Basic Soil Mechanics

Pearson
Education

We work with leading authors to develop the
strongest educational materials in engineering,
bringing cutting-edge thinking and best learning
practice to a global market.

Under a range of well-known imprints, including
Prentice Hall, we craft high quality print and
electronic publications which help readers to
understand and apply their content,
whether studying or at work.

To find out more about the complete range of our
publishing please visit us on the World Wide Web at:
www.pearsoned.co.uk

Basic
Soil Mechanics

Fourth edition

Roy Whitlow

An imprint of Pearson Education

Harlow, England · London · New York · Reading, Massachusetts · San Francisco · Toronto · Don Mills, Ontario · Sydney
Tokyo · Singapore · Hong Kong · Seoul · Taipei · Cape Town · Madrid · Mexico City · Amsterdam · Munich · Paris · Milan

Pearson Education Limited
Edinburgh Gate
Harlow
Essex CM20 2JE
England

and Associated Companies throughout the world

Visit us on the World Wide Web at:
www.pearsoned.co.uk

First published 1983
Second edition 1990
Third edition 1995
Fourth edition 2001

© Pearson Education Limited 1983, 2001

The right of Roy Whitlow to be identified as the author of
this Work has been asserted by him in accordance with the Copyright,
Designs and Patents Act 1988.

ISBN-13: 978-0-582-38109-4

British Library Cataloguing-in-Publication Data
A catalogue record for this book can be obtained from the British Library

Library of Congress Cataloging-in-Publication Data
Whitlow, R. (Roy), 1932–
  Basic soil mechanics / Roy Whitlow.—4th ed.
    p. cm.
  ISBN 0-582-38109-6
    1. Soil mechanics. I. Title.

  TA710.W46 2001
  64.1'5136—dc21   00-036315

10 9
09

Typeset in 10/12 pt Times NR MT by 35
Printed in Great Britain by Henry Ling Limited, at the Dorset Press,
Dorchester, DT1 1HD

Contents

Preface to fourth edition

In the preface to the third edition I remarked that the state of the art of soil mechanics was constantly changing and that learning was becoming more student centred. Both of these remarks are still just as true, especially the latter. Information – on any topic – is becoming more readily available, more comprehensive and more detailed, and one might say more confusing. The purpose of a textbook is to encapsulate specific subject information into an organised and easily navigable text that will provide a permanent, portable and readily accessible learning support tool. On its own, however, a textbook is not very interactive. Teachers and tutors provide vital interaction through lectures, discussions, assignments and even examinations. Computers also can provide rewarding and beneficially interactive interfaces between students and subject matter. While a textbook provides a permanent body of knowledge and reference, computer-based information can provide an interactive learning environment within which a learner can seek, manipulate and explore a subject at his/her own pace. It seems sensible, therefore, to combine the two: a textbook and a computer-based facility.

Included in this fourth edition of *Basic Soil Mechanics* is a CD-ROM which contains learning aids in the form of spreadsheets, electronic reference files and dictionaries, learning assignments and self-testing quizzes. The formats of the various files on the CD are those used by Microsoft™ in Windows95/98 and Microsoft Office, which are most commonly used; although conversion to other formats is relatively simple. The package is designed to assist students and others who are learning about soil mechanics, and should be particularly useful in programmes based on student-paced learning. At the same time, the materials are presented so that they can be utilised by lecturers, either directly or after adaptation. For example, many of the spreadsheets and the study assignments are editable, so that lecturrs can tailor the material to suit their own teaching style and preferences. The author hopes too that students, lecturers and other readers will be encouraged to write similar materials for their own use.

The electronic reference and dictionary files can be accessed directly or from other materials, such as the spreadsheets. These are written in the Microsoft Help style and format, so that methods of navigation, indexing and so on, will be familiar to most users and easily discernible to newcomers.

The changes in content in the fourth edition are not in basic principles, but in approaches to understanding soil behaviour and hence in the presentation of theories and models to be used in solving practical problems and design. Three chapters and some other sections have been substantially rewritten for these reasons.

It no longer seems reasonable to leave the unifying approach of *critical state theory* as an interesting extra – something to note, but not actually use. The behaviour of soils in compression and the associated ideas of volume change and stiffness should be considered *together* with the concepts of strength and yielding, and not as separate sets of ideas. It is now well known that changes in either pore pressure or volume during compression prior to failure fundament-ally affect the parameters that describe failure conditions. Also, the straining behaviour in soil in shear tests can be mapped to provide design routines and aid predictive calculations with respect to settlement and distortion. For all of these, and a few other reasons, the study of volume change and strength should stem from a basic understanding of the soil models proposed in Chapter 6, which has been modified to this end. Chapters 7 and 10 have been completely re-written with the intention underpinning the approaches described above, and also to improve structure and clarity. Some additional material has been incorporated. For example, a number of people now like to model consolidation in terms of parabolic isochrones, as well as, or instead of, the more traditional Fourier series curves (proposed by Karl Terzaghi). Both sets of ideas are now included in Chapter 10. A few traditional ideas have been omitted upon the basis that now they are little used. For example, in Chapter 6, the description and use of Newmark's chart has been removed: the calculation of stresses in the ground due to surface loads is now done almost exclusively using spreadsheets or other computer methods.

Once again I should like to thank all those readers who have written to me, or communicated in some way with me, to offer advice, point out errors, and so on. Many of the comments have been most constructive, some enlightening, and some have been incorporated in this edition. Please keep it up: I really do appreciate hearing from anyone with suggestions.

I wish also to thank the editorial staff at Pearson Education for their patience, understanding and invaluable help, and particularly my publisher, Karen Sutherland, for her inspiration and encouragement. Once again my heartfelt thanks go to my wife Marion whose support as always has been one hundred per cent.

Roy Whitlow
Bristol, February 2000

List of symbols

The symbols used in this text have been chosen in accordance with current conventional usage (e.g. ISSMFE, 1985); those of most importance are listed below.

A, a	area
$A, \bar{A}$	pore pressure coefficient
A_f	pore pressure coefficient at failure
A_v	air voids ratio
a'	intercept of $t':s'$ failure envelope
B	breadth of footing
$B, \bar{B}$	pore pressure coefficient
b	breadth
C_c	compression index (slope of e/log σ' line)
C_g	coefficient of gradation; N correction factor for gravel
C_k	Hazen's permeability coefficient
C_s	compressibility of soil skeleton; slope of swelling/recompression line
C_u	uniformity coefficient
C_v	compressibility of pore fluid
C_α	coefficient of secondary compression
c'	cohesion intercept of the τ_f/σ' envelope
c'_r	residual value of apparent cohesion in terms of effective stress
c_u	undrained shear strength
c_v	coefficient of consolidation
c_s	coefficient of swelling
c_w	adhesion between soil and a structure surface
D	depth; diameter; depth factor
d	depth; diameter; drainage path length
d_{10}, d_{60}, etc.	particle size characteristics
E	Young's modulus of elasticity
E'	drained (effective stress) modulus of elasticity
E'_o	one-dimensional modulus of elasticity
E_u	undrained (total stress) modulus of elasticity

e	void ratio
e_c	critical void ratio
e_0	void ratio intercept of NCL at $\sigma' = 1.0$ kPa
e_κ	void ratio intercept of SRL at $\sigma' = 1.0$ kPa
e_Γ	void ratio intercept of CSL at $\sigma' = 1.0$ kPa
F	factor of safety; force
f_s	skin friction
f_y	yield stress of steel
G'	shear modulus
G_s	specific gravity of soil particles
g	acceleration due to gravity (9.81 m/s^2)
g'	deviator stress intercept of the Hvorslev surface
H	height; layer thickness; slope of Hvorslev surface in $p'{:}q'$ plane
H_c	critical height of unsupported vertical cut
h	height; total, hydrostatic or hydraulic head
h_c	capillary head
$\bar{h}$	height to line of resultant thrust
I_B	brittleness index
I_C	compressibility index (Burland and Burbidge method for settlement)
I_D	density index
I	stress influence coefficient (various subscripts)
I_L	liquidity index
I_P	plasticity index
I_Z	strain influence factor
I_ρ	displacement influence coefficient
i	hydraulic gradient
i_c	critical hydraulic gradient
J	seepage force
j	seepage pressure
K'	bulk deformation modulus
K_a, K_{ac}	coefficient of active earth pressure
K_p, K_{pc}	coefficient of passive earth pressure
K_o	coefficient of earth pressure at rest
K'_{ps}	bulk modulus for plane strain
K_s	coefficient of earth pressure (in piling equation)
k	coefficient of permeability
L, l	length
$\ln$	natural logarithm
M	mass; slope of critical state line in $p'{:}q'$ plane
m	slope stability coefficient
m_v	coefficient of volume compressibility

N	total normal force; slope stability factor; SPT value
N'	effective normal force
N and N_o	specific volume intercept of NCL when $p' = 1.0$ kN/m^3
N_c, N_q, N_γ	bearing capacity factors
N_d	number of equipotential drops in flow net
N_f	number of flow channels in flow net
N_K	cone factor
n	porosity; slope stability coefficient
P	force; resultant thrust
P_A	resultant active thrust
P_P	resultant passive resistance
P_W	resultant thrust due to hydrostatic pressure
p	pressure; mean total normal stress
p'	mean effective normal stress
p_m	historical maximum mean normal stress
p'_y	yield stress
pF	soil suction
pH	acidity/alkalinity value
Q	quantity of flow; total load on pile
q	rate of flow; bearing pressure; deviator stress
q_a	allowable bearing capacity
q_b	end-bearing capacity of pile
q_c	cone penetration resistance
q_f	ultimate bearing capacity
q_{net}	net bearing pressure
q_o	surface surcharge pressure
q_n	net foundation pressure
q_s	skin friction resistance of pile
R	radius; resistance force
R_m, R_o, R_p	overconsolidation ratios
r	radius; radial distance
r_i, r_p, r_s	compression ratios
r_u	pore pressure ratio
S_p	pile coefficient
S_r	degree of saturation
S_s	specific surface; soil suction
S_t	sensitivity of clay soil
s	settlement; pile spacing
s'	mean effective stress in plane strain conditions
s_c	consolidation settlement
s_c, s_q, s_γ	shape factors (bearing capacity equation)
s_i	immediate settlement
$\hat{s}_c$, $\hat{s}_i$	tolerable settlement
s_L	limiting settlement

T	surface tension; torque; tension force
T_v	time factor
t	temperature; time; thickness
t'	max. shear stress in plane strain conditions
U	force due to pore pressure
U, U_z	degree of consolidation
$\bar{U}$	average degree of consolidation
u, u_0, u_1	pore pressure
u_a	pore air pressure
u_w	porewater pressure
V	volume; vertical ground reaction under foundation
V_a	volume of air
V_s	volume of solids
V_v	volume of voids
V_w	volume of water
v	specific volume; velocity
v_s	seepage velocity
W	weight; load; force
w	water content
w_L	liquid limit (or LL)
w_P	plastic limit (or PL)
w_S	shrinkage limit (or SL)
x	distance in x-direction
y	distance in y-direction
z	depth; distance in z-direction
z_o	tension crack depth
α	angle
α'	slope of $t{:}s$ failure envelope ($\tan \alpha' = \sin \phi'$)
α_c	compressibility coefficient for suction
α_f	angle of failure surface
α_r	ring friction coefficient
β	angle; angle of ground slope
β_c	critical ground-slope angle
β_s	skin friction coefficient (pile equation)
Γ	specific volume intercept of CSL when $\sigma' = 1.0$ kPa
γ	bulk unit weight; shear strain
γ'	effective (submerged) unit weight ($\gamma_{sat} - \gamma_w$)
γ_d	dry unit weight
γ_s	unit weight of solid particles
$\gamma_{sat.}$	bulk saturated unit weight
γ_w	unit weight of water (= 9.81 kN/m^3)

β	finite or large increment; finite difference; differential settlement
δ	very small increment; angle of wall friction
ε	linear (normal) strain
ε_a and ε_r	axial and radial strain
$\varepsilon_1, \varepsilon_2, \varepsilon_3$	principal strains
ε_s	triaxial shear strain
ε_v	volumetric strain
$\varepsilon_x, \varepsilon_y, \varepsilon_z$	orthogonal strains
H	slope of Hvorslev surface in $p':q'$ plane
η	dynamic viscosity; effective stress ratio $(= q'/p')$
θ	angle
κ	slope of swelling/recompression $(v/\ln p')$ line
κ_t	temperature coefficient
λ	slope of critical state line in $\ln p':v$ plane
M	slope of critical state line (q'/p')
μ	coefficient of friction
μ_0, μ_1	displacement coefficients
N	specific volume of normally consolidated soil at $p' = 1.0$ kPa
v, v'	Poisson's ratio
ξ	efficiency of pile group; error quantity
ρ	bulk density; settlement (surface displacement)
ρ_d	dry density
ρ_s	density of solid particles
$\rho_{sat.}$	saturated bulk density
ρ_w	density of water (1.00 Mg/m^3)
ρ_α	apparent resistivity
Σ	sum; summation
σ	total normal stress
σ'	effective normal stress
σ_a, σ'_a	total and effective axial stresses
σ_r, σ'_r	total and effective radial stresses
$\sigma_1, \sigma_2, \sigma_3$	principal total stresses
$\sigma'_1, \sigma'_2, \sigma'_3$	principal effective stresses
$\sigma'_{ha}, \sigma'_{hp}$	active and passive lateral pressures
σ_d	deviator stress (also $= q$)
σ_h, σ'_h	horizontal total and effective stress
σ_n, σ'_n	total and effective normal stress
σ'_o	effective overburden stress
σ'_p	preconsolidation stress
σ'_v	vertical effective stress
$\sigma_x, \sigma_y, \sigma_z$	normal stress components in x-, y- and z-directions

τ	shear stress
τ_c	shear strength at critical state
τ_f	shear strength (shear stress at failure); peak shear strength
$\tau_{max.}$	maximum shear strength
τ_m, $\tau_{mob.}$	mobilised shear strength
τ_r	residual shear strength
τ_{yz}, τ_{zx}, τ_{xy}	orthogonal shear stresses
Φ	potential function in flow equation
ϕ	angle of friction
ϕ'	angle of friction in terms of effective stress
ϕ'_c	critical angle of shearing resistance
ϕ'_f	peak angle of shearing resistance
ϕ_m, $\phi_{mob.}$	mobilised angle of shearing resistance
ϕ'_r	residual angle of shearing resistance
$\phi'_{ult.}$	ultimate angle of shearing resistance
ϕ'_y	yield stress
χ	effective stress coefficient for partially saturated soils
Ψ	stream function in flow equation
ψ	angle; angle of dilation

Acknowledgements

We are grateful to the following for permission to reproduce copyright material: British Standards Institution for Fig. 12.7 from Fig. 44 (BSI, 1975), Tables 2.2 and 2.3 from Tables 6 and 8 (BSI, 1981) and Table 11.1 adapted from Table 1 (BSI, 1986); Geological Society Publishing House and the author, J. H. Atkinson for Fig. 7.40 from Figs 3 and 4 (Atkinson and Clinton, 1986) and Table 7.2 from Table II (Atkinson *et al.*, 1986); the Controller of Her Majesty's Stationery Office for Fig. 11.13 from Fig. 21 and Table 11.4 from Table 1 (Burland *et al.*, 1978); Institution of Civil Engineers for Figs 11.14, 11.16 and 11.21 from Figs 1, 2 and 3 (Burland and Burbidge, 1985).

Chapter 1

Origins and composition of soil

1.1 Origins and modes of formation

The term *soil* conveys varying shades of meaning when it is used in different contexts. To a geologist it describes those layers of loose unconsolidated material extending from the surface to solid rock, which have been formed by the weathering and disintegration of the rocks themselves. An engineer, on the other hand, thinks of *soil* in terms of the work he may have to do on it, in it or with it. In an engineering context *soil* means material that can be worked without drilling or blasting. Pedologists, agriculturalists, horticulturalists and others will also prefer their own definitions.

For engineering purposes soil is best considered as a naturally (mostly) occurring particulate material of variable composition having properties of compressibility, permeability and strength (see also *Soil mechanics glossary* on the CD).

All soils originate, directly or indirectly, from solid rocks and these are classified according to their mode of formation as follows:

Igneous rocks, formed by cooling from hot molten material ('magma') within or on the surface of the earth's crust, e.g. granite, basalt, dolerite, andesite, gabbro, syenite, porphyry.

Sedimentary rocks, formed in layers from sediments settling in bodies of water, such as seas and lakes, e.g. limestone, sandstone, mudstone, shale, conglomerate.

Metamorphic rocks, formed by alteration of existing rocks due to: (a) extreme heat, e.g. marble, quartzite, or (b) extreme pressure, e.g. slate, schist.

The processes that convert solid rocks into soils take place at, or near, the earth's surface and, although they are complex, the following controlling factors are apparent:

(a) Nature and composition of the parent rock.
(b) Climatic conditions, particularly temperature and humidity.
(c) Topographic and general terrain conditions, such as degree of shelter or exposure, density and type of vegetation, etc.
(d) Length of time related to particular prevailing conditions.

(e)   Interference by other agencies, e.g. cataclysmic storms, earthquakes, action of humans, etc.

(f)   Mode and conditions of transport.

It is beyond the scope of this book to discuss these factors in detail; the reader is therefore advised to consult a suitable geological text for further information. However, it is worth looking at the effects of some of these factors in so far as they produce particular characteristics and properties in the ultimate soil deposit.

1.2   The effects of weathering

The general term *weathering* embraces a number of natural surface processes which result from the single or combined actions of such agencies as wind, rain, frost, temperature change and gravity. The particular effect of a specific process on a specific type of rock is, to some degree, unique, but some general examples are worth mentioning.

Frost action, in which water within the pore spaces of a rock expands upon freezing, causes *flakes* of rock to split away. The resultant weathered debris is therefore sharp and angular. This contrasts with the action of wind and flowing water, where attrition causes the particles to become rounded. Where the main process is of a chemical nature, certain minerals in the rock will disintegrate and others will prove resistant. Take, for example, the igneous rock *granite*, which comprises essentially the minerals quartz, the feldspars orthoclase and plagioclase and the micas muscovite and biotite. Both quartz and muscovite are very resistant to chemical decomposition and emerge from the process unchanged, whereas the other minerals are broken down (Fig. 1.1).

1.3   The effects of transport

Soils that have not been transported, i.e. have remained at their parent site, are termed *residual soils*. Such soils are found where chemical processes of weathering predominate over physical processes, such as on flat terrain in tropical areas. The soil content will be highly variable, with a wide range of both mineral type

Existing rock minerals	Probable debris minerals	Possible resulting soils
quartz _____	quartz _____	sand
muscovite _____	muscovite _____	micaceous sand
biotite mica _____	chlorite or vermiculite _____ +Mg carbonate solution	clay (dark)
orthoclase feldspar _____	illite or kaolinite _____ +K carbonate solution	clay (light)
plagioclase feldspar _____	montmorillonite _____ +Na or Ca	expansive clay
	carbonate solution _____	lime mud/clay or marl

Fig. 1.1 The weathering of granite

and particle size. In hot climates, weathering may remove some minerals, leaving others of a more resistant nature in a concentrated deposit, e.g. laterite, bauxite, china clay.

The principal effect of transportation is that of sorting. During the processes of movement, separation of the original constituents takes place. This is influenced by both the nature and size of the original rock or mineral grains. In hot arid climates, for example, a fine wind-blown dust known as *loess* may be carried considerable distances before being deposited. The action of flowing water may dissolve some minerals, carry some particles in suspension and bounce or roll others along. The load carried by a river or stream depends largely on the flow velocity. In the upper reaches the velocity is high, so even large boulders may be moved. However, the velocity falls as the river drops down towards the sea, and so deposition takes place: first, gravel-sized particles are deposited in the flood plain and then coarse to medium sands, and finally, in the estuary or delta areas, fine sands and silts. Clay particles, because of their smallness of size and flaky shape, tend to be carried well out into the sea or lake. Thus, river-deposited (*alluvial*) soils are usually well sorted, i.e. poorly or uniformly graded.

During transportation, particles are brought into contact with the stream bed and with each other and so are abraded. The characteristic shape of alluvial soils is rounded or sub-rounded. Even more wear takes place in shore-zone (*littoral*) deposits, producing a fully water-worn rounded shape.

The movement of ice also provides transport for weathered debris. For example, a glacier acts as a slow-moving conveyor belt, carrying sometimes very large boulders considerable distances. The weight of a boulder causes it to sink down through the ice. As it reaches the sole of the glacier, it is ground against the rock base and may be reduced to a very fine rock-flour. Thus, the range of particle sizes in, say, a *boulder clay* is very large indeed. The material deposited as a glacier begins to melt and retreat is termed *moraine*; this also will comprise a wide range of sizes and usually takes the form of a ridge or a series of flat hummocky hills.

1.4   Mineral composition of soil

The large majority of soils consist of mixtures of inorganic mineral particles, together with some water and air. Therefore, it is convenient to think of a soil model which has three phases: solid, liquid and gas (Fig. 1.2).

Rock fragments.   These are identifiable pieces of the parent rock containing several minerals. In general, rock fragments, as opposed to mineral grains, will be fairly large (>2 mm), i.e. sand to gravel size. The overall soundness of the soil will depend on the extent of differential mineral decomposition within individual fragments. For example, the presence of kaolinised granite fragments could influence the crushing strength or shear strength of the soil.

Mineral grains.   These are separate particles of each particular mineral and range in size from gravel (2 mm) down to clay (1 μm). While some soils will contain mixtures of different minerals, a large number will consist almost entirely of one mineral. The best examples of the latter are to be found in the abundance of

Phases   **Constituents**

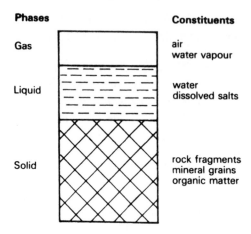

Gas   air
                                    water vapour

Liquid   water
                                    dissolved salts

Solid   rock fragments
                                    mineral grains
                                    organic matter

Fig. 1.2 Three-phase soil model

sand deposits, where the predominant mineral is quartz, due to its previously mentioned enduring qualities. For convenience, it is useful to divide soil into two major groups: *coarse* and *fine* (see also Chapter 2).

(a) *Coarse soils* will be classified as those having particle sizes >0.06 mm, such as SANDS and GRAVELS. Their grains will be *rounded* or *angular* and usually consist of fragments of rock or quartz or jasper, with iron oxide, calcite, mica often present. The relatively equidimensional shape is a function of the crystalline structure of the minerals, and the degree of rounding depends upon the amount of wear that has taken place.

(b) *Fine soils* are finer than 0.06 mm and are typically *flaky* in shape, such as SILTS and CLAYS. Very fine oxides and sulphides, and sometimes organic matter, may also be present. Of major importance in an engineering context is the flakiness of the clay minerals, which gives rise to very large surface areas. This characteristic is discussed in some detail in Section 1.5.

Organic matter.   Organic matter originates from plant or animal remains, the end product of which is known as *humus*, a complex mixture of organic compounds. Organic matter is also a feature of *topsoil*, occurring in the upper layer of usually not more than 0.5 m thickness. *Peat* deposits are predominantly fibrous organic material. From an engineering point of view, organic matter has undesirable properties. For example, it is highly compressible and will absorb large quantities of water, so that changes in load or moisture content will produce considerable changes in volume, posing serious settlement problems. Organic material also has very low shear strength and thus low bearing capacity; furthermore, its presence may affect the setting of cement and therefore provide difficulties in concreting and soil stabilisation processes.

Water.   Water is a fundamental part of natural soil and in fact has a greater effect on engineering properties than any other constituent. The movement of water through a soil mass needs to be studied with care in problems of seepage and permeability, and also, in a slightly different way, when considering compressibility. Water has no shear strength, but is (relatively) incompressible and will therefore transmit direct pressure. For this reason, the drainage conditions in a soil mass are of great significance when considering its shear strength. In addition, water can dissolve and carry in solution a wide range of salts and other compounds, some of which have undesirable effects. For example, the presence of calcium sulphate (and, to a lesser extent, sodium and magnesium sulphates) is fairly common in many clay soils. The presence of sulphate ions has a serious deleterious effect on one of the compounds contained in Portland cement and can therefore be harmful to concrete foundations and other substructures.

Air. Soils may be considered in a practical sense to be perfectly dry or fully saturated, or to be in a condition somewhere between these two extremes. To be exact, the two extremes do not occur. In a so-called 'dry' soil there will be water vapour present, while a 'fully saturated' soil may contain as much as 2 per cent air voids. Air, of course, is compressible and water vapour can freeze, both of which are significant in an engineering context.

1.5 The nature and structure of clay minerals

Clay minerals are produced mainly from the weathering of feldspars and micas (Fig. 1.1). They form part of a group of complex alumino-silicates of potassium, magnesium and iron, known as *layer-lattice minerals*. They are very small in size and very flaky in shape, and so have considerable surface area. Furthermore, these surfaces carry a negative electrical charge, a phenomenon which has great significance in the understanding of the engineering properties of clay soils.

To gain a simple understanding of the properties of clay minerals, it is necessary to examine the essential features of their layer-lattice structure. Figure 1.3

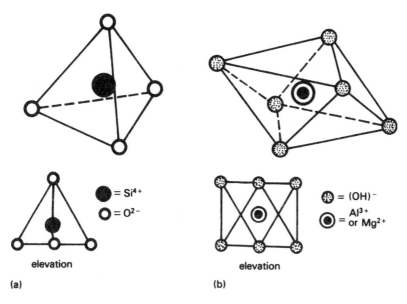

$\bullet = Si^{4+}$
$\bigcirc = O^{2-}$

$\boxtimes = (OH)^{-}$
$\circledcirc = \begin{matrix} Al^{3+} \\ or\ Mg^{2+} \end{matrix}$

elevation   elevation

(a)   (b)

Fig. 1.3 Unit elements of clay minerals
(a) Tetrahedral unit (b) Octahedral unit

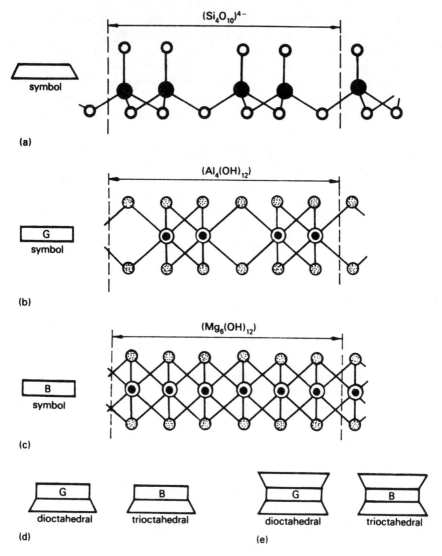

Fig. 1.4 Lattice layer structures
   (a) Silica layer (b) Gibbsite layer (c) Brucite layer (d) Two-layer lattices
   (e) Three-layer lattices

shows the two basic structural units: the *tetrahedral unit*, comprising a central silicon ion with four surrounding oxygen ions, and the *octahedral unit*, comprising a central ion of either aluminium or magnesium, surrounded by six hydroxyl ions. Note that, in both, the metal (with positive valency) is on the inside and the negative non-metallic ions form the outside.

The layer structures are formed when the oxygen ions covalently link between units. Thus, a *silica layer* (Fig. 1.4(a)) is formed of linked tetrahedra, having a general formula of n Si_4O_{10}. The octahedral units also link together at their apices to form a layer, which may be either a *gibbsite layer* $(Al_4(OH)_6)$, in which

Table 1.1   Some layer lattice minerals

	Dioctahedral or gibbsite layer	Trioctahedral or brucite layer
2 layer	Kaolinite Dickite Necrite	Serpentine Chrysotile
3 layer	Pyrophyllite Muscovite Montmorillonite Illite	Talc Biotite Chlorite Vermiculite

only two-thirds of the central positions are occupied by Al^{3+} ions, giving a *dioctahedral* structure (Fig. 1.4(b)), or a *brucite layer* ($Mg_6(OH)_6$), in which all of the central positions are occupied by Mg^{2+} ions, giving a *trioctahedral* structure (Fig. 1.4(c)).

The spacing between the outer ions in the tetrahedral and octahedral layers is sufficiently similar for them to link together via mutual oxygen or hydroxyl ions. Two stacking arrangements are possible, giving either a *two-layer* or a *three-layer* structure. In a two-layer lattice (Fig. 1.4(d)), tetrahedral and octahedral layers alternate, while a three-layer arrangement (Fig. 1.4(e)) consists of an octahedral layer sandwiched between two tetrahedral layers. Mineral particles are built up when the layers are linked together to form stacks. Some of the more common layer-lattice minerals are given in Table 1.1.

Clay minerals are those members of the layer-lattice group commonly encountered in the weathering products of rocks containing feldspars and micas. Depending on the stacking arrangement and type of ions providing linkage between layers, four main groups of clay minerals may be identified: kaolinite, illite, montmorillonite and vermiculite (see Fig. 1.5).

Kaolinite group.   These are the chief constituents of kaolin and china clay derived from the weathering of orthoclase (potash) feldspar which is an essential mineral of granite. Large deposits of china clay occur in Devon and Cornwall; the ball clays of Dorset and Devon have also been formed in this way. The kaolinite structure consists of a strongly bonded two-layer arrangement of silica and gibbsite sheets. Kaolinite itself is a typically flaky mineral, usually with stacks of about 100 layers in a very regular structure.

Another member of this group appearing in some tropical soils is called *halloysite*, in which the layers are separated by water molecules. In contrast to most other clays, which are flaky, halloysite particles are tubular or rod-like. At temperatures over 60 °C halloysite tends to dehydrate, thus care is required in testing procedures on soils containing a significant proportion of this mineral.

Illite group.   The degradation of micas (e.g. muscovite and sericite) under marine conditions results in a group of structurally similar minerals called *illites*. These feature as predominant minerals in marine clays and shales, such as London Clay and Oxford Clay. Some illites are also produced when in the weathering of orthoclase not all of the potassium ions are removed. The structure consists of three-layer gibbsite sheets with K^+ ions providing a bond between adjacent silica

Mineral name	Symbolic structure	Between layers	Approximate size (μm)	Specific surface (m²/g)	Approx. exchange capacity (me/100g)
Kaolinite	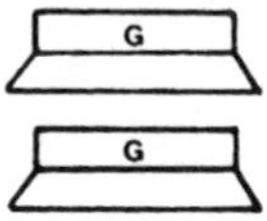	H-bond linkage	$l = 0.2\text{--}2.0$ $t = 0.05\text{--}0.2$	10–30	5
Halloysite	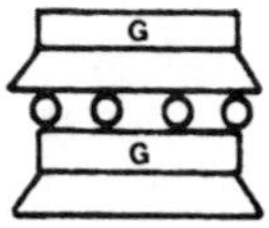	H_2O	(tubular) $l = 0.5$ $t = 0.05$	40–50	15
Illite	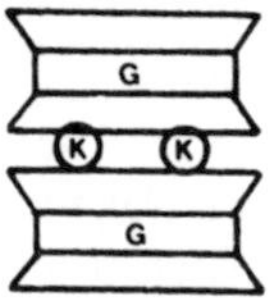	K^+ linkage	$l = 0.2\text{--}2.0$ $t = 0.02\text{--}0.2$	50–100	30
Montmorillonite	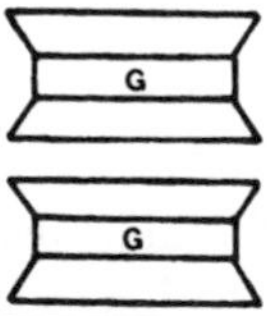	weak cross-linkage between Mg/Al ions	$l = 0.1\text{--}0.5$ $t = 0.001\text{--}0.01$	200–800	100
Vermiculite	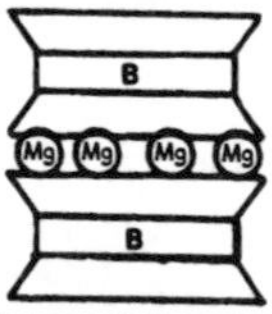	Mg^{2+} linkage	$l = 0.15\text{--}1.0$ $t = 0.01\text{--}0.1$	20–400	150

Fig. 1.5 Structure and size of the main clay minerals

layers (Fig. 1.5). The linkage is weaker than that in kaolinite, resulting in thinner and smaller particles.

Montmorillonite group. The minerals in this group are also referred to as *smectites* and they occur as the chief constituents of bentonite, fuller's earth clays and tropical black cotton soils. Montmorillonite often results from the further degradation of illite, but it is also formed by the weathering of plagioclase feldspar in volcanic ash deposits. Essentially the structure consists of three-layer arrangements in which the middle octahedral layer is mainly gibbsite, but with some substitution of Al by Mg. A variety of metallic ions (other than K^+) provides weak linkage between sheets (Fig. 1.5). As a result of this weak linkage water molecules are easily admitted between sheets, resulting in a high shrinking/swelling potential.

Vermiculite group. This group contains the weathering products of biotite and chlorite. The structure of vermiculite is similar to that of montmorillonite, except that the cations providing inter-sheet linkage are predominantly Mg (Fig. 1.5) accompanied by some water molecules. The shrinkage/swelling potential is therefore similar, but less severe, than that of montmorillonite.

1.6 Important properties of clay minerals

From an engineering point of view, the most significant characteristic of any clay mineral is its extremely flaky shape. A number of important engineering properties are directly attributable to this factor, coupled with others, such as the smallness of particle size and the negative electrical charge carried on the surface. The main properties to be considered in an engineering context are: *surface area, surface charge and adsorption, base exchange capacity, flocculation and dispersion, shrinkage and swelling, plasticity and cohesion.*

Surface area. The smaller and more flaky a particle is, the greater will be its surface area. The ratio of surface area per gram of mass is termed the *specific surface* (S_s) of the soil. Consider a solid cube having a side dimension of d mm and a particle density of ρ_s (g/cm^3 or Mg/m^3).

Surface area $= 6d^2$ mm^2

Mass $= d^3\rho_s \times 10^{-3}$ g

Then specific surface, $S_s = \dfrac{6 \times 10^3}{d\rho_s}$ mm^2/g

$= \dfrac{0.006}{d\rho_s}$ m^2/g

The same expression applies for solid spheres. Thus, a grain of quartz ($\rho_s = 2.65$ g/cm^3) of nominal diameter 1 mm will have a specific surface of about 0.0023 m^2/g. When this is compared with a specific surface of 800 m^2/g for montmorillonite (Table 1.2) the truly enormous surface area of clay minerals will be evident. As a further illustration of this, consider a hypothetical particle of montmorillonite as a single flake having a mass of 1 g. Its thickness would be only 0.001 μm, but to give a specific surface of 800 m^2/g its dimensions (bearing in mind it has two sides) would be 6 m $\times$ 6 m!

Particles of silt size are not as flaky as those in the clay range. Approximate values are:

Silts (0.002–0.06 mm) $S_s = 1$ to 0.04 m^2/g

Sands (0.06–2.0 mm) $S_s = 0.04$ to 0.001 m^2/g

Surface charge and adsorption. The ions forming the platy surfaces of clay minerals are either O^{2-} or $(OH)^-$, and so the surface carries a negative electrical charge. Since water molecules are dipolar, i.e. they have a positive end and a negative end, a layer of these is held against the mineral surface by the hydrogen

Table 1.2   Potential specific surfaces and adsorbed water contents

Mineral	Grain size (mm)		Specific surface	Approx. adsorbed
	d (μm)	t	S_s (m²/g)	water content (%)
Quartz sand	100	d	0.02	0.001
Kaolinite	0.3–2.0	0.2d	20	1
Illite	0.2–2.0	0.1d	80	4
Montmorillonite	0.01–1.0	0.01d	800	40

bond $(H_3O)^+$. Immediately adjacent to the mineral surface water molecules are held in a tightly adhering (*adsorbed*) layer, but further away the bonds become weaker until water becomes more fluid. The properties of this *adsorbed water* layer are markedly different from those of ordinary water: viscosity, density and boiling point are all higher, and freezing point is lower. When determining water contents of clay soils it is advisable to dry them at 105 °C, in order to ensure the expulsion of all adsorbed water.

It is estimated that the thickness of the adsorbed layer is probably about 50 nm. An approximate potential adsorbed water content (w_{AD}) may therefore be calculated:

$$w_{AD} = S_s t \rho_w = 0.05 S_s$$

where t = layer thickness = 50×10^{-9} m
           ρ_w = density of water = 1×10^6 g/m³

The values given in Table 1.2 show the wide range of adsorbed water contents. In addition, certain minerals, such as halloysite and vermiculite, immobilise water between stacked sheets, so that they can remain at low densities with high water contents.

Base exchange capacity.   The total negative charge carried by all clay minerals is neutralised in different ways: partly by internal cations, partly by hydrogen bonds in the adsorbed water and partly due to cations in the adsorbed layer. The balance of the negative surface charge not satisfied internally is termed the *exchange capacity* of the mineral, the units being millequivalents per 100 g (me/100 g). A range of approximate values is shown in Fig. 1.5.

An equilibrium is established generally between the cations in the adsorbed layers and those in the porewater proper. However, the relative proportions of the cations may be varied when different porewater cations are introduced, since some have a greater affinity than others. Those with greater affinity tend to replace in the adsorbed layer those of a lower order. The usual order of replacement ability among commonly occurring cations is:

$$Al^{3+} > (H_3O)^+ > Ca^{2+} > Mg^{2+} > K^+ > Na^+$$

Thus, in tropical climates of high humidity, the soil becomes increasingly acidic as $(H_3O)^+$ replaces Ca^{2+}; in soil surrounding freshly-placed concrete Ca^{2+} tends to replace Na^+. The presence of certain cations tends either to increase or decrease the thickness of the adsorbed layer. For example, twice as many monovalent ions (e.g. Na⁺) are required for balance as divalent ions (e.g. Ca²⁺), resulting in a thicker layer. Cations also tend to be closely surrounded by water molecules held at their negative 'ends'.

Flocculation and dispersion.   The forces acting between two particles close to each other in, say, an aqueous suspension will be influenced by two sets of forces:

(a) Inter-particle attraction due to van der Waals or secondary bonding forces.
(b) Repulsive forces due to the electrically negative nature of the particle surface and adsorbed layer.

The van der Waals attractive force increases if the particles are brought closer together when, for example, the adsorbed layer thickness is decreased due to a base exchange process. In soils where adsorbed layer is thick, the repulsion will be greater and the articles will remain free and dispersed. When the adsorbed layer is thin enough for the attractive forces to dominate, groups of particles form in which edge-to-edge (positive-to-negative) contacts occur; in a suspension these groups will settle together. This process is called *flocculation* and soil exhibiting this phenomenon is termed *flocculent soil* (Fig. 1.6(a)). In marine clays containing a high concentration of cations, the adsorbed layers are thin, resulting in a flocculent structure. By comparison, lacustrine (freshwater) clays tend to possess a *dispersed* structure (Fig. 1.6(b)).

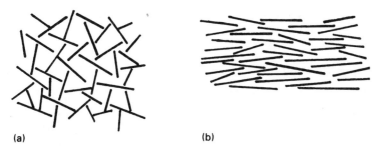

(a)   (b)

Fig. 1.6 Particulate structures in clays
           (a) Flocculent (b) Dispersed

In the laboratory testing of soils a flocculent structure can be dispersed by supplying cations from a suitable salt solution, e.g. sodium hexametaphosphate. Another point worth bearing in mind is that flocculent soils tend to display high liquid limits.

Swelling and shrinkage.   The inter-particle and adsorbed layer forces may achieve equilibrium under constant ambient pressure and temperature conditions due to the movement of water molecules in or out of the adsorbed layer. The moisture content of the soil corresponding to this equilibrium condition is termed the *equilibrium water content* (*ewc*). Any change in the ambient conditions will bring about a change in moisture content. If water is taken in a swelling pressure will be exerted and the volume will tend to increase. Shrinkage will take place if the adsorbed layer is compressed forcing water out, or if suction (e.g. due to climatic evaporation) reduces the moisture content.

The swelling potential of montmorillonite clays is very high. Soils containing a substantial proportion of illite, especially those of marine origin, have fairly high swelling characteristics; while kaolin soils are less susceptible. In soil masses in general, shrinkage manifests itself as a series of polygonal cracks emanating downward from the surface.

Plasticity and cohesion. The most characteristic property of clay soils is their plasticity, i.e. their ability to take and retain a new shape when compressed or moulded. The size and nature of the clay mineral particles, together with the nature of the adsorbed layer, controls this property. Where the average specific surface is high, as in montmorillonite clays, this plasticity may be extremely high and the soil extremely compressible.

The plastic consistency of a clay/water mixture, i.e. a clay soil, varies markedly with the *water content*, which is the ratio of water mass to solid mass. At low water contents, the water present will be predominantly that in the adsorbed layers, thus, the clay particles will be exerting strong attractive force on each other. This binding effect or suction produces a sort of internal tension which is termed *cohesion*. As the water content is increased so the effect of suction is lessened and the cohesion is decreased. When there is sufficient water present to allow the particles to slide past each other without developing internal cracks (i.e. 'crumbling') the soil has reached its *plastic limit*. When the water content is raised to a point where the suction has been reduced to almost nothing and the mixture behaves like a liquid (i.e. flows freely under its own weight) the soil has reached its *liquid limit* (see Chapter 2).

1.7 Engineering soil terminology

As mentioned in Section 1.1, the term *soil* may be construed to mean different things in different areas of study and usage. In the field of engineering the terms used must convey precise information relating to engineering behaviour and processes. The following terms and definitions are commonly encountered in reports, textbooks, research papers, magazine articles and other documents associated with the engineering use and study of soils. More extensive glossaries are available: BS 5930 *Site Investigation*; *Manual of Applied Geology*, Institution of Civil Engineers (1976); *The Penguin Dictionary of Civil Engineering* by John S. Scott, Penguin Books. Also, on the CD-ROM accompanying this book an online glossary is provided: *Soil Mechanics Glossary* and *Basic Geology Glossary*, compiled by Roy Whitlow.

Rock. Hard rigid coherent deposit forming part of the earth's crust, which may be of igneous, sedimentary or metamorphic origin. To a geologist the term *rock* indicates coherent crustal material over about 1 million years old. Soft materials, such as clays, shales and sands, may be described by a geologist as *rock*, whereas an engineer will use the term *soil*. As a general rule, the engineering interpretation of *rock* includes the notion of having to blast it for excavation.

Soil. In engineering taken to be any loose or diggable material that is worked in, worked on or worked with. *Topsoil*, although its removal and replacement are engineering processes, is not normally embraced in the generic term 'engineering' soil. *Subsoil* is essentially an agricultural term describing an inert layer between topsoil and bedrock; its use should be avoided in engineering.

Organic soil. This is a mixture of mineral grains and organic material of mainly vegetable origin in varying stages of decomposition. Many organic soils have their origins in lakes, bays, estuaries, harbours and reservoirs. The presence of organic matter tends to make the soil smoother to the touch; it may also be characterised by a dark colour and a noticeable odour.

Peat. True peat is made up entirely of organic matter; it is very spongy, highly compressible and combustible. Inorganic minerals may also be present and as this increases the material will grade towards an organic soil. From an engineering point of view, peats pose many problems because of their high compressibility, void ratio and moisture content, and in some cases their acidity.

Residual soils. These are the weathered remains of rocks that have undergone no transportation. They are normally sandy or gravelly, with high concentrations of oxides resulting from leaching processes, e.g. laterite, bauxite, china clay.

Alluvial soils (*alluvium*). These are materials, such as sands and gravels, which have been deposited from rivers and streams. Alluvial soils are characteristically well sorted, but they often occur in discontinuous or irregular formations.

Cohesive soils. Fine soils containing sufficient clay or silt particles to impart significant plasticity and cohesion.

Cohesionless soils. Coarse soils, such as sands and gravels, which consist of rounded or angular (non-flaky) particles, and which do not exhibit plasticity or cohesion.

Boulder clay. Sometimes called *till*, this is soil of glacial origin consisting of a very wide range of particle sizes from finely ground rock flour to boulders.

Drift. This is a geological term used to describe superficial unconsolidated deposits of recent origin, such as alluvium, glacial moraine and boulder clay, wind-blown sands, loess, etc.

1.8 Engineering problems and properties

The engineering study of soils involves the application of several scientific disciplines, such as mineralogy, chemistry, physics, mechanics and hydraulics. Also, in some topic areas, mathematics features strongly, and in fact many problems have to be solved in quantitative terms so as to produce numerical 'answers'. The approach and technique required in solving a soil engineering problem varies with the type of problem and with the relative importance of the constraints imposed, but a general approach should include the following considerations:

(a) The nature of the material, including a measure of its relevant engineering properties.
(b) A knowledge of the problem situation or scenario, including an understanding of the behavioural characteristics of the material in those particular circumstances.

(c) Derived from (a) and (b), a model(s) or representation in mathematical or mechanical terms of the behaviour(s) expected.

(d) Application of constraint factors, such as time, safety, aesthetics, planning controls, availability of materials or plant and operational viability, as well as cost factors which often have overriding importance.

(e) The production of solutions which are sound in engineering terms, but which also allow adequately for the other factors.

The common problem areas in soil mechanics may be conveniently summarised as follows:

Excavation.   The digging and removal of soil in order to prepare a site for construction and services. The problems here are closely related to those of support.

Support of soil.   In the case of both natural and built slopes (embankments), it is necessary to determine their intrinsic ability for self-support. Where excavations (e.g. trenches, basements, etc.) or other cuts (e.g. road cuttings) are to be made, it will be necessary to determine the need for, and the extent of, external support. Some of these problems are discussed in Chapters 8 and 9.

Flow of water.   The effects of water in soil masses are dealt with in Chapters 4 and 5. Where a soil is *permeable* and water can flow through it, problems related to the quantity and effects of seepage arise.

Soil as a support medium.   The mass of soil under and adjacent to a structure is part of the foundation system and its behaviour as a support medium must be investigated. Problems of this nature may be divided into two sub-categories:

(a) *Problems of shear failure*, in which possible collapse mechanisms are investigated where rupture surfaces develop due to the shear strength of the soil being exceeded. The concept of shear strength is introduced in Chapter 6 and developed in detail in Chapter 7, together with some applications; further applications are dealt with in Chapters 8 and 11.

(b) *Problems of compressibility.*   A change in volume is induced in all soils when the external loads are increased and in silts, clays and loose sands this can produce a serious *settlement* problem. The processes and mechanics of volume change and compressibility are dealt with in Chapter 10 and the implications in foundations design in Chapter 11.

Building with soils.   Soils are used extensively as construction materials in the building of roads, airfields, dams, embankments and the like. As is the case when using other building materials, such as concrete and steel, the properties of the materials need to be measured and evaluated before use and some measure of quality control introduced to ensure a good and sound product. The soil mechanics aspects of this are included in Chapter 3.

Description and classification.   The starting point in most engineering problems is a good description of the material. This has to be meaningful in an engineering sense, in both qualitative and quantitative terms. It is therefore convenient to commence a detailed study of the engineering behaviour of soils by considering how they might best be described and classified; this is done in Chapter 2.

Exercises

1.1 Describe briefly the origins of soils and summarise the factors which control their formation.

1.2 Discuss the composition of soil in the context of building and civil engineering, pointing out the significance of the presence or absence of notable components.

1.3 Describe the nature and structure of clay minerals and explain their engineering significance as constituents of soil.

1.4 List the common clay minerals and summarise their key properties in an engineering context.

1.5 Summarise the types of engineering problems associated with soils and discuss the nature of possible constraints which may arise from their properties and affect design and construction decisions.

Soil Mechanics Spreadsheets and Reference Assignments and Quizzes (available on the accompanying CD):

Assignments A.1   Soil origins and composition

Quiz Q.1   Soil origins and composition

Chapter 2

Classification of soils for engineering purposes

2.1 Soil classification principles

It is necessary to provide a conventional classification of types of soil for the purpose of describing the various materials encountered in site exploration. The system adopted needs to be sufficiently comprehensive to include all but the rarest of natural deposits, while still being reasonable, systematic and concise. Such a system is required if useful conclusions are to be drawn from the knowledge of the type of material. Without the use of a classification system, published information or recommendations on design and construction based on the type of material are likely to be misleading, and it will be difficult to apply experience gained to future design. Furthermore, unless a system of conventional nomenclature is adopted, conflicting interpretations of the terms used may lead to confusion, rendering the process of communication ineffective.

To be sufficiently adequate for this basic purpose, a classification system must satisfy a number of conditions:

(a) It must incorporate as descriptions definitive terms that are brief and yet meaningful to the user.
(b) Its classes and sub-classes must be defined by parameters that are reasonably easy to measure quantitatively.
(c) Its classes and sub-classes must group together soils having characteristics that will imply similar engineering properties.

Most classification systems divide soils into three main groups: *coarse, fine* and *organic*. The main characteristic differences displayed by these groups are shown in Table 2.1.

In the UK, guidelines and recommendations for both field identification and detailed classification are given in BS 5930 *Site Investigation* (1999). A comprehensive chart (Table 2.2) is given from which identification in the field is possible following the application of several simple tests. Where more data is available, such as from laboratory tests, and especially when the soil is to be used for constructional purposes, the use the soil classification system given in Table 2.3.

Table 2.1 Major classes of engineering soils

	Coarse	Fine	Organic
Inclusive soil types	Stone Gravel Sand	Silt Clay	Peats
Particle shape	Rounded to angular	Flaky	Fibrous
Particle or grain size	Coarse	Fine	–
Porosity or void ratio	Low	High	High
Permeability	High	Low to very low	Variable
Apparent cohesion	None to very low	High	Low
Interparticle friction	High	Low	None to low
Plasticity	None	Low to high	Low to moderate
Compressibility	Very low	Moderate to very high	Usually very high
Rate of compression	Immediate	Moderate to slow	Moderate to rapid

2.2 Field identification

This stage in the process of describing and classifying soils takes place during site exploration; the techniques of site exploration are described in Chapter 12, together with some *in situ* testing procedures. It is important that adequate attempts are made during exploration to describe the nature and formation of all the sub-surface materials encountered. If at all possible, a preliminary classification should be carried out, so that the fullest possible information is passed on. A further and more detailed classification will usually follow a number of laboratory tests.

For the purpose of identification and classification in the field a series of simple tests may be carried out as follows:

Particle size. Identify the main groups by visual examination and 'feel'. Gravel particles (>2 mm) are clearly recognisable; sands (0.06 mm $< d < 2$ mm) have a distinctive gritty feel between the fingers; silts (0.002 mm $< d < 0.06$ mm) feel slightly abrasive, but not gritty; clays (<0.002 mm) feel greasy.

Grading. The *grading* of a soil refers to the distribution of sizes; a *well-graded* soil has a wide distribution of particle sizes, while a *poorly graded* or *uniform* soil contains only a narrow range of sizes.

For a rapid estimate of particle sizes and grading a *field settling test* can be carried out in a tall jar or bottle. A sample of the soil is shaken with water in the jar and then allowed to stand for a few minutes. The coarsest particles settle to the bottom first, followed by progressively smaller sizes. Subsequent examination of the nature and thickness of the layers of sediment will yield approximate proportions of the size ranges.

If over 65 per cent of the particles are greater than 0.06 mm, the soil is classed as *coarse*, i.e. it is a SAND or GRAVEL. The term *fine soil* is used when over 35 per cent of the particles are less than 0.06 mm, i.e. it is a SILT or CLAY. The basis for deciding composite types is given in Table 2.2.

Table 2.2   Field identification and description of soils

	Basic soil type	Particle size (mm)	Visual identification	Particle nature and plasticity
Very coarse soils	BOULDERS		Only seen complete in pits or exposures	Particle shape:
		——— 200		
	COBBLES		Often difficult to recover from boreholes	angular sub-angular sub-rounded
		——— 60		
Coarse soils (over 65% sand and gravel sizes)	GRAVELS	coarse	Easily visible to naked eye; particle shape can be described; grading can be described	rounded flat elongate
		——— 20	Well graded: wide range of grain sizes, well distributed. Poorly graded: not well graded. (May be uniform: size of most particles lies between narrow limits; or gap graded: an intermediate size of particle is markedly under-represented)	
		medium		
		——— 6		
		fine		Texture: rough smooth polished
		——— 2		
	SANDS	coarse	Visible to naked eye; very little or no cohesion when dry; grading can be described	
		——— 0.6		
		medium	Well graded: wide range of grain sizes, well distributed. Poorly graded: not well graded. (May be uniform: size of most particles lies between narrow limits; or gap graded: an intermediate size of particle is markedly under-represented)	
		——— 0.2		
		fine		
		——— 0.06		
Fine soils (over 35% silt and clay sizes)	SILTS	coarse	Only coarse silt barely visible to naked eye; exhibits little plasticity and marked dilatancy; slightly granular or silky to the touch Disintegrates in water; lumps dry quickly; possess cohesion but can be powdered easily between fingers	Non-plastic or low plasticity
		——— 0.02		
		medium		
		——— 0.006		
		fine		
		——— 0.002		
	CLAYS		Dry lumps can be broken but not powdered between the fingers; they also disintegrate under water but more slowly than silt; smooth to the touch; exhibits plasticity but no dilatancy; sticks to the fingers and dries slowly; shrinks appreciably on drying usually showing cracks. Intermediate and high plasticity clays show these properties to a moderate and high degree, respectively	Intermediate plasticity (lean clay) High plasticity (fat clay)
Organic soils	ORGANIC CLAY, SILT or SAND	Varies	Contains substantial amounts of organic vegetable matter	
	PEATS	Varies	Predominantly plant remains usually dark brown or black in colour, often with distinctive smell; low bulk density	

Table 2.2 *Continued*

Basic soil type	Composite soil types (mixtures of basic soil types)			Compactness/strength	
				Term	Field test
BOULDERS	*Scale of secondary constituents with coarse soils*			Loose	By inspection of voids and particle packing
COBBLES	Term		% of clay or silt	Dense	
GRAVELS	slightly clayey slightly silty	GRAVEL or SAND	under 5		
	clayey silty	GRAVEL or SAND	5 to 15	Loose	Can be excavated with a spade; 50 mm wooden peg can be easily driven
	very clayey very silty	GRAVEL or SAND	15 to 35	Dense	Requires pick for excavation; 50 mm wooden peg hard to drive
SANDS	Sandy GRAVEL Gravelly SAND	Sand or gravel and important second constituent of the coarse fraction		Slightly cemented	Visual examination; pick removes soil in lumps which can be abraded
	For composite types described as: clayey: fines are plastic, cohesive; silty: fines non-plastic or of low plasticity				
SILTS	*Scale of secondary, constituents with fine soils*			Soft or loose	Easily moulded or crushed in the fingers
	Term		% of sand or gravel	Firm or dense	Can be moulded or crushed by strong pressure in the fingers
	sandy gravelly –	CLAY or SILT CLAY: SILT	35 to 65 under 35	Very soft Soft	Exudes between fingers when squeezed in hand Moulded by light finger pressure
CLAYS	*Examples of composite types*			Firm	Can be moulded by strong finger pressure
	(Indicating preferred order for description) Loose, brown, sub-angular, very sandy, fine to coarse GRAVEL with small pockets of soft grey clay			Stiff Very stiff	Cannot be moulded by fingers. Can be indented by thumb Can be indented by thumb nail
ORGANIC CLAY, SILT or SAND	Medium dense, light brown, clayey, fine and medium SAND			Firm	Fibres already compressed together
	Stiff, orange brown, fissured sandy CLAY			Spongy	Very compressible and open structure
PEATS	Firm, brown, thinly laminated SILT and CLAY			Plastic	Can be moulded in hand, and smears fingers
	Plastic, brown, amorphous PEAT				

Table 2.2 Continued

Basic soil type	Structure			Colour
	Term	*Field identification*	*Interval scales*	
BOULDERS	Homogeneous	Deposit consists essentially of one type	*Scale of bedding spacing*	Red Pink
COBBLES	Interstratified	Alternating layers of varying types or with bands or lenses of other materials	*Term* *Mean spacing mm*	Yellow Brown Olive Green Blue
GRAVELS		Interval scale for bedding spacing may be used	Very thickly bedded over 2000	White Grey Black etc.
	Heterogeneous	A mixture of types	Thickly bedded 2000 to 600	
	Weathered	Particles may be weakened and may show concentric layering	Medium bedded 600 to 200	Supplemented as necessary with:
			Thinly bedded 200 to 60	
			Very thinly bedded 60 to 20	Light Dark Mottled
SANDS			Thickly laminated 20 to 6	etc.
			Thinly laminated under 6	
SILTS	Fissured	Break into polyhedral fragments along fissures. Interval scale for spacing of discontinuities may be used		and Pinkish Reddish Yellowish
	Intact	No fissures		Brownish etc.
CLAYS	Homogeneous	Deposit consists essentially of one type	*Scale of spacing of other discontinuities*	
	Interstratified	Alternating layers of varying types. Interval scale for thickness of layers may be used	*Term* *Mean spacing (mm)*	
	Weathered	Usually has crumb or columnar structure	Very widely spaced over 2000	
ORGANIC CLAY, SILT or SAND			Widely spaced 2000 to 600	
			Medium spaced 600 to 200	
			Closely spaced 200 to 60	
PEATS	Fibrous	Plant remains recognisable and retains some strength	Very closely spaced 60 to 20	
	Amorphous	Recognisable plant remains absent	Extremely closely spaced under 20	

Reproduced from BS 5930: 1981 *Site Investigations* with permission of the British Standards Institution Now superseded by BS 5930: 1999

Compactness. Compactness or *field strength* may be estimated using a hand spade or pick, or by driving in a small wooden peg; the soil is then reported as being *loose*, *dense* or *slightly cemented*, as appropriate.

Structure. Observations of structural characteristics are most useful and are conveniently made in trial pits, cuttings and other excavations. The following descriptive terms are used:

Homogeneous – *consisting of essentially one soil type*
Inter-stratified – *alternating layers or bands of different materials*; the 'interval spacing' *between* 'bedding planes' *should be reported as indicated in Table 2.2*
Intact – *a non-fissured fine soil*
Fissured – *the direction, size and spacing of fissures should be reported using the scale given in Table 2.2*

Cohesion, plasticity and consistency. If its particles stick together, a soil possesses *cohesion* and, if it can be easily moulded without cracking, it possesses *plasticity*. Both of these behaviours depend on the moisture content of the soil. It is helpful to record in the field the apparent *consistency* of the soil as an indicator of cohesive or plastic behaviour. After removing any particle over 2 mm, squeeze a handful of soil at its natural moisture content and attempt to mould it in the hand and then describe its consistency as follows:

Very soft – *if it exudes between the fingers*
Soft – *if it is very easy to mould and it sticks to the hand*
Firm – *if it moulds easily with moderate pressure*
Very firm – *if it moulds only with considerable pressure*
Hard – *if it will not mould under pressure in the hand*
Crumbly – *if it breaks up into crumbs*

Dilatancy. Remove any particles larger than about 2 mm and moisten the soil sufficiently to make it soft, but not sticky. Place this pat of soil in the open palm of one hand and tap the side of this hand sharply with the other hand; repeat this several times. Dilatancy is exhibited if, as a result of this tapping, a glossy film of water appears on the surface of the pat. If the pat is now pressed gently the water will disappear from the surface and the pat becomes stiff again. Very fine sands and inorganic silts exhibit marked dilatancy, whereas clays and medium-to-coarse sands do not.

Dry strength. The pat of moist soil used for the dilatancy test should now be dried, preferably in an oven, but air-drying will suffice if weather conditions are suitable. The dry strength of the soil is estimated by breaking the dried pat with the fingers. A high dry strength indicates a clay with a high plasticity; inorganic silts exhibit a low dry strength and are powdery when rubbed. The presence of sand reduces the dry strength of silts, and also results in a gritty feel when rubbed.

Weathering. Climatic conditions at the ground surface or on exposed faces can lead to the *weathering* of soil, which can result in a reduction of strength and an increase in compressibility. The degree of weathering evident in newly exposed soil should be reported. Chandler (1969) recommended a system for Keuper marl

Table 2.3   British Soil Classification System for engineering purposes

Soil groups (see note 1)		Sub-groups and laboratory identification				
		Group symbol (see notes 2 and 3)	Subgroup symbol (see note 2)	Fines (% less than 0.06 mm)	Liquid limit (%)	Name
COARSE SOILS — Less than 35% of the material is finer than 0.06 mm						
GRAVELS — More than 50% of coarse material is of gravel size (coarser than 2 mm)	Slightly silty or clayey GRAVEL	G	GW	0 to 5		Well-graded GRAVEL
			GPu GPg	0 to 5		Poorly graded/uniform/gap graded GRAVEL
	Silty GRAVEL	G–M	GWM GPM	5 to 15		Well graded/poorly graded silty GRAVEL
	Clayey GRAVEL	G–F / G–C	GWC GPC	5 to 15		Well graded/poorly graded clayey GRAVEL
	Very silty GRAVEL	GF	GML, etc.	15 to 35		Very silty GRAVEL; subdivide as for GC
	Very clayey GRAVEL	GC	GCL GCI GCH GCV GCE	15 to 35		Very clayey GRAVEL (clay of low, intermediate, high, very high, extremely high plasticity)
SANDS — More than 50% of coarse material is of sand size (finer than 2 mm)	Slightly silty or clayey SAND	S	SW	0 to 5		Well graded SAND
			SPu SPg	0 to 5		Poorly graded/uniform/gap graded SAND
	Silty SAND	S–M	SWM SPM	6 to 15		Well graded/poorly graded silty SAND
	Clayey SAND	S–F / S–C	SWC SPC	6 to 15		Well graded/poorly graded clayey SAND
	Very silty SAND	SM	SML, etc.	15 to 35		Very silty SAND; subdivided as for SC
	Very clayey SAND	SF / SC	SCL SCI SCH SCV SCE	15 to 35		Very clayey SAND (clay of low, intermediate, high, very high, extremely high plasticity)

GRAVEL and SAND may be qualified sandy GRAVEL and gravelly SAND, etc.

FINE SOILS More than 35% of the material is finer than 0.06 mm	**Gravelly or sandy SILTS and CLAYS** 35% to 65% fines	Gravelly SILT	FG	MG	MLG, etc.		Gravelly SILT; subdivide as for CG
		Gravelly CLAY (see note 4)		CG	CLG CIG CHG CVG CEG	<35 35 to 50 50 to 70 70 to 90 >90	Gravelly CLAY of low plasticity of intermediate plasticity of high plasticity of very high plasticity of extremely high plasticity
		Sandy SILT (see note 4)	FS	MS	MLS, etc.		Sandy SILT; subdivide as for CG
		Sandy CLAY		CS	CLS, etc.		Sandy CLAY; subdivide as for CG
	SILTS and CLAYS 65% to 100% fines	SILT (M-SOIL)	F	M	ML, etc.		SILT; subdivide as for C
		CLAY (see notes 5 and 6)		C	CL CI CH CV CE	<35 35 to 50 50 to 70 70 to 90 >90	CLAY of low plasticity of intermediate plasticity of high plasticity of very high plasticity of extremely high plasticity
ORGANIC SOILS		Descriptive letter 'O' suffixed to any group or subgroup symbol					Organic matter suspected to be a significant constituent. Example MHO. Organic SILT of high plasticity
PEAT		Pt					Peat soils consist predominantly of plant remains which may be fibrous or amorphous

Note 1: The name of the soil group should always be given when describing soils, supplemented, if required, by the group symbol, although for some additional applications (e.g. longitudinal sections) it may be convenient to use the group symbol alone.

Note 2: The group symbol or subgroup symbol should be placed in brackets if laboratory methods have not been used for identification, e.g. (GC).

Note 3: The designation FINE SOIL or FINES, F, may be used in place of SILT, M, or CLAY, C, when it is not possible or not required to distinguish between them.

Note 4: GRAVELLY if more than 50% of coarse material is of gravel size. SANDY if more than 50% of coarse material is of sand size.

Note 5: SILT (M-SOIL), M, is material plotting below the A-line, and has a restricted plastic range in relation to its liquid limit, and relatively low cohesion. Fine soils of this type include clean silt-sized materials and rock flour, micaceous and diatomaceous soils, pumice, and volcanic soils, and soils containing halloysite. The alternative term 'M-soil' avoids confusion with materials of predominantly silt size, which form only a part of the group. Organic soils also usually plot below the A-line on the plasticity chart, when they are designated ORGANIC SILT, MO.

Note 6: CLAY, C, is material plotting above the A-line, and is fully plastic in relation to its liquid limit.

Reproduced from BS 5930: 1981 *Site Investigations* with permission of the British Standards Institution. Now superseded by BS 5930: 1999.

which may be generalised to describe the weathered state of originally stiff to hard soils:

Unweathered	– *no visible sign of weathering*
Slightly weathered	– *apparent weakening along joints and fissures, but soil blocks between still intact*
Moderately weathered	– *fabric partly disrupted, with some parts at higher moisture contents than others*
Highly weathered	– *considerable fabric disruption and softening; original structure hardly apparent*
Fully weathered	– *structureless, softened matrix, considerably weaker than original soil*

2.3   Particle size analysis and grading

The range of particle sizes encountered in soils is very wide: from around 200 mm down to the colloidal size of some clays of less than 0.001 mm. Although natural soils are mixtures of various-sized particles, it is common to find a predominance occurring within a relatively narrow band of sizes. When the width of this size band is very narrow the soil will be termed *poorly-graded*, if it is wide the soil is said to be *well graded* (see also Section 2.4). A number of engineering properties, e.g. permeability, frost susceptibility, compressibility, are related directly or indirectly to particle-size characteristics.

Figure 2.1 shows the British Standard range of particle sizes. The particle-size analysis of a soil is carried out by determining the weight percentages falling within bands of size represented by these divisions and sub-divisions. In the case of a coarse soil, from which fine-grained particles have been removed or were absent, the usual process is a *sieve analysis*. A representative sample of the soil is split systematically down to a convenient sub-sample size* and then oven-dried. This sample is then passed through a nest of standard test sieves arranged in descending order of mesh size. Following agitation of first the whole nest and then individual sieves, the weight of soil retained on each sieve is determined and the cumulative percentage of the sub-sample weight passing each sieve calculated.

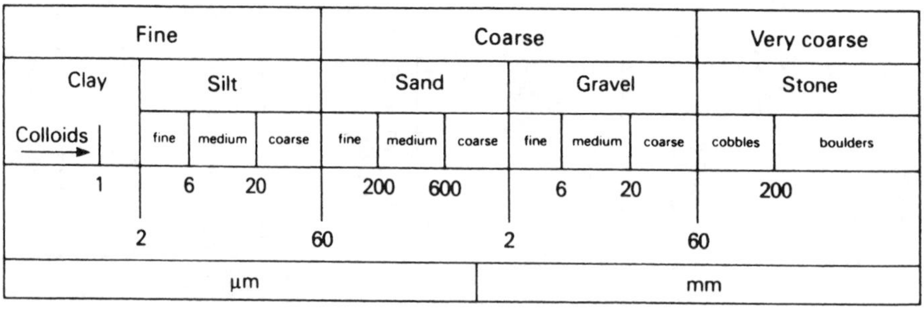

Fine					Coarse					Very coarse	
Clay	Silt			Sand			Gravel			Stone	
Colloids →	fine	medium	coarse	fine	medium	coarse	fine	medium	coarse	cobbles	boulders
1	6	20		200	600		6	20		200	
2		60				2			60		
μm						mm					

Fig. 2.1 British standard range of particle sizes

---

* The laboratory procedures required for this and other tests are detailed in BS 1377 *Methods of Test for Soils for Civil Engineering Purposes*.

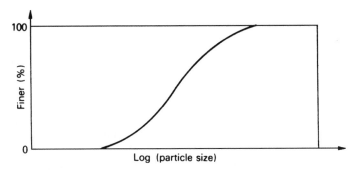

Fig. 2.2 Grading curve

From these figures the particle-size distribution for the soil is plotted as a semi-logarithmic curve (Fig. 2.2) known as a *grading curve*.

Where the soil sample contains fine-grained particles, a *wet sieving* procedure is first carried out to remove these and to determine the combined clay/silt fraction percentage. A suitably sized sub-sample is first oven-dried and then sieved to separate the coarsest particles (>20 mm). The sub-sample is then immersed in water containing a dispersing agent (sodium hexametaphosphate: a 2 g/litre solution) and allowed to stand before being washed through a 63 μm mesh sieve. The retained fraction is again oven-dried and passed through a nest of sieves. After weighing the fractions retained on each sieve and calculating the cumulative percentages passing each sieve, the grading curve is drawn. The combined clay/silt fraction is determined from the weight difference and expressed as a percentage of the total sub-sample weight. The coarsest fraction (>20 mm) can also be sieved and the results used to complete the grading curve.

A further sub-division of particle-size distribution in the fine-grained fraction is not possible by the sieving method. A process of *sedimentation* is normally carried out for this purpose. A small sub-sample of soil is first treated with a dispersing agent and then washed through a 63 μm sieve. The soil/water suspension is then made up to 500 ml, agitated vigorously for a short while and then allowed to settle.

The procedure is based on Stokes' law, which states that the velocity at which a spherical particle will sink due to gravity in a suspension is given by:

$$v = \frac{d^2(\gamma_s - \gamma_w)}{18\eta} \tag{2.1}$$

where d = diameter of particle
             γ_s = unit weight of the grain or particle
             γ_w = unit weight of the suspension fluid (usually water)
             η = viscosity of the suspension fluid

The diameter of those particles that will have settled a given distance in a given time (t) may be obtained by rearranging eqn [2.1]:

$$d = \left[\frac{18\eta h}{(\gamma_s - \gamma_w)t} \right]^{1/2}$$

Usually $h = 100$ mm, giving:

$$d = \left[\frac{1800\eta}{(\gamma_s - \gamma_w)t} \right]^{1/2}$$ [2.2]

Samples taken at a depth of 100 mm, at an elapsed time of t, will not, therefore, include particles of greater size than the diameter d given by eqn [2.2]; but the proportions of particles smaller than d in the suspension will remain unchanged.

The procedure using a hydrometer consists of measuring the suspension density at a depth of 100 mm at a series of elapsed-time intervals. The percentage-finer values corresponding to particular diameters (i.e. particle sizes) are obtained from the density readings, and thus a grading curve for the fine-grained fraction may be drawn. An alternative method of obtaining suspension density values consists of drawing off a small quantity from the prescribed depth using a special pipette (refer to BS 1377 for full details).

The sedimentation method is not particularly accurate in an absolute sense, since errors result from a number of factors: such as the flaky nature of fine-grained particles, the near molecular size of the very fine particles, incomplete dispersion, variations in viscosity due to temperature variations. However, the *equivalent spherical diameter* distribution obtained in this manner provides a sufficiently useful guide for engineering purposes.

2.4   Grading characteristics

The grading curve is a graphical representation of the particle-size distribution and is therefore useful in itself as a means of describing the soil. For this reason it is always a good idea to include copies of grading curves in laboratory and other similar reports. It should also be remembered that the primary object is to provide a descriptive term for the type of soil. This is easily done using the type of chart shown in Fig. 2.3 by estimating the range of sizes included in the most representative fraction of the soil. For example, curve A may be taken to represent a *poorly graded medium SAND: poorly graded* because the curve is steep, indicating a narrow range of sizes, and *medium SAND*, since the largest proportion of the soil (approximately 65 per cent) lies in medium-sand sub-range. Curve B represents a *well-graded* material containing a wide range of particle sizes; from fine sand to medium gravel; this soil is properly described as a *well-graded GRAVEL SAND*, since about half the soil is gravel and the other half sand. Curve C also represents a well-graded material which is predominantly sand, but with a significant silt fraction (about 20 per cent); this soil should be described as a *very silty SAND*, the noun indicating the predominant fraction. Curve D indicates a *sandy SILT*, e.g. an estuarine or deltaic silt; curve E indicates a typical *silty CLAY*, e.g. London Clay or Oxford Clay.

A further quantitative analysis of grading curves may be carried out using certain geometric values known as *grading characteristics*. First of all, three

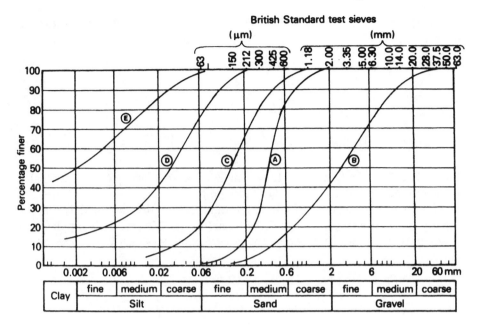

Fig. 2.3 Typical particle size distribution curves

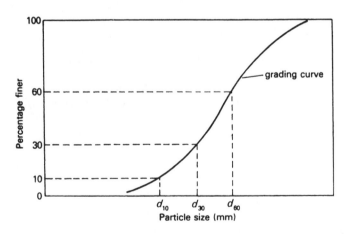

Fig. 2.4 Grading characteristics

points are located on the grading curve to give the following characteristic sizes
(Fig. 2.4):

d_{10} = Maximum size of the smallest 10 per cent of the sample
d_{30} = Maximum size of the smallest 30 per cent of the sample
d_{60} = Maximum size of the smallest 60 per cent of the sample

From these *characteristic sizes*, the following *grading characteristics* are defined:

Effective size $\quad\quad\quad\quad = d_{10}$ mm $\hfill$ [2.3]

Uniformity coefficient, $\quad C_u = \dfrac{d_{60}}{d_{10}}$ $\hfill$ [2.4]

Coefficient of gradation, $\quad C_g = \dfrac{(d_{30})^2}{d_{60} \times d_{10}}$ $\hfill$ [2.5]

Both C_u and C_g will be unity for a single-sized soil, while $C_u < 3$ indicates uniform grading and $C_u > 3$ a well-graded soil.

Most well-graded soils will have grading curves that are mainly flat or slightly concave, giving values of C_g between 0.5 and 2.0. One useful application is an approximation of the coefficient of permeability, which was suggested by Hazen.

Coefficient of permeability $\quad (k) = C_k(d_{10})^2$ m/s

where $\quad C_k$ = a coefficient varying between 0.01 and 0.015.

Worked example 2.1 *The results of a dry-sieving test are given below: plot the particle-size distribution curve and give a classification for the soil.*

Sieve size (mm or µm)	3.35	2.00	1.18	600	425	300	212	150	63
Mass retained (g)	0	2.5	12.5	57.7	62.0	34.2	18.7	12.7	13.1

The quantity passing the 63 µm sieve and collected in the pan was 3.9 g, and the original weighed quantity was 217.2 g.

The retained masses are first expressed as percentages of the total mass and the percentage passing each sieve obtained by successive subtraction. The complete set of results is tabulated below:

	Sieve size	Mass retained (g)	% retained	% passing
mm	3.35	0	0	100.0
	2.0	2.6	1.2	98.8
	1.18	12.5	5.7	93.1
µm	600	57.7	26.6	66.5
	425	62.0	28.6	37.9
	300	34.2	15.7	22.2
	212	18.7	8.6	13.6
	150	12.7	5.8	7.8
	63	13.1	6.0	1.8
Pan		3.9	1.8	
Total		217.4 g	100.0%	

Original total = 217.2 g (i.e. no significant losses)

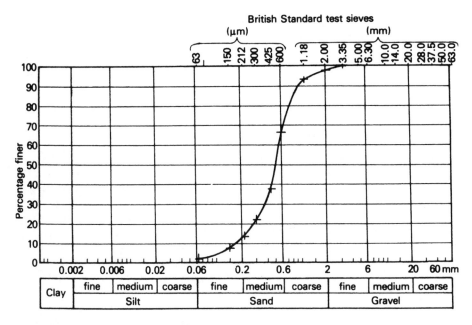

Fig. 2.5 Particle size distribution WE 2.1

The plot of the particle-size distribution curve is shown in Fig. 2.5; an inspection of the curves shows the following approximate proportions:

Coarse sand   33 per cent
Medium sand   54 per cent
Fine sand   13 per cent

The soil may therefore be classified as a *well-graded medium coarse SAND* and the subgroup soil classification symbol will be *SW*.

Worked example 2.2   *A full wet/dry sieve analysis was carried out with the following results:*

Stage 1.   *A total sample of 3274 g was sieved on a 20 mm test sieve:*
            Mass retained = 104.8 g   Mass passing = 3169 g

Stage 2.   *The mass retained on the 20 mm sieve was sieved on larger mesh sizes:*

Sieve size (mm)	50	37.5	28	20
Mass retained (g)	0	45.9	26.2	32.7

Stage 3.   *The mass passing the 20 mm sieve was riffled down to a sub-sample mass of 2044 g and then sieved on a 6.3 mm test sieve:*

            Mass retained = 201.0 g   Mass passing 1843 g

Stage 4.   *The mass retained on the 6.3 mm sieve was sieved on larger mesh sizes:*

Sieve size (mm)	14	10	6.3
Mass retained (g)	40.2	48.8	112.0

Stage 5.   The mass passing the 6.3 mm sieve was riffled down to a sub-sample mass of 212.4 g and then sieved on smaller mesh sizes:

Sieve size (mm or μm)	5.0	3.35	2.0	1.18	600	425	300	212	150	63
Mass retained (g)	22.5	69.4	58.4	43.4	20.9	10.3	9.1	5.6	4.7	7.2

The mass passing the 63 μm sieve was 21.2 g.

Plot the particle-size distribution curve and give a classification for the soil.

The retained masses are first corrected so that they are expressed as proportions of the total sample mass, and the retained- and passing-percentages obtained. The complete set of results and calculations is tabulated below.

Total sample: 3274 g

Sieve size (mm)	(μm)	Mass retained (g)	Correction	Corrected mass retained	% retained	% passing
50.0		0.0		0.0	0.0	100.0
37.5		45.9		45.9	1.4	98.6
28.0		26.2	None	26.2	0.8	97.8
20.0		32.7		32.7	1.0	96.8
Pan		3169	Riffled down to: 2044 g			
14.0		40.2	$\times \dfrac{3169}{2044}$	62.3	1.9	94.9
10.0		48.8		75.6	2.3	92.6
6.3		112.0	$= 1.55$	173.6	5.3	87.3
Pan		1843	Riffled down to: 272.4 g			
5.00		22.5		236.0	7.2	80.1
3.35		69.4	$\times \dfrac{3169}{2044}$	728.0	22.2	57.9
2.00		58.4		612.6	18.7	39.2
1.18		43.4		455.3	13.9	25.3
	600	20.9	$\times \dfrac{1843}{272.4}$	219.2	6.7	18.6
	425	10.3		108.0	3.3	15.3
	300	9.1	$= 10.49$	95.5	2.9	12.4
	212	5.6		58.7	1.8	10.6
	150	4.7		49.3	1.5	9.1
	63	7.2		75.5	2.3	6.8
Pan		21.2		222.4	6.8	

The plot of the particle-size distribution curve is shown in Fig. 2.6: an inspection of which shows the following approximate proportions:

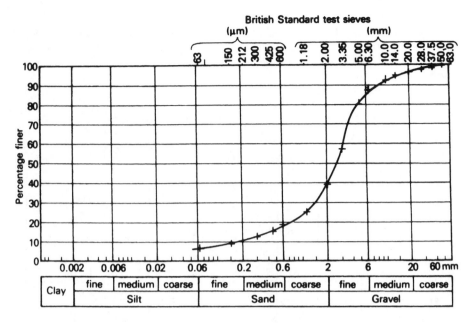

Fig. 2.6 Particle size distribution WE 2.2

Gravel   61 per cent (fine gravel 48 per cent)
Sand   32 per cent
Fines   7 per cent

The soil may therefore be classified as a well-graded *silty sandy GRAVEL*; the Soil Classification symbol will be GW M (a dual symbol being used since the percentage fines is between 5 and 15 per cent).

2.5 Design of filters

In water-pumping operations and in the construction of earth dams, it is often necessary to provide a layer (or layers) of filter material to prevent fine particles being carried into pipes, through mesh screens or into the void space of coarser materials. An effective filter material may be designed using a few simple rules and the grading characteristics of the soil to be protected:

(a) The soil content exceeding a grain size 19 mm should be discounted.
(b) The filter should not contain material of particle size greater than 80 mm.
(c) The filter should have a fines content (particle size <75 μm) of not more than 5 per cent.
(d) The grading curve of the filter should have the same approximate shape as that of the soil.
(e) The d_{15} size of the filter should lie between four times d_{15} for the soil and four times d_{85} for the soil, i.e. $4 \times d_{15}(\text{soil}) < D_{15}(\text{filter}) < 4 \times d_{85}(\text{soil})$.
(f) The d_{85} size of the filter should be not less than twice the inside pipe diameter or screen-mesh size (where applicable).

Worked example 2.3   *Figure 2.7 shows a grading curve for a soil for which a graded filter is required. In pumping operations a pipe is to be used incorporating perforations of diameter 6 mm. Construct a grading curve for a suitable graded filter, labelling key points and sizes.*

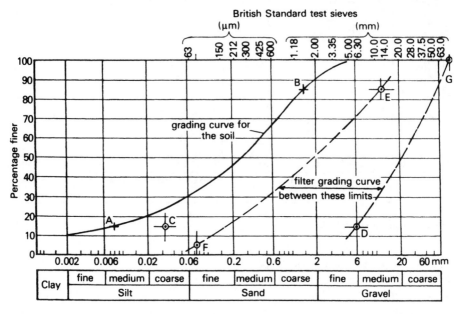

Fig. 2.7 Filter design curve

Grading characteristics of soil:
$d_{15} = 0.008$ mm   (point A)
$d_{85} = 1.50$ mm   (point B)

Required characteristics for filter:
$d_{15} > 4 \times 0.008 = 0.032$ mm   (point C)
$d_{15} < 4 \times 1.50 = 6.0$ mm   (point D)
or $d_{85} > 2 \times 6.0 = 12.0$ mm   (point E)
$d_5 > 75$ μm   (point F)
$d_{100} < 80$ mm   (point G)

2.6   Classification of fine soils

In the case of *fine* soils, it is the shape rather than size of particles that has the greater influence on engineering properties. The combination of very flaky particles and circumstances which may bring about changes in water content results in a material (soil) having properties which are inherently variable. For example, the shear strength of cohesive soils will vary markedly with changes in water content. Also, soils with flaky particles behave as *plastic* material: an increase in applied stress usually brings about an irrecoverable deformation, while the volume remains constant or is reduced and without any signs of cracking or disruption.

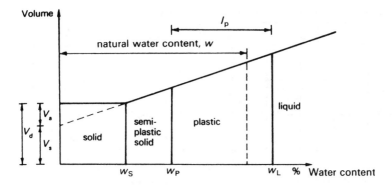

Fig. 2.8   Consistency relationships

Since the *plasticity* of fine soils has an important effect on such engineering properties as shear strength and compressibility, plastic *consistency* is used as a basis for their classification. The consistency of a soil is its physical state characteristic at a given water content. Four consistency states may be defined for cohesive soils: solid, semi-plastic solid, plastic and liquid. The change in volume of a saturated cohesive soil is approximately proportional to a change in water content; the general relationship is shown in Fig. 2.8.

The transition from one state to the next in fact is gradual; however, it is convenient to define arbitrary limits corresponding to a changeover moisture content:

w_L = the *liquid limit*: the water content at which the soil ceases to be liquid and becomes plastic.

w_P = the *plastic limit*: the water content at which the soil ceases to be plastic and becomes a semi-plastic solid.

w_S = the *shrinkage limit*: the water content at which drying-shrinkage at constant stress ceases.

The two most important of these are the liquid and plastic limits, which represent respectively the upper and lower bounds of plastic states; the range of plastic states is given by their difference, and is termed the *plasticity index* (I_P).

$$I_P = w_L - w_P \qquad\qquad [2.6]$$

The relationship between the plasticity index and the liquid limit is used in the British Soil Classification System to establish the sub-groups of fine soil; Fig. 2.9 shows the *plasticity chart* used for this purpose. The A-line provides an arbitrary division between silts and clays, and vertical divisions (of percentage liquid limit) define *five* degrees of plasticity:

Low plasticity:	$w_L < 35\%$
Intermediate plasticity:	$w_L = 35\% - 50\%$
High plasticity:	$w_L = 50\% - 70\%$
Very high plasticity:	$w_L = 70\% - 90\%$
Extremely high plasticity:	$w_L > 90\%$

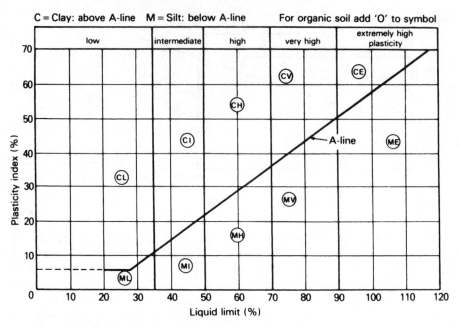

Fig. 2.9 Plasticity chart for the classification of fine soils

Table 2.4   Sub-group symbols in the British Soil Classification System

	Primary letter	Secondary letter
Coarse-grained soils	G = GRAVEL	W = well graded
	S = SAND	P = poorly graded
		Pu = uniform
		Pg = gap graded
Fine-grained soils	F = FINES	L = low plasticity
	(undifferentiated)	I = intermediate plasticity
	M = SILT	H = high plasticity
	C = CLAY	V = very high plasticity
		E = extremely high plasticity
Organic soils	Pt = PEAT	O = organic

A given soil may be located in its correct sub-group zone by plotting a point, having coordinates given by the soil's plasticity index and liquid limit. An explanation of the sub-group symbols is given in Table 2.4.

The relationship between the soil's natural water content and its consistency limits, i.e. its natural or *in situ* consistency, is given by the *liquidity index* (I_L):

$$I_L = \frac{w - w_P}{I_P}$$ [2.7]

where w = natural or *in situ* water content.

Table 2.5 Activity of clays

Minerals	Activity	Soil	Activity
Muscovite	0.25	Kaolin clay	0.4–0.5
Kaolinite	0.40	Glacial clay and loess	0.5–0.75
Illite	0.90	Most British clays including	
		London Clay	0.75–1.25
Montmorillonite	>1.25	Organic estuarine clay	>1.25

Clearly, from Fig. 2.9, the significant values of LI are:

$I_L < 0$: soil is in *semi-plastic solid* or *solid* state.
$0 < I_L < 1$: soil is in *plastic* state.
$I_L > 1$: soil is in *liquid* state.

The consistency limits represent the plasticity characteristics of the soil as a whole. Plasticity, however, is mainly determined by the amount and nature of the clay minerals present. As explained in Section 1.5, the different clay minerals possess different degrees of flakiness. Also, even 'clays' may only comprise 40–50 per cent clay minerals. The degree of plasticity of the clay fraction itself is termed the *activity* of the soil:

$$\text{Activity} = \frac{I_P}{\% \text{ clay particles } (<2 \ \mu m)} \qquad [2.8]$$

Some typical values of activity for some common clay minerals and soils are given in Table 2.5.

2.7 Determination of the consistency limits

The three consistency limits (w_L, w_P and w_S) are determined by arbitrary test routines in the laboratory. Full details of the procedures and apparatus are given in BS 1377. The student may find it useful to refer also to Vickers (1983) and Head (1980). The broad outlines of the main tests are given below.

Determination of liquid limit
The apparatus (Fig. 2.10) consists basically of a stainless steel cone 35 mm long with an apex angle of 30° and having a mass (including the shaft) of 80 g. The cone is mounted on a stand which will allow it to be dropped and then held in position while its vertical movement is measured.

The soil is first dried sufficiently for it to be broken up by a mortar and pestle, with care being taken not to break individual particles. The soil is then sieved and only the material passing a 425 μm mesh sieve taken for testing. This is then thoroughly mixed with distilled water into a smooth thick paste, and stored in an air-tight container for 24 hr to allow full penetration of the water.

At the time of testing, the soil is remixed for 10 min and a portion of it placed in the brass cup. Care must be taken not to entrap air bubbles, and then the surface is struck off level with the top of the cup. After placing the cup on the

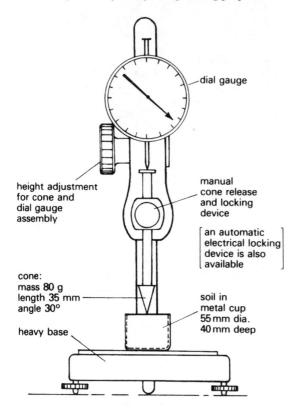

dial gauge

height adjustment
for cone and
dial gauge
assembly

manual
cone release
and locking
device

[an automatic
electrical locking
device is also
available]

cone:
mass 80 g
length 35 mm
angle 30°

soil in
metal cup
55 mm dia.
40 mm deep

heavy base

Fig. 2.10 Cone penetrometer for liquid limit test

base of the stand (Fig. 2.10), the cone is lowered so that it just touches and marks
the surface of the soil paste; the dial gauge is then set and the reading noted. The
cone is released to penetrate the soil paste for exactly 5 s and relocked in its new
position; a second dial gauge reading is now taken. The difference between the
first and second dial readings gives the amount of cone penetration (mm).

The penetration procedure is repeated several times on the same paste mix
and an average penetration obtained, after which a small portion of the soil is
taken and its water content determined. The whole of the penetration procedure
is then repeated with paste mixes having different water contents, five or six
times in all.

A graph is drawn of cone penetration/water content (Fig. 2.11) and the liquid
limit of the soil taken as the water content corresponding to a penetration of
20 mm.

Determination of the plastic limit
It is usual to prepare sufficient natural or air-dried soil as a paste with water
for both the liquid and plastic limit tests. Approximately 20 g of the soil paste is
moulded in the hand until it dries sufficiently for slight cracks to appear. The
sample is then divided into two approximately equal (10 g) portions and each of
these divided into four sub-samples. One of the sub-samples is taken and rolled

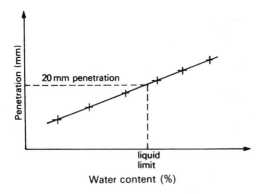

Fig. 2.11

into a ball and then it is rolled on a glass sheet to form a thread of soil. The rolling, using the palm and fingers with light pressure, is continued until the diameter of the thread reaches 3 mm. In this event, the soil is re-formed into a ball; the action of handling the soil has the effect of drying it; it is then re-rolled on the glass sheet (i.e. at a lower water content). This procedure of rolling and re-rolling is continued until the thread starts to crumble just as the diameter of 3 mm is reached; at this point the thread fragments are placed in an airtight container. The same process is carried out on the other three sub-samples, and the crumbled threads of all four gathered together and their combined water content found. The same procedure is followed with the other 10 g portion. The average of the two water contents is reported as the plastic limit. In spite of the seeming arbitrary nature of this test procedure, an experienced technician can obtain plastic limit results with very good reproducibility.

Linear shrinkage test
For soils with very small clay content the liquid and plastic limit tests may not produce reliable results. An approximation of the plasticity index may be obtained in such cases by measuring the linear shrinkage and using the following expression:

$$I_p = 2.13 \times \text{LS} \qquad [2.9]$$

The soil is prepared as for the liquid limit test and a 150 g specimen taken for the linear shrinkage test; this is then thoroughly remixed with distilled water to form a smooth homogeneous paste at approximately the liquid limit of the soil (although the exact water content is not critical). The soil/water paste is placed into a brass mould (Fig. 2.12), taking care not to entrap air, and the surface struck-off level. The soil is air-dried at 60–65 °C until it has shrunk clear of the mould and then placed in an oven at 105–110 °C to complete the drying. After cooling, the length of the sample is measured and the *linear shrinkage* obtained as follows:

$$\text{Percentage linear shrinkage, LS} = \left(1 - \frac{\text{length after drying}}{\text{initial length}}\right)100 \qquad [2.10]$$

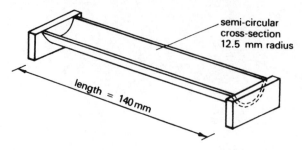

Fig. 2.12 Linear shrinkage mould

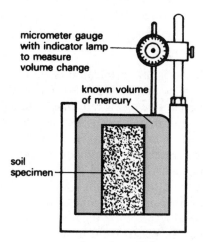

Fig. 2.13 Shrinkage limit test

Determination of shrinkage limit

The shrinkage limit test is not used much, since the w_S is not used directly in soil classification, as are the w_P and w_L. The w_S value can, however, provide some indication of the particulate structure of the soil, since a *dispersed* structure will generally produce a low w_S and a *flocculated* structure a high w_S (see Section 1.6).

A cylindrical specimen of firm plastic soil is taken for the test: usually 76 mm long and 38 mm diameter. At frequent intervals during a slow-drying process, measurements of mass and volume are taken. The volume determination is often done using a mercury displacement vessel (Fig. 2.13): the specimen being immersed in a known volume of mercury and the change in level, measured by a micrometer gauge, used to compute the change in volume.

A volume/water-content graph is plotted and the shrinkage limit obtained as shown in Fig. 2.8.

Worked example 2.4   *In a liquid limit test on a fine-grained soil, using a cone penetrometer, the following results were recorded.*

Cone penetration (mm)	15.9	17.7	19.1	20.3	21.5
Water content (%)	32.6	42.9	51.6	59.8	66.2

In a plastic limit test on the same soil the plastic limit was found to be 25 per cent. Determine the liquid limit and plasticity index of the soil and classify it according to the British Soil Classification System.

The plot of cone penetration/water content is shown in Fig. 2.14, from which the liquid limit is found to be 57 per cent.

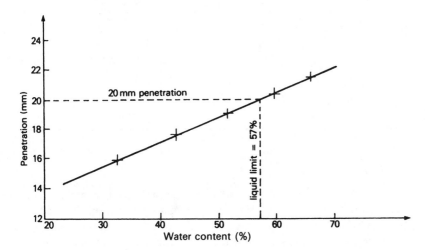

Fig. 2.14

Then the plasticity index, I_P, is given by:

$I_P = w_L - w_P$

$= 57 - 25 = \underline{32}$

From the plasticity chart (Fig. 2.9), a point having the coordinates ($w_L = 57$, $I_P = 32$) falls within the zone labelled CH, i.e. the soil is a *CLAY of high plasticity.*

Worked example 2.5   *After a series of laboratory tests, the following data were established for a fine soil:*

$w_L = 45\%$ $w_P = 18\%$
Clay content = 24.2 % (particles <2 μm)

(a)   Describe the soil according to the British Soil Classification System.
(b)   Calculate the activity of the soil.
(c)   Determine the liquidity index of the soil when its natural moisture content is 29 per cent.

(a)   Using the plasticity chart (Fig. 2.9) the soil is seen to belong to sub-group CL, i.e. *CLAY of low plasticity.*

(b)   Activity = $\dfrac{I_P}{\% \text{ clay particles}} = \dfrac{45 - 18}{24.2} = \underline{1.16}$

(c)   Liquidity index, $I_L = \dfrac{w - w_P}{I_P} \times 100 = \dfrac{29 - 18}{45 - 18} = \underline{41\%}$

Worked example 2.6   *The results of a linear shrinkage test were:*

Length before drying = 140 mm
Length after drying   = 122.4 mm

Calculate the linear shrinkage and obtain an estimate for the plasticity index.

Linear shrinkage, $\text{LS} = \left(1 - \dfrac{122.4}{140.0}\right)100$

$= \underline{12.6\%}$

Plasticity index, $I_\text{P} \approx 2.13 \times \text{LS}$

$= 2.13 \times 12.6 = \underline{27}$

2.8   Soil quality tests

The majority of soils are made up of mineral grains or rock fragments, mixed with relatively clean water, or water containing a small proportion of harmless salts. In some soils, however, the presence of acidic or alkaline compounds and certain salt compounds will produce deterioration of embedded structural components, such as steel pipes and concrete foundations or pipes.

The compounds commonly found to be most undesirable are:

Soluble sulphates, such as those of calcium and magnesium, which react with certain constituents of Portland cement to form sulphoaluminate compounds which inhibit the hardening process and also causing disruption of the cement paste binding the aggregate grains together. The concrete therefore disintegrates and becomes porous, thus allowing further ingress of groundwater and attack; leading also perhaps to the corrosion of steel reinforcement.

The total sulphate content of the soil is determined by first obtaining a solution of the salt by boiling the soil in dilute hydrochloric acid. After filtering, a solution of barium chloride is slowly added to the acid extract; producing a precipitate of barium sulphate, which is insoluble. The precipitate is filtered off, and its mass determined, after burning off the filter paper.

The sulphate content is calculated from the ratio of the mass of precipitate to the mass of the soil sample and expressed as the *percentage of SO_3*.

The sulphate content of groundwater and the *water-soluble* sulphate content of soil are determined using an ion-exchange and titration process, the results being expressed as the *sulphate content in g/litre or parts per 100 000*. Both tests are fully detailed in BS 1377.

The soil may be classified according to either the total or water soluble sulphate content or the sulphate present in groundwater. Such a classification published by the Building Research Establishment (*BRE Digest No. 250*) is summarised in Table 2.6. For each class, recommended precautions are listed, principally advising on the quantity and type of cement to be used in concrete that is to be placed in the soil.

Table 2.6 Sulphates in soil and groundwater

Class	Sulphate (SO₃) content			Recommended cement type	Minimum cement content (kg/m³)
	Total SO₃ (%)	Water-soluble SO₃ (g/litre)	SO₃ groundwater (g/litre)		Maximum water/ cement ratio
1	<0.2	<0.1	<0.3	OPC or RHPC BLENDED CEMENT	$\left[\dfrac{250}{0.70}\right]^{\dagger}$
				PBFC	$\dfrac{300}{0.60}$
2	0.2–0.5	1.0–1.9	0.3–1.2	OPC or RHPC	$\dfrac{300}{0.5}$
				BLENDED CEMENT	$\dfrac{310}{0.55}$
				SRPC	$\dfrac{290}{0.55}$
3	0.5–1.0	1.9–3.1	1.2–2.5	BLENDED CEMENT	$\dfrac{380}{0.45}$
				SRPC	$\dfrac{330}{0.50}$
4	1.0–2.0	3.1–5.6	2.5–5.0	SRPC	$\dfrac{370}{0.45}$
5	>2.0	>5.6	>5.0	SRPC + protective coating*	$\dfrac{370}{0.45}$

OPC: Ordinary Portland cement
RHPC: Rapid-hardening Portland cement
SRPC: Sulphate-resisting Portland cement
PBFC: Portland blast furnace cement

After *BRE Digest No. 250* (June 1981)

BLENDED CEMENT: OPC or RHPC blended with 70–90% ground granulated blast furnace slag or with 25–40% pulverised fuel ash to BS 3892
* See CP 102: 1973
† For plain concrete

Organic acids, which occur naturally in peaty soils, may react with the lime content of Portland cement to form insoluble calcium salts. Very little deterioration is caused to dense impermeable concrete, of low water/cement ratio.

The *percentage organic matter* content may be determined using an oxidising solution such as potassium dichromate, which is then titrated against a standardised solution of ferrous sulphate. The test is fully detailed in BS 1377.

pH value. In some natural soils, especially those containing sulphides (e.g. iron pyrites, galena) or sulphate-reducing bacteria, a high acid content may be found; or a high alkali content in limy soils. The presence of industrial wastes and other pollutants also may produce acidic or alkaline conditions likely to cause corrosion of buried iron and steel and, in some cases, deterioration of concrete.

A measure of acidity/alkalinity is given by determining the *pH value*:

$$\text{pH} = -\log_{10}(\text{hydrogen ion content})$$

A solution with an exact balance of H^+ and OH^- ions is neutral and has a pH value of 7.0; an excess of H^+ occurs in an acid solution so that pH < 7.0; if more OH^- ions are present the solution is alkaline so that pH > 7.0. Tests for the determination of pH value are fully detailed in BS 1377.

Ordinary Portland cement is alkaline resistant (pH > 7), but is not recommended when pH $\leq$ 6; high alumina and supersulphated cement may be used down to pH4; steel corrodes when pH < 9. In cases where the pH value is found to be < 6.5 or > 7.5, a more detailed chemical analysis should be carried out to ascertain the true nature of the active compounds present in the soil.

Soil and groundwater quality investigations
Where there is the possibility of the soil or groundwater conditions providing some form of attack on buried steel, concrete or timber structures, a quality survey should be undertaken. Samples of groundwater and disturbed soil samples should be collected and sent for chemical analysis. Initially, sulphate content and pH tests will be required, with more detailed tests following, if required.

It is also necessary to establish patterns of groundwater movement, by means of standpipe or piezometer observations; observations in trial pits and natural watercourses may also be of use. Since the presence of salts and acid solutions decreases the resistivity of the soil, a resistivity survey can often provide a rapid and economic means of investigation. Care must be taken to carry out investigations at different times of the year and under different climatic conditions; considerable variation in sulphate content, for example, will be noticed after a prolonged dry spell compared with that after a period of heavy rain.

Exercises

2.1–2.3 The following data were recorded during particle size analysis tests in the laboratory. Plot the grading curve in each case and determine the grading characterstics. Using this information classify the soil according to the British Soil Classification System.

Soil 1
Fine-mesh dry sieve analysis weighings:

Sieve size (mm or μm)	1.18	600	425	300	212	150	63
Mass retained (g)	0	1.9	3.2	8.1	10.4	22.7	126.1

Total sample = 212.0 g

Soil 2
Stage 1: Total mass = 2752 g   Coarse-mesh sieving

Sieve size (mm)	37.5	28	20
Mass retained (g)	0	104.6	170.6

Pan contents = 2477 g   Riffled down to 1382 g

Stage 2: Medium-mesh sieving

Sieve size (mm)	14	10	6.3
Mass retained (g)	115.3	127.6	190.6

Pan contents = 949 g   Riffled down to 245.4 g

Stage 3: Fine-mesh sieving

Sieve size (mm or μm)	5	3.35	2.0	1.18	600	425	300	212	150	63
Mass retained (g)	27.0	44.5	35.7	37.3	32.2	13.1	15.9	12.7	9.1	15.9

Pan contents = 1.9 g

Soil 3
Stage 1: Wet and dry sieving on sample of 574.5 g
Medium-mesh fractions:

Sieve size (mm)	14	10	6.3
Mass retained (g)	0	4.6	14.9

Stage 2: Wet and dry sieving on sub-sample of 168.2 g
Fine-mesh fractions

Sieve size (mm or μm)	5.0	3.35	2.0	1.18	600	425	300	212	150	63
Mass retained (g)	4.5	14.5	24.0	27.2	14.1	4.0	3.3	3.1	2.1	7.6

Mass passing 63 μm = 63.7 g

Stage 3: Sedimentation analysis on the 63.7 g passing 63 μm in Stage 2

Size (μm)	40	20	10	6	2
% coarser than	13.1	16.9	25.4	12.0	20.8

2.4 From the grading characteristics given below sketch the grading curves for the three soils.

	Soil		
	A	B	C
d_{10} (mm)	0.28	0.088	0.009
C_u	1.50	19.9	167
C_g	0.87	0.80	0.12
Class	SP	SW	GM

2.5 The following results were recorded during a cone penetrometer test on a cohesive soil:

Average penetration (mm)   15.2   17.3   18.9   21.1   22.8
Average water content (%)   33.4   42.6   49.2   59.4   66.8

(a) Determine the liquid limit of the soil.
(b) If the plastic limit of the soil was found to be 33 per cent, determine its plasticity index and classify the soil.

2.6 Following laboratory tests on four soils, the data given below were determined. For each soil, give the class according to the British Soil Classification System and calculate the activity and liquidity index.

	Soil			
	E	F	G	H
Liquid limit (%)	42	59	26	74
Plastic limit (%)	16	37	20	29
Particles <2 μm (%)	32.5	4.6	9.4	28.8
In situ water content (%)	24	45	18	64

Soil Mechanics Spreadsheets and Reference Assignments and Quizzes (available on the accompanying CD):

Assignments A.2   Soil classification and description
Quiz Q.2   Soil classification and description

Chapter 3

Basic physical properties of soils

3.1 The soil model and basic properties

The basic physical properties of a soil are those required to define its physical state. For the purposes of engineering analysis and design, it is necessary to quantify the three constituent phases (solid, liquid and gas) and to be able to express relationships between them in numerical terms. For example, a soil's water content is simply the ratio of the mass of water to the mass of solid. Densities, i.e. the relationships between mass and volume, are also important measures of a soil's physical state. In a typical soil, the solid, liquid (water) and gas (air) are naturally intermixed, so that relative proportions are difficult to visualise. It is therefore convenient to consider a soil model in which the three phases, while still present in their correct proportions, are separated into distinct amounts.

Several possible phase models can be proposed (Fig. 3.1), each taking its name from that quantity providing a reference amount of unity. For example, the unit *solid volume* model is based on 1 volume unit, e.g. 1 m³, of solid material; the unit *solid mass* model on 1 mass unit, e.g. 1 kg; the unit *total volume* model on 1 volume unit of all three phases combined together. For most purposes in soil mechanics, the *unit solid volume* model is the most convenient, since the solid constituents of soil (with the exception of peaty material) may be considered to

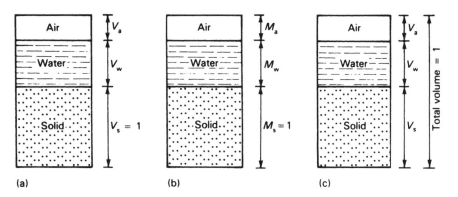

Fig. 3.1 Three-phase soil models
(a) Unit solid volume (b) Unit solid mass (c) Unit total volume

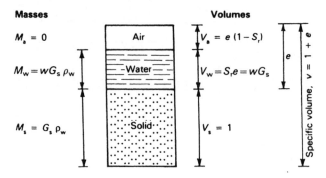

Fig. 3.2   Unit solid volume soil model

be incompressible. The model is therefore constructed, as it were, about 1 unit (1 m³) of solid material, which may be expected to remain constant. All other quantities are now referenced to this amount. A given soil is therefore depicted as a fixed volume of solid material, together with varying amounts of water and air. The amount of volume in the soil not occupied by solids is termed the *voids volume*, the ratio of voids volume to solid volume being *e*. In a perfectly *dry* soil there is no water and the void space is entirely air; in a saturated soil the void space is full of water.

In Fig. 3.2 the soil model is shown in detail with the various mass and volume dimensions indicated. From this basic model several important quantities may now be defined.

Void ratio (*e*)
The volume not occupied by solids is known as *voids volume*: it may be occupied by either water or air, or by a mixture of these.

$$\text{Void ratio, } e = \frac{\text{volume of voids}}{\text{volume of solids}} = \frac{V_v}{V_s} = \frac{V_a + V_w}{V_s} \qquad [3.1]$$

Porosity (*n*)
Another way of expressing the quantity of voids is to relate the voids volume to the *total* volume:

$$\text{Porosity, } n = \frac{\text{volume of voids}}{\text{total volume}} = \frac{V_v}{V}$$

From Fig. 3.2,

$$n = \frac{e}{1 + e} \qquad [3.2]$$

Specific volume (*v*)
The total volume of the soil model is equal to $1 + e$, this quantity is known as the *specific volume* of the soil.

$$\text{Specific volume, } v = 1 + e \qquad [3.3]$$

Degree of saturation (S_r)

The quantity of water in the soil may be expressed as a fraction of the voids volume; this fraction is known as the *degree of saturation*.

Degree of saturation, $\quad S_r = \dfrac{\text{volume of water}}{\text{volume of voids}} = \dfrac{V_w}{V_v}$ [3.4]

Percentage saturation $\quad = 100S_r$

For a perfectly *dry* soil, $\quad S_r = 0$

and for a *saturated* soil, $\quad S_r = 1$

Air-voids content (A_v)

The *air-voids volume* of a soil is that part of the voids volume not occupied by water.

Air-voids volume = volume of voids − volume of water

$$V_a = V_v - V_w$$
$$= e - S_r e$$
$$= e(1 - S_r) \qquad [3.5]$$

The *air-voids content* is the ratio of the air-voids volume to the specific volume of the soil.

$$A_v = \frac{e(1 - S_r)}{1 + e} = n(1 - S_r) \qquad [3.6]$$

or since $S_r = wG_s/e$ (see eqn [3.10] below)

$$A_v = \frac{e - wG_s}{1 + e} \qquad [3.7]$$

Percentage air voids = $100A_v$.

Worked example 3.1 *For a soil having a void ratio of 0.750 and percentage saturation of 85 per cent, determine the porosity and air-voids ratio.*

From eqn [3.2], porosity, $\quad n = \dfrac{0.750}{1 + 0.750} = \underline{0.429}$

The degree of saturation, $\quad S_r = \dfrac{85}{100} = 0.85$

Then from eqn [3.6], air-voids ratio, $A_v = 0.429(1 - 0.85) = \underline{0.064}$ or percentage air voids = $\underline{6.4\%}$

Grain specific gravity (G_s) **and particle density** (ρ_s)

The ratio of the mass of a given volume of a material to the mass of the same volume of water is termed the *specific gravity* of the material. The mass of the one unit of solid volume in the soil model (Fig. 3.2) will therefore be:

$$M_s = G_s\rho_w \qquad [3.8(a)]$$

where ρ_w = the density of water, which may be taken as 1.00 Mg/m^3.

Particle density (ρ_s), or *grain density*, is the mass per unit volume of the solid particles, or grains, and is equal to

$$\rho_s = G_s\rho_w \qquad\qquad [3.8(b)]$$

The mass of water in the soil model will be:

$$M_w = S_r e \rho_w \qquad\qquad [3.9]$$

Water content (w)
The ratio of the mass of water to the mass of solids is termed the *water content* of the soil.

$$\text{Water content, } w = \frac{\text{mass of water}}{\text{mass of solids}} = \frac{M_w}{M_s}$$

From eqns [3.8] and [3.9]:

$$w = \frac{S_r e \rho_w}{G_s \rho_w}$$

$$= \frac{S_r e}{G_s} \qquad\qquad [3.10]$$

or $wG_s = S_r e$ [3.11]

Percentage water content = $100m$.

Worked example 3.2   *An oven tin containing a sample of moist soil was weighed and had a mass of 37.82 g; the empty tin had a mass of 16.15 g. After drying, the tin and soil were weighed again and had a mass of 34.68 g. Determine the void ratio of the soil if the air-voids content is (a) zero (b) 5 per cent (G$_s$ = 2.70).*

$$\text{Water content, } w = \frac{\text{mass of water}}{\text{mass of dry soil}}$$

$$= \frac{37.82 - 34.68}{34.68 - 16.15} = 0.169$$

(a) If the air-voids content is zero, $S_r = 1$

Then from eqn [3.11]: $\qquad e = wG_s$

$$= 0.169 \times 2.70 = \underline{0.456}$$

(b) If the air-voids content is 5 per cent, $A_v = 0.05$

Then rearranging eqn [3.7]: $\qquad e = \dfrac{wG_s + A_v}{1 - A_v}$

$$= \frac{0.169 \times 2.70 + 0.05}{0.95} = \underline{0.533}$$

3.2 Soil densities

The quantities referred to as densities provide a measure of the *quantity* of material related to the *amount of space* it occupies. Several *density* relationships may be defined:

Dry density, $\qquad \rho_d = \dfrac{\text{mass of solids}}{\text{total volume}}$

From Fig. 3.2, $\quad \rho_d = \dfrac{G_s \rho_w}{1+e} = \dfrac{\rho_s}{1+e}$ [3.12]

Bulk density, $\qquad \rho = \dfrac{\text{total mass}}{\text{total volume}} = \dfrac{\text{mass of solids} + \text{mass of water}}{\text{total volume}}$

From Fig. 3.2, $\quad \rho = \dfrac{G_s \rho_w + S_r e \rho_w}{1+e} = \dfrac{\rho_s + S_r e \rho_w}{1+e}$ [3.13]

A useful relationship is obtained from the ratio of these two densities:

$$\frac{\rho}{\rho_d} = \frac{(G_s + S_r e)\rho_w/(1+e)}{G_s \rho_w/(1+e)} = 1 + \frac{S_r e}{G_s}$$

or since $S_r e = w G_s$

$$\rho = (1 + w)\rho_d$$ [3.14]

The *saturated density* is the bulk density of the soil when saturated, i.e. when $S_r = 1$.

Then saturated density, $\rho_{sat.} = \dfrac{G_s + e}{1+e}\rho_w$ [3.15]

The *submerged density* or *effective density* of a soil is the notional effective mass per unit (total) volume, when submerged. When a unit (total) volume of soil is submerged in water it displaces an equal volume of water; the net mass of a unit volume of soil when submerged is then $\rho_{sat.} - \rho_w$. This is referred to as the *submerged density*.

Submerged density, $\rho' = \rho_{sat.} - \rho_w$ [3.16]

Unit weights

The weight of a unit volume of soil is referred to as its *unit weight*. The units of *unit weight* will be *force* per unit volume, whereas the units of *density* are *mass* per unit volume. Unit weights are related to corresponding densities as follows:

Dry unit weight $\qquad\qquad \gamma_d = \rho_d g \text{ kN/m}^3$ [3.17]

Bulk unit weight $\qquad\qquad \gamma = \rho g \text{ kN/m}^3$ [3.18]

Saturated unit weight $\qquad \gamma_{sat.} = \rho_{sat.} g \text{ kN/m}^3$ [3.19]

Unit weight of water $\qquad\quad \gamma_w = \rho_w g \text{ kN/m}^3$ [3.20]

Submerged unit weight $\qquad \gamma' = \gamma_{sat.} - \gamma_w \text{ kN/m}^3$ [3.21]

In eqns [3.17] to [3.21] above the value of gravitational acceleration (g) is taken as 9.81 m/s^2.

Worked example 3.3   *In a sample of moist clay soil, the void ratio is 0.788 and the degree of saturation is 0.93. Assuming $G_s = 2.70$, determine the dry density, the bulk density and the water content.*

Using eqn [3.12], dry density, $\rho_d = \dfrac{2.70 \times 1.00}{1 + 0.788} = \underline{1.51 \text{ Mg/m}^3}$

Using eqn [3.13], bulk density, $\rho = \dfrac{2.70 + 0.93 \times 0.788}{1 + 0.788} \times 1.00 = \underline{1.92 \text{ Mg/m}^3}$

From eqn [3.11], $w = \dfrac{S_r e}{G_s}$

$= \dfrac{0.93 \times 0.788}{2.70} = \underline{0.271}$

Worked example 3.4   *The bulk density of a sand in a drained condition above the water table was found to be 2.06 Mg/m^3 and its water content was 18 per cent. Assume $G_s = 2.70$ and calculate: (a) the drained unit weight, and (b) the saturated unit weight and water content of the same sand below the water table.*

(a)   Using eqn [3.17]:

drained unit weight $\gamma = \rho g = 2.06 \times 9.81 = \underline{20.2 \text{ kN/m}^3}$

(b)   From eqn [3.14]: $e = (1 + w)\dfrac{G_s \rho_w}{\rho} - 1$

$= \dfrac{1.18 \times 2.70 \times 1.00}{2.06} - 1 = \underline{0.547}$

Using eqn [3.15]: $\rho_{sat.} = \dfrac{2.70 + 0.547}{1.547} \times 1.00 = 2.10 \text{ Mg/m}^3$

Then the saturated unit weight $\gamma_{sat} = \rho_{sat.} g$

$= \underline{20.6 \text{ kN/m}^3}$

And the saturated water content, $w = \dfrac{e}{G_s}$

$= \dfrac{0.547}{2.70} = \underline{0.203}$

Worked example 3.5   *After a laboratory compression test, a cylindrical specimen of saturated clay was found to have a mass of 157.28 g and a thickness of 17.4 mm. After then drying to constant weight, its mass was 128.22 g. If the grain specific gravity is 2.68, calculate: (a) the end-of-test water content and void ratio, and (b) the void ratio and water content at the start of the test when the thickness was 18.8 mm, and assuming that the diameter remains constant and the sample remains saturated.*

(a)   End-of-test water content, $w_1 = \dfrac{157.28 - 128.22}{128.22}$

$$= \underline{0.2266 \ (22.7\%)}$$

Since the soil is saturated, $S_r = 1$

Hence the void ratio, $e_1 = wG_s = 0.2266 \times 2.68 = \underline{0.607}$

(b)   If the diameter has remained constant the change in volume (ΔV) is represented by the change in thickness (ΔH), so that:

volumetric strain, $\dfrac{\Delta V}{V_o} = \dfrac{\Delta H}{H_o} = \dfrac{\Delta e}{1 + e_o}$

where V_o, H_o and e_o are the original values of volume, thickness and void ratio respectively.

Original void ratio, $e_o = e_1 + \dfrac{\Delta H}{H_o}(1 + e_1)$

$$= 0.607 + \frac{(18.8 - 17.4)}{17.4}(1 + 0.607) = \underline{0.736}$$

And the original moisture content, $w_o = 0.736/2.68 = \underline{0.275 \ (27.5\%)}$

Worked example 3.6   *A specimen of clay was tested in the laboratory and the following data were collected:*

Mass of wet specimen $M_1 = 148.8 \ g$

Mass of dry specimen $M_2 = 106.2 \ g$

Volume of wet specimen $V = 86.2 \ cm^3$

Specific gravity of particles $G_s = 2.70$

Determine: (a) the water content, (b) the bulk and dry densities, (c) the void ratio and porosity, and (d) the degree of saturation.

(a)   Water content, $w = \dfrac{M_1 - M_2}{M_2} = \dfrac{148.8 - 106.2}{106.2} = \underline{0.401 \ (40.1\%)}$

(b)   Bulk density, $\rho = \dfrac{\text{total mass of wet soil}}{\text{volume}}$

$$= \frac{148.8 \times 10^{-6}}{86.2 \times 10^{-6}} = \underline{1.73 \ Mg/m^3}$$

Dry density, $\rho_d = \dfrac{\text{mass of dry soil}}{\text{volume}}$

$$= \frac{106.2 \times 10^{-6}}{86.2 \times 10^{-6}} = \underline{1.232 \ Mg/m^3}$$

(c)   Void ratio e may be obtained from $\rho_d = G_s\rho_w/(1 + e)$:

$$1 + e = \frac{G_s\rho_w}{\rho_d} = \frac{2.70 \times 1.00}{1.232} = 2.192$$

Hence $e = \underline{1.192}$ and $n = 0.544$

(d)   From $S_r e = wG_s$, $S_r = \dfrac{0.401 \times 2.70}{1.192} = \underline{0.908}$ (91%)

Density index (I_D) – also called relative density
The actual void ratio of a soil lies somewhere between the possible minimum and maximum values, i.e. $e_{min.}$ and $e_{max.}$, depending on the state of compaction. In the case of sands and gravels, there is a good deal of variation between the two extremes. A convenient measure of the state of compaction is provided by a relationship between the void ratio values which is termed the *density index*.

$$I_D = \frac{e_{max.} - e}{e_{max.} - e_{min.}} \qquad\qquad [3.22]$$

Thus for a soil in its densest state, $I_D = 1$, and when in its loosest state, $I_D = 0$. A suggested simple classification of the state of compaction is given in Table 3.1. The void ratio values may be obtained from laboratory tests.

Table 3.1   Relative compaction states

Density index (%)	0–15	15–35	35–65	65–85	85–100
State of compaction	Very loose	Loose	Medium	Dense	Very dense

3.3   Determination of porosity and void ratio

The best practical method of determining porosity, and/or void ratio, of a cohesionless soil consists of filling a suitable mould or container (e.g. compaction mould, gas jar, etc.) with water and then adding the soil to fill the mould. The volume occupied by the soil particles may then be determined by comparing the masses contained in the mould of water and soil + water. The soil should either be oven-dry (when the particles themselves are non-porous), or be in a saturated surface-dry condition.

The minimum void ratio ($e_{min.}$) may be determined by placing a standard compaction mould (mass $= M_1$) under water. The soil is then placed in the mould in three layers of approximately equal thickness, each of which is thoroughly compacted using a vibrating hammer. The collar of the mould is then removed, the surface struck-off level and the mass of mould + soil + water determined (M_2).

If V = volume of mould

Saturated density, $\rho_{sat.} = \dfrac{M_2 - M_1}{V}$

Also, assuming the soil to be saturated, the bulk density is given by eqn [3.15].

So
$$\rho_{sat.(max.)} = \frac{G_s + e_{min.}}{1 + e_{min.}}\rho_w$$

Transposing
$$e_{min.} = \frac{G_s\rho_w - \rho_{sat.(max.)}}{\rho_{sat.(max.)} - \rho_w} \qquad [3.23(a)]$$

And
$$n_{min.} = \frac{e_{min.}}{1 + e_{min.}} \qquad [3.24(a)]$$

The maximum void ratio ($e_{max.}$) may be determined approximately by placing the mould (or other suitable container) under water and quickly pouring the soil into it from just above the top. The collar of the mould (if any) is then removed, the surface struck-off level and the mass of mould + water + soil determined. The values of $e_{max.}$ and $n_{max.}$ can then be obtained using

$$e_{max.} = \frac{G_s\rho_w - \rho_{sat.(min.)}}{\rho_{sat.(min.)} - \rho_w} \qquad [3.23(b)]$$

$$n_{max.} = \frac{e_{max.}}{1 + e_{max.}} \qquad [3.24(b)]$$

Worked example 3.7   *In order to find the density index of a soil, a compaction mould was used having a mass of 5.225 kg and a volume of 944 ml. When the soil was dynamically compacted in the mould, the total mass of soil and mould was 7.289 kg; and when the soil was poured in loosely, the mass was 6.883 kg. If the in situ dry density of the soil is 1.54 Mg/m³ and G_s = 2.70, calculate the density index of the soil.*

The compacted mass will give the maximum saturated density

$$\rho_{sat.(max.)} = \frac{7.289 - 5.225}{944 \times 10^{-3}} = 2.186 \text{ Mg/m}^3$$

From eqn [3.23a]: $e_{min.} = \dfrac{2.70 \times 1.00 - 2.186}{2.186 - 1.00} = 0.433$

The loose mass will give the minimum saturated density

$$\rho_{sat.(min.)} = \frac{6.883 - 5.225}{944 \times 10^{-3}} = 1.756 \text{ Mg/m}^3$$

From eqn [3.23(b)]: $e_{max.} = \dfrac{2.70 \times 1.00 - 1.756}{1.756 - 1.00} = 1.249$

From eqn [3.12]: $e = \dfrac{G_s\rho_w}{\rho_d} - 1$

$$= \frac{2.70 - 1.00}{1.54} - 1 = 0.753$$

Then from eqn [3.22]: $I_D = \dfrac{1.249 - 0.753}{1.249 - 0.433} = \underline{0.61 \ (61\%)}$

3.4   Determination of specific gravity of soil particles

For fine soils, a density bottle of about 50 ml capacity is used. For coarse soils, a 500 ml or 1000 ml container is used, which may be either an ordinary *gas jar* (Fig. 3.3(a)) or a special glass jar, fitted with a conical screw top, which is called a *pycnometer* (Fig. 3.3(b)).

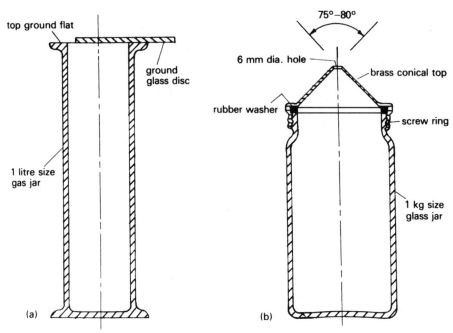

Fig. 3.3   (a) Gas jar (b) Pycnometer

An appropriate quantity of dried soil (depending on the particle size) is placed in the jar and weighed. The jar is then filled with de-aired water and agitated to remove any air bubbles. After carefully topping-up with water, the jar is weighed again. Finally, the jar is emptied and cleaned, and then filled with de-aired water and weighed again.

Now M_1 = mass of empty jar
M_2 = mass of jar + dry soil
M_3 = mass of jar + soil + water
M_4 = mass of jar + water only

Then the specific gravity of the soil particles is obtained thus:

$$G_s = \frac{\text{mass of soil}}{\text{mass of water displaced by soil}}$$

$$= \frac{M_2 - M_1}{(M_4 - M_1) - (M_3 - M_2)} \qquad [3.25(a)]$$

An alternative procedure is to weigh the jar empty and then full of water. A pre-weighed quantity of soil is then poured into the jar and stirred. After carefully topping-up with water, the jar is weighed again. The quantities are now as follows:

M_1 = mass of empty jar
M_s = mass of soil
M_3 = mass of jar + soil + water
M_4 = mass of jar + water only

and $\quad G_s = \dfrac{M_s}{M_4 - M_3 + M_s}$ [3.25(b)]

Worked example 3.8 *A pycnometer used in a specific gravity test was found to have a mass of 524 g when empty and 1557 g when full of clean water. An air-dried sample of cohesionless soil having a mass of 512 g was placed in the jar and stirred to expel any entrapped air. The pycnometer was then carefully filled with clean water, when it had a total mass of 1878 g. Determine the specific gravity of the soil.*

Equation [3.25(b)] may be used in which:

$M_1 = 524$ g $\quad M_s = 512$ g
$M_3 = 1878$ g $\quad M_4 = 1557$ g

Then $G_s = \dfrac{512}{1557 - 1878 + 512} = \underline{2.68}$

3.5 Compaction of soil

The process of *compaction* brings about an increase in soil density, with a consequent reduction of air-voids volume, but with no change in the volume of water. This is usually effected by mechanical means, such as rolling, tamping or vibrating the soil. In the construction of road bases, runways, earth dams and embankments, the soil is placed in layers of specified thickness, each layer being compacted to a specification relating to the type of plant in use. A comparative list of common types of compaction plant is given in Table 3.2.

There are three main objectives in the compaction of soil, namely:

(a) To reduce the void ratio and thus the permeability of the soil (this also has the effect of controlling the water absorption and subsequent changes in water content).
(b) To increase the shear strength and therefore the bearing capacity of the soil.
(c) To make the soil less susceptible to subsequent volume changes and therefore the tendency to settlement under load or under the influence of vibration.

Table 3.2   Suitability of compaction plant

Type of plant	Suitable for	Unsuitable for
Smooth-wheeled roller	Well-graded sand and gravels; silts and clays of low plasticity	Uniform sands; silty sands; soft clays
Grid roller	Well-graded sand and gravels; soft rocks; stony cohesive soils	Uniform sand; silty sands; silty clays
Sheepsfoot roller (tamping roller)	Sands and gravels with more than 20% fines; most fine-grained soils	Very coarse-grained soils; gravels without fines
Pneumatic-tyred roller	Most coarse-grained and fine-grained soils	Very soft clay; soils of highly variable consistency
Vibrating roller	Sands and gravels with no fines; wet cohesive soils	Silts and clays; soils with 5% or more fines; dry soils
Vibrating plates	Soils with up to 12–15% fines; confined areas	Large-volume work
Power rammer	Trench backfill; work in small areas or where access is restricted	Large-volume work

The effectiveness of the compaction process is dependent on several factors:

(a)   The nature and type of soil (i.e. sand or clay; uniform or well graded; plastic or non-plastic).
(b)   The water content at the time of placing.
(c)   The maximum possible state of compaction attainable for the soil.
(d)   The maximum amount of compaction attainable under field conditions.
(e)   The type of compaction plant being used.

Maximum dry density/water content relationship
The state of compaction of a soil is conveniently measured using the dry density, the attainable values of which are related to the water content. As water is added to a dry soil films of adsorbed water form around the particles. As the adsorbed water films increase in thickness the particles become lubricated and are able to pack more closely together, thus the density increases. At a certain point, however, the porewater pressure in adsorbed films tends to push the particles apart and so with further increases in water content the density decreases. The *maximum dry density* therefore occurs at an *optimum water content* w_{opt} as shown in Fig. 3.4.

In order to assess the compaction potential of a soil one of three standard laboratory tests is used, as detailed in BS 1377 and summarised in Table 3.3. The method of test consists essentially of placing the soil in the appropriate mould in either three or five equal layers, each of which is subject to a specified amount of compactive effort. After compacting the final layer, the bulk density of the soil contained in the mould is determined and a sample of soil taken to find its water content. The soil is then removed from the mould, remixed with an

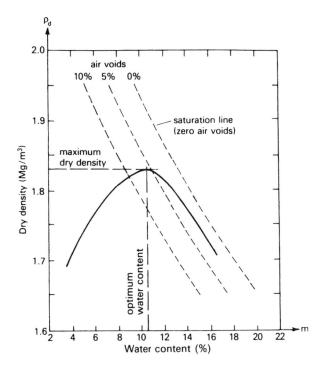

Fig. 3.4 Dry density/water content graph

Table 3.3 BS compaction tests

Name of test	Light compaction		Heavy compaction		Vibrating hammer
Rammer: mass		2.5 kg		4.5 kg	300–400 N
face diameter		50 mm		50 mm	150 mm
drop height		300 mm		450 mm	
Soil size	<20 mm	>20 mm	<20 mm	>20 mm	>20 mm
Soil quantity	5 kg	25 kg	5 kg	25 kg	25 kg
Mould: volume	1000 ml	2300 ml	1000 ml	2300 ml	2300 ml
internal diam.	105 mm	152 mm	105 mm	152 mm	152 mm
internal height	115.5 mm	127 mm	115.5 mm	127 mm	127 mm
No. of layers	3	3	5	5	3
No. blows per layer	27	62	27	62	(60 s)

additional amount of water and the test procedure repeated. A total of at least five tests should be carried out for a given soil. Initially, the soil is air-dried and particles larger than 20 mm removed.

From the values of bulk density and water content obtained, the dry density is calculated (see eqn [3.14]):

$$\rho_d = \rho/(1 + w)$$

and a graph of dry density/water content plotted (Fig. 3.4). From the curve, the *maximum dry density* for that amount of compactive effort is determined. Also reported is the water content value corresponding to the maximum dry density, termed the *optimum water content*.

The maximum possible dry density at a given water content is known as the *saturation dry density*, in which state the soil will have zero air voids, i.e. $A_v = 0$. For values of $A_v > 0$, the maximum attainable dry density is given by the following expression:

$$\rho_d = \frac{G_s \rho_w}{1 + wG_s}(1 - A_v) \qquad [3.26]$$

which is obtained by considering the model soil sample shown in Fig. 3.5 in which:

Total volume, $V = V_a + V_w + V_s$

Substituting and transposing:

$$V(1 - A_v) = V_s(1 + wG_s)$$

Giving
$$\frac{V_s}{V} = \frac{1 - A_v}{1 + wG_s}$$

Now
$$\rho_d = \frac{V_s G_s \rho_w}{V}$$

Then
$$\rho_d = \frac{G_s \rho_w}{1 + wG_s}(1 - A_v) \qquad \text{Q.E.D.}$$

In order to obtain an estimate of the air-voids content in the soil, a set of curves representing 0, 5 per cent and 10 per cent air voids is added to the dry density/water content graph (Fig. 3.4). An increase in compactive effort produces a higher maximum dry density at a lower optimum water content, with the air-voids ratio remaining almost the same (Fig. 3.6).

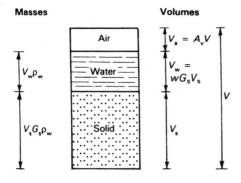

Masses   **Volumes**

Air $V_a = A_v V$

$V_w \rho_w$ Water $V_w = wG_s V_s$

$V_s G_s \rho_w$ Solid V_s

V

Fig. 3.5

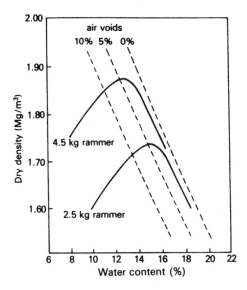

Fig. 3.6 Effect of different compactive effort on ρ_d/w curve

---

Worked example 3.9 *In a BS compaction test the following data were collected:*

Water content (%)	5	8	10	13	16	19
Bulk density (Mg/m³)	1.87	2.04	2.13	2.20	2.16	2.09

$G_s = 2.70$

(a) *Draw the graph of dry density against water content and from it determine the maximum dry density and optimum water content.*

(b) *On the same axes, draw the ρ_d/w curves for zero and 5 per cent air voids, and hence determine the air-voids content at the maximum dry density.*

(a) From eqn [3.14]: $\rho_d = \rho/(1 + w)$

Water content (%)	5	8	10	13	16	19
Dry density (Mg/m³)	1.78	1.89	1.94	1.95	1.86	1.76

Plotting these figures, the curve shown in Fig. 3.7 is obtained.
From the curve: maximum dry density = <u>1.96 Mg/m³</u>
optimum water content = <u>11.9%</u>

(b) Using eqn [3.26], the dry densities corresponding to 0 and 5 per cent air voids will be:

	Water content (%)					
	10	*12*	*14*	*16*	*18*	*20*
ρ_d when: $A_v = 0$	2.13	2.04	1.96	1.89	1.82	1.75
$A_v = 5\%$	2.02	1.94	1.86	1.79	1.73	1.67

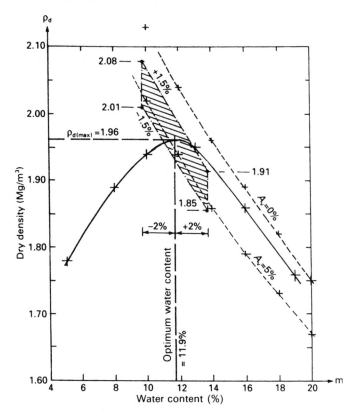

Fig. 3.7

By plotting the two curves (Fig. 3.7) and interpolating at the $\rho_{d(\text{max.})}$ value:

At $\rho_{d(\text{max.})}$, $A_v = \underline{4.0\%}$

Or by calculation using eqn [3.26]:

$$A_v = 1 - \frac{1.96}{2.7 \times 1.00}(1 + 2.70 \times 0.119) = \underline{0.0408}\ (4.1\%)$$

The value of A_v at the maximum dry density is sometimes referred to as the *optimum air-voids ratio or optimum air content.*

Worked example 3.10   *Under field conditions variations in the applied compactive effort may cause the air-voids content to vary by ±1.5 per cent. Also, the field water content may vary above and below the optimum value by 2 per cent. Indicate, therefore, the range of dry densities that may be found after compaction in the field.*

The hatched areas superimposed on the curve in Fig. 3.7 indicate the estimated limits for the dry density values based on $A_v(\rho_{d\text{max}}) = \pm1.5\%$ and $w = w_{\text{opt}} \pm 2\%$

At $w = w_{\text{opt}} - 2\%$ $\rho_d = 2.01$ to 2.08 Mg/m³
At $w = w_{\text{opt}} + 2\%$ $\rho_d = 1.85$ to 1.91 Mg/m³

The minimum control level required to ensure adequate compaction within these limits may be taken from the shaded area: $\underline{1.91\ \text{Mg/m}^3}$.

3.6   *Field density measurements*

The dry density achieved in the field after compaction must be compared with the maximum value obtained in the laboratory. The required quality standard may be specified in terms of the *relative compaction percentage*.

$$\text{Relative compaction} = \frac{\text{achieved } \rho_d}{\text{max. } \rho_d} \times 100 \qquad\qquad [3.27]$$

A summary of specifications for compaction is given in Section 3.7.

In order to measure the achieved field dry density, frequent measurements of bulk density and water content of the placed material are necessary. Samples should be taken along the centre and edges of the compacted area at intervals of 20–50 m, or to give an approximate frequency of one measurement per 1000 m². For layers over 250 mm in thickness, measurements should be taken in the bottom 150 mm. A number of methods are in common use, full details of which are given in BS 1377.

The core cutter method

This method is suitable for fine soils free of stones and consists of driving a steel cylinder, with a hardened cutting edge, into the ground using a specially designed steel rammer and protective dolly (Fig. 3.8). The cutter is then dug out

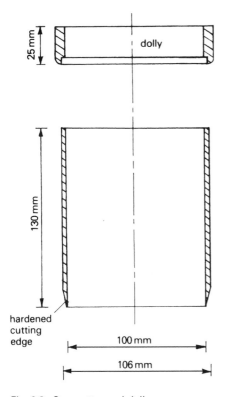

Fig. 3.8 Core cutter and dolly

and the soil trimmed off flush at each end. Since the volume of the cutter is known and the contained mass of soil can be found by weighing, the bulk density may easily be determined. At the same time, small samples of soil are taken from either end from which the water content is determined.

The sand replacement method

This method is suitable for granular soils and involves the use of a *sand-pouring cylinder* as shown in Fig. 3.9. Firstly, a small hole is dug about 100 mm in diameter and not more than 150 mm in depth and the soil removed carefully weighed. The volume of the hole is then determined by pouring sand into it from the pouring cylinder. The sand-pouring cylinder is weighed before and after this operation, and the mass of sand filling the hole determined. Since the density of the sand is known, the volume of the hole can be determined, and hence the bulk density of the *in situ* soil. Two sizes of sand-pouring cylinder are recommended: a small version suitable for fine and medium soils and a large version for fine, medium and coarse soils.

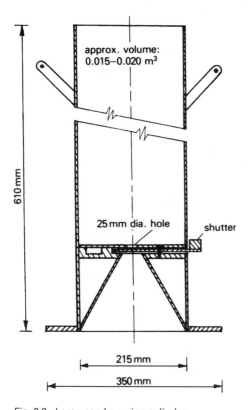

Fig. 3.9 Large sand-pouring cylinder

Worked example 3.11 A sand replacement test was carried out to determine the in situ *bulk density of a soil. From the following recorded data, determine the value of density required.*

Mass of soil removed from hole	*= 2764 g*
Initial total mass of sand-pouring cylinder	*= 5724 g*
Final total mass of sand-pouring cylinder	*= 3172 g*
Volume of cone in sand-pouring cylinder	*= 248 cm³*
Density of pouring sand	*= 1560 kg/m³*

Mass of sand run out of cylinder $= (5724 - 3172)10^{-3} = 2.552$ kg
Mass of sand in cone of cylinder $= 248 \times 10^{-6} \times 1560 = 0.387$ kg
Mass of sand required to fill the hole $= 2.552 - 0.387 = 2.165$ kg

Volume of the hole $= \dfrac{2.165}{1560} = 1.388 \times 10^{-3}$ m³

Bulk density of *in situ* soil $= \dfrac{2764 \times 10^{-6}}{1.388 \times 10^{-3}} = \underline{1.99 \text{ Mg/m}^3}$

The immersion in water method

This method is suitable for cohesive or stabilised soils and where an irregular-shaped intact lump of soil has been obtained. The lump sample is weighed (M_s) and after coating with paraffin wax is weighed again (M_w). The coated sample is then suspended in water from a balance arm and the submerged mass recorded (M_G).

Mass of paraffin wax, $M_P = M_w - M_s$

If the density of the paraffin wax is ρ_p and the density of water is ρ_w,

Volume of specimen, $V_s = \dfrac{M_w - M_G}{\rho_w} - \dfrac{M_P}{\rho_p}$

and bulk density, $\rho = \dfrac{M_s}{V_s}$

The water displacement method

This method is similar to the water-immersion method, except that the volume of the waxed sample is found by lowering it into a container of water with a siphon outlet and measuring the volume of water displaced.

Worked example 3.12 An irregular sample of a firm clay was cut from a trial hole and sent to a laboratory for testing. In order to determine its bulk density the sample was coated with paraffin wax and its volume found by displacement. The following data were collected:

Mass of soil as received	*= 924.2 g*
Mass of soil after coating with paraffin wax	*= 946.6 g*
Volume of water displaced	*= 513.1 ml*
Specific gravity of wax	*= 0.9*

Determine the bulk density of the soil.

Mass of paraffin wax used   = 946.6 − 924.2 = 22.4 g
Volume of paraffin wax used = 22.4/0.9 = 24.9 ml
Volume of soil   = 513.1 − 24.9 = 488.2 ml
Bulk density of soil ρ = mass/volume

$$= \frac{924.2 \times 10^6}{488.2 \times 10^6}$$

$$= \underline{1.89 \ \text{Mg/m}^3}$$

Nuclear methods

Both the bulk density and the moisture content of *in situ* soil can be measured using controlled gamma radiation techniques. The apparatus generally consists of a small shielded radiation source and a detector. Several methods have been devised using surface back-scatter, single and double probes and in-hole devices (Fig. 3.10). The intensity of transmitted or back-scattered radiation varies with density and moisture content. Calibration charts are used which relate detected radiation intensity to values recorded from soils of known intensity (ASSHTO, 1986; ASTM, 1986).

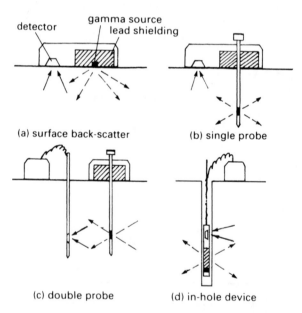

(a) surface back-scatter   (b) single probe

(c) double probe   (d) in-hole device

Fig. 3.10 Nuclear methods to determine *in situ* density and water content

3.7 Specification and quality control of compaction

The amount of compaction achieved on site will depend upon:

(a) *The amount of compactive effort*, i.e. the type and mass of plant used, the number of passes and the layer thickness per pass.

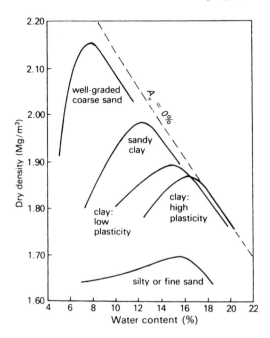

Fig. 3.11 Effect of soil type on compaction

(b) *The field water content*, which may need to be modified in dry conditions or where the shear strength must be reduced for effective compaction (e.g. in clay dam cores and cut-offs).

(c) *The type of soil*: clays of high plasticity may have water contents over 30 per cent and achieve similar strengths to those of lower plasticity with water content below 20 per cent. Well-graded coarse granular soils can be compacted to higher densities than uniform or silty soils (Fig. 3.11).

Specifications for the quality of compaction fall into two distinct groups (Parsons, 1987).

(a) End-result specifications

In end-result specifications a measurable property of the compacted soil is used, such as the dry density or air-voids content. *Relative compaction* may be expressed as the ratio between the achieved field density and an arbitrary laboratory-obtained maximum value (see Section 3.6). Alternatively, the *density index* may be used calculated from either the saturated densities (eqn [3.22]) or the dry densities:

$$\text{Density index, } I_D = \frac{\rho_{dmax}(\rho_{dinsitu} - \rho_{dmin.})}{\rho_{dinsitu}(\rho_{dmax.} - \rho_{dmin.})} \times 100 \qquad [3.28]$$

It is usually necessary to specify a range of working water contents, together with the relative compaction or density index, especially in dry soil conditions (Fig. 3.12). For wetter soils, a specification using the *air-voids content* (eqns [3.6,

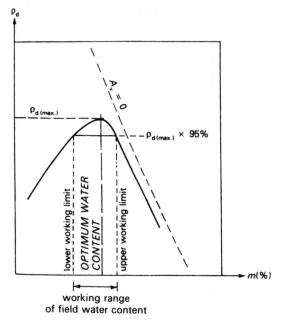

Fig. 3.12   Specification of working water content

3.7 and 3.26]) may be more appropriate. Typical specified values are a maximum of 10 per cent for bulk earthworks and 5 per cent for more important works.

(b)   Method specifications
In a method specification a procedure is laid down, giving a specific type of plant, its mass, the frequency of vibration (if applicable), the thickness of each layer and the number of passes to be made on each layer. This type of specification is particularly suitable for soils wetter than the optimum moisture content, or where conditions vary from day to day, or from one part of a site to another: such conditions are frequently encountered in Great Britain. One widely adopted method specification scheme is published by the Department of Transport and may be obtained from HMSO (Anon, 1986).

3.8   Moisture condition value (MCV)

A disadvantage of compaction tests is the delay involved due to having to determine water contents, although the use of microwave ovens has mitigated this somewhat. A procedure has been developed at the Scottish branch of the Road Research Laboratory to assess the suitability of fill for earthworks without the necessity of measuring water contents (Matheson, 1983, 1988).

In the *moisture condition test* the minimum compactive effort required to produce near-full compaction of a soil is determined. The soil is compacted in a cylindrical mould of internal diameter 100 mm set on a permeable base plate. The rammer used has a flat lower face of diameter 97 mm and a mass of 7.5 kg,

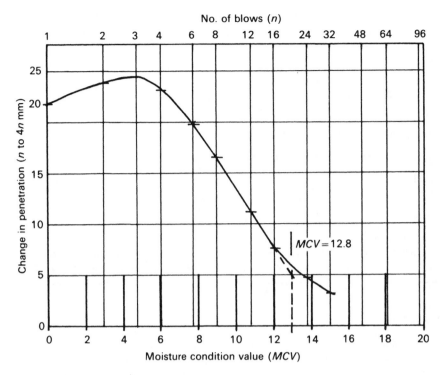

Fig. 3.13 Moisture condition test

and is controlled to fall from a pre-set height. A 1.5 kg sample of soil passing a 20 mm mesh sieve is pressed into the mould and a lightweight separating disc (99 mm diameter) placed on the surface to prevent soil extruding between the mould and the rammer. The rammer is first lowered gently on to the separating disc and the soil allowed to compact under this weight. The release height is then set to 250 mm and one blow applied by releasing the automatic catch; the amount of penetration resulting from this blow is measured to 0.1 mm. Further blows are then applied, the drop height of the rammer being reset each time to 250 mm, and the penetrations recorded.

The change in penetration (Δp) is calculated between that recorded for a given number of blows (n) and that for four times as many blows ($4n$), and a graph plotted of Δp against log n (Fig. 3.13). The steepest possible straight line is then drawn through the points immediately preceding a change in penetration of 5 mm. The **moisture condition value** (**MCV**) is defined as

$$\text{MCV} = 10 \log B$$

where B = number of blows corresponding to $\Delta p = 5.0$ mm.

If a chart similar to that in Fig. 3.13 is used, the MCV can be read off the plot directly using the scale along the bottom. The test is fully detailed in BS 1377. Two other procedures are also described: one to determine the MCV/water-content relationship and the other to assess whether or not a soil is stronger than pre-calibrated standard (MCV).

Worked example 3.13   *The following data were recorded in a moisture condition test. Plot the change in penetration between* n *and* 4n *blows against log* n *and so determine the moisture condition value.*

No. of blows (n)	1	2	3	4	6	8	12	16	24
Penetration (mm)	15.5	25.2	32.1	37.4	44.2	49.1	56.6	60.6	64.0

No. of blows (n)	32	48	64	96	128
Penetration (mm)	65.7	67.8	68.3	68.8	68.9

The changes in penetration Δp between n and $4n$ blows are:

No. of blows (n)	1	2	3	4	6	8	12	16
Δp (n to 4n blows (mm))	21.9	23.9	24.5	23.2	19.8	16.7	11.2	7.7

No. of blows (n)	24	32
Δp (n to 4n blows (mm))	4.8	3.2

Fig. 3.13 shows the plot Δp against log n; the straight-line portion has been extended to intersect with $\Delta p = 5.0$ mm line, from which the MCV can be read off: **MCV = 12.8**.

Exercises

3.1   In a laboratory test a mass of moist soil was compacted into a mould having a volume of 964 cm³. Upon weighing, the mass of soil was found to be 1956 g. The water content was found to be 13 per cent and the grain specific gravity was 2.70.

Calculate: (a) the bulk and dry density, (b) the void ratio and porosity, (c) the degree of saturation, and (d) the air-voids content of the soil.

3.2   Using the model soil sample, show that:

$$e = \frac{wG + A_v}{1 - A_v}$$

$$= \frac{G_s\rho_w - \rho}{\rho - S_r\rho_w}$$

$$= (1 + w)\frac{G_s\rho_w}{\rho} - 1$$

3.3   In a sample of moist soil the porosity is 42 per cent, the specific gravity of the particles 2.69 and the degree of saturation 84 per cent. Determine: (a) void ratio, (b) bulk and dry densities, (c) water content, and (d) saturation bulk density (assuming no swelling takes place).

3.4   A core-cutter cylinder of internal diameter 100 mm and length 125 mm was used to obtain a sample of damp sand from a trial hole. After trimming the ends, the total mass of the cylinder and soil was 3508 g; the mass of the empty cylinder was 1525 g. After oven-drying, the soil on its own weighed 1633 g. If the specific gravity was also found to be 2.71, determine the bulk and dry densities, the water content, the void ratio and the air-voids content of the sample.

3.5   A dry quartz sand has a density of 1.61 Mg/m³ and a grain specific gravity of 2.68. Calculate its bulk density and water content when saturated at the same volume.

3.6 A saturated cylindrical specimen of a clay soil has a diameter of 75.0 mm and a thickness of 18.75 mm and weighs 155.1 g. If the water content is found to be 34.4 per cent, determine the bulk density and void ratio of the specimen. If the original thickness of the specimen was 19.84 mm, what was the initial void ratio?

3.7 A sandy soil has a saturated density of 2.08 Mg/m^3. When it is allowed to drain the density is reduced to 1.84 Mg/m^3 and the volume remains constant. If the grain specific gravity is 2.70, determine the quantity of water in litres/m^2 that will drain from a layer of the sand 2.2 m thick.

3.8 A cohesive soil specimen has a void ratio of 0.812 and a water content of 22.0 per cent. The grain specific gravity is 2.70. Determine: (a) its bulk density and degree of saturation, and (b) the new bulk density and void ratio if the specimen is compressed undrained until it is just saturated.

3.9 A sandy soil has a porosity of 38 per cent and the grain specific gravity is 2.69. Determine: (a) the void ratio, (b) the dry unit weight, (c) the saturated unit weight, and (d) the bulk unit weight at a water content of 17 per cent.

3.10 A laboratory specimen is to be prepared by ramming soil ($G_s = 2.68$) into a cylindrical mould of diameter 104 mm. The finished specimen is to have a water content of 16 per cent and an air-voids content of 5 per cent. Determine: (a) its void ratio and dry unit weight, and (b) the quantities of dry soil and water required to be mixed together in order to form a specimen 125 mm long.

3.11 The following data were recorded during a sand replacement test:

Mass of soil removed from hole	= 1.914 kg
Mass of soil after oven-drying	= 1.664 kg
Initial total mass of sand-pouring cylinder	= 3.426 kg
Mass of sand-pouring cylinder after running sand into hole	= 1.594 kg
Density of pouring sand	= 1.62 Mg/m^3
Mass of sand in cone in sand-pouring cylinder	= 0.248 kg

Determine the bulk and dry unit weights of the soil *in situ*.

3.12 The following data were recorded during a BS compaction test:

Volume of mould = 0.945 × 10^{-3} m^3, $G_s = 2.70$

Mass of wet soil in mould (kg)	1.791	1.937	2.038	2.050	2.022	1.985
Water content (%)	8.4	10.6	12.9	14.4	16.6	18.6

(a) Plot the curve of dry density against water content and from it obtain the maximum dry density and optimum water content for the compacted soil.
(b) On the same axes, plot the dry density/water content curves for zero and 5 per cent air voids and so obtain the air-voids content of the soil at the optimum water content.

3.13 The results below were obtained in a BS compaction test:

Bulk density (Mg/m^3)	1.866	2.019	2.112	2.136	2.105	2.065
Water content (%)	9.4	11.5	13.5	15.1	17.1	19.9

(a) Plot the curve of dry density against water content and from it obtain the maximum dry density and optimum water content for the compacted soil.

(b) Calculate the air-voids ratio, void ratio and degree of saturation of the soil at the optimum water content.

(c) Under field compaction conditions, the air-voids content and optimum water content may both vary up to ±2 per cent. Calculate, therefore, the lower field limit of the maximum dry density.

Soil Mechanics Spreadsheets and Reference Assignments and Quizzes (available on the accompanying CD):

Assignments A.3   Basic physical properties of soils

Quiz Q.3   Basic physical properties of soils

Chapter 4

Groundwater, pore pressure and effective stress

4.1 Occurrence of groundwater

Less than 1 per cent of the earth's water occurs as liquid fresh water associated with land masses, the rest is either saline water in the oceans and seas, or water vapour in the atmosphere. This land-based water is derived from rainfall and subsequently flows under the influence of gravity. On the surface, streams, rivers and lakes are formed; but a very large proportion percolates into the crustal rocks and soils. The depth of penetration of percolating water depends on the porous nature of the rocks; it is suggested by geologists that little or no sub-surface water exists below a depth of 8 km.

From an engineering mechanics point of view, groundwater in soil may be one of two types, occurring in two distinct zones separated by the *water table* or *phreatic surface* (Fig. 4.1).

(a) *Phreatic or gravitational water*, which:
- (i)   is subject to gravitational forces;
- (ii) saturates the pore spaces in the soil below the water table;
- (iii) has an internal pore pressure greater than atmospheric pressure;
- (iv) tends to flow laterally.

(b) *Vadose water*, which may be:
- (i)   transient percolating water, moving downwards to join the phreatic water below the water table;
- (ii) capillary water held above the water table by surface tension forces (with internal pore pressure less than atmospheric).

4.2 Water table or phreatic surface

Since soils are made up of masses of discrete particles, the pore spaces are all interconnected, so that water may pass from zones of high pressure to zones of low pressure. The level at which the porewater pressure is equal to that of the atmosphere is termed the *water table* or *phreatic surface* (Fig. 4.1). In the case of unconfined bodies of groundwater, the water table corresponds to the free water surface; such as may be found at a river or lake or in an excavation (Fig. 4.2).

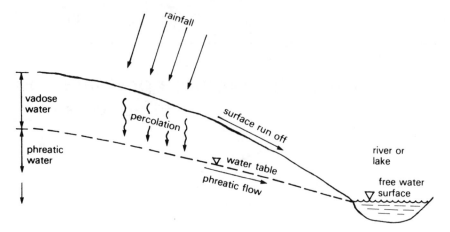

Fig. 4.1   Occurrence of groundwater

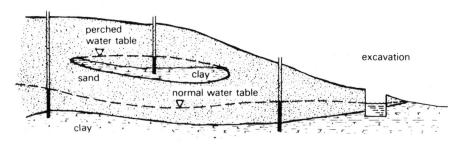

Fig. 4.2   Normal and perched water tables

Where groundwater lies above isolated bodies of soils, such as clay, which have very low permeability, a *perched water table* may occur (Fig. 4.2).

Where a stratum of reasonably high permeability, termed an *aquifer*, is confined above and below by strata of low permeability, the water table will not exist as such. However, the water level in standpipes or wells sunk into the aquifer will indicate the *piezometric surface* level. *Artesian* conditions are said to exist when the piezometric surface lies *above* ground level; *sub-artesian* conditions exist when it is between ground level and the aquifer (Fig. 4.3). The porewater pressure in a confined aquifer is governed by the conditions at the place where the layer is unconfined; if the water table here rises (e.g. due to rainfall), then the artesian porewater pressure is likely to rise also.

4.3   Capillary water

Capillary water is held above the water table by surface tension which is the attractive force exerted at the interface or surface between materials in different physical states, i.e. liquid/gas, solid/liquid. For example, a water/air surface exhibits an apparently elastic molecular skin due to the sub-surface water molecules (which are more dense than air) exerting a greater attraction than the

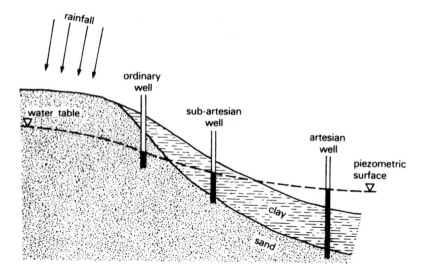

Fig. 4.3 Confined groundwater

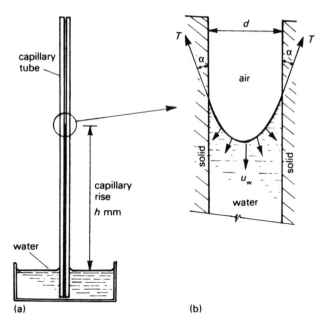

Fig. 4.4 Capillary rise
(a) Capillary tube (b) Detail at surface

air molecules. Similarly, water is attracted towards a solid interface because of the greater density and therefore attraction of the solid.

Consider a glass tube of small diameter (d) standing with its lower end in water (Fig. 4.4).

At the triple interface, the water surface is pulled up into a meniscus, with the surface tension resultant force (T) acting around its perimeter at angle α to the wall of the tube. The water is drawn upwards by this force until, at a height h, the weight of water in the column is in equilibrium with the magnitude of the surface tension force.

Then if atmospheric pressure = 0, equating vertical forces at the surface:

$$T \cos \alpha \pi d = \frac{\pi d^2}{4} u_w$$

Giving capillary pore pressure, $u_w = \dfrac{4T \cos \alpha}{d}$ [4.1]

or putting $u_w = \gamma_w h_c$ (γ_w = unit weight of water)

Capillary rise, $h_c = \dfrac{4T \cos \alpha}{\gamma_w d}$ [4.2]

As an approximation for soils put

$T = 0.000074$ kN/m $\gamma_w = 9.81$ kN/m³ $\alpha = 0$

and $d \simeq e d_{10}$ (where d_{10} = effective size (Section 2.4))

Giving $h_c \simeq \dfrac{4 \times 0.000074 \times 10^6}{e d_{10} \times 9.81} = \dfrac{30}{e d_{10}}$

This estimate may be improved to allow for the effect of grading and grain shape characteristics, such as irregularity and flakiness (Terzaghi and Peck, 1948):

$$h_c = \frac{C}{e d_{10}}$$ [4.3]

where C = a value between 10 and 40 mm²

While the value of h_c represents the *maximum capillary rise*, the soil will only be saturated with capillary moisture up to the *capillary saturation level* (h_{cs}). The approximate relationship between both capillary rise and capillary saturation level and soil type is shown in Fig. 4.5. The overall regime of groundwater may therefore be divided into four zones as shown in Fig. 4.6.

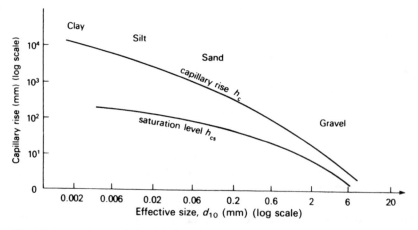

Fig. 4.5 Approximate relationship between capillary rise and soil type

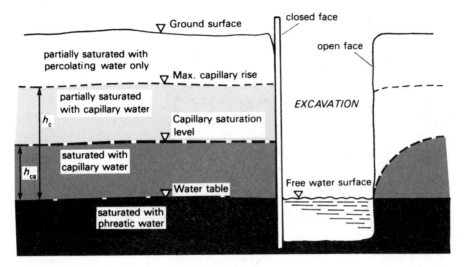

Fig. 4.6 Groundwater zones

4.4 Soil suction

The negative capillary pore pressure (u_c) corresponding to the maximum capillary rise (h_c) is a measure of the *suction* exerted on the porewater by the soil. It is apparent from Fig. 4.5 that the range of suction values is very large. It is therefore convenient to use a logarithmic scale and to define the quantity *soil suction index* or *pF index*:

$$pF = \log_{10} \times \text{(negative capillary suction head in cm)}$$

$$= \log_{10} h_c \qquad [4.4]$$

It has been found experimentally that the *pF* index has a range between 0 and 7, with the higher values relating to oven-dry soils. In sands, the capillary suction head rarely exceeds 50 cm ($pF = 1.7$). It is also apparent that there is a continuous inverse relationship between suction and water content, together with a slight variation depending on whether soil is wetting or drying. A typical curve showing the *pF*/water content relationship is given in Fig. 4.7.

In the case of an undisturbed soil mass in equilibrium, i.e. not subject to any increase in external stress, the pore pressure will balance the suction:

$$u_c + S_s = 0$$

or $\qquad S_s = -u_c$

where S_s = suction pressure = $\gamma_w h_c$

However, the external stress will reduce the water content and vary the suction, so that for a saturated soil:

$$S_s = -u_c + \alpha_c \sigma_z \qquad [4.5]$$

where σ_z = vertical component of stress at depth z
$\qquad \alpha_c$ = a fractional coefficient depending on the compressibility of the soil

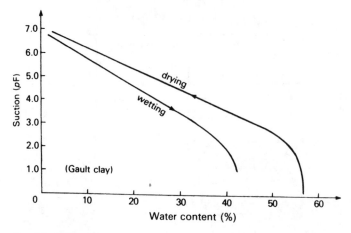

Fig. 4.7 Soil suction and water content

For incompressible rigid materials, such as rocks and compact sands, $\alpha_c = 0$; whereas for a saturated clay, which is highly compressible, $\alpha_c = 1$; for other soils, α_c will vary between 0 and 1. An approximation for the value of α_c may be obtained from the plasticity index of soil: $\alpha_c \simeq 0.025\ I_P$ (for I_P up to 40%).

The potential drying effect of a body of air in contact with soil can be impressively high. Drying of the soil takes place when the suction from the surrounding air is greater than that generated within the soil itself. The suction potential of a mass of air is given approximately by

$$u_s = -150\ 000(1 - R_H)\ kN/m^2 \tag{4.6}$$

where R_H = relative humidity

The equivalent suction index is therefore

$$pF_{(air)} \simeq \log[150\ 000 \times 10^2(1 - R_H)/9.81]$$

$$= 6.2 + \log(1 - R_H) \tag{4.7}$$

The loss of water from exposed ground in hot dry weather leads to desiccation and shrinkage problems, similar problems arise when laboratory samples are left unprotected. Table 4.1 gives some approximate indicative values. Vegetation or inactive surface layers may reduce this effect; nevertheless, the suction potential of clay remains very high. Williams and Pidgeon (1983), investigating soils of the South African high veldt, found pF values of 4.8 and 3.8 at depths of 6 m and 15 m respectively. Driscoll (1983) has suggested that the onset of desiccation (and therefore shrinking) can be estimated from the ratio of the soil water content to its liquid limit:

Start of desiccation: $pF = 2$, $w = 0.5w_L$
Significant shrinkage: $pF = 3$, $w = 0.4w_L$

Table 4.1 *Indicative soil/air suction potentials*

Air conditions	Relative humidity (%)	Potential suction head (m)	Potential suction index (pF)
	0	150 000	6.2
Hot drying oven	15	12 750	6.1
Hot dry day ($T = 40°$)	30	10 500	6.0
Average laboratory	70	4 500	5.7
Cool damp room	90	1 500	5.2
Very wet (internal/external)	99	150	4.2
	99.999	0.15	1.2

4.5 Pore pressure and effective stress

When an external stress is applied to a soil mass that is saturated with porewater, the immediate effect is an increase in the pore pressure. This produces a tendency for the porewater to flow away through adjoining voids, with the result that the pore pressure decreases and the applied stress is transferred to the granular fabric of the soil. At a given time after application, therefore, the applied total stress will be balanced by two internal stress components.

Pore pressure (u). This is the pressure induced in the fluid (either water, or vapour and water) filling the pores. Pore fluid is able to transmit normal stress, but not shear stress, and is therefore ineffective in providing shear resistance. For this reason, the pore pressure is sometimes referred to as *neutral pressure*.

Effective stress (σ'). This is the stress transmitted through the soil fabric via intergranular contacts. It is this stress component that is *effective* in controlling both volume change deformation and the shear strength of the soil since both normal stress and shear stress is transmitted across grain-to-grain contacts. Terzaghi (1943) showed that, for a saturated soil, effective stress may be defined quantitatively as the difference between the total stress and the pore pressure:

$$\sigma' = \sigma - u \qquad [4.8]$$

It should be noted, however, that the effective stress is not the actual grain-to-grain contact stress, but the *average* intergranular stress on a plane area within the soil mass. However, it has been confirmed experimentally that where the grains themselves are relatively incompressible and area of contact between grains is small, the expression provides a high degree of reliability. In the case of soil, only insignificant levels of error will be incurred; although in the case of rocks, which have a rigid structure, some modification is required.

The hydrostatic pore pressure under natural field conditions of no flow is represented by the water table or phreatic surface level. If the water table lies at depth d_w below the surface, then at depth z, the hydrostatic pore pressure is given by $u_z = 9.81(z - d_w)$. When $z > d_w$, u_z will have a positive value; but when $z < d_w$ and there is capillary-held water above the water table, u_z will have a negative value (i.e. suction, Section 4.4).

In many problems, it is necessary to calculate the static effective overburden stress at a given depth; from eqn [4.8]:

$$\sigma'_z = \sigma_z - u_z$$

For examples of effective overburden stress calculations and effective stress/total stress profiles, see Worked Examples 4.2–4.6 at the end of this chapter.

4.6   Pore pressure in partially saturated soils

In the case of partially saturated soils, the pore fluid will consist of water liquid, which is virtually incompressible, and air/water vapour, which is highly compressible. There are two components of pore pressure: the porewater pressure (u_w) and the pore-air pressure (u_a). Due to surface tension the presence of air reduces pore pressure; Bishop (1955) suggested the following relationship:

$$u = u_a - \chi(u_a - u_w)$$

Where χ is a parameter dependent mainly on the degree of saturation and, to a lesser extent, on the soil fabric structure.

It is possible to determine χ experimentally and it seems to vary almost linearly from 0 for $S_r = 0$ (dry soil) to 1 for $S_r = 1$ (saturated soil). However, for soils wetter than their optimum water content, S_r tends to be 0.9 or more, so that χ will be very near to 1. In such cases, the small amount of air present will be in the form of occluded bubbles, which affect the compressibility of the pore fluid without significantly lowering the pore pressure. Thus, eqn [4.8] will not produce unacceptable errors when $w > w_{opt}$.

If compression or shear strength tests are carried out on soils which are drier than the optimum water content, the effect of the compressible pore fluid will distort results. It is not practicable to attempt accurate measurements of χ in routine tests and other factors can also be significant, such as the soil fabric structure and whether the soil is wetting or drying. As a general rule, it is safer to conduct compression and shear strength tests only on naturally saturated soils or soils that have been brought to saturation prior to testing.

4.7   Pore pressure coefficients

Pore pressure coefficients are used to express the increase in pore pressure in relation to the increase in total stress. In order to establish this relationship for the majority of soil problems it is convenient to consider the total stress regime as being made up of two components:

(a)   An *isotropic* or all-round stress; and
(b)   A uniaxial *deviatoric* stress.

Consider an element of soil, of volume V and porosity n, which, from equilibrium and under undrained conditions, is subject to an increase in total stress such that $\Delta\sigma_2 = \Delta\sigma_3$ and $\Delta\sigma_1 > \Delta\sigma_3$ (Fig. 4.8(a)). It will be assumed that there will be a corresponding increase in pore pressure of Δu, and that this is made up of two additive components: Δu_0 caused by the isotropic increase $\Delta\sigma_3$ (Fig. 4.8(b)) and Δu_1 caused by the uniaxial increase $(\Delta\sigma_1 - \Delta\sigma_3)$ (Fig. 4.8(c)).

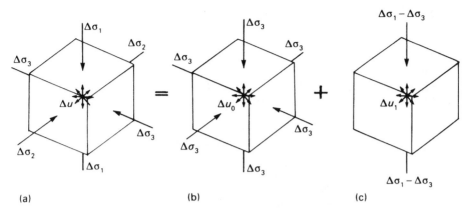

Fig. 4.8 Stresses on a soil element
  (a) General three-dimensional stress system (b) Isotropic stress
  (c) Uniaxial or deviatoric stress

Isotropic increase in stress
Since the increase in stress is uniform, $\Delta\sigma_1 = \Delta\sigma_2 = \Delta\sigma_3$, the corresponding increase in pore pressure = Δu_0 and the increase in isotropic effective stress = $\Delta\sigma_3' = \Delta\sigma_3 - \Delta u_0$.

Now let C_s = compressibility of the soil skeleton
       C_v = compressibility of the pore fluid
Then decrease in volume of soil skeleton $\Delta V_s = C_s V(\Delta\sigma_3 - \Delta u_0)$
and decrease in volume of pore space $\Delta V_v = C_v n V \Delta u_0$

If the soil grains are assumed to be incompressible and as no drainage takes place, then $\Delta V_s = \Delta V_v$

$$C_s V(\Delta\sigma_3 - \Delta u_0) = C_v n V \Delta u_0$$

Giving $$\Delta u_0 = \frac{1}{1 + nC_v/C_s}\Delta\sigma_3$$

Putting $$\frac{1}{1 + nC_v/C_s} = B \quad (\textit{pore pressure coefficient } B)$$

$$\Delta u_0 = B\Delta\sigma_3 \tag{4.9}$$

In rocks, the stiffness of the fabric structure is significant, whereas the fabric of soils is more compressible and the value of B is more dependent on the degree of saturation as shown in Fig. 4.9. Table 4.2 shows typical values calculated for saturated rocks and soils. For practical purposes, therefore, $B = 1$ for saturated soils.

The value of B may be obtained experimentally using the triaxial test apparatus, by varying the cell pressure under undrained conditions and noting the corresponding changes in pore pressure (see Section 7.15).

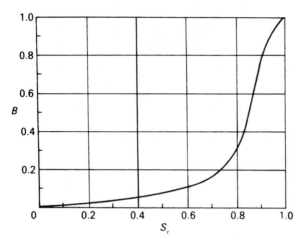

Fig. 4.9 Typical *B/S*r relationship

Table 4.2   *Calculated values of pore pressure coefficient* B

Rock/soil	C_s (m²/kN × 10⁻⁴)	n(%)	B
Bath stone	0.06	15	0.468
Chalk	0.25	30	0.647
Dense sand	15	40	0.9880
Stiff clay	80	42	0.9976
Soft clay	400	55	0.9994

Increase in uniaxial stress

The uniaxial increase in stress is the stress difference or deviatoric component $(\Delta\sigma_1 - \Delta\sigma_3)$ producing a corresponding increase in pore pressure of Δu_1.

The increases in effective stress are therefore:

$$\Delta\sigma_1' = (\Delta\sigma_1 - \Delta\sigma_3) - \Delta u_1$$

$$\Delta\sigma_1' = \Delta\sigma_3' = -\Delta u_1$$

For a perfectly elastic material, the decrease in volume of the soil skeleton would be:

$$\Delta V_s = \tfrac{1}{3}C_s V[\Delta\sigma_1' + \Delta\sigma_2' + \Delta\sigma_3']$$

$$= \tfrac{1}{3}C_s V[(\Delta\sigma_1 - \Delta\sigma_3) - 3\Delta u_1]$$

and the decrease in volume of pore space:

$$\Delta V_v = C_v n V \Delta u_1$$

As before, $\Delta V_s = \Delta V_v$

$$\therefore \quad \tfrac{1}{3}C_s V[(\Delta\sigma_1 - \Delta\sigma_3) - 3\Delta u_1] = C_v n V \Delta u_1$$

Giving $\Delta u_1 = \dfrac{1}{3}\dfrac{1}{1 + nC_v/C_s}(\Delta\sigma_1 - \Delta\sigma_3)$

$$= \tfrac{1}{3}B(\Delta\sigma_1 - \Delta\sigma_3)$$

However, since soils are not perfectly elastic, the volumetric strain constant of $\frac{1}{3}$ must be replaced by a variable coefficient A, so that, in general:

$$\Delta u_1 = AB(\Delta\sigma_1 - \Delta\sigma_3) \qquad [4.10a]$$

$$= \bar{A}(\Delta\sigma_1 - \Delta\sigma_3) \qquad [4.10b]$$

where $\bar{A} = AB$. The pore pressure coefficient A varies with the level of applied stress and with both the rate of strain and whether the stress is increasing or decreasing. In addition, both the drainage conditions and stress history (i.e. whether the soil is normally consolidated or overconsolidated) will influence the value of A. It is therefore necessary to relate a particular measured value of A to a particular stress–strain parameter, such as the maximum deviatoric stress, or the overconsolidation ratio. Data collected during a triaxial compression test may be used to establish values of A (see Section 7.15).

The value of A at the time of shear failure (A_f) is often quoted as a guide value. A summary of some of the ranges obtained for A_f is given in Table 4.3.

Table 4.3 Values of pore pressure coefficient A_f

Soil type	A_f (A at failure)
Highly sensitive clay	1.2–2.5
Normally consolidated clay	0.7–1.3
Lightly overconsolidated clay	0.3–0.7
Heavily overconsolidated clay	−0.5–0
Very loose fine sand	2.0–3.0
Medium fine sand	0–1.0
Dense fine sand	−0.3–0

General pore pressure expression
A general expression for the increase in pore pressure corresponding to a combined increase in isotropic and uniaxial stress is obtained by simply adding eqns [4.9] and [4.10a]:

$$\Delta u = \Delta u_0 + \Delta u_1$$

$$= B\Delta\sigma_3 + AB(\Delta\sigma_1 - \Delta\sigma_3)$$

$$= B[\Delta\sigma_3 + A(\Delta\sigma_1 - \Delta\sigma_3)] \qquad [4.11]$$

4.8 Frost action

In cold conditions, when the surface temperature falls below 0 °C, water in the soil will freeze, causing an increase in volume which is termed *frost heave*. The depth of the frozen layer will increase at a rate dependent on both the temperature and the duration of the freezing conditions. A few centimetres of penetration will result from an overnight frost, but the rate of growth of the frozen layer slows down with increasing penetration. In the United Kingdom, it is unusual for the depth of penetration to exceed 0.5 m, but it may reach half of this in less than a week.

From an engineering point of view, the phenomena associated with frost action are important, in so far as they occur in certain soil types, because of the potential damage that may result. Two forms of problem may be defined:

(a)   The effects of freezing.
(b)   The effects of thawing.

Frost heave
Frost heave is the vertical surface expansion brought about by water freezing in the soil. There are two components of frost heave. Firstly, as water freezes its volume increases by about 9 per cent, thus the volume of the soil increases by about $0.09n$ per unit volume, i.e. from 2 to 6 per cent depending on the void ratio of the soil.

For example, in a soil of void ratio 0.5 with a frost penetration of 0.4 m, the amount of surface heave due simply to freezing expansion would be 12 mm.

Secondly, as water freezes, other phenomena occur. Since the vapour pressure of ice is much lower than that of water, a high degree of suction is created which draws water upwards towards the frozen layer, where it freezes, thus increasing the size of ice crystals. In addition, as it freezes, water gives up latent heat of about 80 J/kg. If the rate of heat dissipation towards the surface is equal to the rate of latent heat released, the thickness of the frozen layer will remain constant. The growth of ice then tends to be highly localised, resulting in the formation of *ice lenses*, which may increase the thickness of the frozen layer by up to 30 per cent.

The pressures set up by frost heave may be sufficient to lift road surfaces, paving and even lightly-loaded foundations. The formation of ice lenses within road bases and sub-grades may also cause damage. Where road surfaces are cracked or crazed by these effects they will become porous; with the result that water will be allowed to penetrate, which in turn may cause a deterioration in the sub-grade.

Thawing effects
If, after a frozen layer containing ice lenses has been formed, a rapid thaw takes place, the soil near the surface will contain an excess of water. The additional water (mainly due to the melting of the ice lenses) will be unable to drain away through the still frozen soil below.

The soil is therefore likely to be in an unstable state, especially in the case of a sloping ground surface. The shear strength and/or compressibility characteristics of the soil will be impaired, so that damage and displacement due to surface loading is probable.

Frost susceptibility
In coarse-grained soils, such as gravels and coarse sands (in which the capillary rise is negligible), expansion can take place within the voids and so increases in overall volume are negligible. In the case of clays, the permeability is so low that the migration of water towards the frozen layer will be too slow to allow a build-up of ice lenses. The frost-susceptible soils therefore tend to be those

Table 4.4 Frost susceptibility of soils and other materials

Frost susceptibility	Coarse soils classified according to the British Soil Classification System (Table 2.3)				
	0–3	3–5	5–10	10–15	15–35
High				GWM GPM GWC GPC SWM SPM SWC SPC	GML,I, etc. GCL,I, etc. SML,I, etc. SMC,I, etc.
Medium			GWM GPM GWC GPC SWM SPM SWC SPC		
Low		GW GP SW SP			
Negligible	GW GP SW SP				

% fines

Other materials	Susceptibility:	When:
Silts	Medium to high	(ALL)
Clays of low to intermediate plasticity: poorly drained	Low to high	$I_p < 15$
well drained	Low to high	$I_p < 20$
Crushed chalk	Medium to high	$w_{sat} > 5\%$
Crushed limestones	Low to high	$w_{sat} > 3\%$
Other crushed rocks and crushed slag	Low to high	% fines > 10%
Burnt colliery shale	Medium to high	
Pulverised fuel ash	Low to high	% fines > 10%

containing a predominance of fine sand or silt. It is also recommended that where possible the water table should be kept at least 600 mm below the formation level.

In a frost design classification system proposed by the US Army Corps of Engineers (1965), the percentage of soil grains finer than 0.02 mm is recommended as a guide, whereas Croney and Jacobs (1967) and subsequently the Road Research Laboratory (1970) have suggested a combination of plasticity index values and the percentage finer than 75 μm. Table 4.4 shows a frost susceptibility classification proposed by the author, based on the above and other sources and on laboratory studies.

4.9 Equilibrium water content

Under constant conditions of stress and drainage, the water content within a soil mass will reach an equilibrium condition. The particular value of water content at a given depth in this equilibrium condition is termed an *equilibrium water content (ewc)*. If in a saturated soil there is a change in conditions of stress of drainage resulting in the movement of water in or out of the soil, a volume change will take place equal to the volume of water added or removed.

The construction of an impervious road surface or pavement will prevent moisture movement through the surface due to infiltration or evaporation. A soil covered in this way in dry weather, when the *in situ* water contents are below the equilibrium water contents, will tend to swell subsequently. It is useful, in attempting to assess the swelling/shrinkage potential, to compare profiles of water content/depth and ewc/depth.

The usual way to calculate the ewc is to calculate the suction and then to use a *pF*/water content curve. In non-arid areas, where the water table is not far below the surface, i.e. not more than 0.7 m clays or 1 to 1.5 m for sands, the suction may be estimated simply from $\gamma_w \times$ height above the water table.

Worked example 4.1   *An impervious paved area is to be constructed on a base of clayey silt during the summer. Observations indicate that the water table will stabilise at 1.7 m below formation level. Using the data given below, prepare an equilibrium water content profile and compare it with the* in situ *water contents.*

Data

$w_L = 66$ per cent $w_P = 35$ per cent   Average $\gamma = 18.7\ kN/m^3$
Contact pressure at formation level = 7.5 kPa
Wetting curve ($pF = 1.0$ to 4.0): ewc = 48 – 9.0 pF

Depth below formation (m)	0.1	0.5	1.5	2.5	4.0	5.0	6.0
In situ water content before construction (%)	24.7	25.3	25.8	25.0	24.5	24.0	23.8

The suction is obtained from eqn [4.5]:

Where $\alpha_c = 0.025\ I_P$ ($I_P = w_L - w_P = 66 - 35 = 31$)

$= 0.025 \times 31 = 0.775$

Vertical stress $\sigma_z = \gamma z + 7.5$ kPa

Suction $S_s = \alpha_c \sigma_z - u_c$

The wetting curve is apparently close to a straight line between $pF = 1.0$ and $pF = 4.0$, hence: ewc = 48 – 9.0 *pF*.

Depth (m)	γz (kPa)	σ_z (kPa)	$\alpha_c\sigma_z$ (m of water)	u_c (m)	S_s (m)	$pF =$ log 100S_s	ewc (%)
0.0	0.0	7.5	0.59	-1.70	2.29	2.36	26.8
0.5	9.35	16.9	1.34	-1.20	2.54	2.40	26.4
1.0	18.70	26.2	2.07	-0.70	2.77	2.44	26.0
1.5	28.05	35.6	2.81	-0.20	3.04	2.48	25.7
2.0	37.40	44.9	3.55	0.30	3.25	2.51	25.4
2.5	46.75	54.3	4.29	0.80	3.49	2.54	25.1
3.0	56.10	63.6	5.02	1.30	3.72	2.57	24.9
4.0	74.8	82.3	6.50	2.30	4.20	2.62	24.4
5.0	93.5	101.0	7.98	3.30	4.68	2.67	24.0
6.0	112.2	119.7	9.46	4.30	5.16	2.71	23.6

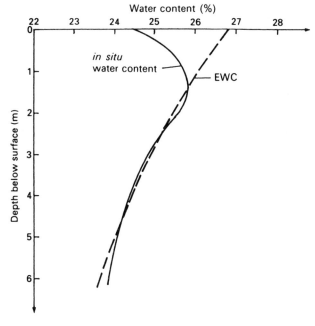

Fig. 4.10

Figure 4.10 shows the comparative plot of *in situ* water content and ewc profiles, from which it will be seen that some expansion may be anticipated in the top 2 m of soil.

Worked examples of effective stress calculations

Worked example 4.2   *The soil layers on a site consist of:*

0–4 m BS gravel-sand ($\gamma_{sat.} = 20.0$ kN/m^3, $\gamma = 19.2$ kN/m^3)
4–9 m BS clay ($\gamma = 18.0$ kN/m^3)

Draw an effective stress/total stress profile between 0 and 9 m BS, when the water table is 1 m above the top of the clay.

The gravel-sand below the water table is saturated and therefore has a bulk unit weight of 20.0 kN/m^3, so that here the increase in total stress with depth is: $\Delta\sigma_z = \gamma_{sat.}\Delta z = 20.0\Delta z$ kPa. Above the water table some water will drain away leaving a unit weight of 19.2 kN/m^3 and $\Delta\sigma_z = \gamma\Delta z = 19.2\Delta z$ kPa. In clays, a combination of low permeability and a high suction will usually ensure saturation above the water tables.

The effective stress at a given depth is: $\sigma'_z = \sigma_z - u_z$, where u_z = the pore pressure due to the static water table. The calculations are tabulated below and the stress profiles plotted in Fig. 4.11.

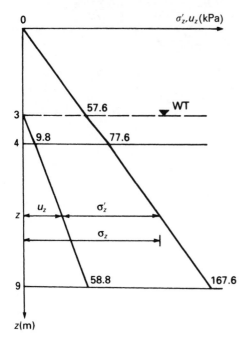

Fig. 4.11

Depth BS (m)	Stresses (kPa)			
	Total stress		Pore pressure	Effective stress
	$\Delta\sigma_z$	σ_z	u_z	$\sigma'_z = \sigma_z - u_z$
0	0	0	0	0
3	$19.2 \times 3 = 57.6$	57.6	0	57.6
4	$20.0 \times 1 = 20.0$	77.6	$9.81 \times 1 = 9.8$	67.8
9	$18.0 \times 5 = 90.0$	167.6	$9.81 \times 6 = 58.8$	108.8

Worked example 4.3   *On a certain site a surface layer of silty sand is 5 m thick and overlies a layer of peaty clay 4 m thick, which in turn is underlain by impermeable rock. Draw effective/total stress profiles for the following conditions: (a) water table at the surface, and (b) water table at a depth of 2.5 m, with the silty sand above the water table saturated with capillary water.*

Unit weights: silty sand = 18.5 kN/m³   clay = 17.7 kN/m³.

(a)   With the water table at the surface, all of the soil is submerged, i.e. unit weight = $\gamma_{sat.}$ and pore pressure = $9.81z$. The calculations are tabulated below and the vertical stress profile is shown in Fig. 4.12(a).

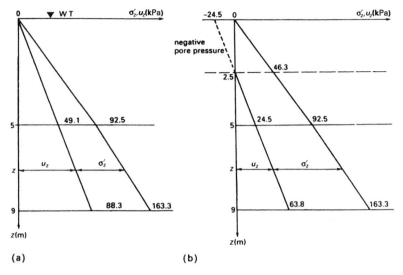

Fig. 4.12

Depth BS (m)	Stresses (kPa)			
	Total stress		Pore pressure	Effective stress
	$\Delta\sigma_z$	σ_z	u_z	$\sigma'_z = \sigma_z - u_z$
0	0	0	0	0
5	$18.5 \times 5 = 92.5$	92.5	$9.81 \times 5 = 49.1$	43.4
9	$17.7 \times 4 = 70.8$	163.3	$9.81 \times 9 = 88.3$	75.0

(b) The soil above the water table is saturated, and the pore pressure will be negative: $u = 9.81(z - 2.5)$. Below the water table the pore pressure will be positive: $u = 9.81(z - 2.5)$.

The calculations are tabulated below and the overburden stress profile shown in Fig. 4.12(b).

Depth BS z(m)	Stresses (kPa)			
	Total stress		Pore pressure	Effective stress
	$\Delta\sigma_z$	σ_z	u_z	$\sigma'_z = \sigma_z - u_z$
0.0		0.0	$-9.81 \times 2.5 = -24.5$	24.5
2.5	$18.5 \times 2.5 = 46.25$	46.3	0	46.3
5.0	$18.5 \times 2.5 = 46.25$	92.5	$9.81 \times 2.5 = 24.5$	68.0
9.0	$17.7 \times 4.0 = 70.8$	163.3	$9.81 \times 6.5 = 63.7$	99.6

Worked example 4.4   *A confined aquifer comprises a 4 m thick layer of sand overlain by a 3 m thick layer of clay and underlain by impermeable rock. The unit weights of the sand and clay respectively are 19.8 kN/m³ and 18.3 kN/m³. Determine the effective overburden stress at the top and bottom of the sand layer, when the levels of the water in a standpipe driven through the clay into the sand layer are: (a) 2 m below the ground surface, and (b) 2 m above the ground surface*

(a)   The water level in the standpipe indicates the piezometric surface; the pore pressures are therefore:

Top of sand	$u_z = 9.81(3 - 2) = 9.8$ kPa
Bottom of sand	$u_z = 9.81(3 + 4 - 2) = 49.1$ kPa

The effective stresses are:

Top of sand	$\sigma'_z = \sigma_z - u$
	$= 3 \times 18.2 - 9.8 = \underline{44.8 \text{ kPa}}$
Bottom of sand	$\sigma'_z = (3 \times 18.2 + 4 \times 19.8 = \underline{84.7 \text{ kPa}}$

(b) Top of sand $u_z = 9.81(3 + 2) = \underline{49.1 \text{ kPa}}$
$\qquad\qquad\qquad\sigma'_z = 3 \times 18.2 - 49.7 = \underline{5.5 \text{ kPa}}$
    Bottom of sand $u_z = 9.81(3 + 4 + 2) = \underline{88.3 \text{ kPa}}$
$\qquad\qquad\qquad\sigma'_z = (3 \times 18.2 + 4 \times 19.8) - 88.3 = \underline{45.5 \text{ kPa}}$

Worked example 4.5   *A sediment settling lagoon has a depth of water of 4 m above the clay base. The clay layer is 3 m thick and this overlies 4 m of a medium sand, which in turn overlies impermeable rock. Calculate the effective stresses at the top of the clay and at the top and bottom of the second layer under the following conditions: (a) initially, before any sediment is deposited, (b) after a 2 m layer of sediment of silty fine sand has been deposited, and (c) after draining the lagoon down to base level, with the same thickness (2 m) of sediment still in place.*

Unit weights: clay = 18 kN/m³; sand = 20 kN/m³; sediment = 16 kN/m³.

(a)   Initially, before deposition of any sediment:

Top of clay	$\sigma_z = 9.81 \times 4 = 39.2$ kPa
	$u_z = 9.81 \times 4 = 39.2$ kPa
	$\sigma'_z = \sigma_z - u_z = 0$
Top of sand	$\sigma_z = 39.2 + 18 \times 3 = 93.2$ kPa
	$u_z = 9.81 \times 7 = 68.7$ kPa
	$\sigma'_z = 24.5$ kPa
Bottom of sand	$\sigma_z = 93.2 + 20 \times 4 = 173.2$ kPa
	$u_z = 9.81 \times 11 = 107.9$ kPa
	$\sigma'_z = 65.3$ kPa

(b)   As the 2 m layer of sediment is deposited it replaces a 2 m layer of water thus increasing the vertical stress:

$$\Delta\sigma'_z = (16.0 - 9.81)2 = 12.4 \text{ kPa.}$$

Top of clay	$\sigma'_z = 12.4$ kPa
Top of sand	$\sigma'_z = 24.5 + 12.4 = 36.9$ kPa
Bottom of sand	$\sigma'_z = 65.3 + 12.4 = 77.7$ kPa

(c) When the water level is reduced by 4 m, a corresponding increase in effective stress takes place: $\Delta\sigma'_z = -\Delta u_z = 9.81 \times 4 = 39.2$ kPa.

Top of clay	$\sigma'_z = 12.4 + 4 \times 9.81 = 51.6$ kPa
Top of sand	$\sigma'_z = 36.9 + 4 \times 9.81 = 76.1$ kPa
Bottom of sand	$\sigma'_z = 77.7 + 4 \times 9.81 = 116.9$ kPa

Exercises

4.1 Explain the differences between *phreatic* and *vadose* groundwater.

4.2 Calculate an approximate value for the maximum capillary rise in soil having the following properties:

$e = 0.800$ effective size, $d_{10} = 0.052$ (assume $C = 25$).

4.3 Calculate the suction pressure required to produce a capillary rise of 1.8 m.

4.4 Define the terms *total stress*, *effective stress* and *pore pressure* and state the relationship that is assumed to exist between them.

4.5 Calculate the value of the pore pressure coefficient B for a soil having a porosity of 45 per cent and a soil skeleton compressibility of 120×10^{-6} m^2/kN. The compressibility of water may be assumed to be 0.454×10^{-6} m^2/kN.

4.6 Describe the effects of frost on the surface layer of soil and explain why some soils are more susceptible than others to the effects of frost.

4.7 Tests on the soil forming a road sub-grade have shown a liquid limit of 54 per cent, a plastic limit of 20 per cent and a bulk density of 18.6 kN/m^3. The water table is expected to remain stable at a depth of 1.5 m below formation level. The water content at a depth of 0.5 m below formation level was found to be 23 per cent immediately prior to construction. The road formation will transmit a contact pressure of 5.8 kPa. Assuming an average water content/suction relationship of $w = 50 + 10pF$, calculate the equilibrium water content for the soil at this point and so determine whether shrinking or swelling might occur after construction.

4.8 On a certain site, a surface layer of sandy gravel is 6 m thick and this overlies a 5 m layer of clay, which in turn overlies impermeable rock. Draw up total stress/effective stress profiles down to the bottom of the clay for the following conditions: (a) water table at the ground surface, and (b) water table at the gravel/clay interface.

Unit weights:	sandy gravel (saturated) $= 21$ kN/m^3
	sandy gravel (drained) $= 18$ kN/m^3
	clay $= 19$ kN/m^3

4.9 In a site reclamation project, 2.5 m of graded fill ($\gamma = 22$ kN/m^3) were laid in compacted layers over an existing layer of silty clay ($\gamma = 18$ kN/m^3) which was 3 m thick. This was underlain by a 2 m thick layer of gravel ($\gamma = 20$ kN/m^3). Assuming that the water table remains at the surface of the silty clay, draw up the total stress/effective stress profiles for: (a) before the fill is placed, (b) immediately after the fill has been placed.

4.10 Figure 4.13 shows the cross-section of an excavation which is to be made alongside a river. Write down an expression for the effective stress at level A–A and use this to establish the depth H to which the water in the trench can be reduced before instability occurs. Assume that in the gravel layer at A–A there is a seepage pressure loss of 30 per cent.

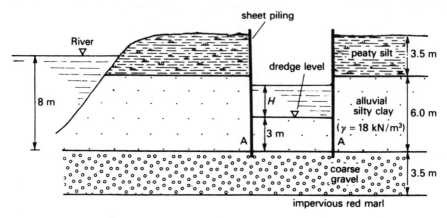

Fig. 4.13

4.11 A wide excavation is to be made on a site which has the following soil conditions:

0–2 m BS medium gravel: $\gamma_{sat.} = 21.8$ kN/m³
                          $\gamma_{drained} = 18.5$ kN/m³
2–6 m BS silty sand: $\gamma_{sat.} = 19.6$ kN/m³
                          $\gamma_{drained} = 18.4$ kN/m³
6–21 m BS heavy clay: $\gamma = 20$ kN/m³
21 m and below, pervious sandstone.

The water table is at a depth of 1.5 m below ground level and the artesian pressure in the sandstone corresponds to a static head of 5 m above ground level.

(a) Calculate the initial effective stress at top and bottom of the clay layer.
(b) To what depth can a pumped-out excavation be taken before a bottom blow-up occurs?
(c) If an excavation of 10 m depth is required and a factor of safety of 1.5 provided against bottom blow-up, calculate the equivalent static head reduction required to the sandstone (by relief pumping).

Soil Mechanics Spreadsheets and Reference Assignments and Quizzes (available on the accompanying CD):

Assignments A.4   Groundwater, pore pressure and effective stress
Quiz Q.4   Groundwater, pore pressure and effective stress

Chapter 5

Soil permeability and seepage

5.1 Water flow

Since soils consist of discrete particles, the pore spaces between particles are all interconnected so that water is free to flow within the soil mass. In such a porous media, water will flow from zones of higher to lower pore pressure. When considering problems of water flow, it is usual to express a pressure as a *pressure head* or *head*, measured in metres of water. Bernoulli's equation states three *head* components, the sum of which provide the *total head* (*h*) causing a water flow:

$$h = h_z + \frac{u}{\gamma_w} + \frac{v^2}{2g} \qquad\qquad [5.1]$$

where h_z = position or elevation head

     $\dfrac{u}{\gamma_w}$ = pressure head due to pore pressure u

     $\dfrac{v^2}{2g}$ = velocity head when the velocity of flow is v

The last component, the velocity head, is usually ignored in problems of ground-water flow, since v is quite small owing to the high resistance to flow offered by the granular structure of the soil. The first two terms, therefore, represent the head tending to cause the flow of water through a mass of soil.

     In saturated conditions, one-dimensional flow is governed by Darcy's law, which states that the flow velocity is proportional to the hydraulic gradient:

$$v \propto i \quad \text{or} \quad v = ki \qquad\qquad [5.2]$$

where v = flow velocity
        k = the flow constant or coefficient of permeability
        i = the hydraulic gradient = $\dfrac{\Delta h}{\Delta L}$ (Fig. 5.1)
     Δh = difference in total head over a flow path length of ΔL

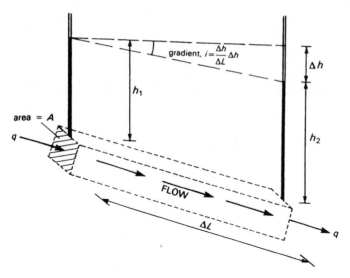

Fig. 5.1   One-dimensional flow in soil

The quantity flowing is therefore given by:

$$q = Av = Aki \tag{5.3}$$

where q = quantity flowing in unit time
          A = area through which flow is taking place

5.2   Coefficient of permeability

The capacity of a soil to allow water to pass through it is termed its *permeability* (or *hydraulic conductivity*). The *coefficient of permeability* (k) may be defined as the flow velocity produced by a hydraulic gradient of unity (see eqn [5.2]). The value of k is used as a measure of the resistance to flow offered by the soil, and it is affected by several factors:

(a)   The porosity of the soil.
(b)   The particle-size distribution.
(c)   The shape and orientation of soil particles.
(d)   The degree of saturation/presence of air.
(e)   The type of cation and thickness of adsorbed layers associated with clay minerals (if present).
(f)   The viscosity of the soil water, which varies with temperature.

The range of values for k is extremely large, extending from 100 m/s in the case of very coarse-grained gravels to almost nothing in the case of clay. In granular materials k varies approximately inversely with the specific surface value (see Section 1.6), but in fine soils the relationships are more complex. In clay soils, such factors as water content and temperature are significant, as also is the presence of fissures when considering the permeability of large masses. Table 5.1

Table 5.1   Range of values of k *(m/s)*

10^2-		
10^1-		
$1-$ Clean gravels		Very good drainage
$10^{-1}-$		
$10^{-2}-$		
$10^{-3}-$ Clean sands		
Gravel–sand mixtures		Good drainage
$10^{-4}-$	Fissured and weathered clays	
$10^{-5}-$		
Very fine sands		
$10^{-6}-$ Silts and silty sands		Poor drainage
$10^{-7}-$		
$10^{-8}-$ Clay silts (>20% clay)		
Unfissured clays		Practically impervious
$10^{-9}-$		

shows the range of average values for k for various soils and also indicates potential drainage conditions.

Approximations of k

A number of approximate empirical relationships have been suggested between k and other soil properties:

$$k \propto d_{10}^2 \quad k \propto (d_{\text{average}})^2 \quad k \propto \frac{e^3}{1+e} \quad k \propto e^2 \quad k \propto \log e$$

It is clear from comparative studies, however, that none are particularly reliable, and that it is far more realistic to obtain estimates for k using field pumping tests or a laboratory method.

The most frequently used approximation is one suggested by Hazen for filter sands:

$$k = C_k d_{10}^2 \text{ (mm/s)} \tag{5.4}$$

where d_{10} = effective size (mm)
   C_k = experimental coefficient dependent on the nature of soil

Experimental evidence (Masch and Denny, 1966; Trenter, 1999) suggests the original Hazen values for C_k are too high, and that the grading of the soil is significant. In Table 5.2 the modified values for C_k are related to the uniformity coefficient C_u (see eqn [2.4]) and to the range of d_{10}.

Table 5.2   Value of modified coefficient C_k

C_u	Range of d_{10} (mm)		
	>0.2	0.2–0.06	<0.06
1	5.0	4.2	3.3
2	3.0	2.6	2.0
3	2.4	2.1	1.7
4	2.0	1.8	1.5

Table 5.3   Values of temperature correction coefficient κ_t

°C	κ_t	°C	κ_t
0	1.779	25	0.906
4	1.555	30	0.808
10	1.299	40	0.670
15	1.133	50	0.550
20	1.000	60	0.468
		70	0.410

Effect of temperature

Since the viscosity and density of water both vary markedly with temperature, it follows that the value of the coefficient of permeability will be affected by changes in temperature. It may be shown theoretically that for a laminar flow condition in a saturated soil mass:

$$k \propto \frac{\gamma_w}{\eta}$$

where γ_w = unit weight of water
η = viscosity of water

A correction for the effect of temperature may, therefore, be obtained as follows:

$$k_t = \kappa_t k_{20} \tag{5.5}$$

where k_t = value of k corresponding to a temperature of t
k_{20} = value of k corresponding to a temperature of 20 °C (i.e. standard room remperature)
κ_t = temperature correction coefficient

Table 5.3 gives values of κ_t, which will have a value of 1.0 at 20 °C since it is at this temperature that most laboratory graduations are standardised.

5.3 Seepage velocity and seepage pressure

The movement of water through a soil mass is generally termed *seepage*. On a microscopic scale the water when flowing follows a tortuous route through the voids in the soil. From a practical point of view, however, it is assumed to follow a straight-line path. In Darcy's equation, the velocity v is interpreted as the *apparent* or *superficial velocity*, i.e. the velocity of flow relative to a soil section area A. The velocity through pores will be greater, and this is termed the *seepage velocity* (v_s).

Consider a soil of porosity: $n = A_v/A$

For a given flow rate: $q = Av = A_v v_s$

where A = section area of soil (perpendicular to flow direction)
A_v = section area of voids

Then $v_s = v\dfrac{A}{A_v} = \dfrac{v}{n} = \dfrac{ki}{n}$ [5.6]

The work done by water during seepage results in a *seepage force* (J) being exerted on the particles. Consider the column of soil shown in Fig. 5.2. When the valve at level A–A is wide open, flow takes place under the influence of a head of h_s, thus an upward-acting seepage force is exerted on the soil particles between C–C and B–B.

When the valve at level A–A is closed, the water level will rise until it reaches O–O where it will remain stationary. At this point there will be no seepage. It may be concluded, therefore, that the seepage force has now been balanced by the additional weight of water between A–A and O–O.

Then seepage force, $(J) = \gamma_w h_s A$

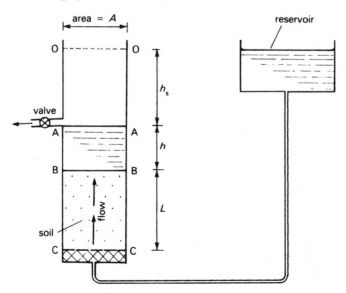

Fig. 5.2 Seepage pressure

But since the flow velocity is constant the seepage force acting on the soil will also be constant between C–C and B–B.

So that seepage force per unit volume, $j = \dfrac{\gamma_w h_s A}{LA}$

and as $\dfrac{h_s}{L} =$ hydraulic gradient i

then $j = i\gamma_w$ [5.7]

The seepage force per unit volume (j) is usually referred to as the *seepage pressure*.

5.4   Quick condition and critical hydraulic gradient

The effect of upward-flowing water in a soil mass in creating a seepage pressure on the particles is to reduce the intergranular or effective stress. If a sufficiently high enough flow rate is achieved the seepage pressure can cancel out the effective stress completely causing a *quick condition*. This is essentially a condition in which the soil has no shear strength, since the intergranular stress has been reduced to zero.

Consider again the situation in Fig. 5.2.

At the quick condition the flow will cause a seepage force at C–C which will be equal and opposite to the effective stress due to the weight of soil.

Then equating forces at C–C:

$$\gamma_w(L + h + h_s)A = (\gamma_{sat}L + \gamma_w h)A$$

giving $\gamma_w h_s = (\gamma_{sat} - \gamma_w)L$

or $\gamma_w i_c = \gamma'$

where i_c is termed the *critical hydraulic gradient*, i.e. the hydraulic gradient at which the quick condition occurs. A numerical value may be obtained for i_c thus:

$$
i_c = \frac{\gamma'}{\gamma_w} = \frac{\gamma_{sat} - \gamma_w}{\gamma_w}
$$

$$
= \frac{(G_s + e)\gamma_w/(1 + e) - \gamma_w}{\gamma_w}
$$

$$
= \frac{G_s - 1}{1 + e}
$$ [5.8]

In cohesionless soils, particularly in medium to fine sands, the quick condition is likely to occur at hydraulic gradients of about 1.0. Typical situations in which this may arise are in alluvial flood plains immediately after a high tide and at the base of spoil tips following heavy rain. The popular term 'quicksand' refers to this condition, being so-named because it is in the range of fine to medium sands where the particular combination of permeability and critical hydraulic gradient gives rise to a rapid occurrence of the condition. Contrary to popular belief, however, this is not a 'sinking sand' condition. 'Quicksand' lacking shear resistance is really a liquid, but its density is roughly twice that of water: a human body would, therefore, float half-submerged.

In soils possessing a significant amount of cohesion, such as silts and clay, the critical hydraulic gradient criterion does not apply, since they have some shear strength even at zero normal stress.

5.5 Determination of k in the laboratory

The coefficient of permeability k can be measured using field tests, or tests conducted in a laboratory. The main categories of problems associated with the reliability of laboratory tests are: (a) those concerned with obtaining good representative samples, (b) those concerned with the reproducibility of laboratory measurements, and (c) those concerned with reproducing field conditions.

Reliability of sample
The permeability of a soil mass is dependent on both its microstructure (i.e. particle size, shape, arrangement, etc.) and its macrostructure (i.e. whether or not stratified, presence of fissures, pipes, lenses, etc.). For obvious practical reasons the size of samples taken for laboratory tests is quite small, and therefore unlikely to be satisfactorily representative in soils with significant macrostructure characteristics. To some extent this deficiency may be overcome by obtaining carefully selected groups of samples.

Reliability of laboratory measurements
The aim in any laboratory procedure should be to reproduce similar results using the same procedure and for these to compare favourably with results using different procedures. In laboratory permeability tests disparities may occur due to:

(a) The presence of air bubbles in the permeant (water).
(b) Variations in sample density and porosity.
(c) Variations in temperature and therefore viscosity of the permeant.

Reproduction of field conditions
It is generally difficult to obtain truly undisturbed samples and particularly difficult to simulate in the laboratory the true field conditions of flow and stress. In the main, disparities between laboratory and field conditions will be:

(a) Variations in density and porosity.
(b) Variation in flow direction with respect to bedding.
(c) Limited ability of a small sample to simulate anisotropic conditions.
(d) Variation in pore pressure and effective stress condition.

For these reasons laboratory tests should, wherever possible and appropriate, be augmented with field tests.

Since the range of values of k is so large the choice of method and equipment varies with the type of soil. The most common types of test in present use are:

(a) *Constant head test*: suitable for gravels and sands with values of $k >$ 10^{-5} m/s.

(b)   *Falling head test*: suitable for fine sands, silts and clays with values of k between 10^{-4} and 10^{-7} m/s.

(c)   *Hydraulic cell test*: suitable for soils with very low permeabilities; both vertical and horizontal permeability can be determined.

5.6   The constant head test

The *constant head test* is used to determine the coefficients of permeability (k) of coarse-grained soils such as gravels and sands having values of k above 10^{-4} m/s.

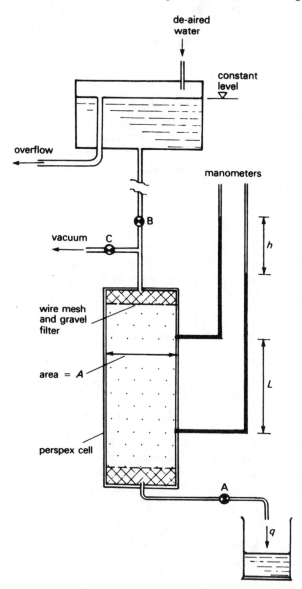

Fig. 5.3   Constant head test apparatus

The apparatus used is called a constant head *permeameter* and is shown diagrammatically in Fig. 5.3.

The soil sample is contained in a perspex cylinder with wire mesh and gravel filters above and below. In the side of the cylinder a number of manometer connection points are provided to enable pairs of pressure head readings to be taken (only one pair is shown in the diagram). Water is allowed to flow through the sample from a tank or reservoir designed to maintain a constant head and the quantity of water measured by weighing a collecting vessel.

Since the presence of air bubbles can seriously affect the results, it is imperative to ensure as little air as possible in the system. This may be effected by first of all supplying only de-aired water to the constant-head tank and secondly by applying a vacuum to the sample prior to commencement of the test. The test is then started with valves A and B open and valve C closed, with the valve A being used to control the rate of flow. Flow is allowed to continue until a *steady state* has been established, i.e. the levels in the manometer tubes remain constant. Once a steady state has been reached, the quantity flowing during a given time interval is measured, and the two manometer levels noted. The flow rate is then varied and the procedure repeated. Several tests at varying flow rates and heads should be carried out and the average value of *k* established which is then corrected for temperature (Table 5.3). The dry density and void ratio should also be reported.

From eqn [5.3]: $q = k\,Ai$

$$\therefore \quad k = \frac{q}{Ai} = \frac{QL}{Aht} \ \text{mm/s} \qquad\qquad [5.9]$$

where Q = quantity of water collected in time t s
$\quad\quad = Q(\text{ml}) \times 10^3 \ (\text{mm}^3)$
$\quad A$ = cross-sectional area of sample (mm^2)
$\quad h$ = difference in manometer levels (mm)
$\quad L$ = distance between manometer tapping points (mm)

Worked example 5.1   *During a test using a constant-head permeameter, the following data were collected. Determine the average value of* k.

Diameter of sample = 100 mm   Temperature of water = 17 °C
Distance between manometer tapping points = 150 mm

Quantity collected in 2 min. (ml)	541	503	509	474
Difference in manometer levels (mm)	76	72	68	65

Cross-sectional area of sample $= 100^2 \times \dfrac{\pi}{4} = 7854 \ \text{mm}^2$

Flow quantity, $Q = Q(\text{ml}) \times 10^3 \ \text{mm}^3$

Flow time, $t = 2 \times 60 = 120$ s

Then from eqn [5.9]: $k = \dfrac{QL}{Aht} = \dfrac{Q \times 10^3 \times 150}{7854 \times h \times 120}$

$\qquad\qquad\qquad = 0.159 Q/h \ \text{mm/s}$

Tabulated results:

Flow quantity Q (ml)	Head difference h (mm)	k = 0.159Q/h (mm/s)	Temperature correction (Table 5.3-interpolating): $\kappa_T = 1.09$
541	76	1.13	
503	72	1.11	
509	68	1.19	
474	65	1.16	

Average   k = 1.15 mm/s   Corrected $k = 1.15 \times 1.09$
= 1.25 mm/s

5.7   The falling head test

The *falling head test* is used to determine the coefficient of permeability of fine soils, such as fine sands, silts and clays. For these soils, the rate of flow of water through them is too small to enable accurate measurements using the constant head permeameter. The *falling head permeameter* is shown diagrammatically in Fig. 5.4.

An undisturbed soil sample is obtained in a 100 mm diameter (usually) cylinder, which may be a U100 sample tube or a core-cutter tube as used in field density tests. Samples can also be prepared by compaction in a standard mould. A wire mesh and gravel filter is provided at the top and bottom of the sample. The base of the cylinder is stood in a water reservoir fitted with a constant-level overflow and the top connected to a glass standpipe of known diameter.

The test is conducted by filling the standpipe with de-aired water and allowing seepage to take place through the sample. The height of water in the standpipe is recorded at several time intervals during the test, and the test repeated using standpipes of different diameter. Following computation of the results, the average value for *k* is determined. It is usual also to report the initial and final unit weights and water contents of the sample.

Commencing with Darcy's law (eqn [5.3]):

$q = kAi$

But, referring to Fig. 5.4, if the level in the standpipe falls d*h* in a time of d*t* then

$$q = -a\frac{dh}{dt}$$

and the hydraulic gradient $i = \dfrac{h}{L}$

so that $q = -a\dfrac{dh}{dt} = kA\dfrac{h}{L}$

where   a = cross-sectional area of the standpipe
A = cross-sectional area of the sample

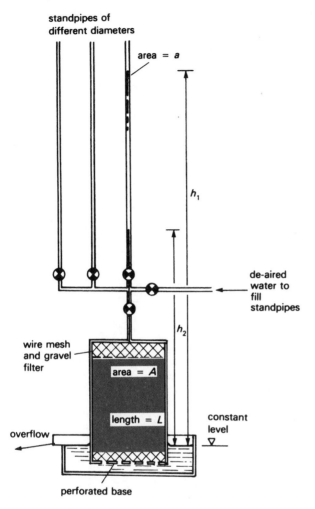

standpipes of
different diameters

area = a

h_1

de-aired
water to
fill
standpipes

wire mesh
and gravel
filter

area = A

length = L

constant
level

overflow

h_2

perforated base

Fig. 5.4   Falling head test apparatus

Rearranging and integrating:

$$-\int_{h_1}^{h_2}\frac{\mathrm{d}h}{h} = \frac{kA}{aL}\int_{t_1}^{t_2}\mathrm{d}t$$

$$\therefore \qquad -\ln\frac{h_2}{h_1} = \frac{kA}{aL}(t_2 - t_1)$$

Giving

$$k = \frac{aL\,\ln(h_1/h_2)}{A(t_2 - t_1)} \qquad\qquad [5.10(a)]$$

or

$$k = \frac{2.3aL\,\log_{10}(h_1/h_2)}{A(t_2 - t_1)} \qquad\qquad [5.10(b)]$$

Worked example 5.2   *During a test using a falling-head permeameter the following data were recorded. Determine the average value of* k.

Diameter of sample = 100 mm
Length of sample   = 150 mm

Cross-sectional area of sample, $A = 100^2 \times \dfrac{\pi}{4}$

Cross-sectional area of standpipe, $a = d^2 \times \dfrac{\pi}{4}$

From eqn [5.10(a)]:

$$k = \frac{aL \ln(h_1/h_2)}{A(t_2 - t_1)}$$

$$= \frac{d^2 \times 1.50 \ln(h_1/h_2)}{100^2(t_2 - t_1)} = \frac{0.015d^2 \ln(h_1/h_2)}{t_2 - t_1} \text{ mm/s}$$

Recorded data				*Computed*	
Standpipe diameter d (mm)	*Level in standpipe*		*Time interval $(t_2 - t_1)$ (s)*	$\ln\dfrac{h_1}{h_2}$	*k (mm/s)*
	Initial h_1 (mm)	*Final h_2 (mm)*			
5.00	1200	800	82	0.4055	1.854×10^{-3}
	800	400	149	0.6931	1.744×10^{-3}
9.00	1200	900	177	0.2877	1.975×10^{-3}
	900	700	169	0.2513	1.807×10^{-3}
	700	400	368	0.5596	1.847×10^{-3}
12.50	1200	800	485	0.4055	1.959×10^{-3}
	800	400	908	0.6931	1.789×10^{-3}
			Average	$k = 1.85 \times 10^{-3}$ mm/s	
				$= \mathbf{1.85 \times 10^{-6}}$ **m/s**	

5.8   Hydraulic cell test

The *Hydraulic cell* was introduced by Rowe and Barden in 1966 for the purpose of carrying out consolidation tests. Its use has also been developed for permeability. Either vertical or horizontal permeabilities may be measured, with a high degree of reliability. The test is of the constant head type, with the field values of void ratio, pore pressure and effective stress closely simulated. This is the preferred method and is described fully in BS 1377.

Vertical permeability
The arrangement of the apparatus for the determination of vertical permeability is shown in Fig. 5.5. An undisturbed sample of soil is fitted into the

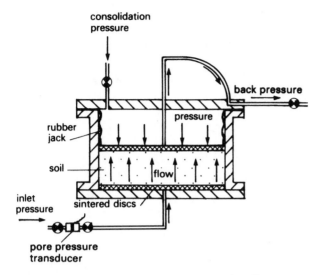

Fig. 5.5 Vertical permeability using a hydraulic cell

body of the cell between porous discs and consolidated to an effective pressure relating to site conditions using the hydraulic pressure jack. The outlet in the base is then connected to a constant pressure system. This can be set to simulate field pore-pressure levels. A second constant pressure system is connected to the upper drain and set at a pressure slightly lower than that at the outlet. The pressure difference is monitored using a differential pressure gauge or a pair of pressure transducers between the inlet and outlet tubes. This is usually maintained at less than 10 per cent of the effective pressure exerted on the sample.

The quantity of flow is obtained from observations taken of two calibrated *volume-change indicators*, one on the inlet side and the other on the outlet side. When the flow quantities for both inlet and outlet tubes are within 10 per cent of each other and steady-state conditions are assumed to prevail the test measurements are taken.

Applying Darcy's law (eqn [5.3]):

$$q = kAi$$

$$k = \frac{q}{Ai}\kappa_T = \frac{qL}{Ah}\kappa_T \qquad [5.11]$$

where q = flow quantity (mm³/s)
        L = length of sample
        i = thickness after initial consolidation (mm)
        A = cross-sectional area of sample (mm²)
        h = differential pressure head (mm of water)
        κ_T = temperature correction coefficient (Table 5.3)

Worked example 5.3   *During a test using a hydraulic cell to determine the vertical permeability of a soil, the following data were collected. Determine the value of* k_v.

Thickness of sample after consolidation = 73.00 mm
Diameter of sample = 254 mm
Difference between inlet and outlet pressures = 65.2 cm of water
Flow rate = 2.39 mm³/s
Temperature = 15 °C

Differential pressure head, $h = 652$ mm of water

Area of sample, $A = 254^2 \times \dfrac{\pi}{4} = 50\,671$ mm^2

Length of sample, $L = 73.00$ mm

Using eqn [5.11], $k_v = \dfrac{qL}{Ah}\kappa_T$

$$= \dfrac{2.39 \times 73.00}{50\,671 \times 652}1.13 = \underline{6.0 \times 10^{-6}} \text{ mm/s}$$

Horizontal permeability

The arrangement of the apparatus for the determination of horizontal permeability is shown in Fig. 5.6. In this set-up the flow is radial, with the cylindrical outer inlet surface being surrounded by a porous drainage material and a central porous drain inserted as the outlet. The overall diameter of the sample is slightly less than that used in the vertical permeability test due to the thickness (about 3.5 mm) of the inlet drain. The centre drain usually takes the form of a sintered bronze spindle.

As with the vertical permeability test, the flow is measured by means of volume-change indicators and the pressure difference by means of either a differential pressure gauge or a pair of pressure transducers.

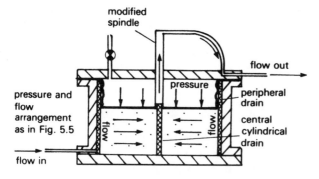

Fig. 5.6 Horizontal permeability using a hydraulic cell

Then the horizontal coefficient of permeability may be obtained from:

$$k_H = \frac{q}{2\pi L}\frac{\log_e(D/d)}{h}\kappa_T \qquad [5.12]$$

where q = flow quantity (mm³/s)
$\quad L$ = length of sample
$\quad$ = thickness after initial consolidation (mm)
$\quad D$ = outside diameter of the sample (mm)
$\quad d$ = inside diameter of the sample
$\quad$ = diameter of centre drain spindle (mm)
$\quad h$ = differential pressure head (mm of water)
$\quad \kappa_T$ = temperature correction coefficient (Table 5.3)

5.9 Field permeability tests

Because of the problems associated with the reliability of laboratory tests, as discussed in Section 5.5, field methods should be used when permeability values are of significant importance. Comprehensive multiple-well pumping tests can be expensive to carry out, but offer a high level of reliability owing to the inclusion of a wide range of macro-structural characteristics. The use of site investigation boreholes can be economically advantageous, providing the pumping/observation sequences are carefully planned and controlled.

Steady state pumping tests
Pumping tests involve the measurement of a pumped quantity from a well, together with observations in other wells of the resulting *drawdown* of the groundwater level. A *steady state* is achieved when, at a constant pumping rate, the levels in the observation wells also remain constant. The pumping rate and the levels in two or more observation wells are then noted. The analysis of the results depends on whether the aquifer is *confined* or *unconfined*.

(a)   Pumping test in a confined aquifer
In a confined aquifer, the pumping rate must not be high enough to reduce the level in the pumping well below the top of the aquifer. The interface between the top of the aquifer and the overlying impermeable stratum therefore forms the top stream line.

An arrangement of a pumping well and two observation wells is shown in Fig. 5.7. The piezometric surface is assumed to be above the upper surface of the aquifer and the hydraulic gradient is assumed to be constant at a given radius.

Then, in steady state conditions, the flow is considered through an elemental cylinder having radius r, thickness dr and height h.

Hydraulic gradient (outside to inside), $i = \dfrac{dh}{dr}$

Area through which flow takes place, $A = 2\pi rD$

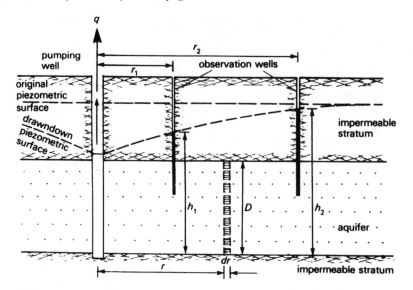

Fig. 5.7 Pumping test in a confined aquifer

Then, starting with Darcy's equation (eqn [5.3]):

$$q = Aki$$

$$= 2\pi r Dk \frac{\mathrm{d}h}{\mathrm{d}r}$$

or

$$\frac{\mathrm{d}r}{r} = \frac{2\pi}{q} Dk\mathrm{d}h$$

Integrating, $\ln(r_2/r_1) = \dfrac{2\pi Dk}{q}(h_2 - h_1)$

Giving $$k = \frac{q}{2\pi D} \frac{\ln(r_2/r_1)}{h_2 - h_1}$$ [5.13]

When only one observation well is available, the pumping well may be utilised as an observation well. Because of soil disturbance around the well, an effective radius of 1.2 × actual radius should be used. The observed pumping well drawdown must also be corrected, since the piezometric level at the well will generally be higher during pumping than the actual level in the well. Stepped drawdown tests may be used to evaluate the pumping level loss, or a loss assumed which is based on experience (e.g. 20 per cent). The following substitutions may then be made in eqn [5.13]:

$r_1 = 1.2 \times r_w$ (r_w = actual pumping well radius)

$h_1 = h_o - d_w$

where d_w = corrected pumping well drawdown (e.g. $0.8 \times$ observed drawdown).

Another, albeit more crude, approximation may be derived from a consideration of the *radius of influence* (r_o) of the pumping. It may be assumed that no draw-down of the piezometric head takes place outside the radius of influence, i.e. when $r = r_o$. Thus, eqn [5.13] may be used with the following substitutions:

$$r_2 = r_o \quad \text{and} \quad h_2 = h_o$$

In the absence of positive observations the location of r_o must be estimated using a rule-of-thumb, e.g. $r_o = 10h_o$; $r_o = 10D$ or $r_o = 10r_1$.

Another simple rule-of-thumb may be used to produce a rough estimate from only the pumping rate (q), the corrected drawdown in the pumping well (d_w) and the aquifer thickness (D). Suppose that $r_o/r_w = 2000$, then from eqn [5.13]:

$$k = \frac{q}{2\pi D} \frac{\ln(2000)}{h_o - d_w} = \frac{1.21q}{D(h_o - d_w)}$$

$$\text{If} \quad r_o/r_w = 5000, \quad k = \frac{1.36q}{D(h_o - d_w)}$$

$$\text{Thus a useful approximation would be:} \quad k = \frac{1.25q}{D(h_o - h_w)} \tag{5.14}$$

Worked example 5.4 *A permeability pumping test was carried out from a well sunk into a confined stratum of dense sand. Figure 5.8 shows the arrangement of the pumping well and observation wells, together with relevant dimensions. Initially, the piezometric surface was located at a depth of 2.5 m. When a steady state was achieved at a pumping rate of 37.4 m³/hr, the following drawdowns were observed:*

Pumping well:　　　　　$d_w = 4.46$ m
Observation well 1:　$d_1 = 1.15$ m
Observation well 2:　$d_2 = 0.42$ m

(a) *Calculate a value for the coefficient of permeability of the sand using (i) the observation well data, and (ii) a corrected drawdown level in the pumping well.*
(b) *Estimate the radius of influence at this pumping rate.*

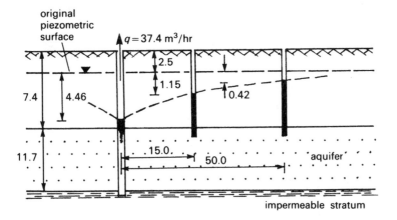

Fig. 5.8

(a) (i) Observation well data: $r_1 = 15$ m $r_2 = 50$ m

$h_o = 11.7 + 7.4 - 2.5 = 16.6$ m

$h_1 = h_o - d_1 = 16.6 - 1.15 = 15.45$ m

$h_2 = h_o - d_2 = 16.6 - 0.42 = 16.18$ m

Also $q = 37.4/3600$ m^3/s   and $D = 11.7$

Then $k = \dfrac{q}{2\pi D} \dfrac{\ln(r_2/r_1)}{h_2 - h_1} = \dfrac{37.4 \times \ln(50/15)}{2\pi \times 11.7 \times 3600(16.18 - 15.45)}$

$= \underline{2.33 \times 10^{-4} \text{ m/s}}$

(ii) Assuming a 20 per cent loss, the corrected pumping well drawdown is

$d_w = 0.8 \times 4.46$ m

Therefore, $h_1 = h_w = 16.6 - 0.8 \times 4.46 = 13.03$ m

and assuming also that $r_1 = r_w = 0.1$ m

$k = \dfrac{37.4 \times \ln(50/0.1)}{2\pi \times 11.7 \times 3600(16.18 - 13.03)} = \underline{2.79 \times 10^{-4} \text{ m/s}}$

(b) When $r = r_0$ (i.e. the radius of influence), there is no drawdown and therefore $h = h_0$. Then, putting $r_1 = 50$ m and $h_1 = 16.18$ m,

$\dfrac{\ln(r_o/50)}{16.60 - 16.18} = \dfrac{2.33 \times 10^{-4} \times 2\pi \times 11.7 \times 3600}{37.4}$

From which, $r_0 = \underline{100 \text{ m}}$

(b)   Pumping test in an unconfined aquifer

An unconfined aquifer is a free-draining surface layer underlain by an impervious base. Under conditions of steady state pumping the hydraulic gradient at a given radius is assumed to be constant in a homogeneous medium. An arrangement of a pumping well and two observation wells is shown in Fig. 5.9. Consider the inflow through an elemental cylinder of radius r, thickness dr and height h.

Hydraulic gradient (outside to inside), $i = \dfrac{dh}{dr}$

Area through which flows take place, $A = 2\pi r$

Then, starting with Darcy's equation (eqn [5.3]):

$$q = Aki$$

$$= 2\pi rhk \dfrac{dh}{dr}$$

or

$$\dfrac{dr}{r} = \dfrac{2\pi}{q} khdh$$

Integrating, $\ln(r_2/r_1) = \dfrac{\pi}{q} k(h_2^2 - h_1^2)$

Giving $k = \dfrac{q}{\pi} \dfrac{\ln(r_2/r_1)}{(h_2^2 - h_1^2)}$ [5.15]

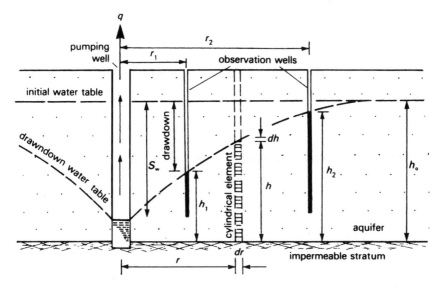

Fig. 5.9 Pumping test in an unconfined aquifer

When the drawdown is a significant fraction of the original saturated thickness, higher flow velocities develop and friction losses increase. Corrected drawdown values should therefore be computed as follows:

Corrected drawdown, $\quad d_c = d - \dfrac{d^2}{2h_o}$ $\hfill$ [15.16]

where $\quad d$ = observed drawdown

$\qquad h_o$ = initial saturated height of the aquifer

When only one observation well is available, eqn [5.15] may be used with substitutions in respect of r_o and h_o, or r_w and h_w, as explained for the confined aquifer pumping test.

A very approximate estimate may be obtained from the pumping well data only, using an assumption similar to those given for eqn [5.14]

$$k \simeq \frac{2.4q}{h_w(2 - h_w)}$$ $\hfill$ [5.17]

Worked example 5.5 *A permeability test was carried out from a well sunk through a surface layer of medium dense sand. Figure 5.10 shows the arrangement of the pumping well and observation wells, together with relevant dimensions. Initially, the water table was located at a depth of 2.5 m. When a steady state was achieved at a pumping rate of 23.4 m^3/hr, the following drawdowns were observed:*

Pumping well: $\qquad$ $d_w = 3.64$ m
Observation well 1: $\quad$ $d_1 = 0.96$ m
Observation well 2: $\quad$ $d_2 = 0.48$ m

(a) Calculate a value for the coefficient of permeability of the sand using (i) the observation well data, and (ii) using a corrected drawdown level in the pumping well.

(b) Estimate the radius of influence at this pumping rate.

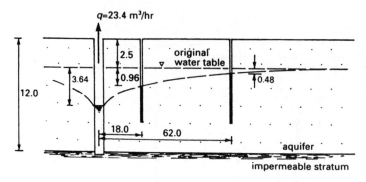

Fig. 5.10

(a) (i) Observation well data: $r_1 = 18$ m $r_2 = 62$ m
$$h_o = 12.0 - 2.5 = 9.50 \text{ m}$$
$$h_1 = h_o - d_1 = 9.50 - 0.96 = 8.54 \text{ m}$$
$$h_2 = h_o - d_2 = 9.50 - 0.48 = 9.02 \text{ m}$$

Also $q = 23.4/3600$ m^3/s

Then $k = \dfrac{q \ln(r_2/r_1)}{\pi\ h_2^2 - h_1^2} = \dfrac{23.4 \times \ln(62/18)}{3600\pi \times (9.02^2 - 8.54^2)}$

$$= \underline{3.04 \times 10^{-4} \text{ m/s}}$$

(ii) The corrected pumping well drawdown is given by eqn [5.16]:

$$d_c = d_w - \frac{d_w^2}{2h_o} = 3.64 - \frac{3.64^2}{2 \times 9.50} = 2.94 \text{ m}$$

Therefore, $h_1 = h_w = 9.50 - 2.94 = 6.56$ m

and assuming also that $r_1 = r_w = 0.1$ m

$$k = \frac{23.4 \times \ln(62/0.1)}{3600\pi \times (9.02^2 - 6.56^2)} = \underline{3.47 \times 10^{-4} \text{ m/s}}$$

(b) When $r = r_o$ (i.e. the radius of influence), there is no drawdown and therefore $h = h_o$. Then, putting $r_1 = 62$ m and $h_1 = 9.02$ m,

$$\frac{\ln(r_o/62)}{9.50^2 - 9.02^2} = \frac{3.04 \times 10^{-4} \times 3600\pi}{23.4}$$

From which, $r_o = \underline{229 \text{ m}}$

Borehole tests

Several methods have been suggested whereby a measure of the coefficient of permeability can be obtained using one borehole only. A *rising head* test is carried out by pumping water out of the hole, and then observing the rate of rise of level. In an *inflow* test water is introduced into the hole; a *variable head* test then involves observation of the rate of fall of level, while in a *constant head* test the head is maintained at a given level by adjusting and measuring the flow rate at intervals from the start of the test.

It is generally held that constant head tests give more accurate results, providing the water pressure is less than that which would cause fracturing or other disturbance of the soil. A useful rule of thumb here is to keep the increase in water pressure to below one-half of the effective overburden pressure. Where the flow rate is likely to be high ($k > 10^{-3}$ m/s) the errors increase and thus the field pumping test would be more suitable.

In the simplest form of test, the hole is prepared by cleaning out any loose soil or other debris. For more accurate results, a perforated tube or a piezometer tip (see Section 12.6) is installed in a gravel filter.

A number of formulae have been suggested for calculating the coefficient k, including the following, which are recommended in BS 5930.

(a) Variable head test

(i) $k = \dfrac{A}{FT}$ [5.18]

(ii) $k = \dfrac{A}{F(t_2 - t_1)} \ln(H_1 - H_2)$ [5.19]

(b) Constant head test

(i) Hvorslev's time lag analysis: $k = \dfrac{q}{FH_c}$ [5.20]

(ii) Gibson's root-time method: $k = \dfrac{q_\infty}{FH_c}$ [5.21]

where A = cross-sectional area of the standpipe or borehole casing
           F = an *intake factor* dependent on conditions at the bottom of the borehole (Figs. 7 and 8 of BS 5930)
           T = basic time lag (Fig. 9 of BS 5930)
    H_1, H_2 = variable heads measured at elapsed times of t_1 and t_2 respectively
         H_c = constant head
          q = rate of inflow
        q_∞ = steady state of inflow, obtained from a graph of q against $1/\sqrt{t}$, at $1/\sqrt{t} = 0$ (Fig. 10 of BS 5930)

5.10 Rapid field determination of permeability

A rapid form of falling-head test may be carried out in the field on coarse soils using the apparatus shown in Fig. 5.11. This consists of a glass tube of diameter 50 mm and length 500 mm, or other near dimension, and a water container such as a large bucket. Two graduation marks are made about 200–250 mm apart on the upper length of the tube, and the lower end covered with a close wire mesh.

With the tube held in place in the water vessel as shown (Fig. 5.11), a layer of 50–100 mm of the soil is carefully deposited over the mesh using a funnel with

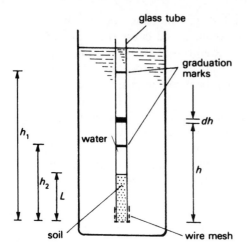

Fig. 5.11 Rapid falling head test

a rubber-tube extension. When the glass tube is withdrawn from the vessel, the level in it begins to fall. The time taken for the level to fall between the two graduation marks is recorded. The average of several tests may be taken for a good approximation of k.

Consider a change in level of dh taking place in time dt

Quantity flowing in unit time, $$q = d^2 \frac{\pi}{4} \frac{dh}{dt}$$

Cross-sectional area of sample, $$A = d^2 \frac{\pi}{4}$$

Hydraulic gradient, $$i = \frac{h}{L}$$

Then starting with Darcy's law (eqn [5.3]):

$$q = Aki$$

$$d^2 \frac{\pi}{4} \frac{dh}{dt} = d^2 \frac{\pi}{4} k \frac{h}{L}$$

or $$\frac{dh}{h} = \frac{k}{L} dt$$

Integrating $$\ln(h_1/h_2) = \frac{k}{L}(t_2 - t_1)$$

Then putting elapsed time, $$t = t_2 - t_1$$

$$k = \frac{L \times \ln(h_1/h_2)}{t} \qquad [5.22]$$

Worked example 5.6   *A glass tube of diameter 50 mm was used in a rapid falling-head test. The sand layer in the bottom of the tube was of thickness 75 mm and the two graduation marks were respectively 200 mm and 100 mm from the bottom of the tube (see Fig. 5.11). In five tests the times required for the water level in the tube to fall between the two marks were 66, 68, 65, 69 and 67 s.*

From eqn [5.22], $k = \dfrac{75 \times \ln(200/100)}{t} = \dfrac{52}{t}$

Then the values of k obtained are: 0.788, 0.765, 0.800, 0.754 and 0.776, giving an average

$k = \underline{0.78 \text{ mm/s}}$

Another rapid field method of determining the approximate permeability of a coarse-grained soil involves the measurement of the average *seepage velocity*. Two boreholes or trial pits are sunk in a situation where a natural hydraulic gradient exists between them (Fig. 5.12). A quantity of tracer dye is introduced into the up-gradient hole and the time (t) required for it to appear in the other hole noted.

Then the seepage velocity, $v_s = L/t$
and the hydraulic gradient, $i = h/L$
So starting with Darcy's law: ($v = ki$)

$$k = \frac{v}{i} = \frac{nv_s}{i}$$

Substituting $k = \dfrac{nL^2}{ht}$ [5.23]

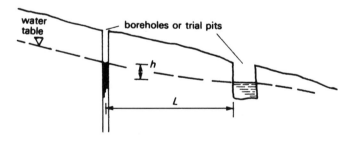

Fig. 5.12   Field seepage test

5.11   Two-dimensional flow

The seepage taking place around sheet-piling, dams, under other water-retaining structures and through embankments and earth dams is two-dimensional. That is to say, the vertical and horizontal velocity components vary from point to point within the cross-section of the soil mass (Fig. 5.13). To begin with, the general case will be considered of two-dimensional flow in a *homogeneous* and *isotropic* (i.e. $k_H = k_V$) soil mass, and then the graphical representation known as a *flow net* will be introduced.

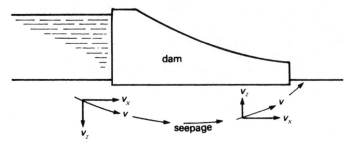

Fig. 5.13 Two-dimensional flow

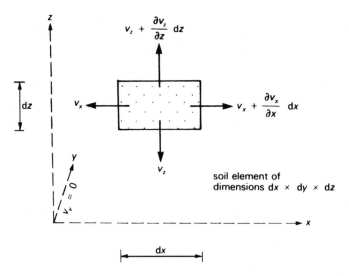

Fig. 5.14 General two-dimensional flow condition

General flow equation
In Fig. 5.14 an element of saturated soil having dimensions dx, dy and dz is shown with flow taking place in the xz plane only. The velocity gradients in the x- and z-directions are $\partial v_z/\partial z$ and $\partial v_x/\partial x$ respectively.

Assuming water to be incompressible and no volume change in the soil skeleton:

Quantity flowing into element = quantity flowing out of element

Substituting $v_x dy dz + v_z dy dx = \left(v_x + \dfrac{\partial v_x}{\partial x}dx\right)dy dz + \left(v_z + \dfrac{\partial v_z}{\partial z}dz\right)dy dx$

Therefore $\dfrac{\partial v_x}{\partial x} + \dfrac{\partial v_z}{\partial z} = 0$ [5.24]

This is known as the flow *continuity equation*.

Now from Darcy's law $(v = ki)$: $\quad v_x = -k\dfrac{\partial h}{\partial x} \quad$ and $\quad v_z = -k\dfrac{\partial h}{\partial z}$

Two Laplace equations can be derived to represent the flow condition based on two functions defined as follows:

Let $\quad v_x = -k\dfrac{\partial h}{\partial x} = \dfrac{\partial \Phi}{\partial x} = \dfrac{\partial \Psi}{\partial z}$

and $\quad v_z = -k\dfrac{\partial h}{\partial z} = \dfrac{\partial \Phi}{\partial z} = \dfrac{\partial \Psi}{\partial x}$

in which $\quad \Phi_{(x,z)}$ = the *potential function* such that $\Phi = -kh$
and $\qquad \Psi_{(x,z)}$ = the *stream function*

So substituting into eqn [5.24]:

$$\frac{\partial^2 \Phi}{\partial x^2} + \frac{\partial^2 \Phi}{\partial z^2} = 0 \qquad\qquad [5.25]$$

Also it can be shown that:

$$\frac{\partial^2 \Psi}{\partial x^2} + \frac{\partial^2 \Psi}{\partial z^2} = 0 \qquad\qquad [5.26]$$

Graphical representation of the flow equations
Comparing the definitions of the potential function Φ and the stream function Ψ it will be seen that:

differentiating $\qquad \Phi_{(x,z)}: d\Phi = \dfrac{\partial \Phi}{\partial x}dx + \dfrac{\partial \Phi}{\partial z}dz$

$$= v_x dx + v_z dz$$

and differentiating $\qquad \Psi_{(x,z)}: d\Psi = \dfrac{\partial \Psi}{\partial x}dx + \dfrac{\partial \Psi}{\partial z}dz$

$$= -v_z dz + v_x dz$$

Then if Φ is constant, $\qquad d\Phi = 0 \quad$ and $\quad \dfrac{dz}{dx} = \dfrac{-v_x}{v_z}$

and if Ψ is constant, $\qquad d\Psi = 0 \quad$ and $\quad \dfrac{dz}{dx} = \dfrac{v_z}{v_x}$

Thus a curve representing a constant value of $\Phi_{(x,z)}$ (and therefore constant head h) will intersect at right angles a curve representing a constant value of $\Psi_{(x,z)}$. Also the tangent to the $\Psi_{(x,z)}$ curve represents the direction of the resultant seepage velocity (Fig. 5.15).

Thus the Laplace equation may be seen to describe two sets of orthogonal curves: those of constant Φ called *equipotentials* and those of constant Ψ called *flow lines*. A graphical construction of equipotentials and flow lines is called a *flow net*.

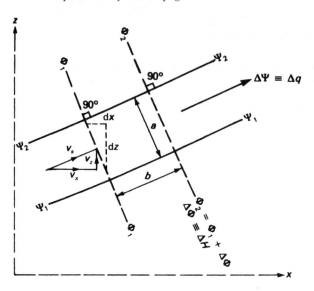

Fig. 5.15 Graphical interpretation of flow

Properties of a flow net

Once a *flow net* has been constructed its graphical properties may be used in the solution of seepage problems, e.g. determination of seepage quantities and seepage pressures. To this end flow nets are constructed so that the intervals between adjacent equipotentials represent a *constant difference in head* (ΔH) and the intervals between adjacent flow lines represent a *constant flow quantity* (Δq) (Fig. 5.15).

Then the total head lost, $\quad H = \Delta H \times$ no. of equipotential drops
$$= \Delta H \times N_d$$
and the total seepage flow, $\quad q = \Delta q \times$ no. of flow intervals (channels)
$$= \Delta q \times N_f$$

Starting from Darcy's law (eqn [5.3]), for unit-thickness in the y direction:

$$\Delta \Psi = \Delta q = \Delta A k i$$

$$= ak\frac{\Delta H}{b} = \frac{a}{b}\Delta \Phi$$

However, the scales may be chosen so that when the flow net is drawn, $\Delta \Phi = \Delta \Psi$, i.e. $a/b = 1$. This is achieved in practice by ensuring that the fields bounded by equipotentials and flow lines are 'square' (or as near square as possible).

Then the total flow, $\quad q = k \cdot \Delta H N_f$

but $\quad\quad\quad\quad\quad\quad \Delta H = \dfrac{H}{N_d}$

Giving $\quad\quad\quad\quad\quad\quad q = kH\dfrac{N_f}{N_d}$ $\quad\quad\quad\quad\quad\quad\quad$ [5.27]

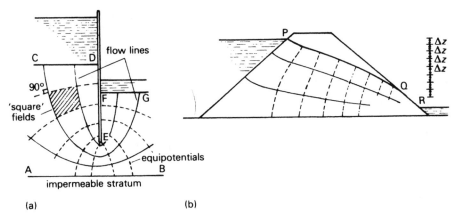

Fig. 5.16 Flow net boundaries

5.12   Flow nets – construction rules and boundary conditions

In the previous section, flow nets were introduced as diagrammatic constructions representing two-dimensional flow conditions. Once a flow net has been
drawn, quantities such as the amount of seepage and pressure distribution may
be evaluated. In order to produce a correctly constructed flow net a number of
'rules' must be observed (see Fig. 5.16).

Square fields: the areas bounded by equipotentials and flow lines must be as near
square as possible.

Right-angle intersections: the intersection of an equipotential with a flow line
must occur at 90°.

Impermeable boundary: since no flow takes place *across* an impermeable boundary,
here Ψ = constant and so such a boundary is a flow line (AB and DEF).

Permeable boundary: a submerged permeable boundary along which the head is
constant will be an equipotential (CD and FG).

Phreatic surface: along a phreatic surface (PQ) (also known as the top *streamline*)
the pore pressure $u = 0$, so that $\Delta\Phi = -k\Delta H = -k\Delta z$, also since Ψ is constant it is
a flow line.

Seepage surface: a seepage surface occurs where the phreatic surface intersects
tangentially with the ground surface (QR); it has the same boundary properties
as the phreatic surface.

To construct a flow net, first of all a scaled cross-section is drawn defining all the
boundaries due to the site, structure, etc. Then, following the rules given above,
a few trial flow lines and equipotentials are drawn in pencil. As further 'square'
fields are added the flow net takes shape. A series of successive trials is necessary,
involving the judicious use of pencil and eraser, with the ultimate aim of drawing
a flow net in which *all parts* comply with the rules. A good deal of practice will
be required, so that students should attempt as many flow nets as possible to
gain proficiency in the sketching process.

Theoretically, any number of flow lines may be drawn and the greater the number, the more accurate should be the calculations that follow. However, from a practical point of view the task is simplified by drawing only a few flow lines; it is not often that more than five or six will be necessary.

Worked example 5.7   *Figure 5.17(a) shows the cross-section of a line of sheet-piling driven to a depth of 7 m into a stratum of homogeneous sandy soil which has a thickness of 12 m and is underlain by an impermeable stratum. From an original depth of 5.5 m the water level on one side of the piles is reduced by pumping to a depth of 0.5 m. Draw a flow net for the seepage conditions and from it determine: (a) the quantity of seepage under the piles per metre run, and (b) the pore pressure in the soil at points P and Q.*

The coefficient of permeability $k = 7.2 \times 10^{-3}$ mm/s

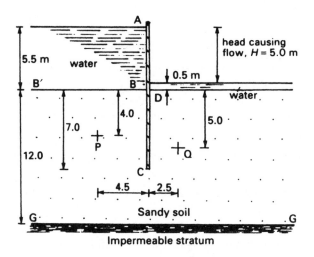

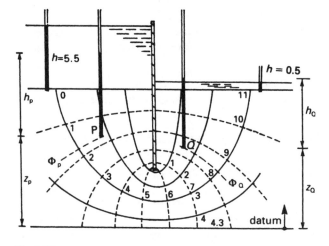

Fig. 5.17

The area in which seepage is taking place and therefore in which the flow net is to be drawn is bounded as follows.

Impermeable boundaries: along the sheet piling BCD and along the impermeable stratum GG; BCD and GG are, therefore, flow lines.

Permeable boundaries: along B′B the pressure head is a constant 5.5 m and along DD′ the pressure head is 0.5; B′B and DD′ are, therefore, equipotentials of value $h = 5.5$ and $h = 0.5$ respectively.

By successive trials a flow net such as that shown in Fig. 5.17(b) may be drawn, although the number of flow lines could be different. It should also be noted that the number of flow channels need not necessarily be a whole number. For the flow net shown:

No. of equipotential intervals $N_d = 11$
No. of flow channels $N_f = 4.3$

(a) The quantity of seepage beneath the piling is given by eqn [5.27]:

$$q = kH\frac{N_f}{N_d}$$

$$= 7.2 \times 10^{-6} \times 5.0 \times \frac{4.3}{11} = \underline{14.07 \times 10^{-6} \text{ m}^3/\text{s per m}}$$

or

$$q = 14.07 \times 10^{-6} \times 3600 = \underline{0.0507 \text{ m}^3/\text{h per m}}$$

(b) At any point i, the total head above the datum = pressure head + position head

i.e. $H_i = h_i + z_i$

At the entry boundary B′B, $H_{B'B} = 5.5 + 12.0 = 17.5$ m

Within the flow net the total head is reduced by ΔH at each equipotential line:

$$\Delta H = \frac{H}{N_d} = \frac{5.0}{11} = 0.455 \text{ m}$$

By numbering the equipotentials it can be seen that the equipotential passing through point P would have a value of $\Phi_P = 1.75$, so that:

Total head at P,	$H_P = 17.5 - 1.75 \times 0.455$
Also,	$H_P = h_P + z_P$
Therefore, the pressure head at P,	$h_P = 17.5 - 8.0 - 1.75 \times 0.455$
	$= 8.71$ m
and pore pressure at P,	$u_P = 9.81 \times 8.71 = \underline{85.4 \text{ kPa}}$
Similarly, at Q: pressure head,	$h_Q = 17.5 - 7.0 - 8.4 \times 0.455$
	$= 6.68$ m
pore pressure,	$u_Q = 9.81 \times 6.68 = \underline{65.5 \text{ kPa}}$

Worked example 5.8   *Figure 5.18 shows the cross-section of a long coffer dam formed by driving two parallel rows of sheet piles into a stratum of sand. Within the coffer dam the sand is excavated down to the level shown and the water maintained at this level by pumping. Draw a suitable flow net and use it to determine the pumping rate required per metre run ($k = 4.5 \times 10^{-5}$ m/s).*

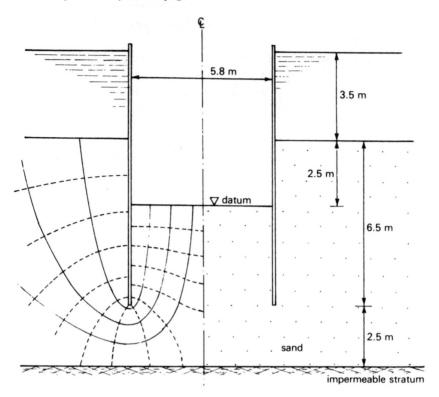

Fig. 5.18

Since the flow net will be symmetrical about the centre-line ($\mathcal{L}$) only one-half needs to be drawn. In order to satisfy the boundary conditions in this case an odd number of flow channels is chosen with the centre of the middle one emerging at the centre-line of the coffer dam. A flow net comprising seven flow channels and thirteen equipotential intervals is shown in the figure.

Then $N_f = 7.0$ and $N_d = 13$

The seepage quantity and hence the pumping rate will be:

$$q = kH\frac{N_f}{N_d} = 4.5 \times 10^{-5} \times 6.0 \times \frac{7}{13}$$

$$= 14.54 \times 10^{-5} \text{ m}^3/\text{h per m}$$

or $q = 14.54 \times 10^{-5} \times 3600 = \underline{0.52 \text{ m}^3/\text{hr per m}}$

Worked example 5.9   *Figure 5.19 shows the cross-section of a dam founded on a permeable stratum, which in turn is underlain by an impermeable stratum. A row of sheet piles has been inserted near the upstream face of the dam in order to reduce the quantity of seepage. Draw a flow net to represent the seepage conditions and from it: (a) determine the seepage quantity per metre run (k = 5.2 × 10⁻⁵ m/s), and (b) plot the distribution of uplift pressure acting on the base of the dam.*

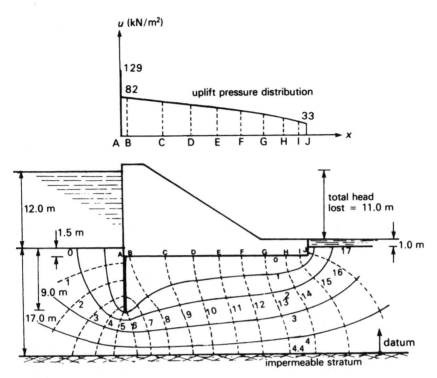

Fig. 5.19

The flow net shown in the figure comprises 4.4 flow channels and 17 equipotential intervals.

$N_f = 4.4$ and $N_d = 17$ Also, $\Delta h = 11.0/17 = 0.647$

(a)   The seepage quantity will, therefore, be:

$$q = kH\frac{N_f}{N_d}$$

$$= 5.2 \times 10^{-5} \times 11 \times \frac{4.4}{17} = 1.48 \times 10^{-4} \text{ m}^3/\text{s per m}$$

or $q = 0.533 \text{ m}^3/\text{hr per m}$

(b)   At any point i on the base where the potential value is Φ_i, the total head will be given by:

$$H_i = 17.0 + 12.0 - \Phi_i\Delta h = h_i + z_i$$

Thus, uplift pressure, $u_i = 9.81 \, h_i$

$$= 9.81 \, (29.0 - z_i - \Phi_i\Delta h)$$

Then, obtaining values of Φ_i and z_i from the flow net, the uplift pressures will be those tabulated below.

Position ($z_i = 15.5$ m for each)									
A	B	C	D	E	F	G	H	I	J
0.5	8.0	9.0	10.0	11.0	12.0	13.0	14.0	15.0	15.7
129	82	75	69	63	56	50	44	37	33

First column labels: Φ_i and u_i (kPa).

5.13 Seepage through earth dams and embankments

Whereas the flow under impermeable structures, such as sheet piles and concrete or masonry dams, is confined, the seepage taking place through a permeable structure, such as an earth dam, is unconfined. In such problems the upper boundary of the seepage zone is the phreatic surface, this being the top flow line and along which the pressure is that of the atmosphere. The first step in drawing a flow net for seepage through a dam or embankment is therefore to locate and draw the phreatic surface or top flow line.

As already established in Section 5.11, since the pressure head along the phreatic surface is zero (i.e. atmospheric), equal drops in total head along it must correspond to equal vertical position intervals (Fig. 5.16(b)). It can be shown mathematically that the basic shape of the phreatic surface is therefore that of a parabola. Although this solution is largely acceptable, some modification is required to meet inconsistencies which may occur at the intersection of the phreatic surface with the entry and exit surfaces of the dam. These modifications will be summarised first and then the basic method of constructing the parabolic phreatic surface will be described.

Entry and exit conditions
The upstream surface of the dam is the entry surface to the seepage zone and since it is the equipotential $\Phi_{max.}$ representing the maximum pressure head, the flow lines must intersect it at right angles. This is true providing the slope of $\Phi_{max.} \leq 90°$ (Fig. 5.20(a)). In certain cases, where an upstream coarse filter is provided, the slope of $\Phi_{max.}$ may be $> 90°$. The phreatic surface at entry is then horizontal, since water at zero pore pressure cannot flow upwards (Fig. 5.20(b)).

At the downstream or exit surface, the theoretical parabola may have to be modified depending on the conditions at the toe. When the exit surface is horizontal (Fig. 5.21(a)) no correction to the basic parabola is required. Where an angular toe filter of coarse material forms the exit surface and $\beta < 180°$ (Fig. 5.21(b)) a correction is made by relocating the exit point of the phreatic surface using a method proposed by Casagrande.

If the basic parabola cuts the exit surface at point K and the exit surface intersects the impermeable base at F, the corrected position of the phreatic surface is located at point J using the ratio $\Delta a/a$:

where $a = FK$

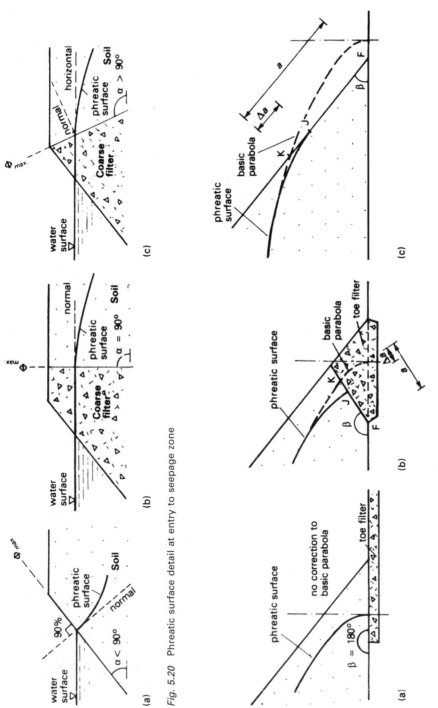

Fig. 5.20 Phreatic surface detail at entry to seepage zone

Fig. 5.21 Phreatic surface detail at exit from seepage zone

and $\Delta a/a$ is obtained from Table 5.4

Then $KJ = \Delta a$

In the case where the base of the toe is impermeable (i.e. no toe filter is provided), the phreatic surface exits tangentially along the downstream slope (Fig. 5.21(c)). Its point of exit can also be located by the method described above.

Table 5.4   Correction factors for earth dam flow net

β	30°	60°	90°	120°	150°	180°
$\Delta a/a$	(0.36)	0.32	0.26	0.18	0.10	0

After Casagrande (1940)

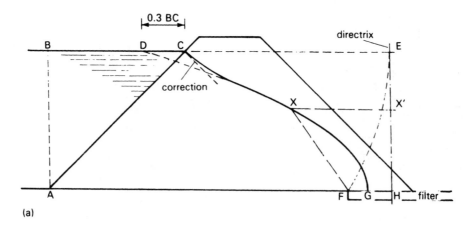

(a)

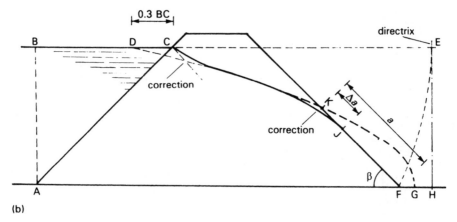

(b)

Fig. 5.22 Phreatic surface construction for an earth dam
(a) Horizontal surface exit (b) Exit tangential to downstream face

Construction of basic parabola

All of the flow lines and the equipotentials are parabolic curves with a common focus. The first step in the drawing of the flow net is to construct the basic parabola which is the phreatic surface. A graphical method of doing this, suggested by Albert Casagrande in 1937, consists of drawing a geometric parabola and then correcting the entry and exit ends as described above.

Figure 5.22(a) shows a typical earth dam with a horizontal surface. The parabola is assumed to start at D (where CD = 0.3 BC) and have a focus at F. The directrix is located by striking an arc of radius DF about point D (i.e. DE = DF). The vertical tangent ($\overline{EH}$) to this arc is the directrix. Since all points on a parabola are equidistant from the directrix and the focus then:

$$\overline{FG} = \overline{GH}$$

and for all points X, $\overline{XX'} = \overline{FX}$

Thus the parabola may be constructed between D and G. The entry detail at point C is corrected as explained above.

A similar procedure is adopted when the phreatic surface exits tangentially to the downstream face of the dam (Fig. 5.22(b)); except in this case, the focus is the toe of the downstream slope. Also, a correction to the exit point from point K to point J is required, the procedure for which has already been described.

Worked example 5.10   *The section of a homogeneous earth dam is shown in Fig. 5.23(a), the soil permeability being 6.0×10^{-6} m/s. Construct the flow net and from it: (a) calculate the estimated seepage quantity, and (b) plot the distribution of pore pressure along the circular arc AS.*

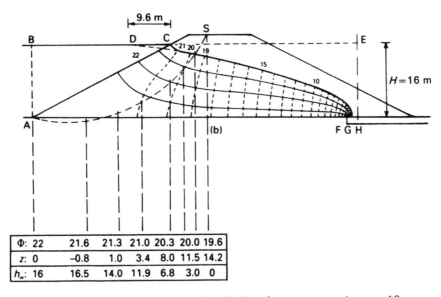

Φ: 22	21.6	21.3	21.0	20.3	20.0	19.6
z: 0	−0.8	1.0	3.4	8.0	11.5	14.2
h_w: 16	16.5	14.0	11.9	6.8	3.0	0

Fig. 5.23   (a) Dimensions (b) Flow net (c) Distribution of pore pressure along arc AS

Firstly, the phreatic surface must be drawn (Fig. 5.23(b)). Draw the vertical line AB and then locate point D.

$$\overline{CD} = 0.3\overline{CB} = 0.3 \times 32 = 9.6 \text{ m}$$

Strike an arc of radius DF to intersect the horizontal projection of BDC in E. Draw the vertical line EH: the directrix of the parabola. Now construct the basic parabola from the focus F. Modify the phreatic surface at the entry surface to pass through C. The other flow lines and equipotentials can now be drawn – remember they are all (basically) parabolas.

From the flow net shown in Fig. 5.23(b), $N_d = 22$ and $N_f = 3.5$.

(a)   Therefore the seepage quantity $q = kH \dfrac{N_f}{N_d}$

$$= 6.0 \times 10^{-6} \times 16 \times \frac{3.5}{22} = 15.3 \times 10^{-6} \text{ m}^3/\text{s}$$

$$= \underline{5.5 \times 10^{-2} \text{ m}^3/\text{hr}}$$

(b)   Estimate the potential value at each point where the arc AC crosses a flow line.

Then at these points, pore pressure head $h_w = H \dfrac{\Phi_h}{N_d} - z$

$$= 0.727\Phi_h - z$$

Figure 5.23(c) shows the plot of the distribution of pore pressure head along AS.

5.14   Seepage in anisotropic soils

Horizontal and vertical flow in stratified soils
Where a soil consists of a number of layers of different soil types each with different permeability, the horizontal and vertical permeabilities will be different. The presence of thin sandy layers (high permeability) in thick layers of fine soil (low permeability) can very often produce relatively high horizontal seepage rates. This principle is exploited in the construction of embankments where sand drains are incorporated to speed up the drainage of excess porewater and hence improve the consolidation rate.

Consider the soil mass shown in Fig. 5.24 which consists of three layers, each of which has a different coefficient of permeability.

Horizontal flow (i.e. tangential to strata)   The head lost between the entry and exit faces will be the same for each layer:

$$h_1 = h_2 = h_3 = h$$

Hence the hydraulic gradients are the same:

$$i_1 = i_2 = i_3 = i$$

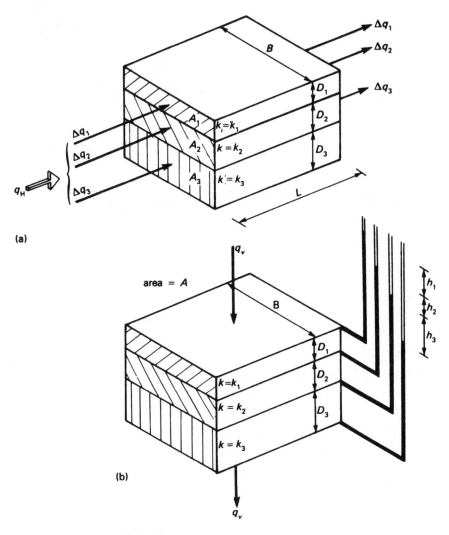

Fig. 5.24 Flow in stratified soil
(a) Horizontal flow (b) Vertical flow

Starting with Darcy's law ($q = Aki$) the flows in the layers will be:

$\Delta q_1 = A_1 k_1 i_1 \quad \Delta q_2 = A_2 k_2 i_2 \quad Aq_3 = A_3 k_3 i_3$

Also $\quad A_1 = BD_1 \quad A_2 = BD_2 \quad A_3 = BD_3 \quad$ and $\quad \bar{A} = B(D_1 + D_2 + D_3)$

The total flow $\quad q_H = \Delta q_1 + \Delta q_2 + \Delta q_3 = \bar{A} k_H i$

where $\quad k_H$ = average horizontal coefficient of permeability

Substituting $\quad BD_1 k_1 i_1 + BD_2 k_2 i_2 + BD_3 k_3 i_3 = B(D_1 + D_2 + D_3)k_H i$

Giving
$$k_H = \frac{D_1 k_1 + D_2 k_2 + D_3 k_3}{D_1 + D_2 + D_3}$$
[5.28]

Vertical flow (i.e. normal to strata)   The rate of flow will be the same through each layer:

$$\Delta q_1 = \Delta q_2 = \Delta q_3 = q_v$$

The heads lost in each layer will be h_1, h_2 and h_3 giving hydraulic gradients of

$$i_1 = h_1/D_1 \quad i_2 = h_2/D_2 \quad i_3 = h_3/D_3$$

Starting with Darcy's law $(q = Aki)$

The total flow,

$$q_v = k_v \bar{A} i = k_v A \frac{\bar{h}}{L}$$

But total head lost, $\bar{h} = h_1 + h_2 + h_3$ and $L = D_1 + D_2 + D_3$

Also

$$h_1 = \frac{qD_1}{Ak_1}, \text{ etc.}$$

Therefore

$$q_v = \frac{k_v A \left(\dfrac{qD_1}{Ak_1} + \dfrac{qD_2}{Ak_2} + \dfrac{qD_3}{Ak_3} \right)}{D_1 + D_2 + D_3}$$

Giving

$$k_v = \frac{D_1 + D_2 + D_3}{D_1/k_1 + D_2/k_2 + D_3/k_3} \qquad [5.29]$$

In both eqns [5.28] and [5.29], the number of components in both the numerator and denominator is equal to the number of layers.

Worked example 5.11   *A stratified soil consists approximately of alternating layers of sand and silt. The sand layers are generally 150 mm in thickness and have a permeability of $k = 6.5 \times 10^{-1}$ mm/s, the silt layers are 1.80 m thick and have a $k = 2.5 \times 10^{-4}$ mm/s. Assuming that within each layer flow conditions are isotropic, determine the ratio of the horizontal permeability to that of the vertical.*

One strata 'cycle' will consist of a layer of sand (1) and a layer of silt (2).

Then $k_1 = 6.5 \times 10^{-1}$ mm/s $k_2 = 2.5 \times 10^{-4}$ mm/s
$\quad\quad\quad D_1 = 150$ mm $\quad\quad\quad D_2 = 1800$ mm

From eqn [5.28]: $k_H = \dfrac{D_1 k_1 + D_2 k_2}{D_1 + D_2}$

$$= \frac{150 \times 6500 + 1800 \times 2.5}{150 + 1800} \times 10^{-4}$$

and from eqn [5.29]: $k_v = \dfrac{D_1 + D_2}{D_1/k_1 + D_2/k_2}$

$$= \frac{150 + 1800}{150/6500 + 1800/2.5} \times 10^{-4}$$

Hence ratio $\dfrac{k_H}{k_v} = \dfrac{(150 \times 6500 + 1800 \times 2.5)(150/6500 + 1800/2.5)}{(1.50 + 1800)^2} = \underline{186}$

Flow net in anisotropic soil

In an anisotropic soil $k_x \neq k_z$ and, in the average direction of flow, k_f has a value between k_x and k_z. The Laplace equation for two-dimensional flow (see eqn [5.25]) therefore becomes:

$$k_x \frac{\partial^2 h}{\partial x^2} + k_z \frac{\partial^2 h}{\partial z^2} = 0 \qquad [5.30]$$

or $\quad \dfrac{\partial^2 h}{(k_z/k_x)\partial x^2} = 0$

But putting $\quad x^2 \dfrac{k_z}{k_x} = x_T^2 \quad$ i.e. $\quad x_T = x \sqrt{\left(\dfrac{k_z}{k_x}\right)} \qquad [5.31]$

the continuity equation becomes: $\quad \dfrac{\partial^2 h}{\partial x_T^2} + \dfrac{\partial^2 h}{\partial z^2} = 0 \qquad [5.32]$

Equation [5.31] gives a scale factor by which the actual anisotropic flow region may be transformed into a notional isotropic region. The coefficient of permeability in the direction of flow is equal to the equivalent isotropic value:

$$k_f = k_x \sqrt{\left(\frac{k_z}{k_x}\right)} = \sqrt{(k_x k_z)} \qquad [5.33]$$

Hence the seepage quantity, derived from a transformed flow net, is:

$$q = k_f \frac{N_f}{N_d} = \frac{N_f}{N_d} \sqrt{(k_x k_z)} \qquad [5.34]$$

The first stage in constructing the flow net is to draw the cross-section of the flow region using a normal vertical (z-axis) scale and a transformed horizontal (x-axis) scale (Fig. 5.25(a)). On this section is drawn a flow net assuming isotropic conditions, i.e. with 'square' fields, and 90° intersections (Fig. 5.25(b)). Finally the true flow net is obtained by redrawing the section using the same scale in both x- and z-directions (Fig. 5.25(c)). The horizontal distances including the network lines are altered by dividing by the factor $\sqrt{(k_z/k_x)}$. The true flow net is the correct representation of the anisotropic conditions, although the intersections are no longer at 90°, nor are the fields square.

Flow across a soil interface

Where the anisotropic conditions include a boundary or interface between soils of different permeability which is inclined to the direction of the flow, the flow lines will be refracted. In other words, at such an interface the direction of flow changes abruptly. If the flow takes place into a less permeable soil ($k_1 > k_2$), the flow lines are refracted towards the normal at the interface, and away from the normal when $k_1 < k_2$.

In Fig. 5.26, the interface is shown between two soils of permeabilities k_1 and k_2 respectively. Consider two flow lines Ψ_1 and Ψ_2 intersecting the boundary to make angles to the normal of α_1 and α_2 respectively. The corresponding equipotentials at the intersection are Φ_1 and Φ_2.

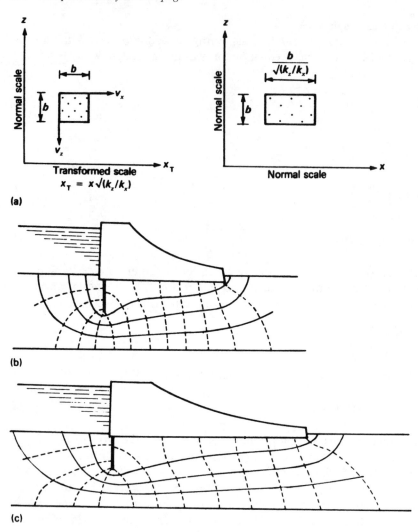

Fig. 5.25 Flow net construction in anisotropic soil
(a) Transformation of flow element (b) Flow net drawn to transformed cells (c) Flow net redrawn to normal scale

Now the head loss, $\Delta h = \Phi_1 - \Phi_2$
and the flow across the interface between A and B $= \Delta q$
For continuity   flow in = flow out
Therefore $A_1 k_1 i_1 = A_2 k_2 i_2$

or $$BCk_1 \frac{\Delta h}{\frac{1}{2}AC} = ADk_2 \frac{\Delta h}{\frac{1}{2}BD}$$

But $BC = AB \cos \alpha_1$ and $AD = AB \cos \alpha_2$

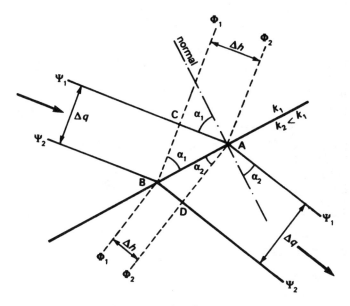

Fig. 5.26 Flow net across a soil interface

Substituting $\quad \dfrac{AB}{AC}\cos\alpha_1 k_1 = \dfrac{AB}{BD}\cos\alpha_2 k_2$

giving $\qquad\qquad \dfrac{k_1}{\tan\alpha_1} = \dfrac{k_2}{\tan\alpha_2}$

or $\qquad\qquad \dfrac{k_1}{k_2} = \dfrac{\tan\alpha_1}{\tan\alpha_2}$ [5.35]

5.15 Instability due to seepage ('piping')

The term *piping* is used to describe the unstable condition which can occur when the vertical component of seepage pressure acting in an upward direction exceeds the downward weight of the soil. When the upward seepage force becomes equal to the submerged weight of the soil, no frictional resistance can be developed between the particles. The soil/water mixture therefore has no shear strength and acts as a liquid.

If the upward seepage forces exceed the submerged weight, the particles may be carried upwards to be deposited at the ground surface. Thus a 'pipe' is formed in the soil near the surface. Piping failure can lead to the complete failure of a foundation or to the collapse of a supporting structure, such as the toe of a dam, or part of a coffer dam. It is necessary, therefore, to check this potential instability condition during the design of water-retaining structures.

Where a flow net has been drawn to represent the seepage conditions, a simple rule-of-thumb method may be used to determine a factor of safety against the occurrence of piping. This is done by considering a prism of soil adjacent to

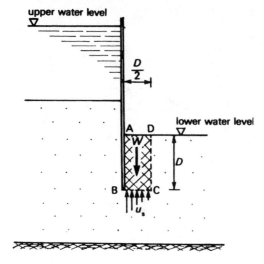

Fig. 5.27 Factor of safety against piping

the downstream face of the structure. Consider the case of a sheet pile wall as shown in Fig. 5.27. The effective weight of the prism of soil ABCD will be

$$W = (\gamma_{\text{sat.}} - \gamma_{\text{w}})D \times \frac{D}{2}$$

The distribution of seepage pressure on the base BC of the prism is obtained from the flow net: suppose the average value to be u_s.

Then the upward seepage force on BC $= u_s \times \dfrac{D}{2}$

Since piping will occur when the upward seepage force is equal to W, the *factor of safety against piping* is given by:

$$F_{\text{(piping)}} = \frac{\text{downward weight}}{\text{upward seepage force}} = \frac{W}{u_s D/2} = \frac{(\gamma_{\text{sat.}} - \gamma_{\text{w}})D}{u_s} \qquad [5.36]$$

The method of calculating u_s is illustrated in worked example 5.7. The factor of safety against piping may be increased in a number of ways. For example, in the case of a coffer dam the depth of penetration of the piles could be increased, or a layer of coarse filter material could be laid on the downstream side before pumping down to the final level.

   In the case of dams, both an increase in the factor of safety against piping and a reduction in the quantity of seepage can be obtained by increasing the length of the flow path. This may be done by driving a row of sheet piles, preferably at or near the upstream face, or by laying an apron of impermeable paving in front of the upstream face. A layer of coarse filter material laid on the downstream side is another possibility (see Fig. 5.28).

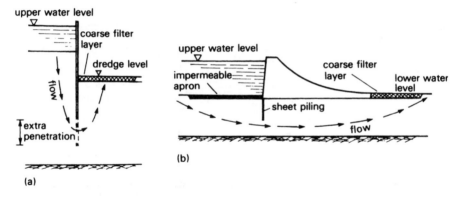

Fig. 5.28 Methods of improving seepage conditions
(a) Coffer dam (b) Concrete or masonry dam

5.16 *Dewatering excavations*

It is frequently necessary to reduce the groundwater level or piezometric level in or around excavations in order to achieve one or more of the following objectives:

(a) to produce 'dry' working conditions
(b) to prevent heave or uplift at the bottom of excavations
(c) to reduce lateral pressures on temporary side supports
(d) to improve the stability of temporary cut side slopes
(e) to reduce the water content of soils that are to be excavated, e.g. from borrow pits.

Dewatering operations are mostly temporary arrangements designed to facilitate construction or repair work. Permanent systems involving pumping tend to be expensive to maintain and often difficult to manage over long periods. Current dewatering methods fall into three main groups:

(a) *Internal sump pumping.* Suitable for small and short-term operations and to supplement other methods. Sump layouts and pump sizes can be altered quickly offering great flexibility.
(b) *External well pumping.* Deep wells, for use with lift pumps or submersible pumps, are drilled and often cased; *wellpoints* are shorter and are sunk into position using high-pressure water jets. The main objective with well pumping is to lower the ground water level (or piezometric surface over confined aquifers).
(c) *Cut-offs.* For more secure and for permanent control vertical impermeable barriers are installed, using grouting, diaphragm walls, chemical/resin injection, etc. These are useful in the control of groundwater from contaminated areas, such as tips, old industrial sites, etc.

Figure 5.29(a) shows a simplified diagrammatic arrangement illustrating different methods. For more detailed information concerning the installation and operation of dewatering systems a specialised text on ground engineering should be consulted, e.g. Powers (1976), Tomlinson (1995), Cedergren (1987), Corbett (1987).

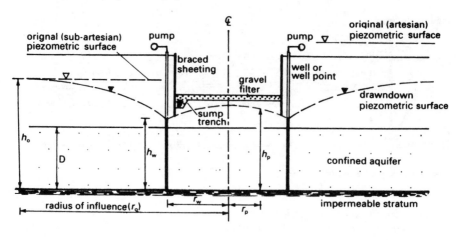

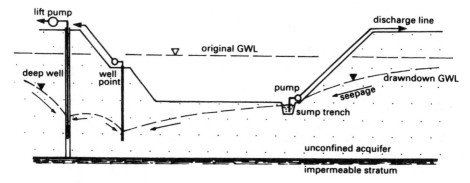

Fig. 5.29   Dewatering excavations
(a) Reducing artesian pressure in a confined aquifer (b) Alternative methods in an unconfined aquifer

In the context of structural design and the operational aspects of construction, it is usually necessary to evaluate the pumping capacity required and to provide estimates of reduced piezometric levels in the vicinity of an excavation. Assuming that appropriate tests have been carried out to ascertain average values of the coefficient of permeability (k), a reasonable estimate of the rate of inflow may be obtained using either eqn [5.13] or eqn [5.15], with the following substitutions:

$r_1 = \bar{r}_w$ = average radius of the pumping wells from the centre of the excavation
$h_1 = \bar{h}_w$ = average piezometric level in the pumping wells
$r_2 = \bar{r}_o$ = average radius of influence of the pumped zone from the centre of the excavation
$h_2 = h_o$ = original piezometric level

For excavations over confined aquifers (Fig. 5.29(b)) it is particularly important to evaluate the uplift pressures in order to establish factors of safety against base heave. At a given point P, at radius r_p, it may be assumed that the total drawdown is a simple summation of the drawdowns induced at that point by individual wells.

Thus, from eqn [5.13], for n wells the total drawdown at P is given by:

$$d_{\mathrm{P}} = h_{\mathrm{o}} - h_{\mathrm{P}} = \frac{1}{2\pi Dk} \sum_{i=n} [q_i \ \ln[(r_{\mathrm{o}}/r_i)]] \qquad [5.37]$$

where q_i = rate of inflow to well i
  r_i = radius of well i from point P

It should be noted that the relief of uplift pressure depends not only on the achieved drawdown, but also on the original piezometric level, which itself is affected by climatic, tidal or other changes in environmental conditions. Piezometric levels should therefore be monitored at several strategic points within or alongside the excavation. Pumping rates may then be adjusted to suit changes or anomalies in site conditions.

***

Worked example 5.12   *A deep excavation is required in an alluvial silty clay overlying a confined aquifer of sand. Since the artesian pressure in the sand layer will produce unacceptable uplift pressures during excavation, the dewatering system shown in Fig. 5.30 has been proposed. From site permeability tests the average coefficient of permeability of the sand has been found to be 4.3×10^{-3} m/s and the expected radius of influence of the pumped zone has been estimated at 750 m.*

(*a*) *Calculate the minimum capacity of pump required for each well, assuming the same model of pump for each.*

(*b*) *Obtain an estimate of the anticipated piezometric drawdown at the centre and at a corner of the excavation when the pumping rate from all 20 wells is equal to the value calculated in (a).*

(*c*) *Calculate the drawdown similarly at other points and comment on the location of the wells.*

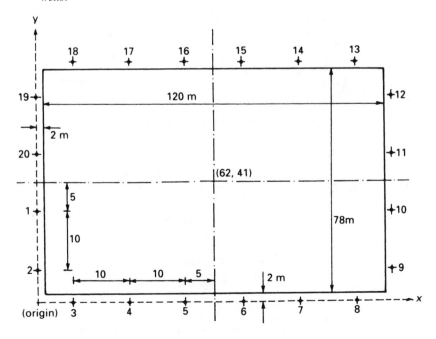

Fig. 5.30

(a)   The average radius $\bar{r}_w$ may be taken as the radius of a circle having the same area as the excavation:

$$\bar{r}_w = \sqrt{(124 \times 82/\pi)} = 56.9 \text{ m}$$

However, in this case, since drawdowns are to be calculated, a true average radius can be obtained. The calculations are tabulated below, the layout being a reproduction of a computer spreadsheet analysis.

WE5/12 Well point system analysis

Radius of influence (m)	750	Flow rate (litres/min)	64 674
Min. drawdown reqd. (m)	7.5	Pump capacity reqd. (l/min)	3234
Coeff. of permeability (m/s)	0.0043	Drawdown at centre (m)	7.94
Thickness of aquifer (m)	13.6		

Well no.	(from origin)		(from centre)					Drawdown at point P = 7.51						
	X	Y	x	y	r	$\ln(r_o/r_i)$	q	$\ln(\ ) \times q$	x_p	y_p	r_p	$\ln(r_o/r_p)$	$\ln(\ ) \times q$
Origin	0	0	62	41					2	2			
1	72	82	10	41	42.20	2.88	3400	9784	70	80	106.30	1.95	6 643
2	92	82	50	41	50.80	2.69	3400	9153	90	80	120.42	1.83	6 219
3	112	82	60	41	64.66	2.45	3400	8333	110	80	136.01	1.71	5 805
4	124	74	62	33	70.24	2.37	3400	8052	122	72	141.66	1.67	5 667
5	124	52	62	11	62.97	2.48	3400	8423	122	50	131.85	1.74	5 911
6	124	30	62	−11	62.97	2.48	3400	8423	122	28	125.17	1.79	6 807
7	124	8	62	−33	70.24	2.37	3400	8052	122	6	122.15	1.81	6 170
8	112	0	50	−41	64.66	2.45	3400	8333	110	−2	110.02	1.92	6 526
9	92	0	30	−41	50.80	2.69	3400	9153	90	−2	90.02	2.12	7 208
10	72	0	10	−41	42.20	2.88	3400	9784	70	−2	70.03	2.37	8 062
11	52	0	−10	−41	42.20	2.88	3400	9784	50	−2	50.04	2.71	9 205
12	32	0	−30	−41	50.80	2.69	3400	9153	30	−2	30.07	3.22	10 937
13	12	0	−50	−41	64.66	2.45	3400	8333	10	−2	10.20	4.30	14 613
14	0	8	−62	−33	70.24	2.37	3400	8052	−2	6	6.32	4.78	16 237
15	0	30	−62	−11	62.97	2.48	3400	8423	−2	28	28.07	3.29	11 170
16	0	52	−62	11	62.97	2.48	3400	8423	−2	50	50.04	2.71	9 205
17	0	74	−62	33	70.24	2.37	3400	8052	−2	72	72.03	2.34	7 966
18	12	82	−50	41	64.66	2.45	3400	8333	10	80	80.62	2.23	7 583
19	32	82	−30	41	50.80	2.69	3400	9153	30	80	85.44	2.17	7 386
20	52	82	−10	41	42.20	2.88	3400	9784	50	80	94.34	2.07	7 049

Average radius, $\bar{r}_w = 58.2$ m

Total pump capacity, $q_{reqd} = 2\pi Dk \dfrac{h_2 - h_1}{\ln(r_2/r_1)}$

However, $h_2 - h_1 = h_o - \bar{h}_w = d_w$ (drawdown required) $= 7.5$ m

$r_2 = \bar{r}_o = 750$ m   and $r_1 = \bar{r}_w = 58.2$ m

Therefore, $q_{reqd} = \dfrac{2\pi \times 13.6 \times 0.0043 \times 60\,000 \times 7.5}{\ln(750/58.2)}$

$= 64\,674$ litres/min

For 20 pumps of equal capacity, q per pump = 3234 litres/min. Therefore use pumps of 3400 litres/min capacity.

(b)   The centre of the excavation lies at coordinates (62,41). Assuming that each pump is working at 3400 litres/min, the summation $\Sigma\,[q_i\,\ln(r_o/r_i)]$ may be obtained so that

Drawdown at centre, $d_c = \dfrac{1}{2\pi Dk}\displaystyle\sum_{i=n}[q_i\,\ln(r_o/r_i)]$

$$= \frac{3400\Sigma\,[\ln(750/r_i)]}{2\pi\times13.6\times0.0045\times60\,000} = \underline{7.94\ m}$$

At the corner of the excavation nearest to origin the coordinates are (2,2); the well radius from this corner is shown in column r_P.

Drawdown at corner P, $d_p = \dfrac{3400\Sigma\,[\ln(750/r_P)]}{2\pi\times13.6\times0.0045\times60\,000} = \underline{7.51\ m}$

The same value will be obtained for the other three corners. This is the minimum drawdown achieved with this pumping regime and it satisfies the required critical value.

(c)   The drawdowns (not well levels) at the pumping wells may be calculated in a similar way by inserting the values of the well coordinates from the origin at the head of the x_P and y_P columns.

e.g.   Well 1 is at $x = 0$ m and $y = 15$ m, giving a drawdown here of 8.17 m.

Figure 5.31 shows a quarter-plan of the excavation giving the calculated drawdowns at each well and also at points on a 10 m × 10 m grid. Drawdown contours have also been added to emphasise the pattern of values. The lowest (and therefore worst) drawdown values occur near the corners, while the highest values occur at the middle wells on the longest sides. The siting of the wells could be improved by increasing their spacing along

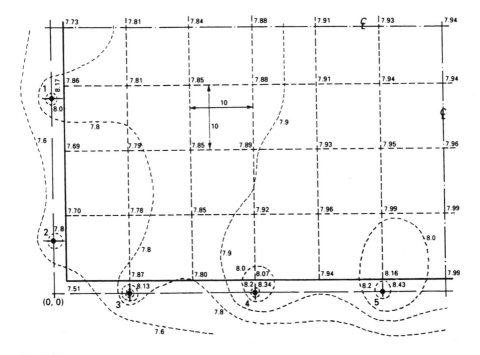

Fig. 5.31

the longest sides, thus bringing wells 3, 8, 13 and 18 closer to the corners and so increasing the corner drawdowns. To provide effective monitoring, observation standpipes should be sunk at each corner.

Thus, if the original piezometric level was, say, 1.3 m below the ground surface and the excavation at a corner was taken down to 7.2 m, the reduced piezometric level at this point would be 8.8 m below ground surface and 1.6 m below excavation level.

Exercises

5.1   (a)   Explain the terms *quick condition* and *critical hydraulic gradient.*
   (b)   Derive an expression for the critical hydraulic gradient (i_c) of a granular soil in terms of the grain specific gravity (G_s) and void ratio (e). What will be the value of i_c, when $G_s = 2.70$ and $e = 0.60$?

5.2   The following data were recorded during a constant-head permeability test:

Internal diameter of permeameter   = 75 mm
Head lost over a sample length of 180 mm = 247 mm
Quantity of water collected in 60 s   = 626 ml

Calculate the coefficient of permeability for the soil.

5.3   A constant head permeameter has an internal diameter of 62.5 mm and is fitted in the side with three manometer tappings at points A, B and C. During tests on a specimen of sand the following data were recorded:

Test no.	Quantity of water collected in 5 min.	Manometer level (mm) above datum		
		A	B	C
1	136.2 ml	62	90	117
2	184.5 ml	84	122	164
3	309.4 ml	112	175	244

Length between tapping points: A–B = 120 mm
                                 B–C = 125

Determine the coefficient of permeability of the soil (average of six values).

5.4   In a falling-head permeability test the following data were recorded:

Internal diameter of permeameter = 75.2 mm
Length of sample   = 122.0 mm
Internal diameter of standpipe   = 6.25 mm
Initial level in standpipe   = 750.0 mm
Level in standpipe after 15 min. = 247.0 mm

Calculate the permeability of the soil.

5.5   A falling-head permeameter has an internal diameter of 75 mm and is connected to a standpipe of diameter 12.5 mm. A specimen of fine soil of length 80 mm is to be tested and is held in place between two discs of fine wire mesh. The head of water in the standpipe is allowed to fall from 950 mm to 150 mm and the times noted as follows:

Time taken when mesh discs only are in place = 4.4 s
Time taken when soil specimen is in place   = 114.8 s

Calculate the coefficient of permeability of the soil, making due allowance for the permeability of the wire mesh discs.

5.6 In a rapid falling head test, a glass tube of 37.5 mm internal diameter was used, with a layer of sand of 60 mm thickness at the bottom end. The average time for the water level in the tube to fall between graduation marks respectively 200 mm and 100 mm from the bottom end was 84.6 s. Calculate the coefficient of permeability of the sand.

5.7 For a field pumping test a well was sunk through a horizontal layer of sand which proved to be 14.4 m thick and to be underlain by a stratum of clay. Two observation wells were sunk, respectively 18 m and 64 m from the pumping well. The water table was initially 2.2 m below ground level. At a steady state pumping rate of 328 litres/min., the drawdowns in the observation wells were found to be 1.92 and 1.16 m respectively. Calculate the coefficient of permeability of the sand.

5.8 A horizontal layer of sand of 6.2 m thickness is overlain by a layer of clay with a horizonal surface of thickness 5.8 m. An impermeable layer underlies the sand. In order to carry out a pumping test a well was sunk to the bottom of the sand and two observation wells sunk through clay just into the sand at distances of 14 m and 52 m from the pumping well. At a steady-state pumping rate of 650 litres/min., the water levels in the observation wells were reduced by 2.31 m and 1.82 m respectively. Calculate the coefficient of permeability of the sand if the initial piezometric surface level lies 1.0 m below the ground surface.

5.9 On a certain site there are three horizontal soil layers down to an impermeable rock head:

Layer A: thickness = 3.5 m; $k = 2.5 \times 10^{-5}$ m/s
Layer B: thickness = 1.8 m; $k = 1.4 \times 10^{-7}$ m/s
Layer C: thickness = 4.2 m; $k = 5.6 \times 10^{-3}$ m/s

Calculate the average horizontal and vertical permeability of the soil between the surface and rock head.

5.10 In the arrangement shown in Fig. 5.32 steady-state flow conditions can be maintained with the tank level either at A–A or B–B. The saturated unit weight of the soil is 20 kN/m^3. Calculate the seepage pressure and the effective stress at level C–C for both positions of the tank, with the water level at E–E remaining constant.

5.11 Sketch a flow net for the seepage under the line of sheet piling shown in Fig. 5.33 and calculate the estimated seepage loss in m^3/hr per metre run.

5.12 A long coffer-dam is to be formed in a tidal estuary by driving parallel rows of sheet piles as shown in Fig. 5.34. Sketch a flow net for the seepage occurring at high tide and estimate the pumping rate (m^3/hr metre run) required to keep the excavation dry at the dredge line. The sand is isotropic and homogeneous and has a coefficient of permeability of 4.5×10^{-5} m/s.

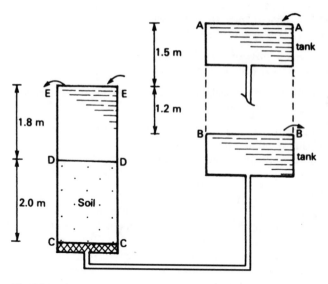

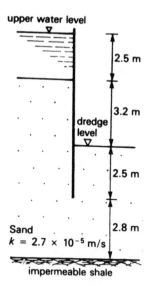

Fig. 5.32

upper water level

Fig. 5.33

5.13 Figure 5.35 shows the cross-section of a long coffer dam formed by parallel rows of sheet piles driven into an isotropic layer of sand.

(a)   Sketch a flow net (half only required) for the seepage and estimate the seepage quantity in m³/hr per metre run if $k = 6.5 \times 10^{-4}$ m/s.

(b)   Calculate the factor of safety against there being a base failure in the excavation due to heaving. (Saturated unit weight of soil = 20.4 kN/m³.)

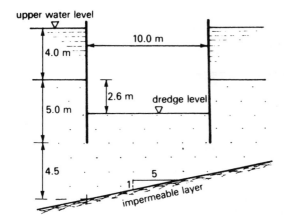

Fig. 5.34

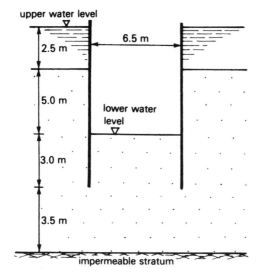

Fig. 5.35

5.14 A sheet pile wall is driven into a horizontal bed of silty sand and shown in Fig. 5.36, in order to impound a maximum depth of 5 m of water. The concrete apron is to be laid to reduce the seepage loss. The coefficient of permeability of the silty sand is 7.5×10^{-6} m/s.

(a) Sketch a flow net and from it estimate the maximum seepage quantity in m³/hr per metre run.

(b) Plot a curve showing the maximum hydrostatic uplift pressure distribution on the concrete apron.

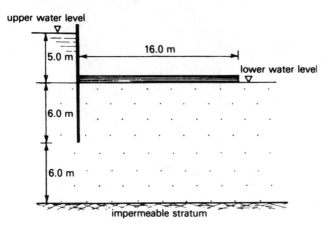

Fig. 5.36

5.15   Figure 5.37 shows the cross-section of a concrete dam founded on an isotopic layer of silty sand, which is underlain by an impermeable layer.

(a)   Sketch a flow net and estimate the seepage quantity under the dam on its own.

(b)   Sketch two further flow nets for the seepage, after a sheet piling cut-off has been installed: (i) at A–A, and (ii) at B–B.

(c)   Comment on the effects of installing the sheet piling cut-off at C–C.

5.16   Redo part (a) of Exercise 5.15 for the case of anisotropic soil conditions in which $k_H = 2.8 \times 10^{-5}$ m/s and $k_V = 7.6 \times 10^{-6}$ m/s.

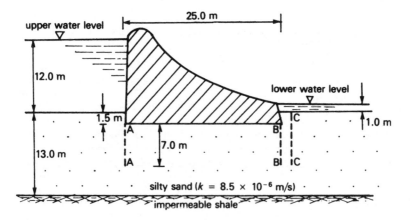

Fig. 5.37

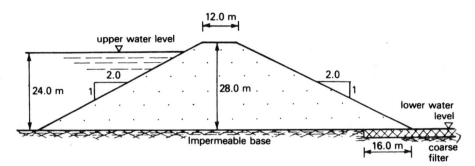

Fig. 5.38

5.17 The cross-section of a homogeneous earth dam is shown in Fig. 5.38. The coefficient of permeability is 5.0×10^{-6} m/s. Sketch a flow net and estimate the quantity of seepage in m^3/day per metre run: (a) with the toe filter as shown, and (b) without a toe filter.

Soil Mechanics Spreadsheets and Reference Assignments and Quizzes (available on the accompanying CD):

Assignments A.5   Soil permeability and seepage

Quiz Q.5   Soil permeability and seepage

Chapter 6

Stresses and strains in soils

6.1 Basic mechanics: equilibrium and compatibility

Many of the problems encountered in soil engineering require an understanding of applied mechanics, i.e. the study of the action of forces on bodies, together with consequential movements, displacements or distortions. Mathematical descriptions and models of mechanical behaviour are required in *four* main branches of mechanics:

(a) *The mechanics of solids*: concerning applied forces or stresses and *internal* strains or displacements, such as changes in shape or volume, e.g. settlement problems, yielding.

(b) *Rigid body mechanics*: concerning forces applied to rigid (non-deforming) bodies and consequential movement mechanisms, e.g. shear slips in slopes, retaining walls.

(c) *Structural mechanics*: concerning deformable frameworks, beams, columns, etc., and consequential bending, shearing, twisting, etc., e.g. piles, anchored walls.

(d) *Fluid mechanics*: concerning forces and pressures associated with both static and flowing water, e.g. hydrostatic pore pressure, seepage pressure, water flow (permeability).

There are two basic principles of mechanics upon which all other relationships are founded, and these are *equilibrium* and *compatibility*.

The principle of equilibrium
Any body that is not accelerating must be in directional and rotational equilibrium, i.e. the net force acting on the body in any direction must be zero and the net moment about any point must be zero. For example, in Fig. 6.1(a) the equilibrium of the forces can be written $0 = \Sigma V_i = V_1 + V_2 + V_3 + V_4$ and $0 = \Sigma H_i = H_1 + H_2 + H_3 + H_4$, where V and H are signed directional components of force. The moment equilibrium about (say) point A is written $0 = \Sigma M_A = $ sum of the moments of forces about A.

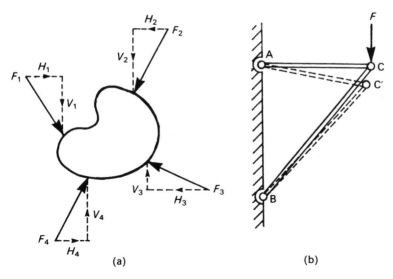

Fig. 6.1 Equilibrium and compatibility

The principle of compatibility

Any displacements, movements or changes in shape or volume must be *compatible*, i.e. no material can be gained or lost. For example, in Fig. 6.1(b), the movement of joint C (from C to C′) must be compatible with the movements due to the change in length of the members AC and BC.

Soils differ from other engineering materials, such as steel and concrete, because of their particulate and porous nature: the application of loading to bodies of soil can result in changes in volume. Changes in volume are often accompanied by changes in water/air content, density, stiffness and strength. Such changes depend on the intensity and rate of loading, and on the loading history of the soil. A study of soil behaviour under load commences with basic solid mechanics.

6.2   Stress and strain: principles and definitions

Material behaviour and mechanical characteristics can be demonstrated and (with numerical data) quantified using stress–strain curves. Figure 6.2 shows typical curves representing behaviour under tensile loading. *Brittle* materials (e.g. cast iron, high-carbon steels, concrete, hard rocks, etc.) are characterised by a steep linear *elastic* curve, followed almost immediately by the breaking point, i.e. *brittle failure*. A *ductile* material (e.g. structural steel, aluminium and other alloys) is shown in Fig. 6.2(c): after an initial fairly steep linear *elastic* phase, the material yields and elongation increases with only a relatively small increase in stress before brittle failure takes place. In this latter phase, the deformation is *plastic* and not fully recoverable upon unloading.

Stiffness and elasticity

The relationship between the stress (i.e. the intensity of loading) and the strain that it causes is termed the *stiffness* of the material. A material is said to be

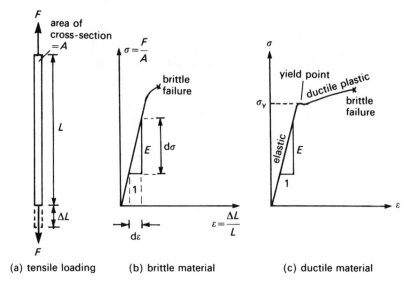

(a) tensile loading　　　(b) brittle material　　　(c) ductile material

Fig. 6.2　Brittle and ductile materials in tension

elastic when the same amount of strain is caused at the same level of stress regardless of whether or not the material has been loaded or unloaded beyond this point. The slope of the stress/strain curve is termed the *stiffness modulus*, *modulus of elasticity* or *Young's modulus*.

$$\text{Modulus of elasticity, } E = \frac{\delta\sigma}{\delta\varepsilon} \qquad\qquad [6.1]$$

If the stress–strain curve is a straight line the stiffness is constant for a range of stress–strain values; this is known as *Hookean* behaviour, i.e. obeying *Hooke's law* ($\varepsilon \propto \sigma$). In the case of non-Hookean materials (e.g. some non-ferrous metals, plastics, soils), the stress–strain curve is not straight and the material behaviour may or may not be elastic. The slope *E* therefore varies with stress and must be quoted as either the *tangent* value or the *secant* value (Fig. 6.3).

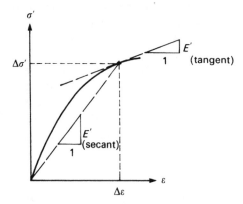

Fig. 6.3　Tangent and secant stiffness moduli for inelastic materials

Tangent stiffness, $E' = \dfrac{\delta\sigma'}{\delta\varepsilon}$ [6.2(a)]

Secant stiffness, $E'_{sec} = \dfrac{\Delta\sigma'}{\Delta\varepsilon}$ [6.2(b)]

Note that a prime has been added to the quantities of stress to denote that this is a value of *effective stress* – this is important in soil mechanics.

Yield point
In materials exhibiting elastic or near-elastic ductile behaviour the change from elastic to plastic straining is often marked by an abrupt change in the slope of the stress–strain curve. This point is called the *yield point* and the stress at which it occurs is called the *yield stress* (σ_y) (Fig. 6.2(c)).

Volumetric strain and bulk modulus
Figure 6.4(a) shows a three-dimensional element subject to an *isotropic* (i.e. equal-all-round) stress or hydrostatic pressure. If the initial volume of the element is V_0 and this changes after loading by dV, the volumetric strain is

$$\varepsilon_v = \frac{dV}{V_0}$$

The volumetric stiffness modulus is termed the *bulk modulus* (Fig. 6.4(b)).

Bulk modulus, $K' = \dfrac{\delta p}{\delta\varepsilon_v}$ [6.3]

In compressible materials (e.g. soils), the bulk modulus increases with the applied isotropic stress.

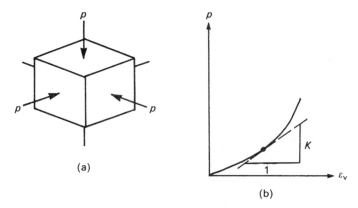

(a)

(b)

Fig. 6.4 Volumetric strain

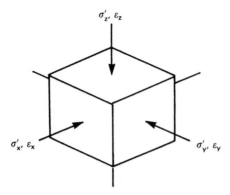

Fig. 6.5 General three-dimensional (triaxial) compression

Triaxial stress and strain

In many practical problems, stress–strain conditions must be considered in terms of three orthogonal axes. Consider a cubical element subject to effective *triaxial stresses*, σ'_x, σ'_y, σ'_z (Fig. 6.5). The general *triaxial strains* will be:

$$\left.\begin{aligned}
\varepsilon_x &= \left(\frac{\sigma'_x}{E'_x} - v'\frac{\sigma'_y}{E'_y} - v'\frac{\sigma'_z}{E'_z}\right) \\[2mm]
\varepsilon_y &= \left(-v'\frac{\sigma'_x}{E'_x} + \frac{\sigma'_y}{E'_y} - v'\frac{\sigma'_z}{E'_z}\right) \\[2mm]
\varepsilon_z &= \left(-v'\frac{\sigma'_x}{E'_x} - v'\frac{\sigma'_y}{E'_y} + \frac{\sigma'_z}{E'_z}\right)
\end{aligned}\right\} \qquad [6.4]$$

where v' = Poisson's ratio (referred to effective stress).

In materials having isotropic stiffness, i.e. $E'_x = E'_y = E'_z = E'$:

$$\left.\begin{aligned}
\varepsilon_x &= \frac{1}{E'}(\sigma'_x - v'\sigma'_y - v'\sigma'_z) \\[2mm]
\varepsilon_y &= \frac{1}{E'}(-v'\sigma'_x + \sigma'_y - v'\sigma'_z) \\[2mm]
\varepsilon_z &= \frac{1}{E'}(-v'\sigma'_x - v'\sigma'_y + \sigma'_z)
\end{aligned}\right\} \qquad [6.5]$$

and volumetric strain,

$$\varepsilon_v = \varepsilon_x + \varepsilon_y + \varepsilon_z = \frac{1}{E'}(1 - 2v')(\sigma'_x + \sigma'_y + \sigma'_z) \qquad [6.6]$$

Shear stress and strain

Figure 6.6(a) shows how a two-dimensional element distorts when subject to a shear stress (τ); the shear strain is represented by the angle γ (strictly this is tan γ,

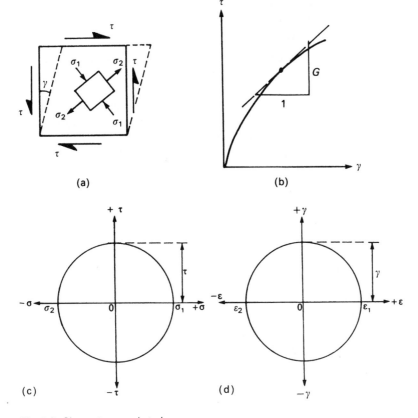

Fig. 6.6 Shear stress and strain

but since γ is very small, $\tan \gamma \approx \gamma$). The slope of the shear–stress/shear–strain curve (Fig. 6.6(b)) is termed the *modulus of rigidity* or *shear modulus*.

$$\text{Modulus of rigidity, } G' = \frac{\delta \tau}{\delta \gamma} \qquad \qquad [6.7]$$

The state of strain produced by the simple shear stress (τ) can also be produced by the principal stresses σ_1 and σ_2 (Fig. 6.6(a)). Figure 6.6(c) and (d) show the corresponding Mohr circles for stress and strain, from which it can be seen that

$$\tau = \tfrac{1}{2}(\sigma_1 - \sigma_2) \qquad \qquad [6.8(a)]$$

$$\gamma = (\varepsilon_1 - \varepsilon_2) \qquad \qquad [6.8(b)]$$

It can be shown that in isotropically elastic materials the following relationships exist between the stiffness moduli:

$$G' = \frac{E'}{2(1 + v')} \qquad \qquad [6.9]$$

$$K' = \frac{E'}{3(1 - 2v')} \qquad \qquad [6.10]$$

The stress–strain characteristics of soil are more complex
In principle, soils behave like other solids when subject to changes in loading, but there are significant differences to, say, steel or concrete:

(a) Except for some partially cemented types, soils cannot sustain tension.
(b) When loaded, soils will generally undergo a change in volume or an increase in pore fluid pressure.
(c) Saturated soils can only undergo a change in volume as porewater is squeezed out (or lost by drying, etc.); the rate of water loss (drainage) is controlled by the permeability of the soil.
(d) Some (hard or stiff) soils will exhibit brittle failure by shearing, while others will simply distort plastically.
(e) Once a shear slip has occurred the problem changes from one of solid mechanics to one of rigid body mechanics.

A number of different stress states can exist in soil masses and these need to be carefully examined. The one-dimensional tension state illustrated in Fig. 6.2 does not exist; in addition, the loading of soil masses must be considered in terms of both *undrained* and *drained* conditions (see Section 6.3).

(a)   Isotropic stress
Isotropic means 'equal in all directions' (Fig. 6.7), a practical example being the first stage in a triaxial test (Section 7.6), another is hydrostatic pressure.

$$\sigma_x = \sigma_y = \sigma_z = p \quad \text{and} \quad \varepsilon_v = \frac{3p'}{E'}(1 - 2v')$$

Therefore

$$K' = \frac{p'}{\varepsilon_v} = \frac{E'}{3(1 - 2v')} \qquad\qquad [6.11]$$

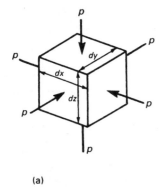
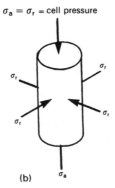

(a)   (b)

Fig. 6.7   Isotropic stress
          (a) On a rectilinear element (b) On a triaxial test specimen

(b)   Biaxial symmetry
When the two horizontal stresses are equal, a state of *biaxial symmetry* exists. For example, in the axial loading stage of the triaxial test (Fig. 6.8(a)), or the stresses beneath a circular loaded area (Fig. 6.8(b)).

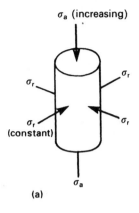

σ_a (increasing)

σ_r

σ_r (constant)

σ_r

σ_r

σ_a

(a)

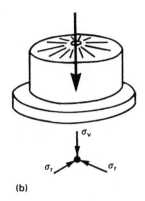

σ_v

σ_r

σ_r

(b)

Fig. 6.8 Stresses with biaxial symmetry
(a) On a triaxial specimen during axial loading (b) Beneath the centre of a circular loaded area

Putting vertical axial stress, $\sigma_z = \sigma_a$ and horizontal radial stress, $\sigma_x = \sigma_y = \sigma_r$. Then from equations [6.5]:

$$\varepsilon_a = \frac{1}{E'}(\sigma'_a - 2v'\sigma'_r)$$

$$\varepsilon_r = \frac{1}{E'}(-v'\sigma'_a + (1 - v')\sigma'_r)$$

$$\varepsilon_v = \varepsilon_a + 2\varepsilon_r = \frac{(1 - 2v')}{E'}(\sigma'_a + 2\sigma'_r)$$

or writing *mean normal stress,* $p' = \frac{1}{3}(\sigma'_a + 2\sigma'_r)$

$$\varepsilon_v = \frac{3(1 - 2v')p'}{E'} = \frac{p'}{K'}$$

Thus $K' = \dfrac{E'}{3(1 - 2v')}$ as in eqn [6.9(b)]

In the context of triaxial tests, it is convenient to define the following parameters:

Deviatoric stress,	$q' = \sigma'_a - \sigma'_r$	[6.12]
Mean normal stress,	$p' = \frac{1}{3}(\sigma'_a + 2\sigma'_r)$	[6.13]
Shear strain,	$\varepsilon_s = \frac{2}{3}(\varepsilon_a - \varepsilon_r)$	[6.14]
Volumetric strain,	$\varepsilon_v = \dfrac{3(1 - 2v')p'}{E'} = \dfrac{p'}{K'}$	[6.15]

Note that ε_s defines the change in shape of the soil mass, whereas ε_v defines the change in volume. The stiffness moduli are then:

$$G' = \frac{\delta q'}{\delta\varepsilon_s} \qquad\qquad [6.16]$$

$$K' = \frac{\delta p'}{\delta\varepsilon_v} \qquad\qquad [6.17]$$

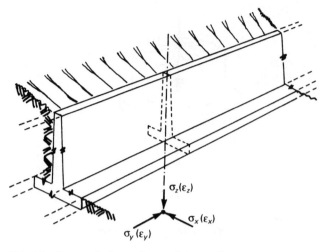

Fig. 6.9 Plane strain beneath a retaining wall

(c)   Plane strain

Under very long structures, such as retaining walls and strip foundations (Fig. 6.9), the strain in the longitudinal direction will be zero (except at each end). Strain therefore takes place only in the plane of the cross-section, producing conditions of *plane strain*. Putting

$$\varepsilon_y = 0 \quad \text{and} \quad \sigma'_x = \sigma'_h, \quad \sigma'_y = v'(\sigma'_z - \sigma'_h)$$

Then

$$\varepsilon_z = \frac{1 + v'}{E'}[(1 - v')\sigma'_z - v'\sigma'_h]$$

$$\varepsilon_h = \frac{1 + v'}{E'}(-v'\sigma'_z + (1 - v')\sigma'_h)$$

$$\varepsilon_v = \varepsilon_z + \varepsilon_h = \frac{(1 + v')(1 - 2v')}{E'}(\sigma'_z + \sigma'_h)$$

or putting

$$s' = \tfrac{1}{2}(\sigma'_z + \sigma'_h)$$

$$\varepsilon_v = \frac{2(1 + v')(1 - 2v')s'}{E'}$$

Thus, the volumetric modulus for plane strain is

$$K'_{ps} = \frac{E'}{2(1 + v')(1 - 2v')} \qquad\qquad [6.18]$$

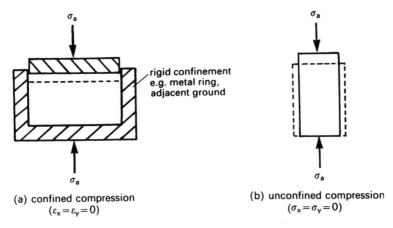

(a) confined compression
$(\varepsilon_x = \varepsilon_y = 0)$

(b) unconfined compression
$(\sigma_x = \sigma_y = 0)$

Fig. 6.10 One-dimensional compression

(d)   Confined one-dimensional compression
Below large loaded areas, such as raft foundations, the strain in both lateral directions may be considered to be zero. A similar state of one-dimensional strain exists in the *oedometer test* (Section 10.3), wherein the specimen is confined laterally by metal ring (Fig. 6.10(a)). Putting

$$\varepsilon_x = \varepsilon_y = \varepsilon_h = 0 \quad \text{and} \quad \sigma'_x = \sigma'_y = \sigma'_h$$

Then

$$\varepsilon_h = 0 = \frac{1}{E'}(-v'\sigma'_z + \sigma'_h - v'\sigma'_h)$$

Therefore

$$\sigma'_h = \frac{v'}{1 - v'}\sigma'_z \qquad\qquad [6.19]$$

Also $\varepsilon_v = \varepsilon_s$ and the *confined stiffness modulus* may be written:

$$E'_0 = \frac{\sigma'_z}{\varepsilon_z} \quad \left(\text{or more strictly: } \frac{\delta\sigma'_z}{\delta\varepsilon_z}\right) \qquad\qquad [6.20]$$

Then, for an isotropically elastic material:

$$\varepsilon_z = \frac{1}{E'}(\sigma'_z - 2v'\sigma'_h)$$

$$= \frac{1}{E'}\left(\sigma'_z - 2v'\frac{v'}{1 - v'}\sigma'_z\right) = \frac{\sigma'_z}{E'}\left[\frac{(1 + v')(1 - 2v')}{(1 - v')}\right]$$

So that

$$E'_0 = E'\frac{1 - v'}{(1 + v')(1 - 2v')} \qquad\qquad [6.21]$$

Note that for a saturated soil with drainage prevented (i.e. undrained conditions), $v = v_u = 0.5$; in which case $E'_0 = \infty$, in other words no volume change can occur.

(e)   Unconfined compression
This state rarely occurs *in situ*, but an *unconfined compression test* may be carried out on specimens of saturated clay. The lateral stresses are zero: $\sigma'_x = \sigma'_y = 0$.

Then $\varepsilon_z = \dfrac{\sigma'_z}{E'}$ and $\varepsilon'_x = \varepsilon'_y = \dfrac{v'\sigma'_z}{E'}$

and $\varepsilon_v = \varepsilon_z + \varepsilon_x + \varepsilon_y = \dfrac{1 - 2v'}{E'}\sigma'_z$

Thus $K' = \dfrac{\sigma'_z}{\varepsilon_v} = \dfrac{E'}{1 - 2v'}$ [6.22]

(f)   Shear box test
During the application of the normal load there is no shear stress and confined one-dimensional compression is taking place; the stiffness moduli are:

$E'_{0(tan)} = \dfrac{\delta\sigma'_n}{\delta\varepsilon_n}$ or $E'_{0(sec)} = \dfrac{\Delta\sigma'_n}{\Delta\varepsilon_n}$ [6.23(a)]

$G'_{(tan)} = \dfrac{\delta\tau}{\delta\gamma}$ or $G'_{(sec)} = \dfrac{\Delta\tau}{\Delta\gamma}$ [6.23(b)]

6.3   Soil compression and volume change

If a body of soil is loaded it will be compressed; if it is unloaded it will swell. In a saturated soil mass the immediate response to loading is an increase in pore pressure, and if drainage is possible water flows out of the soil as the soil compresses or back into it as it swells. The rate of flow and therefore of volume change depends on the permeability of the soil: in gravels and sands it is rapid, but in silts and clays it is very slow. Such changes in volume are related to the *effective stress* – see Section 4.5.

Undrained loading conditions
Undrained conditions occur when either drainage is prevented (e.g. as in an undrained triaxial test) or when the rate of loading is too rapid to allow any significant outflow of water (e.g. immediately after the construction of a foundation on a clay soil). The deformation of an undrained soil mass is related to the stiffness of both the porewater and the soil solids.

$K_{(water)} \approx 2.2 \times 10^6 \text{ kPa}$ $K_{(mineral\ grains)} \approx 50 \times 10^6 \text{ kPa}$

Since the stresses normally encountered in soil engineering are below 1000 kPa the resulting volumetric strain will be less than 10^{-4}, so there will be no significant undrained volume change due to an increase in stress and the undrained stiffness moduli are:

Isotropic compression: $K_u = \infty$

One-dimensional compression: $E_u = \infty$

Thus, $\varepsilon_v = 0 = \dfrac{\sigma}{E_u}(1 - 2v_u)$,   giving $v_u = 0.5$

Under undrained conditions stress–strain behaviour is defined in terms of *total stresses.* Since $\sigma_1 - \sigma_3 = \sigma'_1 - \sigma'_3$, then $G = G'$ and from eqn [6.9]:

$$E_u = E' \frac{1.5}{1 + v'} \qquad\qquad [6.24]$$

Drained conditions
When loading is applied slowly, such that the water drains away without any increase in pore pressure, the volume will decrease and stress–strain behaviour must be defined in terms of *effective stresses.* The principal constants are:

$$K' = \frac{\delta p'}{\delta\varepsilon_v}, \quad G' = \frac{\delta\tau}{\delta\gamma}, \quad E' = \frac{\delta\sigma'_a}{\delta\alpha_a} \text{ and } v' \quad (\text{see also Section 6.2})$$

Isotropic compression and swelling
If the loading is isotropic and drained, the soil compresses and becomes more dense, and therefore stiffer, so that the specific-volume/stress line is curved and getting less steep. Figure 6.11(a) shows a v/p' curve for a loading–swelling–reloading sequence. From an initial isotropic stress of p'_0 at O, the load is increased to stress of p'_y at A. The loading is then reduced to p'_0 and the soil swells back to B; then as the soil is reloaded the curve moves to C and then D. Note that the swelling–reloading line is flatter (indicating increased stiffness) and almost straight. If this behaviour is plotted on $v/\ln p'$ axes (Fig. 6.11(b)) the curves become straight lines and provide a good representation of the behaviour of the majority of soils.

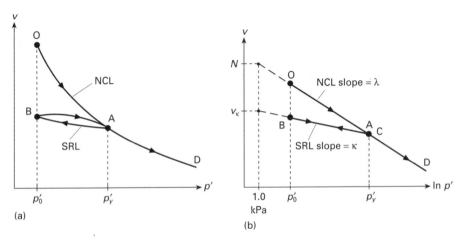

Fig. 6.11 Isotropic compression and swelling

The line OACD represents loading to stresses higher than any previous values and is called the **normal compression line (NCL)**. This corresponds to the natural deposition of soil as it compresses slowly from a mud of high specific volume to its present condition. If at some time it is *unloaded* (due to erosion, disturbance, excavation, etc.) it swells along the swelling line, and if reloaded it compresses along the same (almost) straight line ABC; this is the **swelling–recompression line (SRL)**. The previous highest stress p'_y is the *yield stress* and as the stress is increased beyond this the compression (CD) is again along the NCL.

The following expressions can be deduced:

NCL: $v = N - \lambda \ln p'$ $\hspace{4cm}$ [6.25(a)]

SRL: $v = v_\kappa - \kappa \ln p'$ $\hspace{4cm}$ [6.25(b)]

where N = the specific volume (v) at $p' = 1.0$ kPa
$\hspace{1.4cm}$ λ = slope of the CSL
$\hspace{1.4cm}$ v_κ = the specific volume (v) at $p' = 1.0$ kPa
$\hspace{1.4cm}$ κ = slope of the SRL

At the intersection of the two lines v has a common value

$$v = N - \lambda \ln p'_y = v_\kappa - \kappa \ln p'_y$$

Then $\ln p'_y = \dfrac{N - v_\kappa}{\lambda - \kappa}$ $\hspace{4cm}$ [6.26]

The ratio κ/λ lies in the range 0.1–0.4, but depends also on the extent of particle distortion during compression: with very flaky clay particles κ/λ tends to be large, e.g. 0.25–0.4. In hard (e.g. quartz) sands $\kappa/\lambda = 0.05$–0.1; in softer sands (with crushable particles) $\kappa/\lambda < 0.03$.

Isotropic compression can also be represented by the bulk modulus (K') (see eqn [6.3]), but this is not a constant value:

For normal compression: $K' = \dfrac{vp'}{\lambda}$ $\hspace{3cm}$ [6.27(a)]

For swelling or reloading: $K' = \dfrac{vp'}{\kappa}$ $\hspace{3cm}$ [6.27(b)]

A soil that has been loaded and then unloaded is in a state of **overconsolidation**, which lies below the NCL and is defined by an **overconsolidation ratio (R_p)**.

$$R_p = \frac{p'_y}{p'_0} \hspace{4cm} [6.28]$$

where p'_y = the yield stress (i.e. the intersection of the NCL and the SRL)
$\hspace{1.4cm}$ p'_0 = the stress at the initial state prior to loading

The initial state of a soil can therefore be described quantitatively by the parameters v, p' and R_p. States on the NCL have an $R_p = 1.0$ and states with the *same* $R_p > 1$ value lie on the same straight line parallel to the NCL. In Fig. 6.12 the states B_{1-3} all have the same R_p value, each initial state (B) corresponding to

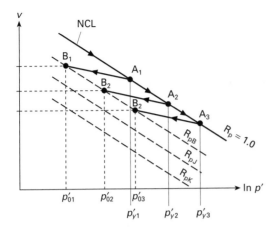

Fig. 6.12 Overconsolidation and R_p lines

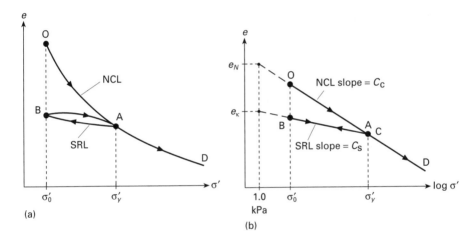

Fig. 6.13 One-dimensional compression and swelling

a unique yield point (A). A family of lines exists therefore for different values of R_p, each parallel to the NCL.

One-dimensional compression

Drained behaviour during a one-dimensional compression (e.g. during natural deposition, the oedometer test) is always close to the idealised compression curve shown in Fig. 6.13(a). The vertical strain can be expressed in terms of a change in thickness (Δh), a change in void ratio (Δe) or a change in specific volume (Δv):

$$\frac{\Delta h}{h_0} = \frac{\Delta e}{1 + e_0} = \frac{\Delta v}{v_0} \quad \text{or} \quad d\varepsilon = \frac{dh}{h} = \frac{de}{1 + e} = \frac{dv}{v} \tag{6.29}$$

When e is plotted against $\log \sigma_z'$ (Fig. 6.13(b)) the normal compression and swelling–recompression lines are basically the same as for isotropic compression, but with different parameters

NCL: $e = e_N - C_c \log \sigma'_z$ [6.30(a)]

SRL: $e = e_\kappa - C_s \log \sigma'_z$ [6.30(b)]

where e_N = the void ratio (e) at $\sigma'_z = 1.0$ kPa
C_c = slope of the CSL, referred to as the **compression index**
e_κ = the specific volume (v) at $\sigma'_z = 1.0$ kPa
C_s = slope of the SRL, referred to as the **swelling index**

At the intersection of the two lines e has a common value

$$e = e_N - C_c \log \sigma'_y = e_\kappa - C_s \log \sigma'_y$$

Then $\log \sigma'_y = \dfrac{e_N - e_\kappa}{C_c - C_s}$ [6.31]

The ratio C_s/C_c usuallly lies in the range 0.2–0.9, but depends also on the extent of particle distortion during compression: with very flaky clay particles C_s/C_c tends to be large, e.g. 0.25–0.4. In hard (e.g. quartz) sands $C_s/C_c = 0.05$–0.1; in softer sands (with crushable particles) $C_s/C_c < 0.03$.

The slopes of the NCL and SRL in isotropic and one-dimensional compression are respectively parallel and, since $\ln x = 2.3 \log_{10} x$, $C_c = 2.3\lambda$ and $C_s = 2.3\kappa$.

One-dimensional compression can also be represented by a stiffness modulus (E'_0), but this is not a constant value for normally consolidated soils:

For normal compression: $E'_0 = \dfrac{d\sigma'}{d\varepsilon} = (1 + e)\dfrac{d\sigma'}{de} = (1 + e)\dfrac{\sigma'}{C_c}$ [6.32(a)]

For swelling or reloading: $E'_0 = (1 + e)\dfrac{\sigma'}{C_s}$ [6.32(b)]

For overconsolidated soils the value of E'_0 can be estimated from the SRL. For example, if a clay is found to have a void ratio (e) of 1.00 and $C_s = 0.05$, then $E'_0 = 36\sigma'_z$. Also, since σ' is usually nearly proportional to the depth below the surface, it follows that E'_0 also increases almost linearly with depth.

An alternative parameter is the **coefficient of volume compressibility (m_v)**, which is the reciprocal of the stiffness modulus,

$$m_v = \dfrac{d\varepsilon_z}{d\sigma'_z} = \dfrac{1}{E'_0}$$ [6.33]

For **overconsolidation** states which lie below the NCL, the **overconsolidation ratio** is

$$R_o = \dfrac{\sigma'_y}{\sigma'_0}$$ [6.34]

where σ'_y = the yield stress (i.e. the intersection of the NCL and the SRL)
σ'_0 = the stress at the initial state prior to loading

The initial state of a soil can therefore be described quantitatively by the parameters e, σ' and R_o. States on the NCL have an $R_o = 1.0$ and states with the *same* $R_o > 1$ value lie on the same straight line parallel to the NCL (as in Fig. 6.12).

Table 6.1 Values of compression and swelling parameters

Soil type	Plasticity			NCL slope	
	Class	w_L	I_p	λ	C_c
Clays	Extremely high	>90	>50	>0.29	>0.72
and	Very high	70–90	>36	0.21–0.29	0.54–0.72
Silty clays	High	50–70	>22	0.13–0.21	0.36–0.54
	Intermediate	35–50	>10	0.09–0.11	0.22–0.36
Sandy clays	Low	<35	<10	<0.09	<0.22
Sands	Hard			0.1–0.2	0.2–0.4
	Crushable			0.2–0.4	0.4–0.9

For clays: $\lambda \approx I_p/170$ or $C_c \approx 0.009 \, (w_L - 10)$ and $\kappa = 0.25\lambda - 0.35\lambda$
For some sands λ is relatively large due to particle crushing

In one-dimensional compression the horizontal strain (ε_h) is zero; therefore the horizontal stress (σ'_h) must change, and from eqn [6.19]

$$\frac{\sigma'_h}{\sigma'_z} = \frac{v'}{1 - v'} = K_0 \quad \textbf{(coefficient of lateral earth pressure)} \qquad [6.35]$$

But v' cannot easily be measured, therefore an approximate value is usually taken for K_0:

Normally consolidated soil: $K_{0nc} = 1 - \sin \phi'_c$ \qquad [6.36(a)]

Overconsolidated soil: $K_{0oc} = K_{0nc}\sqrt{R_o}$ \qquad [6.36(b)]

A connection can be made between one-dimensional and triaxial compression using K_0. Putting $\sigma'_a = \sigma'_z$ and $\sigma'_r = \sigma'_h$ and referring to equations [6.12] and [6.13]:

Deviatoric or shear stress, $\quad q' = \sigma'_z(1 - K_0)$ \qquad [6.37(a)]

Mean normal stress, $\quad p' = \frac{1}{3}\sigma'_z(1 + 2K_0)$ \qquad [6.37(b)]

6.4 *Analysis of stress using Mohr's circle*

The Mohr circle of stress provides a convenient method of analysing two-dimensional stress states. The reader will find fuller details and proof of the circle construction in most elementary texts on the strength of materials. For the purposes required here the following simplified procedure will be sufficient.

Figure 6.14 shows a typical case: a soil element below a cutting is intersected by a trial failure surface. The analysis requires values and directions for the principal stresses (σ_1, σ_3) and the normal and shear stresses (σ'_n, τ) on the failure surface. A generalised version of this problem is shown in Fig. 6.15(a). The Mohr circle of stress (Fig. 6.15(c)) may be defined in terms of the orthogonal stresses (σ_z, σ_x, τ_{xz}, τ_{zx}) in the location of points P and Q. When the circle has been drawn, points A and B represent the minor and major principal stresses respectively.

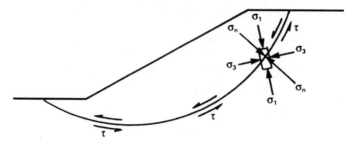

Fig. 6.14 Principal stresses on a soil element

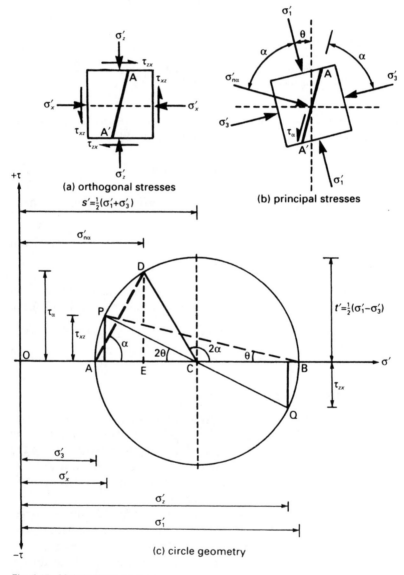

Fig. 6.15 Mohr's circle of stress

$OA = \sigma_3$ and $OB = \sigma_1$

and the angle of inclination of the principal planes is given by $\angle CPB = \theta$.

In many problems – for example, in the analysis of triaxial test results – the Mohr circle is constructed directly with values of the principal stresses. The objective in such cases may be to obtain values for the normal and shear stresses on a particular plane, perhaps a plane of shear slip (shear failure).

Consider the plane AA' passing through the element at angle α to the minor principal stress. Point D on the Mohr circle represents the stresses on this plane:

Normal stress, σ'_n = abscissa at D

Shear stress, τ = ordinate at D

The shear stress (τ) therefore varies from zero when $\alpha = 0$, to a maximum value when $\alpha = 45°$, to zero again when $\alpha = 90°$. Its value may be obtained as follows:

$$\tau = DE = CD \sin(180° - 2\alpha) = CD \sin 2\alpha$$

but $CD = \frac{1}{2}(\sigma'_1 - \sigma'_3)$

Hence $\tau = \frac{1}{2}(\sigma'_1 - \sigma'_3) \sin 2\alpha$ [6.38]

Similarly, the normal stress will be:

$$\sigma'_n = OE = OA + AE = \sigma'_3 + AD \cos \alpha$$

but $AD = 2\,AC \cos \alpha = AB \cos \alpha = (\sigma'_1 - \sigma'_3) \cos \alpha$

Hence $\sigma'_n = \sigma'_3 + (\sigma'_1 - \sigma'_3) \cos^2\alpha$

$$= \frac{1}{2}(\sigma'_1 + \sigma'_3) + \frac{1}{2}(\sigma'_1 - \sigma'_3) \cos 2\alpha \qquad [6.39]$$

The main advantage of using the Mohr circle construction lies in the ease with which the statements for shear stress and normal stress can be written to correspond to special stress regimes.

Figure 6.16 shows how the deviator stress (q') is represented by the diameter of the Mohr circle:

$$q' = (\sigma'_1 - \sigma'_3)$$

Point T on the circle represents the maximum shear stress $(\alpha = 45°)$ and has the coordinates (s', t'):

$$s' = \frac{1}{2}(\sigma'_1 + \sigma'_3) \qquad [6.40(a)]$$

$$t' = \frac{1}{2}(\sigma'_1 - \sigma'_3) \qquad [6.40(b)]$$

When the corresponding circle is constructed for total stresses, it will have the same diameter, but is shifted to the right along the normal stress axis by an amount equal to the pore pressure (u) (Fig. 6.17).

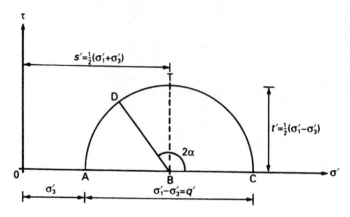

Fig. 6.16

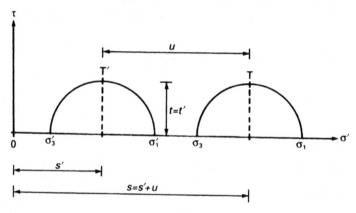

Fig. 6.17

Total stress: $\sigma_1 = \sigma_1' + u$

$$\sigma_3 = \sigma_3' + u$$

Subtracting, $\sigma_1 - \sigma_3 = \sigma_1' - \sigma_3'$

or $q = q'$ [6.41]

Also $s = s' + u$

and $t = t'$ [6.42]

Similarly $p = p' + u$ [6.43]

6.5 The Mohr–Coulomb failure theory

In the case of a shear slip failure or continuous yielding, the Mohr circle containing the normal and shear stresses on the slip plane is clearly a limiting circle. Limiting circles at different values of normal stress will all touch a common tangent which is called a *failure envelope* (Fig. 6.18). The equation of this failure envelope (usually referred to as *Coulomb's equation*) is:

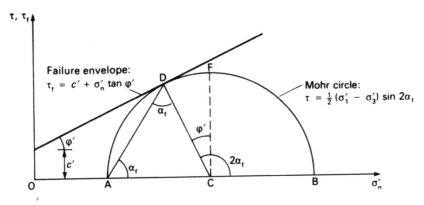

Fig. 6.18 Mohr–Coulomb failure theory

$$\tau'_f = c' + \sigma'_n \tan \phi' \qquad\qquad\qquad\qquad\qquad [6.44]$$

where ϕ' = angle of friction or the angle of shearing resistance
$\quad\quad\quad c'$ = apparent cohesion

From the geometry of the Mohr–Coulomb construction the angle of the plane of failure is:

$$\alpha_f = \tfrac{1}{2}(90° + \phi') = 45° + \phi'/2 \qquad\qquad\qquad [6.45]$$

If a number of samples of the same soil can be brought to a state of *shear slip failure* or *continuous yielding* and the principal stresses (σ'_1 and σ'_3) are measured, the Mohr–Coulomb construction may be used to determine the failure envelope and thus values for the parameters c' and ϕ'.

6.6   Analysis using stress paths

In an elastic body the deformation caused by a change in loading is predictable from the value of E' and the *total change* in load. The final value of strain is not affected by intermediate variations in the pattern of loading, but only with the overall change. Soil masses, however, demonstrate *elasto-plastic* behaviour, so that the exact pattern of loading or unloading may significantly affect the final result.

In an analysis of elasto-plastic behaviour it is instructive to plot the stress changes that take place throughout the entire pattern of loading. Diagrams or graphs of stress changes are referred to as *stress path diagrams*. These will take a number of forms dependent on type of analysis required.

(a)   Stress paths in σ/ε space
Figure 6.19 shows idealised stress–strain graphs representing the behaviour of different metals in tension. The arrows indicate the direction of stress change and the lines O → F are *stress paths*. Typically, elastic behaviour is indicated when the strain (ε) corresponds to the same value of stress (σ) irrespective of any intermediate change in loading. The stress path O → C is a straight line, thus demonstrating Hooke's law, i.e. $d\sigma/d\varepsilon$ = constant.

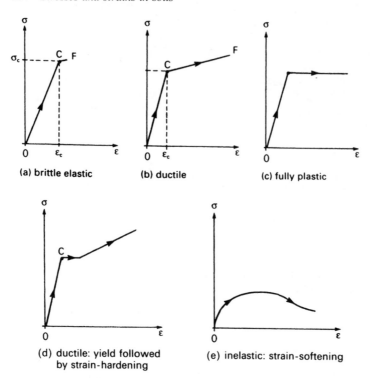

(a) brittle elastic   (b) ductile   (c) fully plastic

(d) ductile: yield followed
by strain-hardening   (e) inelastic: strain-softening

Fig. 6.19 Stress–strain behaviour

At point C (the *yield point*), a sharp increase in strain occurs and the material ceases to behave entirely elastically. Further increase in loading causes *plastic* straining which is not recovered with unloading. In *brittle* materials an ultimate failure occurs soon after the yield point (Fig. 6.19(a)), whereas *ductile* materials exhibit a significant amount of plastic straining before breaking (Fig. 6.19(b)). In a *fully plastic* state the material continues to yield without any increase in stress (Fig. 6.19(c)). In materials such as mild steel and copper strain hardening (increase in yield point stress) occurs (Fig. 6.19(d)). In materials such as lead, aluminium and many plastics the straining is largely inelastic and strain softening (reduction of the yield point stress) may occur (Fig. 6.19 (e)).

Loose or soft soils in compression generally exhibit strain-hardening characteristics, i.e. they contract and become stiffer. The shear behaviour of soil is more complex and much depends on the density. In compact sands and over-consolidated clays a brittle failure in the form of a shear slip is likely to occur at the *peak stress* (Fig. 6.20). In loose or soft soils contraction takes place up to the yield point and then continuous straining occurs at a constant or decreasing *ultimate stress*. Where very large strains (>1 m) occur, e.g. in hillside or embankment landslips, the ultimate stress may further decrease to a lower *residual stress*, which is a form of strain-softening behaviour.

The use of stress paths to study this sort of behaviour is considered further in Section 7.18.

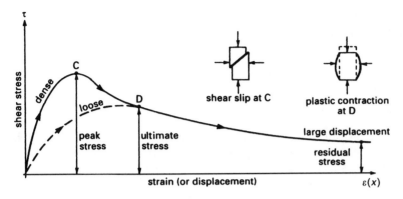

Fig. 6.20 Stress–strain curve of a shear test on soil

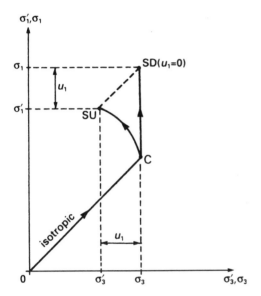

Fig. 6.21 Stress paths in σ_1/σ_3 space

(b) Stress paths in σ_1/σ_3 space

For many problems and in the interpretation of shear test results comparisons are often required between *drained* and *undrained* behaviour and between *effective* stresses and *total* stresses. Stress paths plotted on principal stress axes may be used. In Fig. 6.21, O → C represents an isotropic increase in stress ($\Delta\sigma_1 = \Delta\sigma_2 = \Delta\sigma_3$) under drained conditions ($\Delta u = 0$). If σ_1 is now increased uniaxially ($\sigma_2 = \sigma_3 = $ constant), the stress path for drained conditions will be C → SD, whereas with drainage prevented the pore pressure will rise and the stress path will be C → SU, since $\sigma' = \sigma - u$.

(c)　Stress paths in t'/s' space

Stress states can be conveniently represented by the Mohr circle and this can also be related to a failure criterion. The coordinates of the maximum shear stress point of a Mohr circle are given by:

$$s' = \tfrac{1}{2}(\sigma_1' + \sigma_3')$$

$$t' = \tfrac{1}{2}(\sigma_1' - \sigma_3')$$

Figure 6.22 shows drained and undrained stress paths for a uniaxial increase in σ_1. From an initial stress state $\sigma_1' = \sigma_3' = OC$ and with full drainage the total and effective stress paths follow the same line: $C \rightarrow ST(SE)$. Under undrained conditions (in a saturated soil) the pore pressure increases with the increase in σ_1.

The total stress path (TSP) is again at 45°; $O \rightarrow ST$. However, the effective stress path (ESP) curves backwards, since $s' = s - u$: $C \rightarrow SE$.

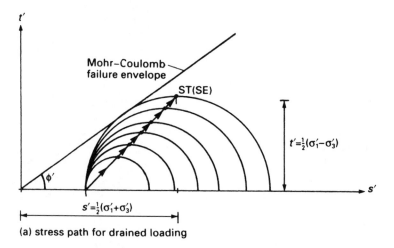

(a) stress path for drained loading

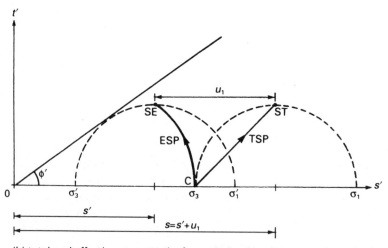

(b) total and effective stress paths for undrained loading

Fig. 6.22　Stress paths in t'/s' space

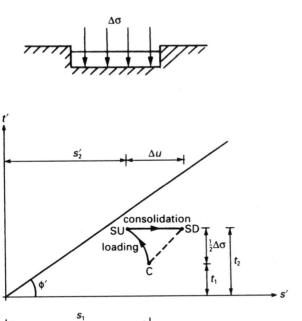

Fig. 6.23 Stress path for loading from a strip footing

The comparative behaviour of soil under increasing and decreasing stress can be illustrated using t'/s' stress paths. In Fig. 6.23 the stress path for a point below a strip footing in a clay soil is shown. During construction the stresses increase rapidly from their original values of σ_v and σ_h under virtual undrained conditions and the pore pressure rises by Δu. In the course of time, the pore pressure dissipates due to consolidation drainage (SU → SD). It is important to note that the final stress point is further away from the failure envelope, leading to the conclusion that in the short term undrained loading is more critical than drained loading in these circumstances, i.e. under a foundation. In Fig. 6.24 the stress path is shown for undrained loading at a point adjacent to an excavation. In this case, a reduction in stress induces a negative (suction) pore pressure ($-\Delta u$), thus the immediate or undrained stress path is C → SU. In response to the suction the water content gradually rises until the negative excess pore pressure has been reduced to zero, the stress path being SU → SD. The final stress point is nearer to the failure envelope and thus the long-term drained strength is the more critical in problems relating to the stability of excavations and cut slopes.

At the point of failure, the Mohr circle touches the Mohr–Coulomb failure envelope. The stress point (s_f', t_f') in this circle is therefore an alternative parameter of the failure condition. A line drawn through a series of such stress points in failure circles may be defined as the *stress point failure envelope* and utilised as an alternative failure criterion (Fig. 6.25):

$$t_f' = a' + s_f' \tan \alpha' \qquad\qquad [6.46]$$

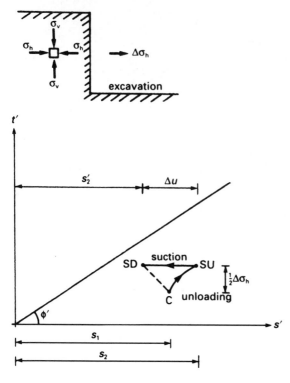

Fig. 6.24 Stress for unloading adjacent to an excavation

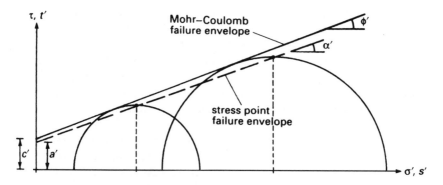

Fig. 6.25 Stress point (t'/s') failure envelope

The parameters of the stress point failure envelope are related to those of the Mohr–Coulomb criterion as follows:

$$\sin \phi' = \tan \alpha' \qquad\qquad\qquad\qquad\qquad\text{[6.47(a)]}$$

$$c' \cos \alpha' = \alpha' \qquad\qquad\qquad\qquad\qquad\text{[6.47(b)]}$$

(d) Stress paths in q'/p' space

While the stress path methods described above are useful in problems involving plane strain, they are somewhat limited in a general sense since they cannot easily represent true triaxial conditions. If the mean stress p' and the deviator stress q' are used instead of s' and t' then plane strain, biaxially symmetrical and true triaxial stress states can be represented with equal facility.

For true triaxial stress ($\sigma_1 \neq \sigma_2 \neq \sigma_3$):

Mean normal stress, $p' = \frac{1}{3}(\sigma'_1 + \sigma'_2 + \sigma'_3)$

and $\qquad\qquad\qquad p = p' + u$

Deviator stress, $\qquad q' = \sigma'_1 - \sigma'_3$

and $\qquad\qquad\qquad q = q'$

For triaxial stresses with biaxial symmetry (e.g. triaxial test):

$p' = \frac{1}{3}(\sigma'_1 + 2\sigma'_3)$ and $p = p' + u$

$q' = \sigma'_1 - \sigma'_3$ and $q = q'$

Figure 6.26 shows a q'/p' plot typical of an undrained triaxial test. The isotropic consolidation stage follows the path $O \rightarrow C$ and at C;

$\sigma_1 = \sigma_2 = \sigma_3$ and $u_o = 0$

giving $p'_o = p_o = \sigma'_3 = \sigma_3$

As σ_1 is increased uniaxially, the total stress path is $C \rightarrow SD$ and has a slope $dq'/dp' = \frac{1}{3}$.

$p' = \frac{1}{3}(\sigma'_1 + 2\sigma'_3) = \frac{1}{3}(\sigma'_1 - \sigma'_3 + 3\sigma'_3)$

$\quad = \frac{1}{3}q' + \sigma'_3$

Differentiating, $\dfrac{dp'}{dq'} = \dfrac{1}{3}$

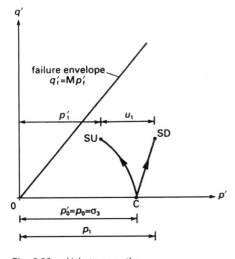

Fig. 6.26 q'/p' stress paths

As the sample is undrained the pore pressure increases during the uniaxial stage from zero to u_1 and the effective stress path will be $C \rightarrow SU$.

An envelope may be defined corresponding to the values of q' and p' at failure:

$$q'_f = Mp'_f \qquad\qquad [6.48]$$

The relationship between M (M) and the angle of friction ϕ' defined by the corresponding Mohr–Coulomb failure envelope can be obtained as follows:

From the Mohr circle (Fig. 6.18), when $c' = 0$:

$$\sin \phi' = \frac{CD}{OC} \frac{\frac{1}{2}(\sigma'_1 - \sigma'_3)}{\frac{1}{2}(\sigma'_1 + \sigma'_3)}$$

Transposing,
$$\frac{\sigma'_3}{\sigma'_1} = \frac{1 - \sin \phi'}{1 + \sin \varphi'}$$

From eqn [6.48],
$$M = \frac{q'}{p'} = \frac{\sigma'_1 - \sigma'_3}{\frac{1}{3}(\sigma'_1 + 2\sigma'_3)}$$

Substituting,
$$M = \frac{3\left(\sigma'_1 - \dfrac{1 - \sin \phi'}{1 + \sin \phi'}\sigma'_1\right)}{\sigma'_1 + \dfrac{2(1 - \sin \phi')\sigma'_1}{1 + \sin \phi'}}$$

$$= \frac{3(1 + \sin \phi' - 1 + \sin \phi')\sigma'_1}{(1 + \sin \phi' + 2 - 2 \sin \phi')\sigma'_1}$$

$$= \frac{6 \sin \phi'}{3 - \sin \phi'}$$

Transposing,
$$\sin \phi' = \frac{3M}{6 + M} \qquad\qquad [6.49]$$

6.7 The yielding of soils

Stress path diagrams can be usefully applied to the study of yielding phenomena in soils. Figure 6.19 illustrates the dramatic changes that follow when a material is strained beyond the yield point. In structural components made of steel, concrete, timber, etc. yielding does not (normally) occur, but in many situations in soils engineering the applied loading can take the stress–strain state past the yield point. In soil masses, yielding can be brought about by different combined stress regimes, e.g. isotropic, triaxial, plane strain, one-dimensional compression, etc. Figure 6.27 shows some typical stress–strain curves which demonstrate yielding behaviour in clay. The value of yield stress (p'_y, q'_y, σ'_y, etc.) at the first yield point Y_1 depends on the combined stress regime under which it occurred, and yet the same soil ought to yield each time at a characteristic *yield criterion* – which is clearly not just a simple stress (or strain) value.

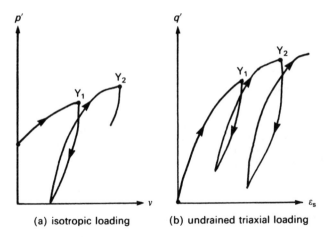

(a) isotropic loading   (b) undrained triaxial loading

Fig. 6.27 Loading–unloading of clay (after Wood (1990) and others)

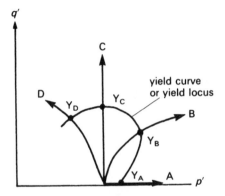

Fig. 6.28 Location of yield curve from yield points

The nature of the yield criterion can be demonstrated by conducting a number of different tests on specimens of the same soil. Suppose the following tests were carried out in triaxial apparatus on specimens of a non-sensitive clay soil, each commencing at the same cell pressure.

(A) isotropic compression: yield point Y_A obtained from $v/\ln p'$ plot
(B) one-dimensional compression: yield point Y_B obtained from $v/\ln p'$ or q'/p' plot
(C) constant mean stress: yield point Y_C obtained from q'/ε_s plot
(D) undrained triaxial compression: yield point Y_D obtained from q'/ε_s plot

Figure 6.28 shows how the four yield points would occur in a q'/p' plot. A *yield curve* (or *yield locus*) can be drawn through the points Y_A, Y_B, Y_C, Y_D (and through other points from different tests). Thus, yield can be represented by a yield curve, which is similar to a failure envelope. Yield curves for subsequent yielding (after unloading and reloading) can be obtained by plotting results from appropriately devised test routines. Finally, a curve can be added passing through the *brittle failure points* (Fig. 6.29). Figure 6.30 shows the generalised set of yield curves in

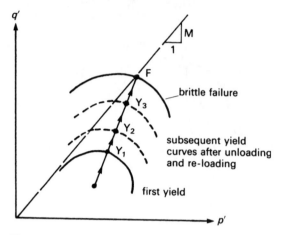

Fig. 6.29 Successive yield curves in q'/p' space

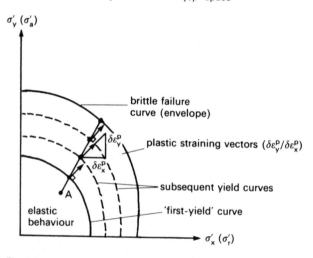

Fig. 6.30 Yield curves in orthogonal stress plane

orthogonal stress space defining the elastic, plastic and brittle behaviour of the soil. A *yield surface* can be constructed by plotting the plastic strain (i.e. intervals between subsequent yield points) along a third axis (Fig. 6.31).

The plastic deformation (or flow) of a material is associated with the yield surface. The vector of plastic straining at a given yield point will be perpendicular to the yield curve, the direction $\delta\varepsilon_y^p/\delta\varepsilon_x^p$ (or $\delta\varepsilon_q^p/\delta\varepsilon_p^p$, etc.); this is referred to as the *plastic strain increment ratio* (or the *plastic dilatancy*). At the same point, the *stress ratio* is $\delta\sigma_y'/\delta\sigma_x'$ (or $\delta q'/\delta p'$, etc.). The relationship between the stress ratio and the plastic strain increment ratio is known as the *flow rule*.

Stress ratio, $\eta = \delta q'/\delta p$ [6.50(a)]

Plastic strain increment ratio, $\tan\beta = \delta\varepsilon_q^p/\delta\varepsilon_p^p$ [6.50(b)]

The flow rule for a soil can be defined with a stress/dilatancy (η/b) plot (Fig. 6.32). Note that for isotropic compression $\eta = 0$ and at the critical state $\eta = M$.

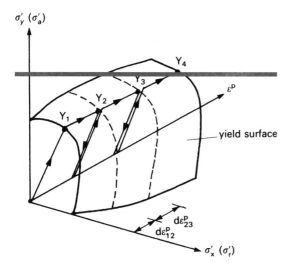

Fig. 6.31 Stress path below and across the yield surface

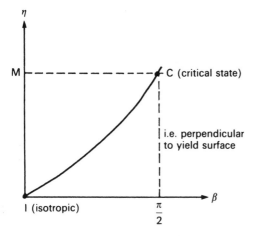

Fig. 6.32 Stress–dilatancy plot illustrating the flow rule

6.8 Critical state theory

When saturated drained soils are loaded they are compressed, causing the volume and water content to decrease; when they are unloaded they swell, with the volume and water content increasing. If undrained the volume remains constant and pore pressures change. The rate of compression or swelling depends on the permeability: in sands it is very quick, whereas in clays it is very slow. Also, as soil is compressed its shear strength increases: the volume change behaviour and strength of soil are related. In the Mohr–Coulomb failure theory volume changes are ignored, and thus it is difficult to model soil behaviours accurately for over-consolidated and undrained soil. And yet there is a unique relationship between applied stress, shear strength and volume – this occurs at the critical state.

The *critical state theory* provides a unified model of soil behaviour in which stress and volume states are interrelated. The concept was first proposed in 1958 by Roscoe, Schofield and Wroth in a paper on the yielding of soils; further work followed, mainly in the University Engineering Department at Cambridge, leading to several subsequent publications: Parry (1960), Roscoe and Burland (1968), Scholfield and Wroth (1968), Atkinson and Bransby (1978), Atkinson (1981) and Muir Wood (1990).

It is proposed that soil will yield, and reach its *critical state strength*, at a critical specific volume and shear stress. Yielding or shear slipping is then considered to be occurring at a point on a *state boundary surface*, which is a three-dimensional 'failure' envelope equivalent to the Mohr–Coulomb envelope (which is two-dimensional).

Critical state line in triaxial compression

Consider a series of six triaxial compression tests on specimens of the same normally consolidated clay in which pairs of specimens are consolidated to the same value of isotropic stress (p') before the axial stress is increased with full drainage up to the yield point. Figure 6.33(b) shows the stress paths for the six tests in q'/p' space.

Consolidation stages: $O \rightarrow C_1, O \rightarrow C_2, O \rightarrow C_3$

Undrained specimens: $O \rightarrow U_1, O \rightarrow U_2, O \rightarrow U_3$

Drained specimens: $O \rightarrow D_1, O \rightarrow D_2, O \rightarrow D_3$

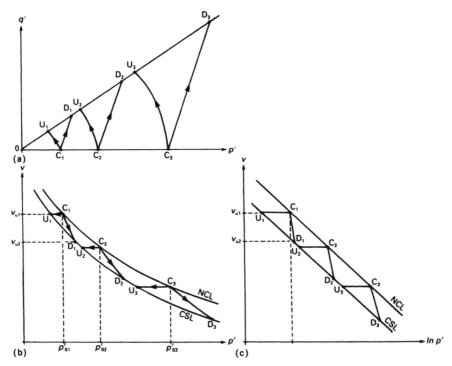

Fig. 6.33 $q'/v/p'$ plots of triaxial test results

At the respective yield points (U or D) the stress paths each terminate on the *same failure envelope* ($q'_f = Mp'_f$). During the uniaxial (shearing) stages of the drained tests a change in volume takes place, whereas in the undrained tests the volume remains constant. For a complete model of the stress–strain behaviour, therefore, changes in volume associated with changes in stress must be incorporated.

During consolidation under isotropic stress (p') the volume change path (O → C) will move along the *normal consolidation line* (NCL). The volume/stress paths are drawn in v/p' space (Fig. 6.33(b)), where v = specific volume (= $1 + e$). The *drained* paths C → D indicate a decrease in volume and the *undrained* paths C → U indicate constant volume. The curve passing through points U_1, D_1, U_2, D_2, U_3 and D_3 represents the failure criterion in v/p' space.

Thus, Figs. 6.33(a) and 6.33(b) are respectively an elevation and a plan of a three-dimensional failure criterion line in $q'/v/p'$ space; this is called the **critical state line (CSL)**. The critical state line is a curve drawn on a three-dimensional *state boundary surface* which represents the yielding of soil, i.e. it is the boundary between elastic and plastic behaviour. For convenience in mathematical interpretations the plan view of the CSL is often shown as $v/\ln p'$ (Fig. 6.33 (c)).

The critical state model was developed using remoulded saturated clays, but it may be assumed sufficiently representative of naturally occurring soils to provide a generalised model of behaviour. The defining equations and other relevant relationships must now be established. The CSL is shown in a three-dimensional projection in Fig. 6.34.

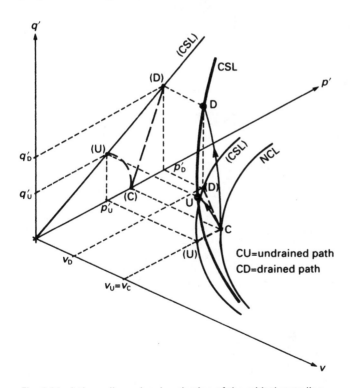

Fig. 6.34 A three-dimensional projection of the critical state line

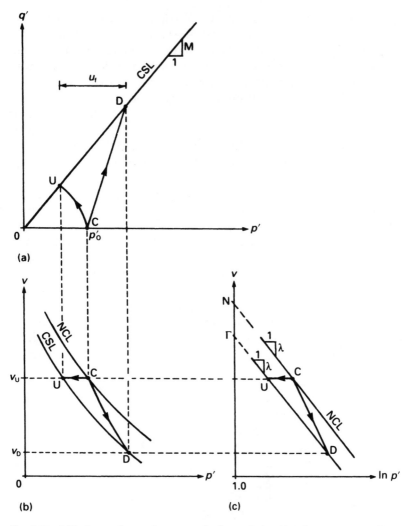

Fig. 6.35 Critical state line and stress paths for undrained loading on a normally consolidated clay

Defining the critical state line (CSL)

Deviator stress, $q' = \sigma'_1 - \sigma'_3$ [or in triaxial test $= \sigma'_a - \sigma'_r$]

Mean normal stress, $p' = \frac{1}{3}(\sigma'_1 + \sigma'_2 + \sigma'_3)$ [or in triaxial test $= \frac{1}{3}(\sigma'_a - 2\sigma'_r)$]

Specific volume, $v = 1 + e$

Equations of the CSL (Fig. 6.35):

$q' = Mp'$ [6.51(a)]

$v = \Gamma - \lambda \ln p'$ [6.51(b)]

where M = slope of the CSL in q'/p' plane
        Γ = the specific volume (v) at $p' = 1.0$ kPa
        λ = slope of the CSL in the $v/\ln p'$ plane

Rearranging eqn [6.51(b)]:

$$p' = \exp\left(\frac{\Gamma - v}{\lambda}\right)$$

[6.52]

$$\text{Then,} \quad q' = Mp' = M \exp\left(\frac{\Gamma - v}{\lambda}\right)$$

[6.53]

The equations of the compression lines are (eqns [6.25(a) and (b)] and Fig. 6.11(b)):

NCL: $v = N - \lambda \ln p'$

SRL: $v = v_\kappa - \kappa \ln p'$

The state boundary surface
Referring again to Fig. 6.33, the families of stress paths C → U and C → D are seen to have similar shapes. These paths in fact traverse a three-dimensional surface whose boundaries are the CSL and the NCL. This is clearly part of the state boundary surface and is called the *Roscoe surface* (after the late Professor K. H. Roscoe). The position of the stress path on the Roscoe surface is determined by the consolidation pressure (p'_0).

The Roscoe surface is applicable to normally consolidated soil where the stress path commences on the NCL, but for *lightly overconsolidated* soil, the stress path commences on the SRL at a point (L) located between the NCL and the CSL (Fig. 6.36). The specific volume at (L) is greater than critical and the water content wetter. The undrained stress path (volume constant) is L → U and for drained loading (volume decreasing) the path is L → D.

A *heavily overconsolidated* soil will have been consolidated to a (p', v) point (H) on the SRL *below* the CSL (Fig. 6.37). Under undrained loading (volume constant), the stress path is H → UH, where UH is a point above the CSL on the q'/p' elevation. After yielding, the stress path will continue with further straining along a straight line (TS) to meet the CSL at S. The critical state is only likely to be reached in part of the soil adjacent to slip surfaces that may develop. The greater the degree of overconsolidation, the greater is the strain required to bring the soil to its critical state.

Under drained loading heavily overconsolidated soil will expand and the volume will continue to increase after yielding. The stress path will be H → DH, where DH is a failure point also on the line TS. After yielding, the increase in volume causes the stresses to fall back to residual value RH which may be on or below the CSL. The soil adjacent to slip planes will be affected to a much greater degree and will thus become weaker. The line TS therefore represents that part of the state boundary surface which governs the yielding of heavily

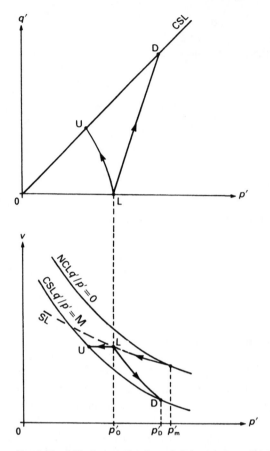

Fig. 6.36 Critical state plot for a lightly overconsolidated soil

overconsolidated soils and is called the *Hvorslev surface* (after Professor M. J. Hvorslev).

The third part of the state boundary surface lies between origin 0 and T in the q'/p' plane. This is called the *no tension cut-off* and represents the condition of zero tensile stress ($\sigma'_3 = 0$), which is an assumed limit for soils. Figure 6.38 shows a q'/p' constant volume section of the complete state boundary surface, the defining equations of which are:

No-tension cut-off (OT): $q' = 3p'$ [6.54]

Hvorslev surface (TS): $q' = Hp' + (M - H) \exp \dfrac{\Gamma - v}{\lambda}$ [6.55]

Roscoe surface (SC): $q' = Mp' \left[1 + \dfrac{\Gamma - v - \lambda \ln p'}{\lambda - \kappa} \right]$ [6.56]

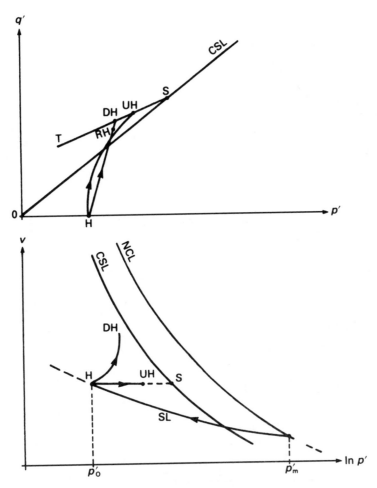

Fig. 6.37 Critical state plot for a heavily overconsolidated soil

Equations for plane strain states

When examining plane strain stress states it is convenient to redefine the appropriate parameters as follows (Fig. 6.39):

CSL:
$$t'_f = s' \sin (\phi'_c) \qquad [6.57]$$

$$v_f = \Gamma_{ps} - \lambda \ln s'_f \qquad [6.58]$$

NCL:
$$v = N_{ps} - \lambda \ln s' \qquad [6.59]$$

SRL:
$$v = v_\kappa - \kappa \ln s' \qquad [6.60]$$

Hvorslev surface:
$$t' = s' \sin \phi'_h + (\sin \phi'_c - \sin \phi'_h) \exp \frac{\Gamma_{ps} - v}{\lambda} \qquad [6.61]$$

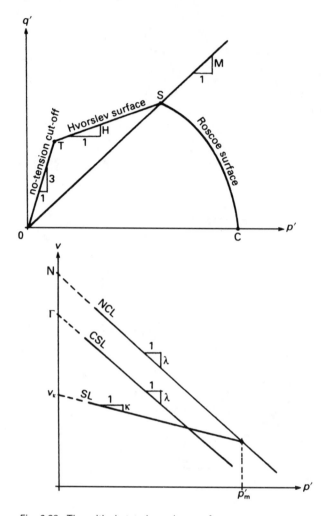

Fig. 6.38 The critical state boundary surface

Roscoe surface: $t' = s' \sin \phi'_c \left[1 + \dfrac{\Gamma_{ps} - v - \lambda \ln s'}{\lambda - \kappa} \right]$ [6.62]

A generalised three-dimensional view of the complete state boundary surface is shown in Fig. 6.40. SS is the critical state line, NN is the normal consolidation line and the three component surfaces are:

VVTT = no tension cut-off

TTSS = Hvorslev surface

SSNN = Roscoe surface

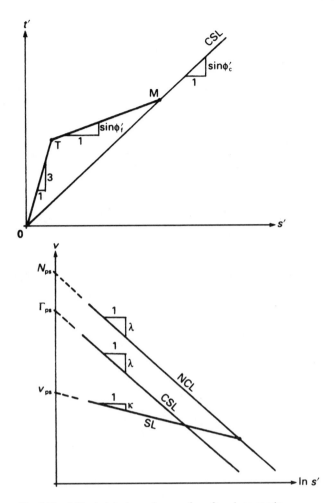

Fig. 6.39 Critical state boundary surface for plane strain

Normalizing and equivalent stresses

The stress paths for a normally consolidated soil will lie on the Roscoe surface, whereas stress paths for overconsolidated soil lie below it and progressively further away as the degree of consolidation increases. Each curve has the same shape, but is a different size, depending on the degree of overconsolidation at the start of a test. In order to obtain easily comparable and non-dimensional plots adjusted for stress history, test results can be *normalised* by dividing by an equivalent stress value that occurs on either the CSL or the NCL. The CSL may be preferable, since it is unique for a given soil.

The normalising parameters for different types of test are shown in Fig. 6.41. Point T represents a current stress-strain state. For direct shear or one-dimensional compression, the normalising parameters are either σ_c' (the equivalent stress on the CSL) or σ_e' (the equivalent stress on the NCL). For triaxial test

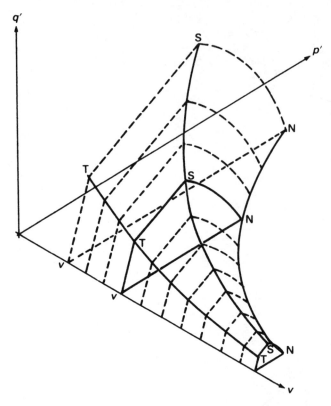

Fig. 6.40   A three-dimensional view of the complete state boundary surface

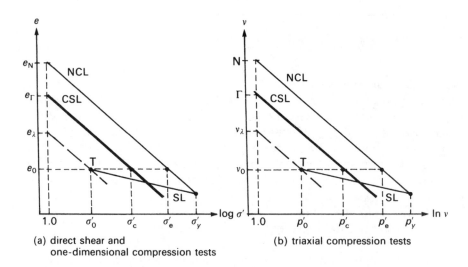

(a) direct shear and
    one-dimensional compression tests

(b) triaxial compression tests

Fig. 6.41   Parameters for normalising test results

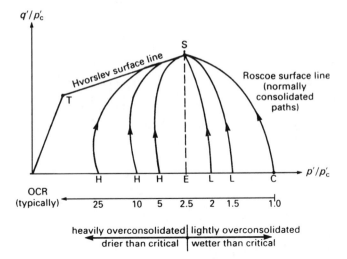

Fig. 6.42   Effect of OCR on undrained stress paths

results plotted in $q'/\ln p'/v$ space, the corresponding parameters are p'_c and p'_e. Figure 6.42 shows a normalised plot, i.e. q'/p'_c against p'/p'_c, representing a constant volume section of the state boundary surface. For a normally consolidated soil ($p'_0 = p'_c$) the stress paths traverse the Roscoe surface to reach the CSL at S. With overconsolidation ($p'_0 < p'_c$), so the stress paths start between E and C. Lightly overconsolidated soils are less dense and wetter than critical and their stress paths (L → S) reach the CSL from below. Heavily overconsolidated soils are more dense and drier than critical and their stress paths commence between 0 and E, before curving slightly in the opposite direction as they rise toward the Hvorslev surface. They then follow the Hvorslev surface if straining continues undrained or fall back slightly when drained.

It is important to recognise the three significant stress states in the case of heavily overconsolidated soils. The *peak stress* shear is reached when the stress path reaches the Hvorslev surface, whereas the *critical stress* occurs at the CSL. After large strains, especially along slip surfaces, the yield stress state will fall back to a lower *residual* value. This concept is an essential consideration in the interpretation of shear tests and in the application of measured parameters in design problems. For numerical examples and interpretative examples see Chapter 7.

6.9   Contact pressure

Contact pressure is the intensity of loading transmitted from the underside of a foundation to the soil. The distribution of contact pressure depends on both the rigidity of the footing and on the stiffness of the foundation soil.

Consider a footing carrying concentrated column loads such as that shown in Fig. 6.43. When supported on hard soil or rock, which has a high stiffness

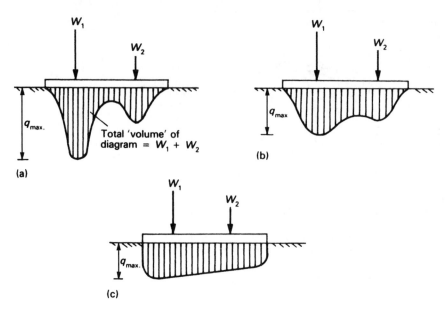

Fig. 6.43   Effect of soil type on contact pressure
(a) Rock (b) Stiff soil (c) Soft soil

modulus, the load is transmitted to a relatively small area, since a high intensity of stress can develop (Fig. 6.43(a)). On a less stiff foundation medium the loading is distributed laterally, producing lower values of contact pressure, until in a soft soil it may be almost uniform (Figs. 6.43(b) and 6.43(c)).

When there is a relatively thick layer of compressible soil beneath a footing the settlement profile tends to become dish-shaped. A uniformly loaded footing of perfect flexibility will theoretically transmit a uniform contact pressure in order to produce this dish shape (Fig. 6.44(a)). A perfectly rigid footing will settle uniformly across its breadth. Thus, in a compressible soil a rigid footing, in settling uniformly, will transmit a higher contact pressure near the edges. Extremely high edge stresses cannot occur, however, since as the soil yields some of the load is transferred inward, producing a distribution such as that shown in Fig. 6.44(b).

In sands, the contact pressure near the edge will be lower (tending to zero under shallow footings) and that under the centre will be higher because of the higher confining pressure (Fig. 6.44(c)). Under deeper footings the confining pressure and therefore the contact pressure is more uniform.

Most footings are neither perfectly flexible, nor perfectly rigid, so that actual distributions are somewhere between these extremes. In designing wide rein-forced concrete foundations a sensible distribution of contact pressure should be sought, bearing in mind the stiffness of both the footing and the foundation soil. It should also be remembered that the contact pressure distribution will also affect the bending moments in the footing. For the purpose of calculating stresses and displacements within the soil mass sufficient accuracy may be obtained by assuming a uniform distribution of contact pressure.

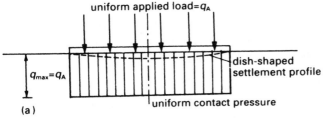

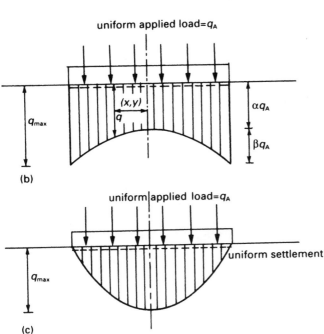

Fig. 6.44 Effect of footing rigidity of contact pressure
(a) Flexible footing (b) Rigid footing on cohesive soil (c) Rigid footing on cohesionless soil

6.10   Stresses in a soil mass due to applied loading

Soil masses in practical situations mostly have a horizontal surface from which the depth (z) is measured downwards, with x and y being lateral dimensions. The magnitudes of x, y and z are usually large compared with the size of the structure. The conceptual model of the soil mass is therefore a semi-infinite elastic half-space, the boundaries of which are sufficiently remote not to affect the analysis. The soil is assumed to be homogeneous and isotropically elastic, thus a straightforward application of elastic theory is the normal approach. Of course soils do not comply with such ideal conditions, but providing the stresses stay well below the yield point, the magnitude of errors will be small, and well within natural variations of other properties.

In 1885 Boussinesq published solutions for the stresses beneath a point load applied at the surface of such a soil mass. Subsequently, many other solutions have been developed for both stresses and displacements relating to different types of loading, layers of finite thickness, multi-layered masses and internally loaded masses (Ahlvin and Ulery, 1962; Giroud, 1970; Newmark, 1942; Poulos and Davis, 1974; and others). Some of the more commonly used expressions are given below.

(a)   Stresses due to a vertical point load
The original work of Boussinesq related to a point load. Using polar coordinates (r, θ, z), the increases in stress at a given point (Fig. 6.45) due to a point load (P) at the surface are:

$$\Delta\sigma_z = \frac{3Pz^3}{2\pi R^5}$$

$$\Delta\sigma_r = \frac{P}{2\pi R^2}\left(\frac{3zr^2}{R^3} - \frac{R(1-2v')}{R+z}\right)$$

$$\Delta\sigma_\theta = \frac{P}{2\pi R^2}(2v'-1)\left(\frac{z}{R} - \frac{R}{R+z}\right) \qquad\qquad [6.63]$$

$$\Delta\tau_{rz} = \frac{3Pz^2r}{2\pi R^5} = \Delta\sigma_z\frac{r}{z}$$

$$\Delta\tau_{\theta z} = \Delta\tau_{\theta r} = 0$$

where v' = Poisson's ratio referred to effective stress.

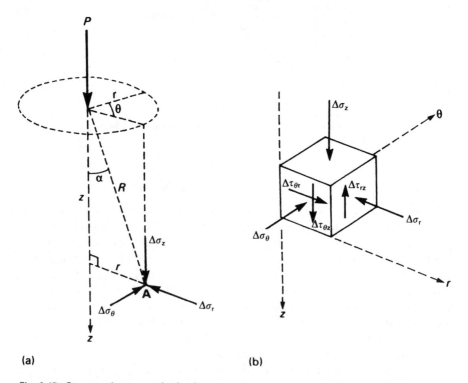

(a)   (b)

Fig. 6.45 Stresses due to a point load

In undrained conditions, $v_u = 0.5$, and so $\Delta\sigma_\theta = 0$ and $\Delta\sigma_r = \Delta\sigma_z(r^2/z^2)$. The increase in vertical stress can be written:

$$\Delta\sigma_z = \frac{P}{z^2} \cdot I_p \tag{6.64}$$

where I_p = the point load *influence* factor

$$= \frac{3}{2\pi}\left(\frac{z}{R}\right)^5 = \frac{3}{2\pi}\left(\frac{1}{1 + (r/z)^2}\right)^{5/2} \tag{6.65}$$

Values of I_p can be tabulated against r/z for use in practical calculations, but it is quicker to use a computer spreadsheet, or even a hand-held calculator. Figure 6.46 shows the variation of $\Delta\sigma_z$ vertically with depth and horizontally offset from the load point. The point load expression can be used in problems

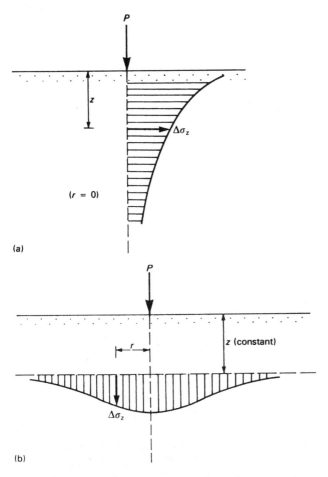

Fig. 6.46 Variation of stress due to a point load
(a) Variation with depth (z) (b) Variation with radial offset (r)

where the contact pressure is non-uniform. The loaded area is first simplified to a set of small rectangular areas, each transmitting an equivalent point load at its centre. Applying the principle of superposition, the stresses at the point are calculated by summation – see worked example 6.1.

(b) Stresses due to a long line load
A line load is assumed to have length, but not breadth, and to be uniformly loaded along the length. Then at point A in Fig. 6.47:

$$\Delta\sigma_z = \frac{2Q}{\pi} \cdot \frac{z^3}{(x^2 + z^2)^2}$$

$$\Delta\sigma_x = \frac{2Q}{\pi} \cdot \frac{x^2 z}{(x^2 + z^2)^2}$$

$$\Delta\tau_{xz} = \frac{2Q}{\pi} \cdot \frac{xz^2}{(x^2 + z^2)^2} \qquad\qquad [6.66]$$

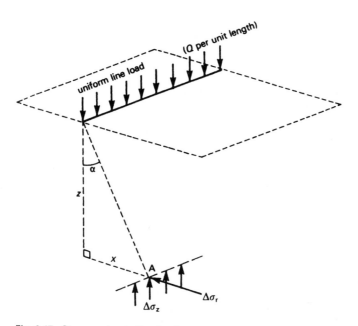

Fig. 6.47 Stresses due to line load

Equation [6.66] can be used to determine the increase in horizontal stress on earth-retaining structures, such as sheet-piles and concrete walls. It is assumed that the wall will be rigid and that there are (in effect) two line loads, one on each side of the wall at equal distance (Fig. 6.48). The increases in horizontal stress may then be determined from:

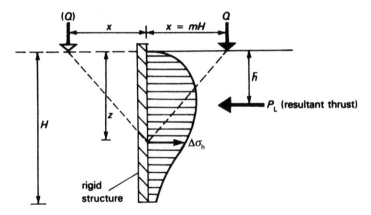

Fig. 6.48 Horizontal thrust on a rigid structure due to a line load

$$\Delta\sigma_h = 2\Delta\sigma_x = \frac{4Q}{\pi} \cdot \frac{x^2 z}{(x^2 + z^2)^2} \qquad [6.67]$$

or, putting $x = mH$ and $z = nH$ (where H = height of the wall):

$$\Delta\sigma_h = \frac{4Q}{\pi H} \cdot \frac{m^2 n}{m^2 + n^2} \qquad [6.68]$$

The resultant horizontal thrust force can be found by integrating this expression over the height (H) of the wall:

$$P_L = \int_0^H \Delta\sigma_h dz = \frac{4Q}{\pi} \int_0^H \frac{x^2 z}{(x^2 + z^2)^2} dz$$

from which $\qquad P_L = \frac{2Q}{\pi} \cdot \frac{H^2}{x^2 + H^2} \qquad [6.68(a)]$

or $\qquad\qquad P_L = \frac{2Q}{\pi} \cdot \frac{1}{m^2 + 1} \qquad [6.68(b)]$

The position of P_L (depth below top of wall) can be obtained by taking area moments.

(c) Stresses due to a uniform strip load

The length of a *strip footing* is very long compared with its breadth. Most usually, strip footings are uniformly loaded, or at least approximately so. The problem is therefore two-dimensional. The orthogonal stresses at a point (z, x) from the centreline (Fig. 6.49) can be expressed in the following simplified form:

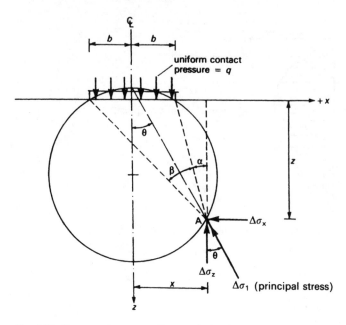

Fig. 6.49 Stresses due to a uniform strip load

$$\Delta\sigma_z = \frac{q}{\pi}[\beta + \sin\beta\cos(2\alpha + \beta)]$$

$$\Delta\sigma_x = \frac{q}{\pi}[\beta - \sin\beta\cos(2\alpha + \beta)]$$

$$\Delta\tau_{xz} = \frac{q}{\pi}[\sin\beta\cos\beta(2\alpha + \beta)] \qquad\qquad [6.69]$$

The values of angles α and β can be determined from the cross-sectional dimensions by simple trigonometry – see worked example 6.2.

(d)   Stresses due to a triangular strip load
A 'triangular' load occurs when the contact pressure varies linearly across the breadth of the strip from zero to a maximum value (Fig. 6.50), e.g. below the sloping sides of an embankment. The orthogonal stresses are:

$$\Delta\sigma_z = \frac{q}{\pi}\left[\frac{x}{c}\beta - \tfrac{1}{2}\sin 2\alpha\right]$$

$$\Delta\sigma_x = \frac{q}{\pi}\left[\frac{x}{c}\beta + \tfrac{1}{2}\sin 2\beta - \frac{z}{c}\ln\left(\frac{x^2 - z^2}{x^2 - c^2 - z^2}\right)\right]$$

$$\Delta\tau_{xz} = \frac{q}{\pi}\left[1 + \cos 2\beta - \frac{2z}{c}\alpha\right] \qquad\qquad [6.70]$$

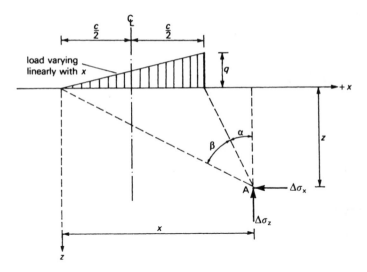

Fig. 6.50 Stresses due to a triangular load

The values of angles α and β can be determined from the cross-sectional dimensions by simple trigonometry – see worked example 6.3.

(e) Stresses due to a uniformly loaded circular area
For uniformly loaded *areas*, e.g. raft foundations or tank bases, the Boussinesq expressions are integrated over the area. For example, in the case of vertical stress below the centre of a circular foundation of radius a (Fig. 6.51):

The load on the small element $= q \times r \mathrm{d}\theta \mathrm{d}r$

Then, using eqns [6.64] and [6.65] and integrating over the circular area:

$$\Delta\sigma_z = \int_0^{2\pi}\int_0^a \frac{q r \mathrm{d}\theta \mathrm{d}r}{z^2} \frac{3}{2\pi}\left(\frac{1}{1+(r/z)^2}\right)^{5/2}$$

Which solves to:

$$\Delta\sigma_z = q\left\{1 - \left[\frac{1}{1+(a/z)^2}\right]^{3/2}\right\}$$ [6.71]

Unfortunately, an analytical solution for the general vertical stress increase, i.e. at point (r, z), has not yet been found. Some solutions have been demonstrated using numerical methods (e.g. Ahlvin and Ulery, 1962), and from these it is possible to obtain values of reasonable accuracy.

$$\Delta\sigma_z = q(A + B)$$
$$\varepsilon_z = (1 + v')[(1 - 2v')A + B]$$ [6.72]

where A and B are partial influence factors as given in Table 6.2.

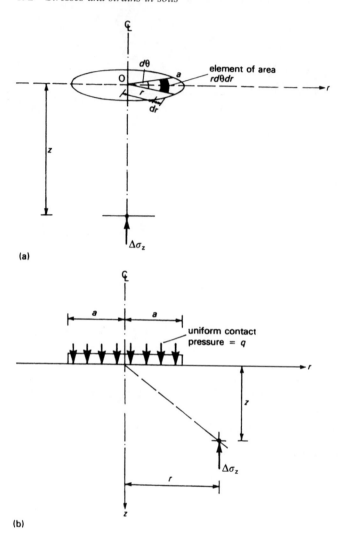

Fig. 6.51 Vertical stress due to a uniformly loaded circular area
(a) Stress beneath centre of circle (b) General vertical stress case

(f) Stresses due to a uniformly loaded rectangular area

This application is perhaps the most widely used in soil engineering design. Expressions for the component stresses can be obtained by integrating the Boussinesq expressions. Several types of solution have been proposed involving formulae, tables and charts. The principal expression is for the increase in vertical stress beneath one corner of a flexible rectangular area carrying a uniform load (q):

$$\Delta\sigma_z = qI_R \qquad\qquad\qquad\qquad\qquad\qquad\qquad\qquad\qquad \text{[6.73(a)]}$$

where I_R is an influence factor dependent on the length (L) and breadth (B) of the loaded area, and the depth (z) of the point at which the stress is required. It is usual to express I_R in term of the parameters $m = B/z$ and $n = L/z$.

Table 6.2 Influence factors (A and B) for vertical stress due to a uniformly loaded circular area

z/a	r/a 0	0.2	0.4	0.6	0.8	1.0	1.2	1.5	2.0	3.0
0	1.0	1.0	1.0	1.0	1.0	0.5	0.0	0.0	0.0	0.0
	0.0	0.0	0.0	0.0	0.0	0.0	0.0	0.0	0.0	0.0
0.2	0.804	0.798	0.779	0.735	0.630	0.383	0.154	0.053	0.017	0.004
	0.188	0.193	0.208	0.235	0.260	0.085	-0.078	-0.044	-0.016	-0.004
0.4	0.629	0.620	0.592	0.538	0.443	0.310	0.187	0.086	0.031	0.008
	0.320	0.323	0.327	0.323	0.269	0.124	-0.008	-0.045	-0.025	-0.008
0.6	0.486	0.477	0.451	0.404	0.337	0.256	0.180	0.100	0.041	0.011
	0.378	0.375	0.363	0.382	0.254	0.144	0.045	-0.021	-0.025	-0.010
0.8	0.375	0.368	0.347	0.312	0.266	0.213	0.162	0.102	0.048	0.014
	0.381	0.374	0.351	0.307	0.238	0.153	0.075	0.006	-0.018	-0.010
1.0	0.293	0.288	0.270	0.247	0.215	0.179	0.143	0.098	0.052	0.017
	0.353	0.346	0.321	0.278	0.220	0.154	0.092	0.028	-0.010	-0.011
1.2	0.232	0.228	0.217	0.199	0.176	0.151	0.126	0.092	0.053	0.019
	0.315	0.307	0.285	0.248	0.201	0.149	0.100	0.044	0.000	-0.010
1.5	0.168	0.166	0.159	0.148	0.134	0.119	0.103	0.080	0.051	0.021
	0.256	0.250	0.233	0.207	0.174	0.137	0.102	0.057	0.014	-0.007
2.0	0.106	0.104	0.101	0.096	0.090	0.083	0.075	0.063	0.045	0.022
	0.179	0.181	0.166	0.152	0.134	0.113	0.093	0.064	0.028	0.000
3.0	0.051	0.051	0.050	0.049	0.047	0.045	0.042	0.038	0.032	0.020
	0.095	0.094	0.091	0.086	0.080	0.073	0.066	0.054	0.035	0.011
4.0	0.030	0.030	0.029	0.028	0.028	0.027	0.026	0.025	0.022	0.016
	0.057	0.057	0.056	0.054	0.051	0.048	0.045	0.040	0.031	0.015
5.0	0.019	0.019	0.019	0.019	0.019	0.018	0.018	0.018	0.016	0.012
	0.038	0.038	0.037	0.036	0.035	0.034	0.031	0.028	0.025	0.015
10.0	0.005	0.005	0.005	0.005	0.005	0.005	0.005	0.005	0.004	0.004
	0.010	0.009	0.009	0.009	0.009	0.009	0.009	0.009	0.008	0.008

Top line = A; bottom line = B $\Delta\sigma_z = q(A + B)$ (see Fig. 6.51) $\Delta\varepsilon_z = q(1 + v)[(1 - 2v)A + B]/E$

$$I_R = \frac{1}{4\pi}\left[\frac{2mn\sqrt{(m^2 + n^2 + 1)}}{m^2 + n^2 + m^2n^2 + 1}\left(\frac{m^2 + n^2 + 2}{m^2 + n^2 + 1}\right)\right.$$

$$\left. + \tan^{-1}\left(\frac{2mn\sqrt{(m^2 + n^2 + 1)}}{m^2 + n^2 + m^2n^2 + 1}\right)\right] \qquad [6.73(b)]$$

Table 6.3 gives values of I_R with respect to m and n, but eqns [6.73] are easily incorporated into a spreadsheet for easier application (Whitlow, 2001, CD). Fadum (1948) plotted a set of curves for I_R against m and n (Fig. 6.52); these are much used in textbooks and examinations.

Any foundation which has a rectilinear plan may be considered as a series of rectangles, each with a corner coincident with the point beneath which the increase in stress is required; the value of stress increase at the point is then found using the principle of superposition – see worked example 6.5.

Table 6.3 Influence factors (I_R) for vertical stress under one corner of a uniformly loaded rectangular area

B/z	L/z 0.1	0.2	0.3	0.4	0.5	0.6	0.7	0.8	0.9	1.0	1.4	2.0	3.0	5.0	∞
0.1	0.0047	0.0092	0.0132	0.0168	0.0198	0.0222	0.0242	0.0258	0.0270	0.0279	0.0301	0.0311	0.0315	0.0316	0.0316
0.2	0.0092	0.0179	0.0259	0.0328	0.0387	0.0435	0.0474	0.0504	0.0528	0.0547	0.0589	0.0610	0.0620	0.0620	0.0620
0.3	0.0132	0.0259	0.0374	0.0474	0.0560	0.0630	0.0686	0.0731	0.0766	0.0794	0.0856	0.0887	0.0898	0.0901	0.0902
0.4	0.0168	0.0328	0.0474	0.0602	0.0711	0.0801	0.0873	0.0931	0.0977	0.1013	0.1094	0.1134	0.1150	0.1154	0.1154
0.5	0.0198	0.0387	0.0560	0.0711	0.0840	0.0947	0.1034	0.1104	0.1158	0.1202	0.1300	0.1350	0.1368	0.1374	0.1375
0.6	0.0222	0.0435	0.0629	0.0801	0.0947	0.1069	0.1168	0.1247	0.1310	0.1361	0.1475	0.1533	0.1555	0.1561	0.1562
0.7	0.0240	0.0474	0.0686	0.0873	0.1034	0.1168	0.1277	0.1365	0.1436	0.1491	0.1620	0.1686	0.1711	0.1719	0.1720
0.8	0.0258	0.0504	0.0731	0.0931	0.1104	0.1247	0.1365	0.1461	0.1537	0.1598	0.1739	0.1812	0.1841	0.1849	0.1850
0.9	0.0270	0.0528	0.0766	0.0977	0.1158	0.1311	0.1436	0.1537	0.1619	0.1684	0.1836	0.1915	0.1947	0.1956	0.1958
1.0	0.0279	0.0547	0.0794	0.1013	0.1202	0.1361	0.1491	0.1598	0.1684	0.1752	0.1914	0.1999	0.2034	0.2044	0.2046
1.4	0.0301	0.0589	0.0856	0.1094	0.1300	0.1475	0.1620	0.1739	0.1836	0.1914	0.2102	0.2206	0.2250	0.2263	0.2266
2.0	0.0311	0.0610	0.0887	0.1134	0.1350	0.1533	0.1686	0.1812	0.1915	0.1999	0.2206	0.2325	0.2378	0.2395	0.2399
3.0	0.0315	0.0618	0.0898	0.1150	0.1368	0.1555	0.1711	0.1841	0.1947	0.2034	0.2250	0.2378	0.2420	0.2461	0.2465
5.0	0.0316	0.0620	0.0901	0.1154	0.1374	0.1561	0.1719	0.1849	0.1956	0.2044	0.2263	0.2395	0.2461	0.2486	0.2491
∞	0.0316	0.0620	0.0902	0.1154	0.1375	0.1562	0.1720	0.1850	0.1958	0.2046	0.2266	0.2399	0.2465	0.2492	0.2500

$\Delta\sigma_z = qI_R$ (See Fig. 6.52)

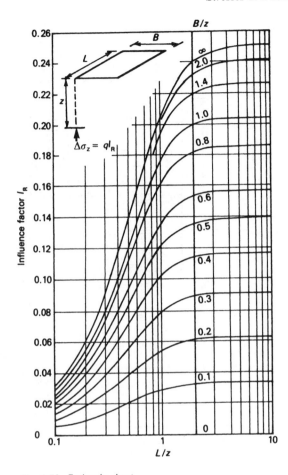

Fig. 6.52 Fadum's chart

Worked examples of stresses due to applied loads

Worked example 6.1 *The footing shown in Fig. 6.53(a) carries an inclusive uniform loading of 250 kPa over the shaded area and 150 kPa over the unshaded area. Determine the increase in vertical direct and shear stresses at a point 3.0 m below the corner A.*

A solution can be obtained using point-load influence factors. First, the footing is divided into squares of 1 m side length and the uniform loading resolved into a series of point loads occurring at centres of the squares (Fig. 6.53(b)).

Point load on each shaded square   = 250 kN
Point load on each unshaded square = 150 kN

The calculations are tabulated below. The x,y,z coordinates are calculated for each square centre from a common origin at point A.

z = depth $R^2 = x^2 + y^2 + z^2$ I_p is obtained from eqn [6.65]

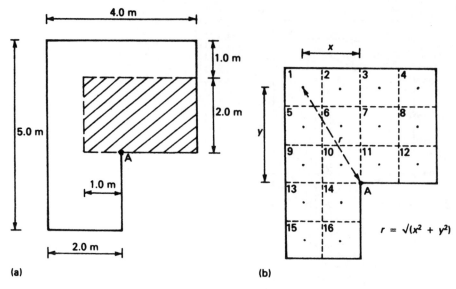

Fig. 6.53

Squares	No.	Coordinates			Depth	Radial	Influence	Load	Vertical	Shear
		x (m)	y (m)	z (m)	z (m)	R	factor I_p	P (kN)	stress (kPa)	stress (kPa)
1, 4	2	1.5	2.5	3.0		4.1833	0.0906	150	3.02	2.93
2, 3	2	0.5	2.5	3.0		3.9370	0.1227	150	4.09	3.47
5, 15	2	1.5	1.5	3.0		3.6742	0.1733	150	5.78	4.08
9, 13, 16	3	0.5	1.5	3.0		3.3912	0.2587	150	12.94	6.82
14	1	0.5	0.5	3.0		3.0822	0.4171	150	6.95	1.64
8	1	1.5	1.5	3.0		3.6742	0.1733	250	4.81	3.40
6, 7, 12	3	0.5	1.5	3.0		3.3912	0.2587	250	21.56	11.36
10, 11	2	0.5	0.5	3.0		3.0832	0.4171	250	23.17	5.46

The summations of the last two columns give the increase in stress required:

Increase in vertical stress, $\qquad \Delta\sigma_z = \underline{82.32 \text{ kPa}}$

Increase in vertical shear stress, $\Delta\tau_{xz} = \underline{39.16 \text{ kPa}}$

Worked example 6.2 *A continuous strip footing of breadth 4.3 m carries a uniform load of 100 kPa. (a) Plot the distribution of vertical stress occurring on a horizontal plane at a depth of 3.0 m below the footing, and (b) compare this distribution with that obtained when a 30° 'load spread' is assumed and comment on any 'errors'.*

(a) Since the loading is uniform, the vertical stress distribution will be symmetrical about the centre-line of the footing. The vertical stress induced at various offset distances is tabulated below and plotted in Fig. 6.54 using eqn. [6.69].

Depth (m)	Offset (m)	Angle α (rad)	Angle β (rad)	Influence factor, I_s	Vertical stress (kPa)
3.0	0	−0.622	1.244	0.697	69.7
	0.5	−0.503	1.226	0.683	68.3
	1.0	−0.366	1.176	0.640	64.0
	1.5	−0.213	1.096	0.571	57.1
	2.0	−0.050	0.995	0.484	48.4
	2.15	0.000	0.962	0.455	45.5
	2.5	0.116	0.882	0.389	38.9
	3.0	0.276	0.767	0.299	29.9
	4.0	0.553	0.564	0.163	16.3

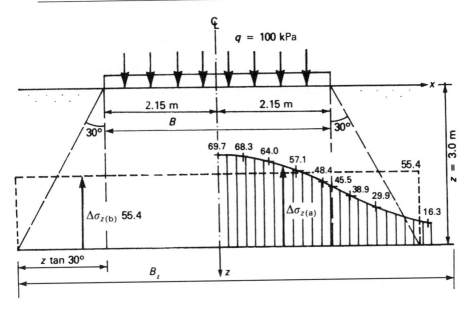

Fig. 6.54

(b) The 30° 'load spread' method is commonly adopted as a means of estimating vertical stress beneath a footing; it does, however, produce substantial errors. The assumption is that the same load over breadth B at the base of the footing is distributed uniformly over breadth B_z at depth z below the base.

Where $\qquad\qquad B_z = B + 2z \tan 30°$

For this footing: $\quad B_z = 4.3 + 2 \times 3.0 \tan 30° = 7.76$ m

Then $\qquad\qquad \Delta\sigma_z = 100 \times 4.3/7.76 = \underline{55.4 \text{ kPa}}$

This distribution is represented by the broken line in Fig. 6.54.

Assuming the plot derived in part (a) to be correct, the errors incurred when assuming a 30° load spread are:

At the centre: $\qquad\qquad\qquad\quad \zeta = (55.4 - 69.7)100/69.7 = -20.4$ per cent (underestimate)
At edge of footing: $\qquad\qquad\quad \zeta = (55.4 - 45.5)100/45.5 = 21.8$ per cent (overestimate)
At the limit of load spread: $\quad \zeta = (55.4 - 17.6)100/17.6 = 214$ per cent (overestimate)

If the vertical stress increase at a specific point is required the 'load spread' method is best avoided. The average increase in vertical stress can be estimated from the area under the curve in Fig. 6.54; in this case a value of 47.5 kPa is obtained. Thus, the 'load spread' method may be used in problems where an average increase in stress is all that is needed, e.g. in settlement calculations. However, the choice of angle of spread is crucial. Many engineers use a load spread ratio of 1 horizontal:2 vertical (26.6°); this is fairly conservative, producing a higher average stress increase than the Boussinesq method, but the calculation is simpler. A load spread angle of <35° should not be used.

Worked example 6.3   *In Fig. 6.55 the cross-section of a proposed embankment is shown. Calculate the increase in vertical stress that will follow completion of the embankment at a depth of 4.0 m at points A and B as shown. Assume an average rolled unit weight of 20 kN/m³ for the placed soil.*

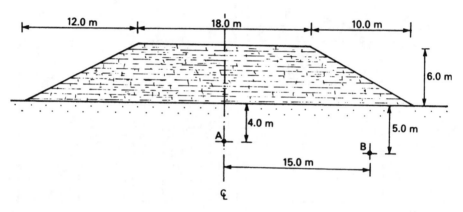

Fig. 6.55

The intensity of uniform load at the base of the central section:

$q = 20 \times 6 = 120$ kPa

The equations to be used: [6.69] for the central portion and [6.70] for the side slopes.

Point A: $z = 4.00$ m   *Offset from ₵ = 0*

(Refer to Figs 6.49 and 6.50)	*Centre*	*Left slope*	*Right slope*
x	0	21.0	19.0
Angle α (rad)	−1.153	1.153	1.153
Angle β (rad)	2.305	0.230	0.211
Influence factor	0.970	0.010	0.009

Vertical stress at point A = 120(0.970 + 0.010 + 0.009) = <u>118.7 kPa</u>

Point B: $z = 5.00$ m   *Offset from* $\mathbb{C} = 15.0$ m

(Refer to Figs 6.49 and 6.50)	Centre	Left slope	Right slope
x	15.0	36.0	4.0
Angle α (rad)	0.876	1.365	−0.876
Angle β (rad)	0.489	0.067	1.551
Influence factor	0.063	0.001	0.354

Vertical stress at point A $= 120(0.063 + 0.001 + 0.354) = \underline{50.1\text{ kPa}}$

Worked example 6.4   *A circular foundation of diameter 10 m transmits a uniform contact pressure of 150 kPa. Plot the following vertical stress profiles induced by this loading: (a) beneath the centre, down to z = 10 m, and (b) on a horizontal plane 6 m below the foundation, between the centre and a distance of 12 m from the centres.*

Referring to Fig. 6.51 and Table 6.2:

$\Delta\sigma_z = q(A + B) = 150(A + B)$

(a)   The vertical stress values at various depths z below the centre of the foundation are tabulated below and plotted in Fig. 6.56(a).

$a = 5.0$ m   At centre, $r/a = 0$

z (m)	0	1	2	3	4	5	6	8	10
z/a	0	0.20	0.40	0.60	0.80	1.00	1.20	1.60	2.00
A	1.00	0.804	0.629	0.486	0.375	0.293	0.232	0.156	0.106
B	0	0.188	0.320	0.378	0.381	0.353	0.315	0.241	0.179
$\Delta\sigma_{zo}$ (kPa)	150	149	142	130	113	97	82	60	43

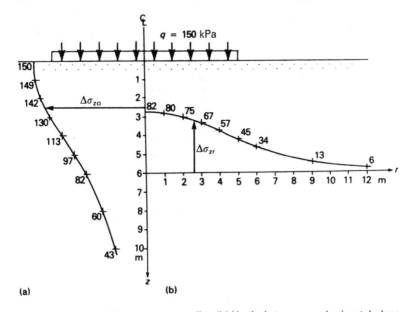

(a)   (b)

Fig. 6.56   (a) Vertical stress at centre-line (b) Vertical stress on a horizontal plane

(b) The vertical stress values on a horizontal plane 6 m below the foundation for various offsets from the centre are tabulated below and plotted in Fig. 6.56(b).

$a = 5.0$ m $z/a = 6.0/5.0 = 1.20$

r (m)	0	1.0	2.0	3.0	4.0	5.0	6.0	9.0	12.0
r/a	0	0.2	0.4	0.6	0.8	1.0	1.2	1.8	2.4
A	0.232	0.228	0.217	0.199	0.176	0.151	0.126	0.069	0.039
B	0.315	0.307	0.285	0.248	0.201	0.149	0.100	0.018	0.000
$\Delta\sigma_{zr}$ (kPa)	82	80	75	67	57	45	34	13	6

Worked example 6.5   *Figure 6.57 shows the plan of a rectangular foundation which trans-mits a uniform contact pressure of 120 kPa. Determine the vertical stress induced by this loading: (a) at a depth of 10 m below point A, and (b) at a depth of 5 m below B.*

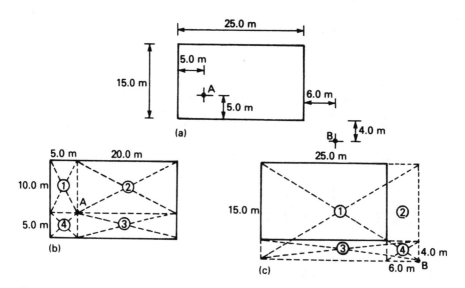

Fig. 6.57

(a) Consider four rectangles (1, 2, 3, 4) each with a corner at A (Fig. 6.57(b)): the vertical stress below A is the sum of the stresses induced by each rectangle:

$$\Delta\sigma_{z(A)} = \Delta\sigma_{z(1)} + \Delta\sigma_{z(2)} + \Delta\sigma_{z(3)} + \Delta\sigma_{z(4)}$$

$$= q(I_{R(1)} + I_{R(2)} + I_{R(3)} + I_{R(4)})$$

The calculations, using eqns [6.73], are tabulated below. Alternatively, either Table 6.3 or Fig. 6.52 can be used.

Rectangle	z = 10.0 m		
	m = B/z	n = L/z	I_R
1	10/10 = 1.0	5/10 = 0.5	0.1202
2	10/10 = 1.0	20/10 = 2.0	0.1999
3	5/10 = 0.5	20/10 = 2.0	0.1350
4	5/10 = 0.5	5/10 = 0.5	0.0840

Then $\Delta\sigma_{z(A)} = 120(0.1202 + 0.1999 + 0.1350 + 0.0840)$

$$= 120 \times 0.5391 = \underline{65 \text{ kPa}}$$

(b) Consider four rectangles (1, 2, 3, 4) each with a corner at point B (Fig. 6.57(c)) – note that for rectangle 1, $L = 31$ m and $B = 19$ m. The vertical stress below B is given by:

$$\Delta\sigma_{z(B)} = q(I_{R(1)} - I_{R(2)} - I_{R(3)} + I_{R(4)})$$

The calculations are tabulated below:

Rectangle	z = 5.0 m		
	m = B/z	n = L/z	I_R
1	19/5 = 3.8	31/5 = 6.2	0.2480
2	19/5 = 3.8	6/5 = 1.2	0.2171
3	4/5 = 0.8	31/5 = 6.2	0.1850
4	4/5 = 0.8	6/5 = 1.2	0.1684

Then $\Delta\sigma_{z(B)} = 120(0.2480 - 0.2171 - 0.1850 + 0.1684)$

$$= \underline{1.7 \text{ kPa}}$$

6.11 Elastic displacements

The assumption that the soil below a foundation behaves as an elastic body may also be adopted in the evaluation of surface displacement, i.e. settlement due to elastic compression. The effect of non-homogeneity only produces significant errors when the inter-stratum differences are considerable. In the case of uniform loading, the vertical surface displacement of a soil layer of infinite depth is given by:

$$s_i = \frac{qB}{E}(1 - v^2)I_\rho \qquad\qquad [6.74]$$

where q = intensity of contact pressure
B = least lateral dimension (breadth or diameter)
v = Poisson's ratio
E = modulus of elasticity
I_ρ = influence factor for vertical displacement

For the application of eqn [6.74] to the computation of foundation settlement see Section 10.2 and worked example 10.1.

Influence factor I_ρ

The vertical displacement influence factor I_ρ is dependent on the shape and stiffness of the foundation. Values of I_ρ obtained using elastic theory for the displacement at the centre or a corner of a uniformly loaded rectangle are given in Table 6.4.

Table 6.4 Influence factors (I_ρ) for vertical displacement due to elastic compression of a layer of semi-infinite thickness

Shape	Flexible*			Rigid[†]
	Centre	Corner	Average	
Circle	1.00	0.64	0.85	0.79
Rectangle				
$\frac{L}{B}$ 1.0	1.122	0.561	0.946	0.82
1.5	1.358	0.679	1.148	1.06
2.0	1.532	0.766	1.300	1.20
3.0	1.783	0.892	1.527	1.42
4.0	1.964	0.982	1.694	1.58
5.0	2.105	1.052	1.826	1.70
10.0	2.540	1.270	2.246	2.10
100.0	4.010	2.005	3.693	3.47

* After Giroud (1968)
[†] After Skempton (1951)

Poisson's ratio (v)

A range of values for v may be obtained using elastic theory. Consider the strain in the direction of σ_3 in a homogeneous element, as shown in Section 7.1, under conditions of confined compression (eqn [6.19]):

$$\sigma_3' = K_0 \sigma_1' = \frac{v'}{1 - v'} \sigma_1'$$

So that $K_0 = \dfrac{v'}{1 - v'}$

giving $v' = \dfrac{K_0}{1 + K_0}$

Thus, for a saturated clay where $K_0 = 1.0$, $v' = 0.5$
and for a sand where $K_0 = 0.5$, $v' = 0.33$

A typical range of values of v' is given in Table 6.5. A value can be found using a volumetric-strain/axial-strain ratio obtained in a triaxial test.

Table 6.5   *Typical values of Poisson's ratio*

Type of soil	v'
Saturated clay	0.4–0.5
Unsaturated or sandy clay	0.2–0.4
Sand: $\phi = 40°$	0.2
$\phi = 20°$	0.5

$$v' = -\frac{d\varepsilon_3}{d\varepsilon_1} \quad \text{and} \quad d\varepsilon_v = d\varepsilon_1 + 2d\varepsilon_3$$

Hence $\quad v' = 0.5(1 - d\varepsilon_v/d\varepsilon_1)$ \hfill [6.75]

Measurement of modulus of elasticity

The value of the undrained modulus of elasticity is not constant, but varies with the level of stress, the void ratio and with the stress history of the soil. Since it varies with stress, it therefore varies with depth. For design purposes, over relatively narrow ranges of depths E_u may be assumed to remain constant in saturated clays under undrained loading. In sands and unsaturated fine soils, however, estimates of E_u must be based on the depth and conditions at the point of interest. Under wide rafts on sands the value E_u increases towards the centre.

Values may be found for E' from the deviator-stress/axial-strain curves obtained from triaxial tests.

$$E' = \frac{d\sigma_1'}{d\varepsilon_1} = \frac{dq'}{d\varepsilon_a} \hfill [6.76]$$

If the slope (κ) of the swelling/reloading line has been obtained (eqn [6.25(b)]), the bulk modulus (K') may be determined.

$$dv = -\kappa \, d(\ln p') = -\kappa \, dp'/p'$$

Therefore $\qquad d\varepsilon_v = -\kappa \, dp'/vp'$

Thus $\qquad\qquad K' = vp'/\kappa$ \hfill [6.77]

and from eqn [6.11], $E' = 3vp'(1 - 2v')/\kappa$ \hfill [6.78]

It is usual to measure the *secant modulus* by drawing a chord from the origin to a reference point on the curve. This point is often located at one-half the maximum stress. Skempton (1951) has suggested that the point should be located at 65 per cent of the maximum stress, this being adjudged equivalent to the probable applied stress, allowing a factor of safety of three. An alternative is to measure E at a reference strain of 1 per cent, since at higher strains the deformation may not be truly elastic.

The effect of sample disturbance is to reduce the value of E_u. Immediate undrained tests therefore yield values which are too low. The *in situ* effective stress condition must be simulated by first consolidating the soil, either under a cell pressure representing the predicted effective overburden stress, or under similar conditions. The sample is then sheared undrained (see also Section 7.2).

Higher, and more reliable, values are obtained from *in situ* tests or from site observations, followed by back analysis (Burland *et al.*, 1966; Marsland, 1971; Hooper, 1973; Crawford and Burn, 1962). Values thus obtained have been reported which are up to four or five times the laboratory value. It is suggested (D'Appolonia *et al.*, 1971) that the true value of E_u will be up to $1\frac{1}{2}$ times the value obtained from the consolidated-undrained test results.

In situ measurements of E_u obtained from pressuremeter tests have been found to correlate well with values obtained from E_u/c_u ratios (Calhoon, 1972) and other methods (Mair and Wood, 1987). Essentially, a pressuremeter enables the measurement of lateral or volumetric strain in a borehole cavity in response to changes in pressure. After applying calibration corrections (see Section 12.5), the results will be presented in the form of either a pressure/cavity-volume or a pressure/cavity-strain curve. The shear modulus (G') may then be obtained.

$$G' = \frac{dp'}{dV}V \quad G' = \frac{1}{2}\frac{dp'}{d\varepsilon_c}$$

where dp'/dV = slope of the pressure/cavity-volume curve
$dp'/d\varepsilon_c$ = slope of the pressure/cavity-strain curve
V = cavity volume

The slopes may be measured directly from the elastic-phase portion of the curve, or from unloading/reloading cycles conducted at higher pressures. Values of E_u may then be determined from eqn [6.9]:

$$E_u = 2(1 + v_u)G' = 3G' \qquad [6.79]$$

(since $v_u = 0.5$)

In anisotropic soils the horizontal (measured) stiffness may be significantly higher than the required vertical value: in the absence of a measured anisotropy ratio, a reduction factor of 25–30 per cent is recommended.

Displacement in a thin layer
In most practical problems the layers are of finite thickness and are very often underlain by a relatively stiff or hard stratum. In such cases, the use of eqn [6.74] will lead to a considerable overestimate; a simple rule of thumb here is to apply eqn [6.72] only to layers, the thickness of which exceeds twice the breadth of the foundation.

For cases where the layer thickness is less than $2B$, and where $v \simeq 0.5$, the following solution (Janbu *et al.*, 1956) may be used:

$$s_i = \frac{qB}{E_u}\mu_0\mu_1(1 - v^2) \qquad [6.80]$$

where μ_0 and μ_1 are coefficients dependent on the breadth and depth of the foundation and also on the thickness of the layer below the foundation (Fig. 6.58).

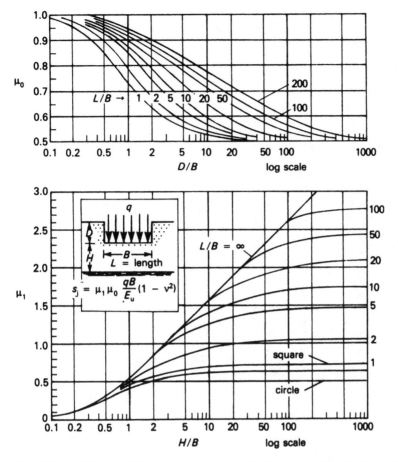

Fig. 6.58 Coefficients of displacement under flexible foundation (after Janbu *et al.*, 1956)

Steinbrenner (1934) presented an approximate general solution for the displacement under a corner of a flexible rectangular foundation over a layer of finite thickness:

$$s_i = \frac{qB}{E}(1 - v^2)I_s \qquad\qquad [6.81]$$

where v = Poisson's ratio

$$I_s = F_1 + \frac{1 - 2v}{1 - v}F_2$$

where F_1 and F_2 are influence factors dependent on ratios of length/breadth and depth/breadth (Table 6.6).

Elastic settlement of foundations
For computations of elastic settlement under foundations using the foregoing method, see Section 10.2.

Table 6.6 Influence factors (F_1 and F_2) for vertical displacement under a corner of a uniformly loaded rectangle over an elastic layer of finite thickness

D/B	L/B				
	1	2	5	10	∞
0.5	0.05	0.05	0.05	0.05	0.05
	0.08	0.09	0.10	0.10	0.10
1	0.16	0.13	0.13	0.13	0.13
	0.09	0.11	0.13	0.13	0.13
2	0.29	0.30	0.27	0.27	0.27
	0.06	0.10	0.14	0.15	0.13
3	0.36	0.40	0.38	0.37	0.37
	0.05	0.08	0.13	0.15	0.16
4	0.41	0.48	0.48	0.47	0.45
	0.04	0.07	0.12	0.15	0.16
5	0.44	0.53	0.55	0.54	0.52
	0.03	0.06	0.11	0.14	0.16
6	0.46	0.57	0.61	0.60	0.58
	0.02	0.05	0.10	0.14	0.16
8	0.48	0.61	0.69	0.69	0.66
	0.02	0.04	0.08	0.12	0.16
10	0.48	0.64	0.75	0.76	0.73
	0.02	0.03	0.07	0.11	0.16

$$s_i = \frac{qB(1-v^2)}{E} I_s$$

$$I_s = F_1 + \frac{1-2v}{1-v} F_2 \qquad \frac{F_1}{F_2}$$

L = length B = breadth D = depth below foundation

After Steinbrenner (1934)

6.12   *Distribution of stress – pressure bulbs*

From the foregoing section in this chapter, it will be seen that the intensity of stress below a foundation induced by the foundation loading decreases both vertically and laterally. It follows, therefore, that at some depth and/or lateral distance away from the foundation the intensity of stress will become relatively insignificant with respect to a particular practical problem.

If equal values of vertical stress are plotted on a cross-section a diagram known as a *pressure bulb* is obtained. Figure 6.59 shows pressure bulbs drawn for different foundation types, in which the stress, given as a fraction of the applied loading intensity, is plotted against the breadth of the foundation. The extent of a pressure bulb of a given value can provide a useful guide when considering which parts of the soil mass below a foundation will be significantly affected by the applied loading. It is particularly useful to note the vertical and lateral extent of the pressure bulbs representing stresses equal to $0.2q$ and $0.1q$. Two extreme 'shape' cases are represented by uniform circular and uniform strip loading; uniformly loaded rectangles will produce intermediate values. Using

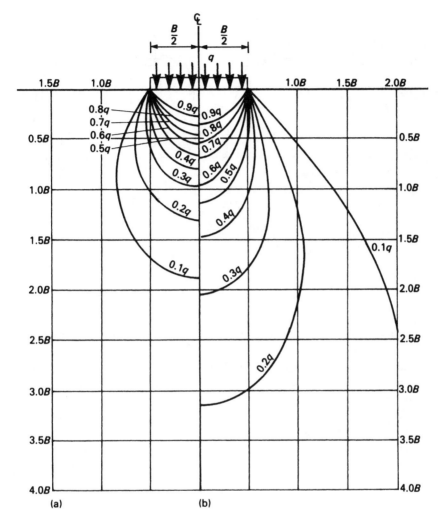

Fig. 6.59 Pressure bulbs for vertical stress
(a) Circular foundation (b) Strip foundation

Table 6.2 and eqn [6.69], the following approximate values may be obtained for the depth and breadth of the $0.2q$ and $0.1q$ pressure bulbs.

	Uniform circular load		Uniform strip load	
	0.2q	0.1q	0.2q	0.1q
Maximum depth of pressure bulb below ℄ of foundation	1.3B	1.9B	3.2B	6.6B
Maximum half-breadth of pressure bulb (i.e. lateral distance from ℄)	0.7B	0.9B	1.1B	2.1B

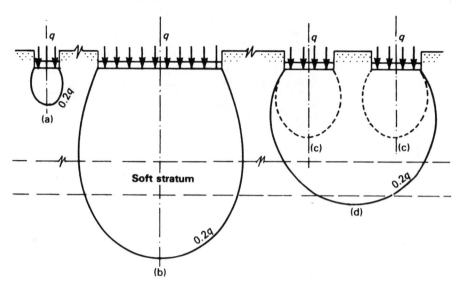

Fig. 6.60 Pressure bulbs indicating depth to which soil is significantly stressed

Pressure bulb dimensions can provide a useful guide when deciding the extent to which site exploration should be carried out. The minimum depth of exploratory boreholes is usually taken as $1.5B$, but in some cases this may prove to be inadequate, as the above figures indicate. It should also be noted that the combined pressure bulbs of adjacent foundations will be deeper than those for the same foundation when isolated. In Fig. 6.60, the effects of both foundation breadth and grouping are shown. The soft stratum is not significantly stressed by foundation (a), but it is subject to vertical stress well over $0.2q$ due to foundation (b) and also due to the combined effect (d) of the adjacent foundations (c).

Exercises

6.1   For the purpose of determining elastic displacements an estimate is required for the elastic modulus of a clay soil. Laboratory tests have shown that the void ratio (e) is 0.750, the compression index (C_c) is 0.25 and Poisson's ratio (v') is 0.30. Obtain estimated values of E' and E_u in terms of the effective overburden stress (σ').

6.2   Commencing with the three-dimensional strain equations applied to the case of one-dimensional compression, show that the principal stress ratio σ_3'/σ_1' is equal to $v'/(1 - v')$ and thence determine the undrained values of v' and E.

6.3   Four column loads of 900 kN, 800 kN, 200 kN and 500 kN respectively are located at the corners of a square of 4 m side on the surface of a soil mass. A culvert passes diagonally across the square, directly under the 900 kN and 200 kN load, and at a depth (to its top) of 4 m. Calculate the vertical stress imposed on the culvert due to the four loads at three points: under the 900 kN load, under the 200 kN load and at a point midway between them.

6.4   The concrete raft foundation shown in Fig. 6.61 is subject to uniform loadings of 250 kPa and 150 kPa on the shaded and unshaded areas respectively. Calculate the

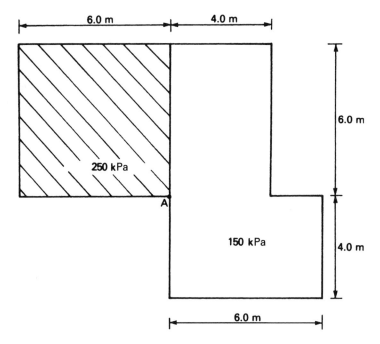

Fig. 6.61

intensity of vertical stress induced at a point in the soil 3 m below corner A: (a) using eqn [6.63], (b) using the factors given in Table 6.3.

6.5 Two parallel line-loads of 100 kN/m and 80 kN/m respectively, and 2 m apart, act vertically on a horizontal soil surface.

   (a) Calculate the vertical stress induced by this loading at a depth of 2 m: directly beneath each load, and at the half and quarter points between them.
   (b) From the resulting distribution, determine the maximum vertical and horizontal direct stresses and the maximum vertical shear stress at this depth and indicate the point where they occur.

6.6 A proposed long strip footing is to transmit a contact pressure of 215 kPa and be 3.5 m wide. The footing will be founded in a layer of sand which is 6 m deep and underlain by a layer of clay 3 m thick. Determine the maximum depth at which the footing may be founded in order that the increase in vertical stress at the centre of the clay layer does not exceed 70 kPa. Assume the groundwater level to be well below the footing and the unit weight of the sand to be 19 kN/m³.

6.7 Three parallel strip foundations, each 3 m wide and 5 m apart centre-to-centre, transmit contact pressures of 200 kPa, 150 kPa and 100 kPa respectively. Calculate the intensity of vertical stress due to the combined loads beneath the centre of each footing at a depth of 3 m.

6.8 A very long spoil heap is to be built up with a base width of 28 m and two side slopes of 8 m width. It is important that the increase in vertical stress at a depth of 4 m below the base does not exceed 120 kPa.

   (a) Determine the maximum uniform depth permitted for the central portion of the spoil heap ($\gamma = 18$ kN/m³).

(b) Calculate the maximum intensity of vertical stress induced (i.e. when the maximum height is reached) at points 4 m below the base, directly under both the top and the toe of the slope.

6.9   Figure 6.62 shows the plan of a large circular raft foundation; the centre (shaded) area transmits a contact pressure of 80 kPa and the outer annular area transmits a contact pressure of 200 kPa. Using the factors in Table 6.2, calculate the intensity of vertical stress induced at points in the soil mass 6 m below A, B and C.

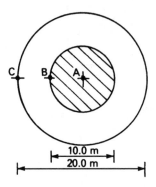

Fig. 6.62

6.10   Figure 6.63 shows the plan of a rectangular raft foundation which transmits a uniform contact pressure of 180 kPa to the soil beneath. The line of a culvert is also shown, which passes under the raft at a depth to the soffit of 3 m (fall ignored). Calculate the intensity of vertical stress on the culvert that will be induced by the raft loading at the points A, B, C, D and E shown.

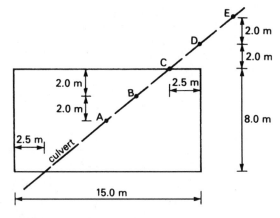

Fig. 6.63

Soil Mechanics Spreadsheets and Reference Assignments and Quizzes (available on the accompanying compact disc):

Assignments A.6   Stresses and strains in soils

Quiz Q.6   Stresses and strains in soils

Chapter 7

Measurement of shear strength

7.1 Shear strength and the friction model

Strength is the measure of the maximum stress state that can be induced in a material without it failing. Although strength can be stated in terms of compressive stress or tensile stress, fundamentally its the ability to sustain shear stress that provides strength. The *shear strength* of a soil is the maximum value of shear stress that may be induced within its mass before the soil yields. In brittle soils, yielding may lead to the formation of shear slip surface, over which sliding movement takes place, e.g. landslips, rotational slope and excavation failures. In softer more plastic soils yielding occurs as a result of internal particle flow. Measures of shear strength are required in the analyses and design of geotechnical structures, such as foundations, retaining walls, earth slopes and road bases.

Essentially, shear strength within a soil mass is due to the development of frictional resistance between adjacent particles, and analyses are based primarily on the friction model. The force transmitted between two bodies in static contact (Fig. 7.1) can be resolved into two components: a normal component N, perpendicular to the interface (RS), and a tangential or shear component T, parallel to it. When shear slipping movement takes place along this surface the

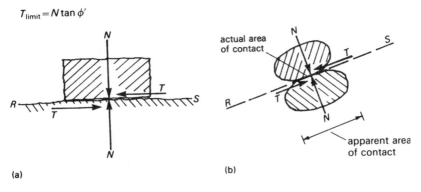

Fig. 7.1 The friction model

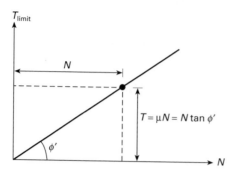

Fig. 7.2 Angle of internal friction

ratio *T/N* will have reached a limiting value termed the *coefficient of friction* (μ), so that

$$T_{\text{limit}} = \mu N$$

The coefficient of friction is a function of the roughness of the slip surface and may be assumed to remain constant for a given material. The limiting value of *T* can therefore be written:

$$T_{\text{limit}} = N \tan \phi' \qquad\qquad\qquad [7.1]$$

where ϕ' is defined as the angle of friction and is the angle given by the graph of *T/N* (Fig. 7.2).

A suggested mechanism of frictional resistance is that at the true points of contact the particles become welded or locked together (Bowden and Tabor, 1950, 1964). For sliding to occur, it is necessary for the material to yield locally at the points of contact. The yield stress of a material may be approximated by measuring the indentation hardness: typical values are 4000 MPa for tool steel and 7000 MPa for quartz. Thus, for a tool steel transmitting a normal stress of 100 MPa the true contact area would be approximately 1/40 of the apparent area. In the case of a typical sand, the normal stress due to 25 m of overburden will be about 0.5 MPa, so that the true contact area between grains may be as little as 1/14 000 of the superficial area (Fig. 7.1(b)).

The friction model provides a useful mathematical basis for descriptions of soil shear strength; however, a number of important factors peculiar to soils must also be considered. For example, loading a soil mass will induce changes in its volume or the internal pore pressure. Water may seep out of the soil at a rate controlled by the permeability, but the permeability range of sands to clays encompasses some ten orders of magnitude, so that *rates* of volume change are widely different. The limiting shear stress value (*shear strength*) is also affected by both the amount and the rate of strain, as well as by the stress history of the soil (see also Chapter 6).

Of considerable importance is the amount of strain. Figure 7.3 shows a typical shear stress/displacement (strain) curve for the shearing of a prismatic element of soil under constant normal stress (σ_n). The shear stress at first increases quickly while the strain increases slowly, but as the soil yields the strain increases dramatically while the shear stress levels off and then begins to fall. This

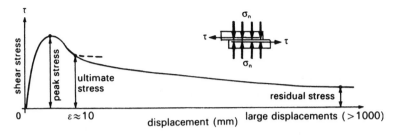

Fig. 7.3 Peak, ultimate and residual limiting stress

maximum value of shear stress at the yield point is termed the *peak stress* and represents the limiting value corresponding to that value of normal stress. The shear stress continues to fall until it levels off again at a lower value, known as the *ultimate stress*. The ultimate stress represents the shear strength of the material at its *critical volume* (see also Section 6.6). The ultimate stress value will usually be reached at strains of between 10 per cent and 20 per cent.

At very much larger strains, e.g. on active landslip surfaces in clay soils, the limiting shear stress falls further, and at displacements of over a metre it may be reduced to values as low as 10 per cent of the peak stress. This very low large-strain value is referred to as the *residual stress*. Residual stress values are related to the gradual rearrangement of clay particles on long slip surfaces and perhaps also to the effects of polishing or slickensliding.

Strength envelopes

A *strength envelope* is a graphical representation of a particular limiting condition of, say, the shear stress/normal stress ratio. Points below the envelope represent stress ratios possible prior to yielding, whereas points on the envelope represent the stress ratio at yielding. Real points above the envelope cannot exist.

Figure 7.4(a) shows the stress–strain graphs resulting from drained shear tests on three specimens of the same dense coarse soil at normal stresses $\sigma'_{n(1)} < \sigma'_{n(2)} < \sigma'_{n(3)}$. When the peak and ultimate stresses (τ_f) are transposed (using the same

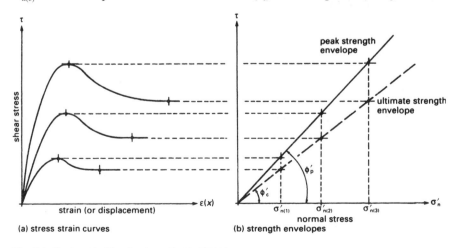

(a) stress strain curves

(b) strength envelopes

Fig. 7.4 Peak and ultimate strength envelopes

stress scale) to intercept ordinates along a normal stress (σ'_n) axis (Fig. 7.4(b)), the stress ratio points form straight lines. Envelopes for peak and ultimate stress ratios are obtained by drawing the best straight line through the plot points. The angle between a strength envelope and the normal stress axis is termed the *angle of shearing resistance* and may be taken as a good estimate of the *angle of friction (ϕ')* specified in the friction model (eqn [7.1]). It is necessary to add a subscript to ϕ' in order to specify the behavioural condition:

ϕ'_p = peak angle of friction
ϕ'_{ult} = ultimate angle of friction
ϕ'_c = critical angle of friction (may be taken as = ϕ'_{ult})
ϕ'_r = residual angle of shearing resistance
σ'_n = normal effective stress (usually written as σ')

Undrained shear strength and cohesion
The graphs shown in Figs. 7.2, 7.3 and 7.4 relate to *drained* loading conditions, i.e. no increase in pore pressure occurs. Under *undrained* conditions, saturated fine soils will apparently display a constant limiting shear stress at all values of normal stress, since the volume remains constant, and from eqn [4.8], $\Delta\sigma' = \Delta\sigma - \Delta u$. Thus an increase in the total normal stress brings about a similar increase in pore pressure, with the effective normal stress therefore remaining constant. A series of undrained tests on saturated specimens of the same soil, carried out at different normal stresses, will result in a strength envelope plot similar to Fig. 7.5.

The angle of friction (ϕ_u) is zero and the strength envelope intercepts the shear stress axis at a value which is termed the *undrained shear strength*, or the *undrained cohesion (c_u)*, of the soil. The undrained shear strength of saturated soil is apparently constant, but the value corresponds to *a particular water content* (and specific volume). At a different water content (or specific volume) a different value of undrained shear strength would be obtained.

The term *cohesion* is apt to be misleading and is often misunderstood. Earlier opinion as to the nature of this apparent force holding soil grains together was centred on interparticle bonding arising out of the electrostatic conditions on the surfaces of clay minerals. In fact, the apparent cohesion is related simply to the suction effects of pore pressure. Drained shear tests carried out on over-

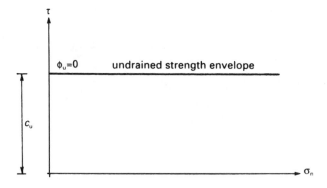

Fig. 7.5 Undrained shear strength envelope

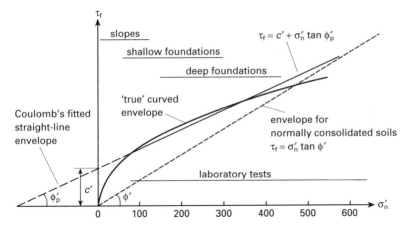

Fig. 7.6 Coulomb's equation and failure envelopes

consolidated clays produce a cohesion intercept (c') which has suggested to some that a true cohesion force may exist between uncemented soil particles (Fig. 7.6). However, drained tests on the same soil under normal consolidation conditions clearly produce an intercept of $c' = 0$. The 'true' drained peak strength envelope is in fact curved (Fig. 7.6), and $c' > 0$ is actually an intercept resulting from fitting a straight line to a small number of plot points. For remoulded clays, which include clays in the vicinity of large-strain slip surfaces (i.e. residual conditions), the value of c' should be taken as zero.

7.2 Critical strength and peak strength

Critical state and critical strength

The shear strength of soil as measured in tests depends mainly on the state of the soil at the start. In hard or dense soils the stress–strain curve is initially steep, reaches a peak and then falls to an ultimate or critical value (Fig. 7.7). In soft or loose soils the peak is lower or even non-existent and the curve flattens out at the ultimate (critical) value. In a drained test the volume changes: increasing in hard/dense soils, decreasing in soft/loose soils. After the ultimate (critical) strength is reached the volume remains constant while shearing continues: the soil is now in the *critical state* and the volume is the *critical volume*. In the **critical state** there is a unique relationship (for this soil) between the shear stress, the normal stress and the volume (or void ratio). A series of tests on the same soil will produce a critical strength envelope that is a straight line passing through the origin ($\sigma' = 0$, $\tau_f = 0$); this is referred to as the *critical state line (CSL)* (see also Section 6.8).

The critical state line is, however, three-dimensional, having also a volume or void ratio axis (Fig. 7.8). For direct shear tests the axial parameters are (τ_f, σ', e), while for triaxial tests they are (q', p', v). Figure 7.8(a) show elevations of the CSL drawn from $\tau_f{:}\sigma'$ and $e{:}\sigma'$ axes respectively; when a log σ' axis

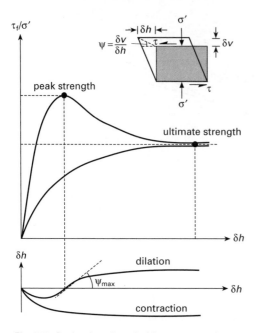

peak strength

ultimate strength

$\psi = \dfrac{\delta v}{\delta h}$

dilation

ψ_{max}

contraction

Fig. 7.7 Drained peak and ultimate strength

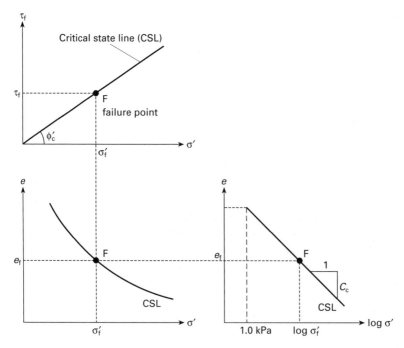

Critical state line (CSL)

τ_f

F
failure point

ϕ'_c

σ'_f

e

F

CSL

e_f

σ'_f

e

F

C_c

CSL

1.0 kPa   log σ'_f log σ'

Fig. 7.8 Critical strength

is used the CSL elevation is a straight line (Fig. 7.8(c)). The relevant expressions are:

For direct shear (Fig 7.8): $\tau_f = \sigma_f' \tan \phi'$ [7.2(a)]

$$e_f = e_\Gamma - C_c \log \sigma'$$ [7.2(b)]

where ϕ' = slope of the CSL in τ_n/σ' plane
  e_Γ = the void ratio at $\sigma' = 1.0$ kPa
  C_c = slope of the CSL in the $e/\log \sigma'$ plane

For triaxial tests: (eqn [6.43(a)]) $q_f' = Mp_f'$

(eqn [6.43(a)]) $v_f = \Gamma - \lambda \ln p_f'$

where M = slope of the CSL in the q'/p' plane
  Γ = the specific volume at $p' = 1.0$ kPa
  λ = slope of the CSL in the $v/\ln p'$ plane

The *critical strength* is the only unique measure of the strength of a soil; other measures, e.g. undrained strength or peak strength, depend on the initial water content/volume state of the soil and are therefore not constant. It is important to note that cohesion plays no part in the description of critical strength when is developed entirely by inter-particle friction. In coarse angular-grained sands and gravels some particle crushing occurs, and may obscure true strength values.

Peak states and peak strengths

The critical strength is reached at strains over 10 per cent. In conditions where the strains are small a *peak strength* is developed provided the initial void ratio is less than the critical void ratio. Figure 7.9 shows stress–strain curves for several samples of a soil of different densities during drained tests. The more dense (or more dry) the soil is initially, the greater will be the peak strength, while those with void ratios below critical do not exhibit peak strength. Note also how the volume changes correspond to the initial void ratio.

Peak strength also depends on the level of normal stress. At zero normal stress the peak strength is zero, since no inter-particle friction is being developed.

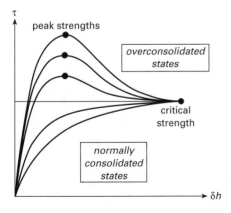

Fig. 7.9 Peak strengths

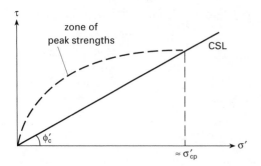

Fig. 7.10 Curved zone of peak strengths

Figure 7.10 shows an envelope of peak strengths: as the normal stresses are increased there is at first a sharp increase in peak strengths, then the envelope flattens slightly; eventually at around the pre-consolidation stress (σ'_{cp}) (see Section 10.3) the peak envelope merges with the critical strength envelope. A peak strength envelope is therefore curved, and there is a separate envelope for each initial water content/volume state.

In the interpretation of shear test results it is convenient (although potentially inaccurate) to assume a straight-line peak strength envelope (Fig. 7.10). The projection of a peak strength envelope gives the cohesion intercept c' on the shear stress axis, the value of which depends on the initial water content/volume state. Thus, if a series of sets of samples of the same soil were subject to drained tests, each set being prepared to different void ratios, then a different value of c' would be evident for each set. The equation of a peak strength envelope is therefore:

For direct shear: (Fig. 7.11(a)) $\tau_f = c' + \sigma'_f \tan \phi'_p$ [7.3]

where c' = the cohesion intercept (dependent on void ratio)
         ϕ'_p = slope of the peak strength envelope in τ_n/σ' plane

For triaxial tests: (Fig. 7.11(a)) $q'_f = g' + Hp'_f$ [7.4]

where g' = the intercept on the q' axis
        H = slope of the peak strength envelope in the q'/p' plane

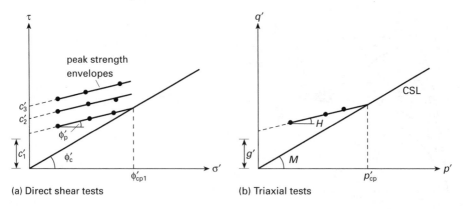

(a) Direct shear tests   (b) Triaxial tests

Fig. 7.11 Straight-line peak strength envelopes

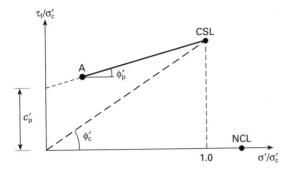

Fig. 7.12 Normalised peak strength envelope

In Fig. 7.11, if the peak strength envelope intercepts the CSL at a stress of σ'_c then the cohesion intercept

$$c' = (\tan \phi'_c - \tan \phi'_p)\sigma'_c \qquad\qquad\qquad\qquad [7.5(a)]$$

or $\quad g' = (M - H)p'_c \qquad\qquad\qquad\qquad\qquad\qquad [7.5(b)]$

A single peak strength expression and envelope can be obtained by normalising the stresses with respect to σ'_c. The normalised stresses are τ_f/σ'_c and σ'/σ'_c (Fig. 7.12(a)) and the resulting peak strength envelope becomes a single line intercepting the CSL at a single point, which is the *critical state point*.

Also, from eqn [7.2(b)],

$$\log \sigma'_c = \frac{e_\Gamma - e}{C_c}$$

Putting the normalised cohesion as $c'_p = c'/\sigma'_c$, then

$$\log\left(\frac{c'}{c'_p}\right) = \frac{e_\Gamma - e}{C_c} \qquad\qquad\qquad\qquad [7.6]$$

Thus c'_p increases as the void ratio decreases.

The equivalent expressions for triaxial test results are normalised with respect to p'_c (Fig. 7.12(b)), giving the single peak strength line:

$$\frac{q'_f}{p'_c} = g'_p + H_p\left(\frac{p'_f}{p'_p}\right) \qquad\qquad\qquad\qquad [7.7]$$

where $\quad g'_p = g'/p'_c$

Now, as the stresses tend towards zero, $\tau_f \approx \frac{1}{2}(\sigma'_a - \sigma'_r)_f = \frac{1}{2}q'_f$, so that $g'_p \approx 2c'_p$.

Peak strength and dilation

Denser (drier) than critical (or overconsolidated) soils dilate prior to reaching a peak strength and the amount of dilation is related to the peak strength achieved. In Fig. 7.7 a prismatic element of soil is shown distorting and dilating during shearing, the horizontal displacement is δh and the increase in volume (dilation) is δv. The angle of dilation is given by $\tan \psi = \delta h / \delta v$. Allowing for the dilation the stress ratio is therefore:

$$\frac{\tau'}{\sigma'} = \tan (\phi'_c - \psi) \qquad [7.8]$$

The diagram shows the stress ratio τ'/σ' plotted against δh and δv plotted against δh. The angle of dilation (ψ) is zero at two points (A and C), and has a maximum value (ψ_p) corresponding to the peak strength.

For triaxial tests the equivalent expression is:

$$\frac{q'}{p'} = M - \frac{d\varepsilon_v}{d\varepsilon_s} \qquad [7.9]$$

where ε_v = volumetric strain

and $\varepsilon_s = \frac{2}{3}(\varepsilon_a - \varepsilon_r)$ (eqn [6.14])

Note that $d\varepsilon_v$ will be negative for dilation.

7.3   Shear failure criteria and parameters

By applying a failure criterion to a set of shear test results it is possible to evaluate parameters which can then be used in design. There are two criteria applicable to the failure of soils: (i) the Tresca criterion, stated in terms of total stresses, and (ii) the Mohr–Coulomb criterion, stated in terms of effective stresses.

The Tresca failure criterion
The Tresca criterion states that failure will occur when the Mohr circle of total stress due to increasing load touches an envelope (Fig. 7.5) defined by

$$\tau_f = c_u \qquad [7.10]$$

where c_u = the undrained strength of the soil (sometimes written as s_u)

Note that this strength is independent of the magnitude of applied stress, but as will be explained later it does depend on the water content of the soil. This criterion only applies in a practical sense to clays and silts: the concept of undrained strength in coarse soils is impractical because of their high permeability.

The Mohr–Coulomb failure criterion
A simple equation and theory relating the shear strength of soil to the applied normal stress was first suggested by Coulomb in 1776. A straight-line equation for the limiting shear stress was given thus:

$$\tau_f = c + \sigma_n \tan \phi \qquad [7.11]$$

where c = apparent cohesion (assumed to be constant)
    σ_n = normal stress on slip surface
    ϕ = angle of friction (or angle of shearing resistance)

The shear strength of soil in fact is related to effective stresses, so the Mohr–Coulomb failure criterion states that the shear strength increases linearly with effective normal stress and is defined by an envelope just touching the Mohr effective stress circles at failure, i.e.

$$\tau_f = c' + \sigma'_n \tan \phi' \qquad [7.12]$$

where c' and ϕ' are shear strength parameters stated in terms of effective stresses
$\quad\quad\quad \sigma'_n$ = effective normal stress on the slip plane (usually written as just σ')
$\quad\quad\quad\quad$ = $\sigma_n - u_f$ (from eqn [4.6])
$\quad\quad\quad u$ = pore pressure acting on the slip surface at failure
$\quad\quad\quad \phi'$ = angle of friction referred to effective stresses

Equation [7.12] corresponds to the expression given above for *peak strength*, eqn [7.3]. It is now recognised that for a statement of *critical strength* the intercept $c' = 0$ and the appropriate expression is eqn [7.2a]: $\tau_f = \sigma' \tan \phi'$. With these provisos in mind, however, the Coulomb equation is still a suitable basis for describing shear strength measured in tests, where only basic parameters (e.g. values of ϕ' and c') are required.

The main disadvantage of the Coulomb equation is that it ignores volume changes and pre-failure behaviour; in such cases the critical state model is preferable. The Mohr–Coulomb failure criterion and other important concepts of soil failure are dealt with in Chapter 6.

Shear strength tests and practical parameters
The purpose of shear strength testing is to establish empirical values for the shear strength parameters. The drainage conditions during the test influence the measured values considerably. Shear strength tests are carried out in two main stages involving the following types of drainage conditions:

(1) *Consolidation stage.* After a test sample has been prepared to size and its mass and water content determined, it is consolidated to a required state. This is done either one-dimensionally in direct shear tests or isotropically in triaxial tests. The objective in this stage is principally to produce an initial stress state, and also to ensure full saturation in triaxial tests. The consolidation of coarse soils in direct shear tests is virtually immediate upon load application, but may require several hours for fine soils.
(2) *Shearing or axial loading stage.* The consolidated test sample is subject either to direct shearing (e.g. shear box test) or to an increase in axial loading (triaxial test) until failure occurs. Readings of vertical (and maybe lateral) strain, axial load, pore pressure and volume change are taken against time.

Undrained tests
In undrained tests no drainage of porewater is allowed during the shearing stage. For fine soils the test sample should be fully saturated; the increase in pore pressure will therefore be equal to the increase in total stress and no increase in effective stress takes place.

If $\Delta u = \Delta\sigma$, then $\Delta\sigma' = 0$, giving the total stress envelope shown in Fig. 7.5:

$$\tau_f = c_u \qquad\qquad\qquad\qquad [7.13]$$

where c_u = undrained cohesion or undrained shear strength

The value of c_u depends on the initial void ratio (or specific volume) which remains constant. Samples tested at the same void ratio or water content should all shear at the same normal stress, giving a total stress envelope parallel to the normal stress axis, i.e. $\phi_u = 0$. If an inclined envelope ($\phi_u > 0$) occurs, the test results are invalid for volume changes have occurred; this may be due to incomplete saturation (air is compressible) or inadvertent drainage (e.g. faulty membranes, valves), inadequate sample preparation, etc.

Effective stress parameters can be obtained by plotting effective stresses, i.e. $\sigma' = \sigma - u$. For normally consolidated soils the failure point will still fall on the CSL and a value obtained for ϕ'. However, the values yielded for peak strength parameters c' and ϕ'_p may not be as reliable as those obtained in drained tests.

Drained tests
The test specimen is first consolidated under conditions of constant one-dimensional or isotropic stress and full drainage. When the consolidation stage is complete, shearing takes place at a rate slow enough to ensure that no increase in pore pressure takes place. A back pressure (in triaxial tests) may be applied to simulate field conditions. The increase in effective stress is equal to the increase in total stress ($\Delta\sigma' = \Delta\sigma$).

For normally consolidated soils the failure points fall on the CSL giving a value for ϕ'. Peak stresses develop in dense or heavily consolidated soil giving rise to the failure envelopes described by eqns [7.3] and [7.4]:

For *residual* conditions involving large strains the failure envelope is curved at low normal stresses and c' tends toward zero. The residual angle of friction is therefore strain-dependent (see also Section 7.13), but the following relationship can be used:

$$\tau_f = \sigma'_r \tan\phi'_r \qquad\qquad\qquad\qquad [7.14]$$

Types of shear test
Below is a summary of the available shear tests: some are laboratory tests and some are *in situ* tests. Those of some importance are described in later sections. They all have one common characteristic in that they are tests on *samples*, whether the tests are carried out *in situ* or on samples brought to the laboratory. In order that reliable results might be produced it is essential that samples are properly representative of the site materials that may be encountered. Great care should be taken in obtaining, packaging and transporting samples from site to laboratory; especially with undisturbed samples, in which the *in situ* structure, density and moisture content must be preserved (see also Section 12.4).

Laboratory tests
Shear box (Section 7.4)
Simple shear

Ring shear (Section 7.15)
Triaxial compression (Section 7.6)
Stress path (Section 7.18)

In situ *tests*
Standard penetration and cone penetration (Sections 11.8 and 12.5)
In situ shear box
Shear vane (Section 7.18)

7.4 The shear box test

This is a *direct* shear test, i.e. the normal and shear stresses on the failure surface
are measured directly. A rectangular prism is carefully cut from a soil sample (or
remoulded, as required) and fitted into a square metal box that is split into two
halves horizontally (Fig. 7.13). In the standard type of apparatus the box is
60 mm × 60 mm, but for testing coarse soils and fissured clay a larger version
is used.

With the halves of the box held together, the soil specimen is sandwiched
within the box between ridged metal plates and porous ceramic stones. A pressure pad is placed on top and the box itself placed in an outer box which runs
horizontally on roller bearings, and which also acts as a water bath. A vertical
load is then applied to the specimen by means of a static weight hanger. After
removing the screws holding the two halves of the box together, the soil is
sheared by applying a horizontal force with a screw jack at a constant rate of
strain. The magnitude of the shearing force is measured by means of a proving
ring or electronic load cell.

The procedure is repeated on four or five specimens of the same soil. Values
of normal stress (σ_n) and shear stress (τ) on the failure plane are computed and
plotted using a computer. The shear strength envelopes corresponding to peak
and ultimate stresses are fitted as the best straight lines through plotted (σ_n, τ)
points.

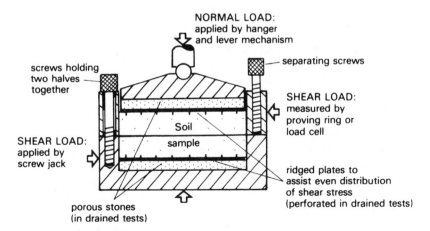

Fig. 7.13 The shear box

Worked example 7.1   *A drained shear box test was carried out on a sandy clay and yielded the following results:*

Normal load (N)	108	202	295	390	484	576
Shear load at failure (N)	172	227	266	323	374	425

Area of shear plane = 60 mm × 60 mm
Determine the apparent cohesion and angle of friction for the soil.

Area of shear plane = $60 \times 60 \times 10^{-4} = 3.6 \times 10^{-3} \text{ m}^2$

So for the first specimen:

Normal stress, $\sigma'_n = \dfrac{108 \times 10^{-3}}{3.6 \times 10^{-3}} = 30.0 \text{ kPa}$

Shear stress at failure, $\tau_f = \dfrac{172 \times 10^{-3}}{3.6 \times 10^{-3}} = 47.8 \text{ kPa}$

Similarly, other values will be:

Normal stress, σ'_n (kPa)	30.0	56.1	81.9	108.3	134.4	160.0
Shear stress at failure, τ_f (kPa)	47.8	63.1	73.9	89.7	103.9	118.1

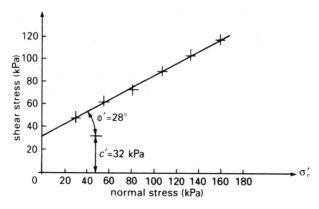

Fig. 7.14

Figure 7.14 shows the peak shear stress plotted against normal stress and the best straight line drawn through the points to give the peak strength envelope. The peak strength parameters measured from the plot are:

Apparent cohesion $c' = \underline{33 \text{ kPa}}$

Peak angle of friction $\phi'_p = \underline{28°}$

Advantages and disadvantages of the shear box test

The shear box test is a direct shear test, i.e. the shear and normal stresses on the plane of failure are measured directly. The most widely used shear strength test, however, is the triaxial compression test (see Section 7.6), which is an indirect test. The advantages of the shear box method in comparison with the triaxial method are summarised below.

Advantages

(a)   Both the shear stress and normal stress on the plane of failure are measured directly; the shear strength parameters (c' and ϕ') are defined in terms of these direct stresses.
(b)   A constant normal stress can be maintained throughout the test.
(c)   It is easier to test cohesionless soils, e.g. sand and gravels, and drained tests can be carried out in a reasonably short time.
(d)   Volume changes can be measured simply.
(e)   Using a *reversible* shear box, tests involving large displacements can be done, e.g. to measure residual strengths of clays (Section 7.14).

Disadvantages

(a)   The distribution of shear stress over the plane of failure is assumed to be uniform, but in fact it is not.
(b)   It is not possible to control drainage from the sample or to measure the pore pressure within the sample. Therefore, only total stress measurements can be made, except when the rate of shearing is kept slow enough to ensure no rise in pore pressure.
(c)   The normal stress cannot easily be varied *during* tests.

7.5   Examining peak and ultimate strength

The relationship between the shear strength of a soil and its initial density state is an important characteristic and it should feature early in the study of soil mechanics. The shear box test provides students with a means of examining this relationship using remoulded sand samples, which are prepared easily to different densities; the test procedure is also quick and simple to monitor.

In dense sands the grains are to some degree interlocked, so that an initial expansion or dilation is necessary in order that shearing can occur. Thus, the shear stress will first rise sharply, with a corresponding increase in volume (Fig. 7.7), until it reaches a *peak* value at a relatively low value of strain. As the interlocking is reduced the displacement continues more quickly and the shear stress falls back and finally levels off at an *ultimate* value.

This behaviour of dense sand contrasts with that of a loosely packed sand, since in the latter a steady increase in shear stress takes place as the displacement increases (Fig. 7.7). The volume decreases and the soil contracts into a more dense state as the shear stress increases towards the same ultimate value. The constant volume (void ratio) achieved at the ultimate stress corresponds to the *critical state* volume (see also Section 7.2). The difference between the peak and ultimate values clearly depends on the original density (void ratio) of the soil: a soil more dense than critical will dilate, whereas a soil looser than critical will contract. The peak stress is therefore a function of the initial volume/density state.

In the case of very loose sands, the volume is much greater than the critical value, so a small amount of displacement will be accompanied by a significant change in volume. Such soils are likely to compact easily under relatively small loads or vibrations, causing contraction and settlement.

The natural angle at which a pile of tipped dry sand or gravel will stand unsupported is often referred to as the *angle of repose* and is equivalent to the ultimate strength value, i.e. ϕ'_c. It should be noted, however, that if the soil is damp the resulting pore suction will increase the effective stress between the grains and thus the tipped slope angle will be greater, as for example in a seaside sandcastle (see also Section 9.2).

Worked example 7.2 *The following results were recorded during shear box tests on specimens of a sand compacted to the same initial density:*

Normal load (N)	110	216	324	432
Ultimate shear load (N)	66	131	195	261
Peak shear load (N)	85	170	253	340

Determine the peak and ultimate angles of friction and the angle of dilation.

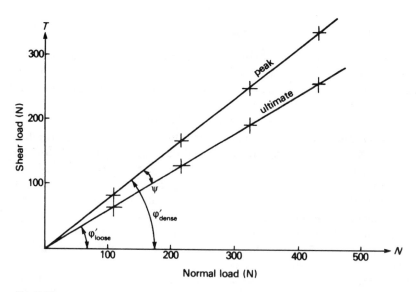

Fig. 7.15

Figure 7.15 shows the plot of these data. Two strength envelopes are drawn:

Peak strength: $\phi'_p = \underline{38°}$ Ultimate strength: $\phi'_c = \underline{31°}$

Angle of dilation $= \underline{7°}$

Note that the ϕ'_p would have had a different value for different initial densities, but ϕ'_c would have remained the same, since the critical volume would be reached each time.

Worked example 7.3   *The readings given below were taken during two shear box tests on remoulded samples of the same sand. In both cases the constant normal stress was 210 kPa In Test 1, the sand was prepared in a loose state; in Test 2, the sand was compacted into a dense state.*

Draw the shear stress/displacement curves for the two tests and determine the peak and ultimate angles of friction. Also, plot the change in volume against horizontal displacement and comment on the observed volume changes.

Test 1: Loose state

Horizontal displacement (10^{-2} mm)	0	50	100	150	200	250	300	350	400	450	500	550
Vertical displacement (10^{-2} mm)	0	−6	−12	−15	−17	−18	−19	−19	−20	−20	−21	−21
Shear stress (kPa)	0	59	78	91	99	106	111	113	114	116	116	116

Test 2: Dense state

Horizontal displacement (10^{-2} mm)	0	50	100	150	200	250	300	350	400	450	500	550
Vertical displacement (10^{-2} mm)	0	−3	1	9	17	23	29	35	39	41	41	41
Shear stress (kPa)	0	73	118	143	152	149	139	133	126	122	120	119

The results are shown plotted in Fig. 7.16. It can be seen that in both tests the sand reached its critical state at a displacement of about 5 mm, after which the volume remained constant. Prior to reaching the critical state the loose sand (Test 1) contracted, whereas the dense sand (Test 2) dilated.

In Test 1, the shear stress increased steadily to an ultimate value of 116 kPa; but in Test 2 there was a rapid rise to a peak value of 152 kPa before a decrease toward an ultimate value similar to that in Test 1.

Ultimate strength: $\tau_f = \sigma'_n \tan \phi'_c$

Then $116 = 210 \tan \phi'_c$ and therefore $\phi'_c = \underline{29°}$

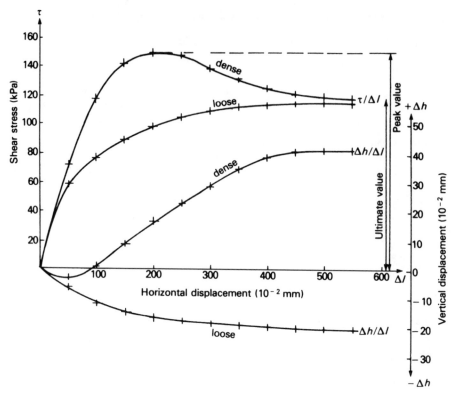

Fig. 7.16

| Peak strength: | $\tau_f = \sigma'_n \tan \phi'_p$ |
| Then | $152 = 210 \tan \phi'_p$ and therefore $\underline{\phi'_p = 36°}$ |

7.6 The triaxial compression test

The triaxial compression test is the most widely used shear strength test; it is suitable for all types of soil except for very sensitive clays and allows a number of different test procedures. Detailed descriptions of triaxial test procedures may be found in Bishop and Henkel (1962), Head (1982), Vickers (1983) and in BS 1377 (1990).

The test is carried out on a cylindrical specimen of soil having a height/ diameter ratio of 2:1; the usual sizes being 76×38 mm and 100×50 mm. The specimen is first accurately cut and trimmed from a field sample, or a remoulded specimen prepared and consolidated in a special device. It is then enclosed between rigid end-caps inside a thin rubber membrane to seal it from the cell water; rubber O-rings are fitted over the membrane at the caps to provide a seal. With the cell disassembled, the specimen is mounted on the pedestal and the end of the rubber membrane stretched over the pedestal and held in place with an O-ring. The cell is then reassembled so that the arrangement is as shown in Fig. 7.17.

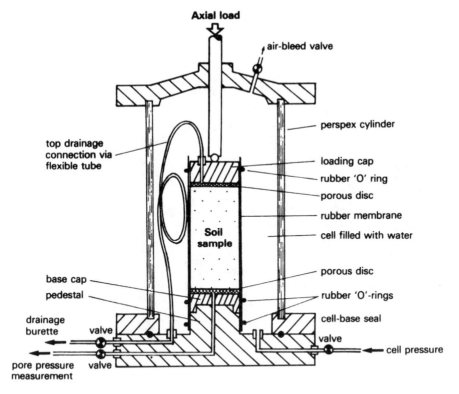

Fig. 7.17 The triaxial cell

Undrained tests

The specimen is subjected to an isotropic stress by filling the cell with water and increasing the pressure to a prescribed value. With the drain open, the sample is allowed to consolidate under the isotropic cell pressure until volume change ceases. (If the specimen is fully saturated initially this stage can be carried out undrained.)

The drain is closed, and with the cell pressure held constant the axial load is then increased by means of a motor-operated screw jack until the sample shears or the ultimate stress is reached. During the axial loading, readings are taken of (a) the change in length of the specimen (using either a dial gauge or a displacement transducer), (b) the axial load (using either a proving ring or an electronic load cell, and (c) the pore pressure within the specimen (using either a pressure transducer or a Bourdon gauge).

Pore pressure measurement

To facilitate pore pressure measurements, perforated end-caps are used, together with porous ceramic discs, at either end of the specimen. The base of the triaxial cell is specially drilled and has a valve arrangement to facilitate a connection to a pressure transducer or equivalent device. Figure 7.18 shows a typical overall arrangement.

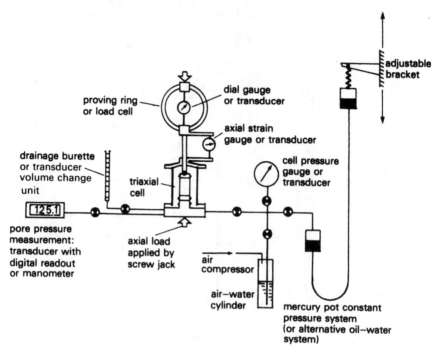

Fig. 7.18 Typical triaxial test arrangement

Drained tests

For a *drained* test, the test specimen is also consolidated isotropically. If site pore pressure conditions are to be simulated an appropriate back pressure is applied. After volume change has ceased, the axial load is applied at a rate slow enough to ensure no further change in pore pressure. Strips of filter paper may be placed vertically around the specimen inside the rubber membrane to assist drainage. A perforated top cap is provided, which is connected to the base inside the cell with a length of plastic tube; from here, via a valve, a connection leads to a volume measurement device (e.g. a calibrated burette or similar).

Sources of error in the triaxial test

Undrained tests

(a) Disturbance or damage during sampling and preparation.
(b) Soil not saturated, i.e. contains air which is compressible.
(c) Air bubbles trapped between the soil and the rubber membrane or end-caps.
(d) Rubber membrane is excessively thick or is punctured.
(e) Poor water seals at ends; air bubbles in pore pressure line.
(f) Lateral stress developed across end-caps (these should be greased).

Drained tests

(a) Rate of loading too fast, thus $\Delta u \neq 0$.
(b) Ineffective seals in volume measurement system.

(c)   Calibration errors in volume measurement system.
(d)   Load loss in axial load piston due to poor lubrication.
(e)   Insensitivity of measurements at low strains due to high early soil stiffness (overcome when constant stress rate control is used instead of constant strain rate).

7.7   Area and volume changes during the triaxial test

As the axial load on the specimen increases a shortening in length takes place, with a corresponding increase in diameter. In addition, when drainage is allowed, the volume of the sample will decrease. A strain dial gauge or transducer indicates the change in length of the specimen (Fig. 7.17).

To record volume changes a drainage connection is made from the top of the specimen. A porous ceramic disc is placed between the soil and the top (perforated) end-cap and a plastic tube connected between the end-cap and a drainage outlet in the cell base (Fig. 7.18). The volume change device may be a simple glass burette (measuring to 0.1 ml) or a transducer-equipped volume gauge with computer monitoring. The volume of water collected represents the change in volume of a saturated specimen.

To calculate the deviator stress from the applied axial load, the changed cross-sectional area of the specimen may be obtained from:

$$A = A_0 \frac{1 - \Delta V / V_0}{1 - \Delta l / l_0} \qquad [7.15]$$

where A_0 = initial cross-sectional area of specimen
V_0 = original volume of specimen
l_0 = original length of specimen
ΔV = change in volume
Δl = change in length
} before increase in axial load but after consolidation

In the case of an *undrained* test, ΔV will be zero, giving

$$A = \frac{A_0}{1 - \Delta l / l_0} = \frac{A_0}{1 - \varepsilon_a} \qquad [7.16]$$

where ε_a = axial strain = $\Delta l / l_0$

7.8   Interpretation of triaxial test results

As the name implies, the specimen in the triaxial test is subject to compressive stresses resolved along three orthogonal axes and applied in the two stages (Fig. 7.19). First, an isotropic consolidating stress σ_c is applied, where σ_c = the cell pressure. During this stage the volume decrease due to consolidation (ΔV_c) should be measured and the final consolidated specific volume of the specimen (v_c) evaluated.

The test is now continued by increasing the axial load while (in the standard test) the cell pressure is held constant; the axial (vertical) compressive stress is

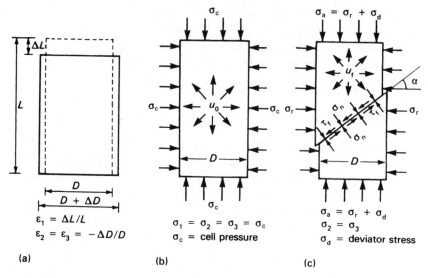

Fig. 7.19   Strains and stresses in the triaxial test
(a) Principal strains (b) Cell pressure only (c) Stresses at shear failure

increased by σ_d, which is termed to deviator stress (σ_d = axial load/area of test specimen). Thus, the final stresses are:

Final axial stress, $\sigma_a = \sigma_c + \sigma_d$

Final radial stress, $\sigma_r = \sigma_c$

When shear failure occurs, the *peak* or *ultimate* deviator stress will have been reached. If the stage is undrained the pore pressure will have risen to u_f. If the stage is drained the specific volume will have fallen from v_c to v_f.

Since there is no shear stress developed on the sides of the specimen, the vertical axial stress and the lateral stresses are *principal stresses*:

Vertical axial stress, $\sigma_a = \sigma_1$ = major principal stress

Lateral stress, $\sigma_r = \sigma_3$ = minor principal stress

At failure, the deviator stress σ_d = the principal stress difference ($\sigma_1 - \sigma_3$). This can also be written as $q_f = \sigma_1 - \sigma_3$ (see Sections 6.6 and 6.8).

Types of failure
As the specimen shortens under the increasing axial load its diameter will increase. In dense or heavily overconsolidated soil the specimen may shear cleanly along a well-defined slip surface as the peak stress is reached; this is a *brittle shear slip* failure (Fig. 7.20(a)). In lightly overconsolidated soil the shear will usually be less definite (Fig. 7.20(b)) and in loose or normally consolidated soils *plastic yielding* will occur without the formation of a slip surface, producing a barrel shaped appearance (Fig. 7.20(c)). In this latter case, a definite ultimate

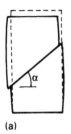 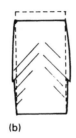 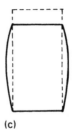

(a) (b) (c)

Fig. 7.20 Types of failure in the triaxial test
(a) Brittle shear failure (b) Partial shear failure (c) Plastic yielding failure

value for the deviator stress may not be discernible; an arbitrary value is therefore taken, corresponding to an axial strain of 20 per cent.

Applying failure criteria to obtain parameters
The triaxial test apparatus and procedures lend themselves to a variety of applications in soil testing which enable the measurement of a number parameters relating to shear strength and soil compression and swelling. To obtain parameters an appropriate criterion must be applied to a set of recorded results. There are four principal criteria applicable to strength testing:

(1) The Tresca criterion: undrained strength in term of total stresses:

$$\tau_f = c_u \quad \text{(discussed in Section 7.3)} \qquad [7.13]$$

(2) The Mohr–Coulomb criterion: drained strength in terms of effective normal stress:

$$\tau_f = c' + \sigma'_n \tan \phi' \quad \text{(discussed in Section 7.3)} \qquad [7.12]$$

Changes in volume are measured in terms of void ratio (e).

(3) The critical state strength criterion: drained or undrained strength in terms of mean normal stress:

$$q_f = Mp' \quad \text{(see Section 6.8)} \qquad [6.43(a)]$$

Changes in volume are measured in terms of specific volume (v).

(4) Hvorslev peak strength criterion: peak strength on a normalized line:

$$\frac{q'_f}{p'_c} = g'_p + H\frac{p'_f}{p'_c} \qquad [7.17]$$

where p'_c = consolidation stress

Values for g' are given by:

$$g' = g'_p p'_c \qquad [7.18]$$

Obtaining Mohr–Coulomb parameters
Many of the problems in geotechnical design require only shear strength parameters relating to normally consolidated soils or to measured initial peak states.

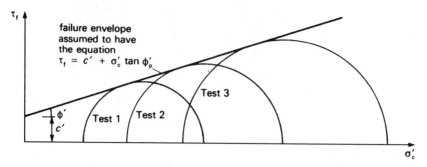

Fig. 7.21 Obtaining the Mohr–Coulomb envelope

In such cases, the Mohr–Coulomb failure criterion (see Sections 6.3 and 6.4) may be used in the interpretation of triaxial test results.

Several (usually three) specimens of the same soil are tested at different cell pressures and a Mohr circle (or usually a semi-circle) is drawn for each peak or ultimate failure stress. A common tangent to these failure circles is then drawn and, providing there is a reasonable goodness of fit, this may be taken as the *strength envelope* for the soil from which values of the angle of friction (ϕ'_c or ϕ'_p) and cohesion (c' or c_u) may be scaled (Fig. 7.21). A minimum of three circles will be required to give a reliable result.

Worked example 7.4   *A drained triaxial compression test carried out on three specimens of the same soil yielded the following results:*

Test No.		1	2	3
Cell pressure (kPa)		100	200	300
Deviator stress at failure (kPa)		210	438	644

Draw the shear strength envelope and determine the peak strength parameters, c' and ϕ'_p, assuming that the pore pressure remains constant during the axial loading stage.

The principal stresses are obtained as follows:

Minor principal stress, $\sigma_3 = \sigma_r =$ cell pressure

Major principal stress, $\sigma_1 = \sigma_a =$ cell pressure + deviator stress

Since $u_f = 0$, $\sigma'_1 = \sigma_1$ and $\sigma'_3 = \sigma_3$

Thus

Test No.	1	2	3
σ'_3 (kPa)	100	200	300
σ'_1 (kPa)	310	638	944

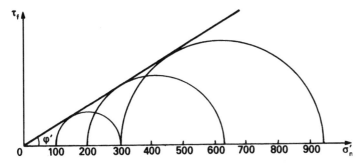

Fig. 7.22

The Mohr circles are shown in Fig. 7.22; when the best common tangent is drawn the Mohr–Coulomb strength envelope is obtained, from which:

$c' = 0$ and $\phi' = 31°$

The unconfined compression test
This is really a special case of triaxial compression, carried out at zero cell pressure ($\sigma_3 = 0$). The Mohr–Coulomb plot of the test results is shown in Fig. 7.23. Since only one circle can be drawn (corresponding to $\sigma_3 = 0$), the test is only applicable to fully saturated non-fissured clays, and only the undrained strength (c_u) can be measured.

The test may be carried out in the laboratory using a standard or slightly modified triaxial apparatus; neither the perspex cell nor the rubber membrane is required. A portable apparatus is also available (but not recommended) for tests carried out in the field (BS 1377:1990).

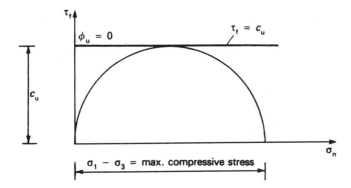

Fig. 7.23   Mohr–Coulomb plot for the unconfined compression test

Obtaining critical state parameters

The principal objective is to plot the critical state line (CSL) from which the parameters, slopes M and λ and intercept Γ, can be determined. The equations are:

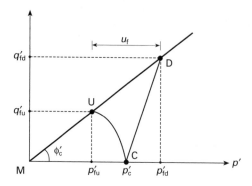

Fig. 7.24 Drained and undrained failure points on the CSL

Deviator stress at failure,	$q_f' = Mp'$
Specific volume,	$v = \Gamma - \lambda \ln p'$
in which	$q' = \sigma_a' - \sigma_r' = \sigma_a - \sigma_r$
mean normal stress,	$p = \frac{1}{3}(\sigma_a + 2\sigma_r)$
and effective mean normal stress,	$p' = p - u_f$

u_f = pore pressure at failure

Γ = specific volume intercept of CSL at $p' = 1.0$ kPa

Note that the critical state line is a common criterion for both drained and undrained tests carried out on soils on the wet side of critical. Figure 7.24 shows the stress paths to failure of two specimens of the same soil both having been consolidated to a mean stress of p_c'. For the drained test the path is CD and point D lies on the CSL. In the undrained test, with volume remaining constant, the pore pressure rises to u_f at failure; the path is CU; point U is also on the same CSL. The effective stresses at failure are:

Drained: p_{fd}' and q_{fd}'

Undrained: p_{fu}' and q_{fu}'

Peak strengths will be reached by soils initially on the dry side of critical, e.g. heavily overconsolidated soils. In the latter case the peak strength envelope is obtained from:

$$q_f' = g' + Hp_f' \qquad\qquad [7.4]$$

A normalized expression is given [eqn 7.17] and from eqn [7.18] the normalized value of the cohesion intercept is:

$$g_p' = g'/p_c'$$

Worked example 7.5   *A series of drained and undrained triaxial compression tests carried out on specimens of the same soil yielded the following results at points of failure:*

Test No. (D = drained, U = undrained)	D1	U1	D2	U2	D3	U3
Cell pressure, σ_r (kPa)	120	120	200	200	400	400
Total axial stress, σ_a (kPa)	284	194	493	320	979	645
Pore pressure at failure, u_f (kPa)	0	69	0	117	0	230
Specific volume, v_f	1.80	1.97	1.70	1.86	1.54	1.72

Plot the critical state line and obtain the critical state parameters M, Γ *and* λ.

First, calculate the following for each test:

Deviator stress, $\qquad\qquad q' = \sigma'_a - \sigma'_r = \sigma_a - \sigma_r$

Total mean normal stress, $\quad p = \frac{1}{3}(\sigma_a + 2\sigma_r)$

Effective mean normal stress, $\;\; p' = p - u_f$

Test No. (D = drained, U = undrained)	D1	U1	D2	U2	D3	U3
Deviator stress, q' (kPa)	164	74	293	120	579	245
Total mean normal stress, p	175	155	298	240	593	482
Effective mean normal stress, p'	175	76	298	123	593	252
$\ln p'$	5.16	4.33	5.70	4.81	6.39	5.53

The CSL is plotted in two elevations. In Fig. 7.25(a) $p':q'$ points are plotted and the best straight line (CSL) drawn through them and through the origin. The slope of the CSL is

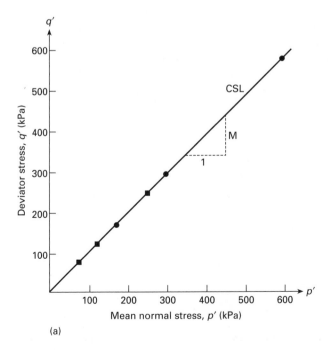

(a)

Fig. 7.25

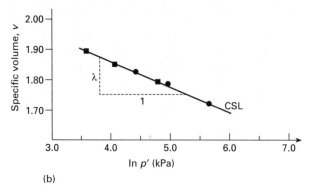

(b)

Fig. 7.25 Continued

measured from the plot:

$M = \underline{0.975}$

Figure 7.25(b) show the plot of v_f against ln p'. The best straight line (the CSL) is drawn through the line and its slope (λ) measured. The intercept (Γ) can be found by extending the straight to cut the ordinate $p' = 1.0$ kPa and reading off the corresponding specific volume.

Slope of CSL, $\lambda = \underline{0.211}$

Intercept, $\Gamma = \underline{2.89}$

Relationship between Mohr–Coulomb and critical state parameters
The critical state line represents the unique drained strength of a soil regardless of which failure criterion used, or even which test is used. For design and other practical purposes it may be convenient to express parameters with respect to one failure criterion having measured them with respect to another. The following equivalance expression can be used.

Critical strength

$$M = \frac{6 \sin \phi'_c}{3 - \sin \phi'_c} \quad \text{or} \quad \sin \phi'_c = \frac{3 M}{6 + M}$$

$$C_c = 2.3 \, \lambda \qquad \text{and} \qquad C_s = 2.3 \, \kappa$$

Peak strengths

$$H = \frac{6 \sin \phi'_p}{3 - \sin \phi'_p} \quad \text{or} \quad \sin \phi'_p = \frac{3 H}{6 + H}$$

$$g'_p \approx 2c'_p \qquad \text{or} \qquad g' \approx 2c'$$

7.9   Comparison of Mohr–Coulomb plots for different types of test

Three possible Mohr circles may be drawn to represent a single triaxial test for a given cell pressure. In an *undrained* test, with pore pressure measured during

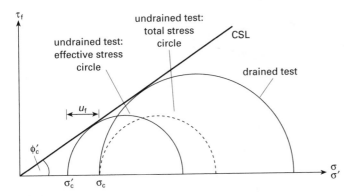

Fig. 7.26 Comparison of Mohr circles for drained and undrained tests

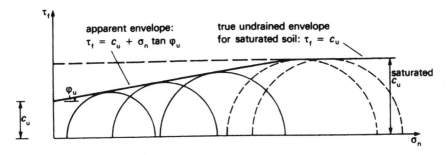

Fig. 7.27 True and fallacious undrained envelopes

axial loading, two circles may be drawn: one in terms of total stresses and the other in terms of effective stresses (Fig. 7.26). A *drained* test carried out on the same soil consolidated at the same cell pressure (σ_c) would yield an effective stress circle of greater diameter. Because the volume is reduced during axial loading in a drained test the soil becomes stiffer, and therefore the deviator stress at failure will be greater than that for the undrained test. For normally consolidated samples the two effective stress circles will both touch the critical state line (CSL), since this defines critical strength however it is measured.

Undrained strength should be related to the water content (or specific volume) of the soil and is measured in terms of total stresses. During undrained shearing in the test the volume should remain constant, so an objective of the consolidation stage is to ensure full saturation in the specimen. A problem arising in the interpretation of an undrained test is further illustrated in Fig. 7.27. At first, it appears that the shear strength of the soil is given by the line $\tau_f = c_u + \sigma_n \tan \phi_u$. However, this is a fallacy, since undrained strength is independent of cell pressure. In fact, if further tests were carried at different cell pressures or at different loading rates and more results added to the plot the total stress envelope would be found to be curved. This problem here is that volume change is occurring, probably due to the presence of air in the specimen which dissolves in the porewater as the stresses increase. Because of the low permeability of fine soils, this internal solution process takes time; at sufficiently high stresses all the

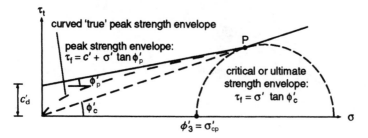

Fig. 7.28   Peak and critical strength plots of drained test results

air would dissolve and the soil would become saturated. The only usable value for c_u is one related to the *in situ* state of vertical and horizontal stress and water content.

In the interpretation of drained tests, if the specimens are normally consolidated at different stress levels before axial loading the critical strength envelope should pass through the origin, i.e. $c' = 0$ (Fig. 7.28).

Then $\tau_f = \sigma' \tan \phi'_c$

However, when the soil is on the dense (dry) side of critical drained test results will give the *peak strength* envelope:

$\tau_f = c'_p + \sigma' \tan \phi'_p$

This will apply to dense sands and to heavily overconsolidated clays, both of which must dilate in order to reach the critical state. The point of convergence (P) of the peak and critical strength envelopes represents the failure condition of a specimen consolidated at a cell pressure equal to the maximum historical consolidation stress (σ'_{cp}). At very low effective stresses the peak strength envelope is clearly curved and tends toward the origin.

Effect of stress history

The stress–strain behaviour of soil, especially clay, is significantly affected by its stress history. At a given depth an element of soil is subject to vertical and horizontal stresses due to the weight of overburden and any superimposed loading on the surface. Under the influence of these stresses the soil will have been consolidated during the course of time since its deposition, so that it is presently in a consolidated state. A *normally consolidated* soil is a material that has never been subject to stresses greater than those presently existing. An *overconsolidated* soil is material that, at some time in its past, has been subject to consolidating stress greater than that presently existing. This could arise, for example, where some of the overburden has been eroded, or where a thick layer of ice once imposed a surface surcharge before melting (see also Section 6.3).

Normally consolidated clay

If drainage is prevented from a saturated test specimen any increase in total stress results in an equal increase in pore pressure, leaving the effective stress unchanged. A number of identical saturated specimens tested undrained at

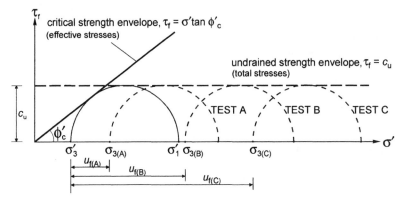

Fig. 7.29 Undrained test envelopes for a normally consolidated clay

different cell pressures will fail at the same deviator stress. The Mohr–Coulomb plot is shown in Fig. 7.29; the total stress circles have the same diameter and the envelope has zero slope ($\phi_u = 0$). In a drained test, as discussed previously, the excess pore pressure remains zero, the effective stress equals the total stress and the failure envelope ideally passes through the origin. The effective stress parameter c' is therefore zero for a normally consolidated clay.

Overconsolidated clay
Heavily overconsolidated clay is on the *dry* side of critical and tends to dilate significantly during shearing. Under undrained conditions the tendency to dilate brings about a decrease in pore pressure and the effective stress is increased. Thus, the undrained strength is greater than the drained strength, which is opposite to the behaviour of normally consolidated clay.

When the test consolidation pressure (p'_c or σ'_c) exceeds the historical pre-consolidation stress the specimen will follow the normal consolidation line. The strength envelope for overconsolidated specimens is curved, but can be assumed straight over a limited range of normal stress to give a cohesion intercept (c') (see Section 7.2).

7.10 Variations in undrained strength

The undrained strength of a soil is not a constant intrinsic property, but it is regulated by its volumetric state, which is a function of its stress history. The volumetric state of a soil can be defined by its specific volume (v), its void ratio (e), its water content (w), or its unit weight (γ). Under natural conditions of deposition, the void ratio and water content of a soil vary with depth and so therefore does its undrained strength; in a homogeneous soil undrained strength increases with depth.

The critical state theory (Section 6.8) can be used to provide evaluations of undrained strength. Following natural deposition, a clay will have consolidated one-dimensionally along the normal consolidation line to arrive at its present specific volume (Fig. 6.11 and eqn [6.25(a)]):

$$v = N - \lambda \ln p'$$

Under undrained conditions, failure will take place, at the same specific volume, at a point on the critical state line at which:

$$q' = Mp' \quad \text{and} \quad v = \Gamma - \lambda \ln p'$$

Rearranging and substituting, $\quad q' = M \exp\left(\dfrac{\Gamma - v}{\lambda}\right)$

Now the undrained strength, $\quad c_u = \frac{1}{2}q'$

Then, for a normally consolidated soil,

$$c_u = \frac{1}{2}M \exp\left(\frac{\Gamma - v}{\lambda}\right) \qquad [7.19]$$

However, for natural one-dimensional consolidation,

$$v = N_0 - \lambda \ln p_z'$$

where $\quad p_z' = $ mean normal effective stress at depth z

Hence $\quad c_u = \frac{1}{2}M \exp\left(\dfrac{\Gamma - N_0 + \lambda \ln p_z'}{\lambda}\right)$

$$= \frac{1}{2}M \exp\left(\frac{\Gamma - N_0}{\lambda} + \ln p_z'\right) \qquad [7.20]$$

Since both M and $(\Gamma - N_0)/\lambda$ are constants, c_u is seen to increase *linearly* with depth in a normally consolidated clay.

The variation in undrained strength with water content may be similarly demonstrated using eqn [7.19]. From eqn [3.11], for a saturated soil, $e = wG_s$. Then $v = 1 + wG$, and

$$c_u = \frac{1}{2}M \exp\left(\frac{\Gamma - 1 - wG_s}{\lambda}\right) \qquad [7.21]$$

In the case of an *overconsolidated* soil, the present *in situ* specific volume will be

$$v = v_{\kappa 0} - \kappa \ln p_z'$$

Hence $\quad c_u = \frac{1}{2}M \exp\left(\dfrac{\Gamma - v_{\kappa 0} + \kappa \ln p_z'}{\lambda}\right) \qquad [7.22]$

i.e. a *non-linear increase* with depth.

Worked example 7.6   *From oedometer and triaxial tests on a normally consolidated recent shallow lake clay the following parameters have been obtained:*

$\gamma = 18.0 \ kN/m^3 \quad M = 1.09 \quad \Gamma = 2.51 \quad e_0 = 1.59 \quad C_c = 0.299$

Using the critical state theory, draw up a predicted c_u/depth profile for depths down to 40 m, assuming the water table to be at the surface.

The vertical and horizontal effective stresses at depth z are:

$$\sigma'_v = (\gamma_{sat} - \gamma)z \quad \text{and} \quad \sigma'_h = K_0\sigma'_v$$

where $K_0 = 0.5$ for undrained conditions.

Then the mean normal stress, $\ p'_z = \tfrac{1}{3}(\sigma'_v + 2\sigma'_h)$

$$= \tfrac{1}{3}\sigma'_v(1 + 2K_0)$$

$$= \tfrac{2}{3}(18.0 - 9.81)z = 5.46z \ \text{kPa}$$

The oedometer test results are given for the NCL in terms of e and log σ'; the corresponding parameters for the v–ln p' line are from Section 6.3:

$\lambda = C_c/2.3 = 0.299/2.3 = 0.130$

$N_0 = 1 + e_0 = 1 + 1.59 = 2.59$

Then, from eqn [7.18]: $\quad c_u = \tfrac{1}{2}M \exp\left(\dfrac{\Gamma - N_0}{\lambda} + \ln p'_z\right)$

$$= \tfrac{1}{2} \times 1.09 \exp\left(\dfrac{2.51 - 2.59}{0.130} + \ln(5.46z)\right)$$

$$= 0.545 \exp[-0.615 + \ln(5.46z)]$$

Tabulated values are:

z (m)	1	2	5	10	15	20	25	30	35	40
c_u (kPa)	1.8	4	9	18	27	36	45	54	63	72

Approximately, $c_u = 1.8z$ kPa.

Worked example 7.7 *From laboratory tests on samples of a non-fissured overconsolidated clay the following parameters have been obtained:*

$M = 0.96 \qquad \lambda = 0.21 \qquad \Gamma = 3.09 \quad \gamma = 18.0 \ kN/m^3$
$\kappa = 0.052 \quad v_{\kappa 0} = 2.17 \quad K_0 = 0.8$

Estimate the change in undrained shear strength that will take place adjacent to an excavation down to a depth of 12 m.

From eqn [7.22],

$$c_u = \tfrac{1}{2}M \exp\left(\dfrac{\Gamma - v_{\kappa 0} + \kappa \ln p'_z}{\lambda}\right)$$

Then before excavation the undrained strength near the surface is:

$$c_u = \tfrac{1}{2} \times 0.96 \times \exp[(3.09 - 2.17)/0.21] = 38 \ \text{kPa}$$

At a depth of 12.0 m, assuming the water table to be below the excavated depth, the mean normal stress is:

$$p'_z = \tfrac{1}{3}\sigma'_v(1 + 2K_0)$$

$$= \tfrac{1}{3} \times 12.0 \times 18.0(1 + 2 \times 0.8) = 187.2 \ \text{kPa}$$

$$c_u = \tfrac{1}{2} \times 0.96 \exp\left(\frac{3.09 - 2.17}{0.21} + \frac{0.052}{0.21} \ln p'_z\right)$$

$$= 0.48 \times \exp[4.38 + 0.248 \times \ln(187.2)] = 140 \text{ kPa}$$

After excavation, the horizontal stress on the cut face is reduced to zero, so that:

$$p'_z = \tfrac{1}{3}\sigma'_v(1 + 2K_0) = \tfrac{1}{3} \times 12.0 \times 18.0 \times 1.80 = 129.6 \text{ kPa}$$

and $c_u = 0.48 \times \exp[4.38 + 0.248 \times \ln(129.6)] = 128 \text{ kPa}$

So the reduction in undrained strength of about 9 per cent.

A further reduction will take place as exposure continues and the water content (and therefore the specific volume) increases. The shear strength just below the centre of the excavation will also be reduced and may be expected to fall to 38 kPa.

7.11   *Further worked examples on the interpretation of triaxial tests*

Worked example 7.8   *The following results were obtained from undrained tests on specimens of a saturated normally consolidated clay.*

Cell pressure (kPa)	100	200	300
Ultimate deviator stress (kPa)	137	210	283
Ultimate pore pressure (kPa)	28	86	147

Determine (a) the effective stress parameters c' *and* ϕ'_c *and (b) the critical state parameter* M.

(a)   The ultimate effective stresses are:

$$\sigma'_a = \sigma_a - u_f$$

$$\sigma'_r = \sigma_r - u_f$$

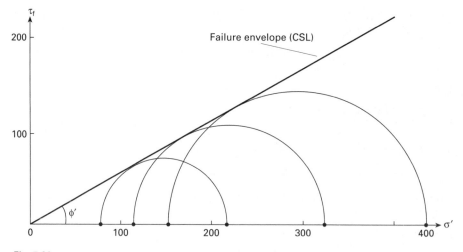

Fig. 7.30

| Effective radial stress, σ_r' (kPa): | 100 | 200 | 300 |
| Effective axial stress, σ_a' (kPa): | 78 | 114 | 153 |

The Mohr–Coulomb plot is shown in Fig. 7.30, from which the following values may be obtained:

Critical angle of friction, $\phi_c' = \underline{29°}$
Cohesion intercept, $c' = \underline{0}$

(b) The parameter M can be obtained from the relationship:

$$M = \frac{6 \sin \phi_c'}{3 - \sin \phi_c'} = \frac{6 \sin 29°}{3 - \sin 29°} = \underline{1.16}$$

Worked example 7.9 *The following results were obtained from undrained tests on specimens of an overconsolidated clay:*

Cell pressure (kPa)	100	250	400
Deviator stress at failure (kPa)	340	410	474
Deviator pore pressure (kPa)	−42	64	177

Determine the effective stress parameters c' and ϕ_p'

Effective radial stress = cell pressure – pore pressure

Diameter of a Mohr circle = deviator stress = $\sigma_a' - \sigma_r' = \sigma_a - \sigma_r$

σ_r (kPa)	100	250	400
u_f (kPa)	−42	64	177
σ_r'	142	186	223
$\sigma_a' - \sigma_r'$	340	410	474

Note that the effect of the negative pore pressure is to give $\sigma_r' > \sigma_r$.

The Mohr–Coulomb plot is shown in Fig. 7.31, from which:

$c' = \underline{31\ kPa}$ and $\phi_p' = \underline{27°}$

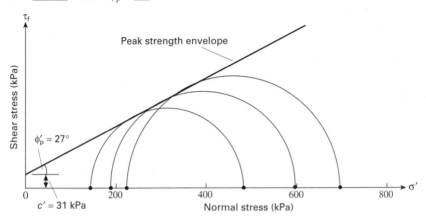

Fig. 7.31

Worked example 7.10   *In a undrained triaxial test on a specimen of clay normally consolid-ated at a cell pressure of 150 kPa, the ultimate deviator stress was 260 kPa and the ultimate pore pressure was 50 kPa. Draw the appropriate shear strength envelope and (a) determine the parameter ϕ', and (b) estimate the undrained strength c_u.*

Figure 7.32 shows the Mohr circle and envelope plot.

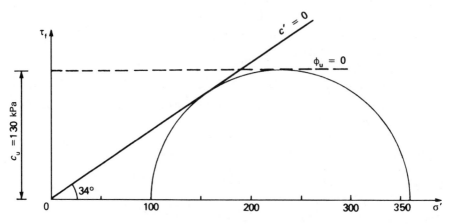

Fig. 7.32

(a)   At failure, $\sigma'_3 = 150 - 50 = 100$ kPa.
The envelope passes through the origin and $\phi'_c = \underline{34°}$.
(b)   The undrained strength is given by the height of the Mohr circle: $c_u = \underline{130\ \text{kPa}}$.
Note: an estimate based on just one reading test would not be acceptably reliable.

Worked example 7.11   *A normally consolidated clay was found to have shear strength parameters of $c' = 0$ and $\phi'_c = 26°$. Triaxial tests were carried out on three specimens of the soil.*

(a)   In Test 1, the specimen was consolidated under an isotropic stress of 200 kPa and the axial loading stage is undrained. Determine the ultimate deviator stress if the final pore pressure is 50 kPa.
(b)   In Test 2, the specimen is consolidated under an isotropic stress of 200 kPa and the axial loading stage is drained with the back pressure remaining at zero. Calculate the ultimate deviator stress.
(c)   In Test 3, both stages are undrained. Determine the pore pressure expected when the specimen reaches an ultimate deviator stress of 148 kPa, assuming that the specimen is saturated throughout.

The envelope plot is shown in Fig. 7.33.

(a) $\sigma'_a - \sigma'_r =$ ultimate deviator stress $= 200 - 50 = 150$ kPa
To obtain the Mohr circle, first draw the chord line AB at angle $\alpha = 45° + \phi'_c/2$ from $\sigma'_r = 150$ kPa. The bisecting normal to AB cuts the σ' axis at the centre of the circle (C), and the circle radius is AC = BC.
Then, $\sigma'_a - $ AD $= 235$ kPa
    Alternatively, the solution may be obtained analytically. In Fig. 7.30,

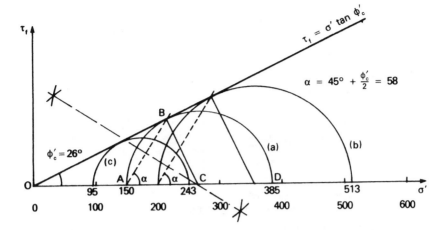

Fig. 7.33

$$\sin \phi'_c = \frac{BC}{OC} = \frac{\sigma'_a - \sigma'_r}{\sigma'_a + \sigma'_r}$$

From which $\dfrac{\sigma'_r}{\sigma'_a} = \dfrac{1 - \sin \phi'_c}{1 + \sin \phi'_c} = \tan^2\left(45° - \dfrac{\phi'_c}{2}\right)$

Then if $\phi'_c = 26°,\ \tan^2(45° - 13°) = 0.390$

and $\sigma'_a = \sigma'_r/0.390 = 150/0.390 = 385$ kPa

giving $\sigma'_a - \sigma'_r = 385 - 150 = \underline{235\ \text{kPa}}$

(b) The pore pressure at failure $= 0$, therefore $\sigma'_r = \sigma_r = 200$ kPa. The Mohr circle is drawn as before, giving: $\sigma'_a - \sigma'_r = \underline{313\ \text{kPa}}$.

Or analytically, $\sigma'_a = 200/0.390 = 513$ kPa, giving $\sigma'_a - \sigma'_r = \underline{313\ \text{kPa}}$.

(c) The ultimate deviator stress is the diameter of the Mohr circle, to which the strength envelope is a tangent, giving:

$\sigma'_a = 243$ kPa and $\sigma'_r = 95$ kPa

So that the final pore pressure will be: $u_f = \sigma_r - \sigma'_r = 200 - 95 = \underline{105\ \text{kPa}}$

Worked example 7.12 *In a undrained test on a specimen of saturated clay using the shear box apparatus, the following data were recorded:*

Ultimate shear stress = 90 kPa
Constant normal stress = 180 kPa

If the peak strength parameters are known to be $c' = 25$ *kPa and* $\phi'_p = 27°$, *what was the ultimate pore pressure in the specimen?*

The shear strength envelope is shown in Fig. 7.34 and is given by:

$\tau_f = c' + \sigma' \tan \phi'_p = 25 + 0.51\sigma'$

Then, for an ultimate shear stress of 90 kPa, the corresponding effective normal stress will be:

$\sigma' = (90 - 25)/0.51 = 128$ kPa

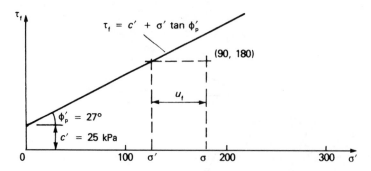

Fig. 7.34

and the ultimate pore pressure, $u_f = \sigma - \sigma'$
$$= 180 - 128 = \underline{52 \text{ kPa}}$$

Worked example 7.13   *The following critical state parameters are known for a normally consolidated clay:*

$M = 0.96$ $\lambda = 0.220$ $N = 3.18$ $\Gamma = 3.06$

In a triaxial compression test, a specimen of the clay was consolidated under an isotropic cell pressure of 300 kPa.

(a) *Calculate the ultimate values expected for the mean normal stress (p'), the deviator stress (q'), the axial stress (σ'_a) and the void ratio (e_f), if the axial loading stage is fully drained.*

(b) *Calculate the ultimate values for* p', q' *and* σ'_a *for a test in which the axial loading stage is undrained. In this case, what will be the ultimate pore pressure (u_f)?*

(a) In the drained axial loading stage:

$du = 0$, $d\sigma_r = 0$ and $dq'/dp' = 3$ (Section 6.4(d))

Thus, at failure, $q'_f = 3(p'_f - p'_c)$, where p'_f = mean normal stress at failure and p'_c = mean normal isotropic cell pressure.
Also, from eqn [6.43(a)]

$q'_f = Mp'_f$

Hence, the ultimate mean normal stress is:

$$p'_f = \frac{3\ p'_c}{3 - M} = \frac{3 \times 300}{3 - 0.96} = \underline{441.2 \text{ kPa}}$$

Ultimate deviator stress, $q'_f = 0.96 \times 441.2 = \underline{423.5 \text{ kPa}}$

Ultimate axial stress, $\sigma'_a = q'_f + \sigma'_a = \underline{723.5 \text{ kPa}}$

The ultimate specific volume will be on the critical state line:

$v_f = \Gamma - \lambda \ln p'_f$

$= 3.06 - 0.220 \ln 441.2 = \underline{1.720}$

(b) In an undrained axial loading stage the volume will remain constant, i.e. equal to that it was brought to on the normal consolidation line:

$v_f = v_c = N - \lambda \ln p' = 3.18 - 0.22 \ln 300 = 1.925$

Then, from eqn [6.45(b)]:

$$p'_f = \exp\left(\frac{\Gamma - v_f}{\lambda}\right) = \exp\left(\frac{3.06 - 1.925}{0.220}\right) = \underline{174 \text{ kPa}}$$

and $\quad q'_f = 0.96 \times 174.0 = \underline{167 \text{kPa}}$

and $\quad \sigma'_a = 167.0 + 300 = \underline{467 \text{ kPa}}$

Now the total stress, $p'_f = p'_c + \frac{1}{3}q'_f = \underline{355.7 \text{ kPa}}$.
Hence the ultimate pore pressure, $u_f = p_f - p'_f = \underline{181.7 \text{ kPa}}$.

Worked example 7.14 *The following data were recorded during the isotropic consolidation of a specimen of clay.*

Mean normal stress, p' (kPa)	25	50	100	200	300	400	600	25
Change in volume (ml)	0	0.67	1.39	2.33	4.75	6.54	8.92	5.69

Physical properties of the specimen at p' = 25 kPa:

$G_s = 2.72 \quad \gamma = 18.6 \text{ kN/m}^3 \quad w = 34\% \quad Volume = 86.19 \text{ ml}$

(a) *Plot the v/ln* p' *curves and hence determine values for the parameters* λ, κ, N *and* v_κ.
(b) *Determine the variation in the volumetric stiffness* (K') *with the mean normal stress* (p').

(a) It is first necessary to establish the specific volume corresponding to $p' = 25$ kPa.

Dry density, $\qquad \rho_d = \dfrac{18.6}{9.81(1 + 0.34)} = 1.415 \text{ Mg/m}^3$

Specific volume, $\qquad v = 2.72/1.415 = 1.922$

Now, volumetric strain, $\quad \varepsilon_v = \dfrac{\Delta V}{V_0} = \dfrac{\Delta v}{v_0}$

Therefore, $\qquad \Delta v = (1.922/86.19)\Delta V$

The specific volumes corresponding to p' can now be calculated:

p'	25	50	100	200	30	400	600	50
Δv	0	0.015	0.031	0.052	0.106	0.146	0.199	0.127
v	1.922	1.907	1.891	1.870	1.816	1.776	1.723	1.795

Figure 7.35 shows the $v/\ln p'$ plot.

The parameter κ is the slope of the first part of the curve:

$$\kappa = \frac{1.922 - 1.891}{\ln(100/25)} = \underline{0.0224}$$

This may be verified from the slope of the unloading curve:

$$\kappa = \frac{1.795 - 1.723}{\ln(600/25)} = 0.0227$$

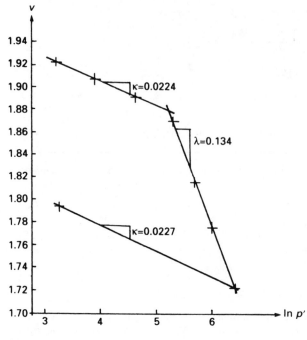

Fig. 7.35

The parameter λ is the slope of the normal consolidation line (NCL):

$$\lambda = \frac{1.870 - 1.723}{\ln(600/200)} = \underline{0.134}$$

N = the intercept of the isotropic NCL at $p' = 1.0$ kPa

$\quad = 1.870 + 0.134 \times \ln 200 = \underline{2.580}$

v_κ = the intercept of the swelling–reloading curve (SRL) at $p' = 1.0$ kPa

$\quad = 1.922 + 0.224 \times \ln 25 = \underline{1.994}$

(b)   The volumetric stiffness, $K' = \dfrac{dp'}{d\varepsilon_v}$

But $\qquad d\varepsilon_v = \dfrac{dv}{v}$ and $dv = -\kappa\, d(\ln p') = -\kappa \dfrac{dp'}{p'}$

Hence $\qquad K' = \dfrac{vp'}{\kappa}$

Thus, for $p' = 100$ kPa, $K' = \dfrac{1.891 \times 100}{0.0224} = \underline{8442 \text{ kPa}}$

$\qquad\qquad p' = 50$ kPa, $K' = \dfrac{1.907 \times 50}{0.0224} = \underline{4257 \text{ kPa}}$

Overconsolidated soil may be assumed to behave elastically, i.e. in this case when $25 < p' < 180$, $K' \approx 85p'$ kPa.

Worked example 7.15 *Drained triaxial tests were carried out on three specimens of a saturated clay and the following data recorded:*

Initial specimen dimensions: length = 76 mm, diameter = 38 mm

Specimen	1	2	3
Cell pressure (kPa)	100	200	400
Ultimate axial load (kN)	0.168	0.344	0.696
Change in length (mm):			
during consolidation, Δl_c	0.73	1.77	2.82
during axial loading, Δl_a.	9.38	12.24	15.38
Change in volume (ml):			
during consolidation, ΔV_c,	2.48	6.02	9.90
during axial loading, ΔV_a.	5.93	6.05	6.07

(a) *Using the Mohr–Coulomb failure criterion, determine the drained shear strength parameter ϕ'_c.*

(b) *Use the data to establish the critical state parameters **M** and **Γ**. Assume that the isotropic normal consolidation line has the parameters $\lambda = 0.17$ and N = 2.98.*

(a) The cross-sectional area at failure may be estimated from eqn [7.15].

$$A = A_0 \frac{1 - \Delta V/V_0}{1 - \Delta l/l_0}$$

where $\Delta V = \Delta V_c + \Delta V_a$ and $\Delta l = \Delta l_c + \Delta l_a$

Original volume, $V_0 = \dfrac{\pi}{2} \times 38^2 \times 76/1000 = 86.19$ ml

Original area, $A_0 = \dfrac{\pi}{2} \times 38^2 = 1134$ mm^2

e.g. specimen l: $A = 1134 \times \dfrac{1 - (2.48 + 5.93)/86.19}{1 - (0.73 + 9.38)/76.0} = 1180$ mm^2

The calculations are tabulated below:

Specimen	1	2	3
$1 - \Delta V/V_0$	0.9024	0.8600	0.8147
$1 - \Delta l/l_0$	0.8670	0.8157	0.7605
Final area, A (mm^2)	1180	1196	1215
Deviator stress, $\sigma'_a - \sigma'_3$ (kPa)	142	288	573

The Mohr–Coulomb plot is shown in Fig. 7.36(a), from which:

$\phi'_c = \underline{25°}$

(b) The mean normal stresses must first be calculated:

Consolidation stress, $p'_c = \frac{1}{3}(\sigma'_1 + \sigma'_2 + \sigma'_3) = \sigma'_c$

Ultimate stress, $p'_f = \frac{1}{3}(\sigma'_a + 2\sigma'_a)$

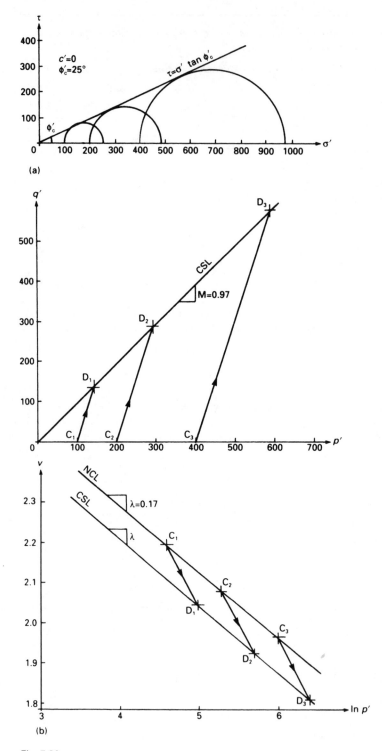

(a)

(b)

Fig. 7.36

The values of specific volume after consolidation (v_c) will lie on the normal consolidation line and on the critical state line at failure (v_f). Thus, for specimen 1:

Axial stress, $\sigma'_a = \sigma'_3 + q' = 242$ kPa

Mean normal stress, $p'_f = \frac{1}{3}(242 + 2 \times 100) = 147$ kPa

Consolidated specific volume, $v_c = N - \lambda \ln p'_c$

$$= 2.98 - 0.17 \ln 147 = 2.197$$

Final specific volume, $v_f = v_c + \Delta v$

Now, volumetric strain during axial loading, $\varepsilon_v = \dfrac{\Delta V}{V_0} = \dfrac{\Delta v}{v_0}$

Thus $\quad \Delta v = \dfrac{2.197 \times 5.93}{86.19 - 2.48} = 0.156$

Hence, $\quad v_f = 2.197 - 0.156 = 2.041$

and the CSL parameters are: $\quad M = q'_f/p' = \underline{0.970}$

$$\Gamma = 2.197 + 0.17 \ln 147 = \underline{2.89}$$

The results are tabulated below and the plot shown in Fig. (7.36(b)).

Specimen		1	2	3
$p'_c = \sigma'_c$	(kPa)	100	200	400
q'_f	(kPa)	142	288	573
σ'_a	(kPa)	242	488	973
p'_f	(kPa)	147	296	591
v_c		2.197	2.079	1.961
Δv		0.156	0.157	0.156
v_f		2.041	1.922	1.805
M		0.970	0.973	0.970
Γ		2.89	2.89	2.89

7.12 Sensitivity of clays

Certain clays, especially those of post-glacial origin, are sensitive to disturbance of their natural internal structure by remoulding. The ratio of the undrained strength after remoulding to the undisturbed undrained strength is known as the *sensitivity* of the clay.

Sensitivity, $S_t = \dfrac{\text{undisturbed undrained strength}}{\text{remoulded undrained strength}}$ [7.23]

Most ordinary clays have sensitivity values up to about 4, while some over-consolidated clays of low liquidity index have very low values and are termed insensitive; some so-called quick clays may have sensitivities of over 100. Bjerrum (1954) found Norwegian clays of low salinity to have sensitivities as high as 500 and the sensitivity of the Canadian Leda clay has been reported at over 700 (Sangrey, 1972). Two possible explanations exist for sensitive behaviour.

Table 7.1   Sensitivity of clays

Sensitivity S_t	Description
1	Insensitive
1–4	Low sensitivity
4–8	Sensitive
8–16	Extra-sensitive
>16	Quick

Where marine clays and clayey silts have been post-glacially uplifted and the original sodium saline porewater has been diluted by rainwater, an open skeletal texture appears to result. A similar open texture is thought to result from a degree of cementation in the original structure, which is then subsequently disrupted. These open textures are fragile and, with shearing disturbance, collapse quickly with an accompanying rapid rise in pore pressure. The sudden decrease in effective intergranular stress, and hence shear strength, causes liquefaction in many sensitive soils, resulting in catastrophic landslips and flow-slides.

A sensitivity classification is given in Table 7.1.

7.13   Measurement of residual strength

When the shear strain is very large clay soils exhibit a residual strength which is much lower than the peak strength (see also Section 7.1):

$$\tau_f = \sigma'_r \tan \phi'_r \qquad\qquad [7.13]$$

In a detailed study of the long-term stability of cuttings in London Clay, Professor A. W. Skempton (1964) observed that the values of shear strength on long-established slip surfaces tended to be considerably less than corresponding laboratory peak strengths. He also noted that the older the slip surface and the greater the displacement, the lower was the shear strength obtained by back analysis.

Further work by Bishop *et al.* (1971), using a newly designed ring shear apparatus, confirmed that in soils with a large clay content (>50 per cent) a dramatic fall in strength occurs with large continuous displacements (>1 m). Typical reported values measured in tests and determined by back-analysis of actual slips are:

	Peak strength		*Residual strength*	
	c'	ϕ'_p	c'_r	ϕ'_r
Brown London clay	12 kPa	22°	1 kPa	11°
Atherfield clay	21 kPa	18°	1 kPa	16°
Weald clay	48 kPa	23°	4 kPa	10°

It is also apparent that in first-time slides the value of ϕ_r' is similar to, or only slightly less than, that of ϕ_p', leading to the conclusion that the *amount* of strain is important. A likely explanation is that the flaky clay particles adjacent to the slip plane become re-orientated into a more parallel arrangement, thus producing a smoother surface. Displacements of several millimetres have only a marginal effect on reducing ϕ', but movements of a metre or more are sufficient to cause a substantial decrease in strength. Plastic yielding probably occurs first at asperity contacts (Calladine, 1971).

Also of significance is the curvature of the residual shear strength envelope (Bromhead and Dixon, 1986; Lupini *et al.*, 1981). Hawkins and Privett (1985, 1986) point out that it is inadvisable to base the t_f/s_r' plot on only three tests, since this may lead to an impression of a straight-line envelope ($\tau_f = \sigma' \tan \phi_r'$). When six or more points are plotted over a wider range of normal stress values, the curvature of the envelope becomes noticeable, especially at low values of σ'. A plot of the residual friction coefficient (tan $\phi_r' = \tau_f/\sigma'$) is a clearer way of presenting residual strength characteristics.

Residual strength tests

(a) Shear box or triaxial tests on natural slip surfaces
Carefully cut specimens that include a portion of an existing slip surface are obtained from the site of a slip and tested in a conventional manner, using a shear box or triaxial test apparatus. Most of the practical difficulties lie in the sampling process and in the subsequent orientation of the slip plane. This method is unsuitable for routine purposes.

(b) Reversible shear box tests
A specially adapted shear box is used in which the relative displacement of the two halves of the specimen may be reversed an unlimited number of times. The potential cumulative displacement is therefore extended from a few millimetres to several metres if need be. The rate of shear must be slow enough to allow full dissipation of the excess pore pressure. At least ten reversals will usually be necessary to develop a realistic residual state and it is recommended that at least six tests are carried out over the appropriate range of normal stresses.

(c) Ring shear apparatus
The suggestion for a ring shear apparatus was first made in the 1930s by Hvorslev and was revived by Bishop *et al.* (1971). In 1979, Bromhead published the design of a relatively inexpensive apparatus, which is now commercially available. A small remoulded specimen of the soil is prepared in the form of a square-sectioned ring. The top half of the ring is made to rotate relative to the bottom half (Fig. 7.37) while the normal stress (σ') and the shear stress (τ) are measured. An initial failure surface is formed by rapidly rotating the ring. The speed is then gradually reduced and rotation maintained until the residual state is developed. By varying the normal stress, several points on the τ_f/σ' envelope can be obtained relatively easily.

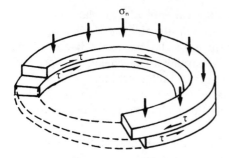

Fig. 7.37   Ring shear test specimen

From a number of studies (Hawkins and Privett, 1985; Bromhead and Dixon, 1986; Bromhead and Curtis, 1983; and others) it is clear that ring shear results correlate well with tests on natural slip surfaces and the results of back analyses. The results of ring shear tests and reversible shear box tests also appear to be in close agreement. Ring samples are simple to prepare from remoulded soil and it is easier to obtain envelope points in satisfactory numbers.

7.14   Development of failure: soils wetter or drier than critical

The Mohr–Coulomb failure theory depends on the assumption that a limiting uniform shear stress develops on a surface of failure. The ordinary interpretation of direct shear and triaxial test results and the analysis of slope stability by a slip circle method are examples of this approach. It is quite apparent, however, that in many cases a definite shear surface does not develop, but that, as the soil yields, contraction takes place.

The reason for this lies in the fundamental difference between the behaviour, while shearing, of soils that are drier than the critical value and those that are wetter (see Section 6.8). Soils on the *dry* or *dense* side of critical tend to dilate and then to fail by discontinuous slipping along a well-defined slip surface. The peak shear strength values associated with this type of failure are predicted reasonably well by the Mohr–Coulomb theory.

Contrastingly, in normally consolidated soils with void ratios on the *wet* or *loose* side of critical, the structure tends to contract and even collapse or flow; slip surfaces do not develop. *Flow slides* tend to develop in this way in clay slopes when the void ratio is increased by freeze/thaw processes. Localised increases in void ratio may occur in spoil tips due to drainage of sub-tip water. Small local displacements resulting in a decrease in shear strength may lead to progressive flow-type failure.

In some overconsolidated *fissured* clays, such as London Clay, especially when they are exposed to sub-aerial weathering, there is a gradual decline in strength. This effect starts adjacent to fissures and may lead to progressive failure. It is suggested that the *critical state strength* ($\tau_f = \sigma' \tan \phi'_c$) should be taken for stability calculations, rather than the *peak strength* ($\tau_f = c' + \sigma' \tan \phi'_p$). In fact, peak strengths should only be invoked when it is known that strains will be small.

7.15   *Measurement of pore pressure coefficients* A *and* B

It was shown in Section 4.7 that the changes in pore pressure induced by changes in isotropic stress or uniaxial stress, or both, can be obtained from the following general expression:

$$\Delta u = B[\Delta\sigma_3 + A(\Delta\sigma_1 - \Delta\sigma_3)] \hspace{3cm} [4.11]$$

in which A and B are termed *pore pressure coefficients.*

The triaxial test apparatus is ideally suited to procedures devised to measure the pore pressure coefficients. The soil specimen is first consolidated and a back pressure maintained to give an initial pore pressure. Very often the back pressure will be chosen to simulate *in situ* conditions.

When the consolidation stage has been completed, the cell pressure is raised ($\Delta\sigma_3$) with drainage prevented and the pore pressure response (Δu_0) is noted. The coefficient B is obtained from:

$$B = \Delta u_0 / \Delta\sigma_3 \hspace{4cm} [7.24]$$

To determine coefficient A, an increment of axial load is applied to produce an increase in deviator stress ($\Delta\sigma_1 - \Delta\sigma_3$). The resulting change in pore pressure (Δu_1) is noted and then:

$$AB = \bar{A} = \Delta u_1 / (\Delta\sigma_1 - \Delta\sigma_3) \hspace{3cm} [7.25]$$

An overall coefficient can also be obtained corresponding to the change in the major principal stress:

$$\bar{B} = (\Delta u_0 + \Delta u_1) / \Delta\sigma_1 = B[1 - (1 - A)(1 - \Delta\sigma_3 / \Delta\sigma_1)] \hspace{2cm} [7.26]$$

where B is obtained at constant lateral strain by maintaining the same ratio σ_3/σ_1 as σ_3 and σ_1 are increased together.

Worked example 7.16   *Using the triaxial apparatus, a soil sample was first of all consolidated at an isotropic cell pressure of 600 kPa, with the back pressure maintained at 300 kPa. Then, with the drains closed, the cell pressure was raised to 720 kPa, resulting in the pore pressure increasing to 415 kPa. Following this, the axial load was raised to give an increase in deviator stress of 550 kPa, while the cell pressure was held constant. The final pore pressure reading was 562 kPa. Calculate the pore pressure coefficients* B, A *and* B̄.

Following an increase in isotropic stress, $\Delta\sigma_3 = 720 - 600$. The pore pressure change was $\Delta u_0 = 415 - 300$. Therefore

$$B = \frac{\Delta u_0}{\Delta\sigma_3} = \frac{415 - 300}{720 - 600} = \underline{0.958}$$

Then an increase in deviator stress ($\Delta\sigma_1 - \Delta\sigma_3$) = 550 kPa produced a further change in pore pressure of $\Delta u_1 = 562 - 415$. Therefore

$$AB = \frac{\Delta u_1}{\Delta\sigma_1 - \Delta\sigma_3} = \frac{562 - 415}{550} = \underline{0.267}$$

Hence $A = \dfrac{0.267}{0.958} = \underline{0.278}$

The overall coefficient $\bar{B}$ is the ratio of the overall change in pore pressure to the overall change in major principal stress:

$$\bar{B} = \frac{\Delta u_f}{\Delta \sigma_1} = \frac{\Delta u_0 + \Delta u_1}{\Delta \sigma_3 + (\Delta \sigma_1 - \Delta \sigma_3)} = \frac{562 - 300}{550 - 120} = \underline{0.391}$$

Worked example 7.17   *During consolidated undrained triaxial tests, the following data were recorded:*

Change in length of specimen (mm)	0	0.75	1.50	3.00	6.00	9.10	10.6
Deviator stress (kPa)	0	120	200	280	360	420	460
Pore pressure (kPa)	0	103	153	192	215	232	248

Cell pressure = 250 kPa   Original length of specimen = 76.0 mm

Determine the relationship between the pore pressure coefficient A and axial strain.

Axial strain, $\varepsilon_a = \Delta l/l_0$.

For $\Delta l = 0.75$ mm, $\varepsilon_a = 0.75/76 = 0.01$

Pore pressure coefficient, $B = 1.0$ for saturated soil

Then, from eqn [7.25], $A = \dfrac{\Delta u_1}{\Delta \sigma_1 - \Delta \sigma_3}$

For $(\Delta \sigma_1 - \Delta \sigma_3) = 120$ kPa, $A = 103/120 = 0.858$

The complete set of calculations is tabulated below.

$(\Delta \sigma_1 - \Delta \sigma_3)$	0	120	200	280	360	420	460
Axial strain, ε_a	0	0.01	0.02	0.04	0.08	0.12	0.14
Pore pressure coefficient, A	0	0.858	0.765	0.686	0.597	0.552	0.539

Figure 7.38 shows a plot of A/ε_a.

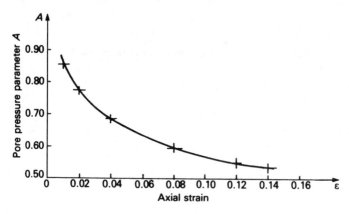

Fig. 7.38

Worked example 7.18   *The following data at failure were recorded during a series of undrained tests on an overconsolidated clay:*

Cell pressure (kPa)	100	200	350	500
Deviator stress (kPa)	286	374	513	652
Pore pressure (kPa)	−43	−12	39	87

Determine: (a) the peak strength parameters c′ *and* ϕ'_p *for the soil, and (b) the relationship between the pore pressure coefficient* A_f *and the consolidation ratio* R_p *if the consolidation stress,* $\sigma'_c = 600\ kPa$.

(a)   The Mohr–Coulomb plot is shown in Fig. 7.39(a), from which:

   c′ = <u>27 kPa</u>   and ϕ'_p = <u>25°</u>

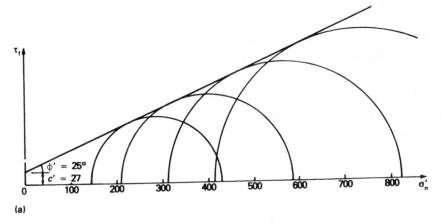

(a)

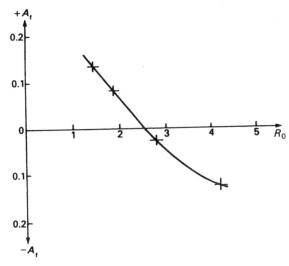

(b)

Fig. 7.39

(b)   From eqn [6.28], the overconsolidation ratio, $R_p = \sigma_c'/\sigma_3'$,

where $\sigma_c' =$ consolidation stress

$\sigma_3' =$ effective radial stress at failure

$=$ cell pressure – pore pressure

Pore pressure coefficient, $B = 1.0$ for saturated soil

Then, from eqn [7.25], $A = \dfrac{\Delta u_1}{\Delta\sigma_1 - \Delta\sigma_3}$

The results are tabulated below and a plot of A_f/R_p shown in Fig. 7.39(b).

σ_3'	143	212	311	413
R_p	4.2	2.8	1.9	1.45
Δu_1	−43	−12	39	87
$\Delta\sigma_1 - \Delta\sigma_3$	286	374	513	652
A_f	−0.15	−0.03	0.08	0.13

Worked example 7.19   *By means of undrained triaxial tests, the peak strength parameters of an overconsolidated clay were found to be:*

$c' = 10\ kPa$ and $\phi_p' = 24°$

The value of A_f for the in situ conditions is estimated at −0.18. In an unconfined compression test the specimen sheared at a compressive stress of 162 kPa. Determine the initial value of the suction pore pressure in the soil.

Maximum compressive stress, $\sigma_1' - \sigma_3' = 162$ kPa

and $\sigma_3' = 0$

Therefore, $u_1 = -0.18 \times 162 = -29.16$ kPa

Let the initial pore pressure $= u_0$

Then, at failure: $\sigma_3' = \sigma_3 - u_0 + 29.16 = 29.16 - u_0$

$\sigma_1' = 29.16 - u_0 + 162 = 191.16 - u_0$

From the Mohr–Coulomb plot (Fig. 8.13):

$$\frac{c' \cot \phi' + \sigma_3'}{c' \cot \phi' + \sigma_1'} = \tan^2\left(45° - \frac{\phi'}{2}\right)$$

Now, $c' \cot \phi' = 10 \times \cot 24° = 22.46$ kPa

and $\tan^2(45° - \phi'/2) = 0.422$

Then $\dfrac{22.46 + 29.16 - u_0}{22.46 + 191.16 - u_0} = 0.422$

giving $u_0 = \dfrac{-90.15 + 51.6}{0.578} = \underline{-66.7\ kPa}$

7.16 Stress path tests

The results of tests carried out to establish general design parameters are usually presented in terms of the final failure condition. No account is taken of the sequence of stress changes throughout either the tests themselves or in the design problems for which the parameters are required. While this is satisfactory for much in routine design work, it is often necessary to obtain specific design parameters for more complex geotechnical problems, e.g. detailed changes in pore pressures or volumetric and shear strains.

The stress–strain/volumetric behaviour of soil depends on the sequence of stress changes that occur, i.e. along a *stress path* (see also Section 6.6). *Stress path tests* are designed to simulate such changes expected to occur in a geotechnical problem. The procedure must be designed to control and monitor throughout the test all changes of stress, strain, pore pressure and volume.

The equipment required for stress path tests is basically that used for conventional triaxial tests on 100 mm diameter samples (Atkinson and Clinton, 1986). An essential modification is the inclusion of a hydraulic axial loading cell to facilitate stress-controlled loading. This enables a more accurate determination of stress–strain behaviour at low strains than is possible in constant-rate-of-strain tests. The axial load is measured using an electronic load cell and the cell and pore pressures using pressure transducers. A displacement transducer is used to measure axial strain, and a similar device is incorporated in the volume-change gauge.

The test variables are monitored and logged at frequent intervals using a computer. The computer is also programmed to respond to monitored data and can make control adjustments using stepper motors to operate hydraulic valves. Thus, predetermined stress paths may be followed and relevant changes continuously monitored and recorded. Subsequently, the computer will process the results and produce VDU display and/or hard copy output.

A number of studies (Atkinson and Clinton, 1986; Atkinson *et al.*, 1986; and others) have shown that while the angle of friction and the stiffness of soil are relatively unaffected, intermediate and especially undrained pore pressure and strain behaviour are significantly stress-path dependent. Therefore, undrained stability, volumetric and shear strains, and pore pressure changes should be assessed using stress path tests. In the recommended method, undisturbed or remoulded specimens (100 mm diameter) are isotropically consolidated at a stress somewhat higher than that occurring *in situ* and then are allowed to swell back to a prescribed value, e.g. *in situ* stress. This approach has also been recommended in the SHANSEP (stress history and normalised engineering properties) method (Ladd and Foott, 1974; Coatsworth, 1986).

Normal consolidation and K_0 line

The procedure may be used simply to determine normal consolidation parameters for either isotropic or one-dimensional loading, as well as evaluating K_0, or as a preliminary to further stress path tests. Remoulded specimens are consolidated in a stress-path cell to a reasonably high maximum stress, e.g. $p'_m = 700$ kPa, and then allowed to swell to the required consolidation stress (p'_c). Figure 7.40 shows

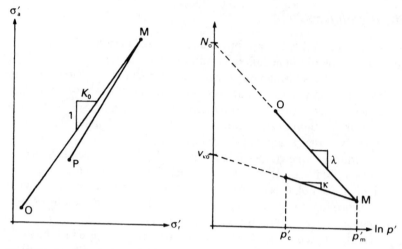

Fig. 7.40 Stress path for one-dimensional consolidation and swelling

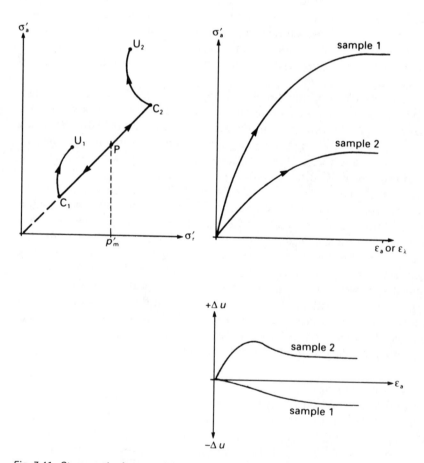

Fig. 7.41 Stress paths for consolidated undrained triaxial tests

a typical stress path plot (OMP) for one-dimensional consolidation and swelling from which λ, κ, N_0 and K_0 would then be determined.

Conventional undrained triaxial compression

In a conventional triaxial procedure the samples are consolidated isotropically before being subjected to triaxial compression by increasing the axial stress (σ_a), while keeping the radial stress (σ_a = cell pressure) constant. In Fig. 7.41, stress paths for two samples are shown. Sample 1 is overconsolidated ($p_0' < p_m'$) and therefore the pore pressure decreases as σ_a increases. Sample 2 is normally consolidated ($p_0' > p_m'$) and therefore so the pore pressure increases.

Problem simulated stress paths

In Fig. 7.42(a), five typical geotechnical design situations are shown. The points of interest all lie originally at the same depth and have the same stress history, but in the proposed constructions (foundation, active pressure, passive pressure, excavation and slope) they will undergo changes along different stress paths (Fig. 7.42(b)). In laboratory tests stress paths may be chosen that will attempt to simulate those envisaged in the problem.

(a)   *Beneath a foundation*
   The first stage, representing relatively fast construction, is undrained (OA), and this is followed by consolidation (AB), where dissipation of excess pore pressure occurs. Finally, the undrained loading is continued to failure (BC).

(b)   *Active side of a retaining wall*
   Excavation in front of the wall reduces the horizontal total stress, but the vertical total stress remains relatively unchanged. In the first stage, the cell pressure is lowered (OD), with the total axial stress held constant under undrained conditions. The sample is then allowed to swell (DE) at the original back pressure until failure occurs.

(c)   *Passive side of a retaining wall*
   Excavation will reduce the total vertical stress, but the tendency for the wall to move will increase the horizontal stress. This is represented by the undrained stress path (OF), after which the excess pore pressure is allowed to dissipate (FG) and then the loading continued until failure (in tension) occurs.

(d)   *As a result of excavation*
   Excavation reduces the total vertical stress, and this is accompanied, to a lesser extent, by a reduction in horizontal total stress. From the initial condition, the total axial stress is reduced under undrained conditions (OI), after which swelling is allowed (IJ). Finally, the stress path is (JK), and, although failure may not be achieved, the pore pressure and strain changes will be of interest.

(e)   *Pore pressure change in a slope*
   Slope failures are often due to an increase in pore pressure, while the total vertical and horizontal stresses remain constant. In the test procedure, therefore, the axial and radial stresses are held constant while the back pressure is slowly raised. The effective stress path is (OL); there is no total stress path.

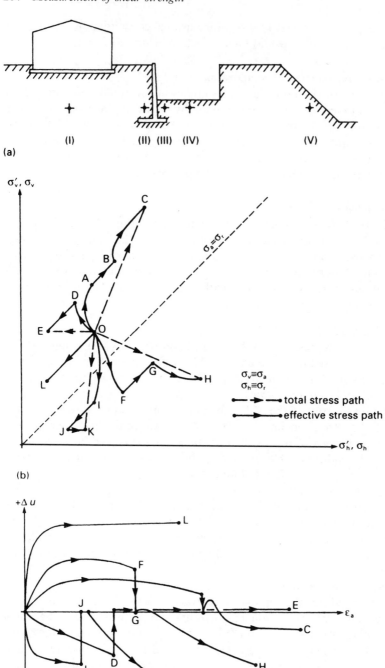

Fig. 7.42 Problem-simulated stress-path tests (reproduced by permission of the Geological Society from Atkinson and Clinton, 1986)

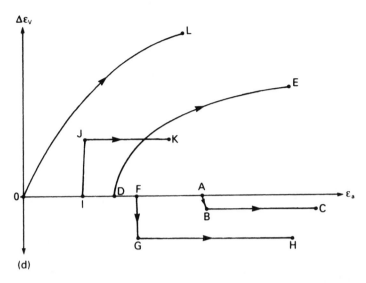

(d)

Fig. 7.42 Continued

Stress path tests demonstrate that the stress–strain behaviour of soil in tension is quite different from that in compression. More significant, however, are the differences in pore pressure and stiffness change with respect to strain. These differences have been demonstrated by Atkinson *et al.* (1986), using stress path tests on reconstituted samples of London Clay (Table 7.2). The five samples were prepared from a slurry and one-dimensionally consolidated to $p'_m = 600$ kPa, before being allowed to swell to $p'_c = 200$ kPa.

Table 7.2 Summary of stress-path tests on London clay (reproduced by permission of the Geological Society from Atkinson et al., 1986)

Test	E_u^* (MPa)	p'_0 (kPa)	v_0 (kPa)	c_u	$\dfrac{E_u}{vp'_0}$	$\dfrac{E_u}{c_u}$	Loading
1	5.2	50	1.80	63	60	135	Triaxial u/d
2	17.5	200	1.73	133	50	130	σ'_a increasing
3	13.6	400	1.70	171	20	80	σ'_r constant
4	15.3	200	1.73	113	45	135	σ'_a constant σ'_r reducing
5	6	200	1.72	120	20	50	σ'_a reducing σ'_r constant

* At $\varepsilon_a = 1\%$

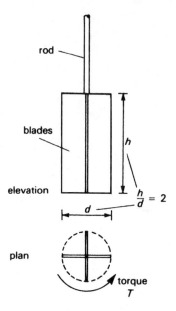

rod

blades

h

elevation

$\dfrac{h}{d} = 2$

d

plan

torque
T

Fig. 7.43   The shear vane

7.17   The shear vane test

In soft silts and clays, especially those of high sensitivity, the effects of disturbance when taking samples can have a considerable effect on subsequent laboratory-measured shear strengths. The shear vane apparatus is used to carry out *in situ* shear strength tests, thus obviating the necessity of obtaining undisturbed samples. Details of the apparatus and procedure are given in Section 12.5.

The shear vane test is suitable for determining the undrained strength of unfissured saturated clays and saturated silts. It is not so reliable for fissured soil or laminated mixtures. The vane consists of four rectangular blades in a cruciform at the end of a steel rod (Fig. 7.43). After the vane has been pushed into the soil, it is rotated by applying a torque at the surface end of the rod. The vane is first rotated at 6–12° per minute to determine the undisturbed shear strength and then the remoulded strength is measured by rotating the vane rapidly. The resulting shear surface comprises the perimeter and ends of a cylinder. The undrained strength is obtained by equating the applied torque to the shear resistance moment:

$$T = c_u \left(\pi h d \times \frac{d}{2} + 2\pi \frac{d^2}{4} \times \frac{d}{3} \right)$$

Hence $c_u = \dfrac{T}{\pi(\frac{1}{2}hd^2 + \frac{1}{6}d^3)} = \dfrac{T}{\frac{1}{2}\pi d^2(h + \frac{1}{3}d)}$ [7.27]

Worked example 7.20 *A shear vane used to test* in situ *a soft clay had a diameter of 75 mm and a length of 150 mm. The average torques recorded after slow and then rapid* *rotations were 64 and 26 Nm respectively. Determine the undrained strength of the clay and also its sensitivity.*

The dimensional characteristic of the vane is:

$$\tfrac{1}{2}\pi d^2 \left(h + \tfrac{1}{3}d\right) = \tfrac{1}{2}\pi \times 0.075^2(0.150 + \tfrac{1}{3} \times 0.075) = 1.546 \times 10^{-3} \text{ mm}^3$$

Undisturbed undrained strength, $c_u = \dfrac{64 \times 10^{-3}}{1.546 \times 10^{-3}} = \underline{41.4 \text{ kPa}}$

Remoulded undrained strength, $c_{uR} = \dfrac{16 \times 10^{-3}}{1.546 \times 10^{-3}} = \underline{10.3 \text{ kPa}}$

Sensitivity, $S_t = \dfrac{41.4}{10.3} = \underline{4.0}$

7.18 Estimates of shear strength parameters from index tests

Soils are generally classified and described using standard index tests, such as the liquid and plastic limit test (see also Section 2.6). The *plasticity index* (I_P) represents the range of water contents over which a soil has a plastic consistency and the *liquidity index* (I_L) gives a measure of the relative *in situ* consistency.

Eqn [2.6]: $I_P = w_L - w_P$

Eqn [2.7]: $I_L = \dfrac{w - w_P}{I_P}$

Preliminary estimates of strength and stiffness parameters can provide a useful basis for early design and feasibility studies; they also enable the realistic planning of a detailed testing programme. A number of quantitative correlations have been proposed between I_P and compressibility (C_c or λ) and between I_L and undrained strength. While all such correlations must be treated with caution and some are sensitive to errors in input data, they may at least be used to establish an early picture of the strength and stiffness nature of the soil. The following examples are not exclusive, nor do they lay claim to a high order of general reliability; however, they do have the advantage of simplicity. When used with good sense and engineering skill born out of experience these estimates will provide useful markers for preliminary design purposes.

Undrained strength estimates

Source	c_{uR} at w_L (kPa)	c_{uR} at w_P (kPa)
Skempton and Northey (1953)	1.7	170
Schofield and Wroth (1978)	1.7	170
Medhat and Whyte (1986)	1.6	110
Youssef *et al.* (1965)	1.7	110

Also Schofield and Wroth (1978): $c_u = 170 \exp(-4.6I_L)$ kPa

Skempton and Bjerrum (1957): $\dfrac{c_u}{\sigma'_{vo}} = 0.11 + 0.37I_p$

Ladd *et al.* (1977): for overconsolidated clay, $\dfrac{(c_u/\sigma'_{vo})_{oc}}{(c_u/\sigma'_{vo})_{nc}} = R_o^{0.8}$

An approximate correlation is: $\log c_u = 0.2 + 2.0I_L$

Also, after Massarch (1979): $K_0 = 0.44 + 0.42I_p$

In the above: σ'_{vo} = effective overburden stress
R_o = one-dimensional overconsolidation ratio
c_{uR} = remoulded undrained strength

Worked example 7.21   *Obtain an estimate for the remoulded undrained strength of a clay which has an* in situ *water content of 23 per cent and the following index properties:*

$w_L = 37$ *per cent* $w_P = 19$ *per cent*

Liquidity index, $I_L = \dfrac{37 - 23}{37 - 19} = 0.78$

$\log c_{uR} = 0.2 + 2.0I_L = 0.2 + 2.0 \times 0.78 = 1.76$

Therefore, estimated $c_{uR} = \underline{58 \text{ kPa}}$

Stiffness estimates

Skempton and Northey (1953), after examining a number of British soils, found that the strength at the plastic limit was one hundred times that at the liquid limit, i.e.

$q'_{PL} = 100 \, q'_{LL}.$

and since $q' = Mp'$, then $p'_{PL} = 100 \, p'_{LL}$, which are points on the critical state line having the specific volume coordinates:

$v_{PL} = \Gamma - \lambda \ln p'_{PL}$

$v_{LL} = \Gamma - \lambda \ln p'_{LL}$

Now, for a saturated soil, $v = 1 + wG_s$, so that

$$w_{LL} - w_{PL} = \frac{\lambda \ln(p'_{PL}/p'_{LL})}{G_s} = \frac{\lambda \ln 100}{G_s} = \frac{4.61\,\lambda}{G_s}$$

But $w_{LL} - w_{PL} = I_P/100$. Hence the slope of the *critical state line* may be estimated from:

$$\lambda = \frac{I_P G_s}{461} \qquad\qquad [7.28]$$

Similarly, the compressibility index is estimated from:

$$C_c = \lambda \ln 10 = \frac{I_p G_s}{200}$$ [7.29]

where I_P is in percentage units.

Exercises

7.1 (a) Sketch a stress–strain graph to show what is meant by *peak strength, critical strength* and *residual strength*.

(b) What is meant by the term *critical state*? How does the behaviour of a soil differ during shearing when it is *looser/wetter* as opposed to being *denser/drier* than the critical state?

(c) Explain the term *angle of repose*.

7.2 The readings given below were recorded during shear box tests on samples of a sand compacted to the same density. The shear surface measures 60×60 mm. Determine (a) the critical angle of friction and (b) the peak angle of friction.

Normal load (N)	Shear load (N)	
	Peak	Ultimate
110	97	61
230	198	128
350	301	198

7.3 The readings given below were taken during two shear box tests on samples of the same sand. In both cases, the constant normal stress was 160 kN/m. In Test 1, the sand was in a loose state; in Test 2, it was in a dense state. Draw the shear stress/displacement graphs for the two tests and determine the peak and ultimate angles of friction.

Test 1

Horizontal displacement $(10^{-2}$ mm$)$	0	50	100	150	250	350	450	550
Vertical displacement $(10^{-2}$ mm$)$	0	−11	−17	−21	−26	−28	−29	−29
Shear stress (kPa)	0	34	53	64	78	86	89	90

Test 2

Horizontal displacement $(10^{-2}$ mm$)$	0	50	100	150	250	350	450	550
Vertical displacement $(10^{-2}$ mm$)$	0	−5	−4	1	29	47	53	56
Shear stress (kPa)	0	59	90	110	121	111	97	93

7.4   The following parameters are known for a normally consolidated clay:

$M = 1.17$ $\lambda = 0.18$ $N_0 = 2.82$ $\Gamma = 2.73$ $G_s = 2.72$

In triaxial tests, two samples of the soil were each isotropically consolidated at a cell pressure of 300 kPa. In the case of sample A the axial load was increased under drained conditions, whereas sample B was tested undrained.

(a)   Calculate the expected ultimate mean normal stress, deviator stress, void ratio and moisture content at the end of the drained test.

(b)   Calculate the expected ultimate mean normal stress, deviator stress and pore pressure at the end of the undrained test.

(c)   Calculate the equivalent value of the critical angle of friction.

7.5   Draw a Mohr circle and failure envelope showing the Mohr–Coulomb failure criterion for a peak strength line ($\tau_f = c' + \sigma' \tan \phi'_p$). Show that

$$\frac{c' \cot \phi'_p + \sigma'_3}{c' \cot \phi'_p + \sigma'_1} = \tan^2 \left(45° - \frac{\phi'_p}{2} \right)$$

7.6   Given that the shear strength parameters for a soil are $c' = 18$ kPa and $\phi'_p = 30°$, calculate its peak strength within a saturated mass of the soil on a plane upon which the total normal stress is 278 kPa and the pore pressure is 94 kPa.

7.7   The following results were recorded during consolidated undrained triaxial tests on a clay soil:

Cell pressure (kPa)	150	300	500
Deviator stress at failure (kPa)	192	385	638
Pore pressure at failure (kPa)	70	139	234

(a)   Determine the peak strength parameters c' and ϕ_p.

(b)   Do the test results indicate the clay to be normally consolidated or overconsolidated?

7.8   A consolidated undrained triaxial test on an overconsolidated clay yielded the following results:

Cell pressure (kPa)	100	300	500
Deviator stress at failure (kPa)	208	410	590
Pore pressure at failure (kPa)	−18	68	168

Determine the peak strength parameters c' and ϕ'_p.

7.9   A series of three triaxial tests was carried out on samples of a saturated clay soil consolidated at a cell pressure of 200 kPa, which was held constant during axial loading. The shear strength parameters were found to be $c' = 0$ and $\phi'_c = 24°$.

(a)   In an undrained test the pore pressure at failure was 125 kPa. Determine the ultimate deviator stress.

(b)   In another undrained test the maximum deviator stress was 160 kPa. Determine the ultimate pore pressure.

(c)   In a drained test the back pressure was kept constant at 80 kPa. Determine the ultimate deviator stress.

7.10 The following parameters have been obtained for an overconsolidated clay.

$M = 0.96$ $\Gamma = 3.15$ $H = 0.66$ $\lambda = 0.18$
$\kappa = 0.049$ $v_a = 2.30$ $\gamma = 18.4$ kN/m^3

(a) Calculate the undrained strength of the soil at depths of 1.0 m, 2.0 m and 5.0 m, assuming that the water table is at the surface.

(b) Calculate an estimate of the historical pre-consolidation stress for the soil.

7.11 Four specimens of a clay soil were consolidated and then swelled to an isotropic stress of 650 kPa, while another four specimens of the same soil were consolidated and swelled to 290 kPa. Shear tests on the specimens then revealed the following peak state stresses.

Series A: Consolidation stress, $\sigma'_c = 650$ kPa

Normal stress at failure, σ'_f (kPa)	120	220	340	480
Shear stress at failure, τ_f (kPa)	150	181	211	254

Series B: Consolidation stress, $\sigma'_c = 290$ kPa

Normal stress at failure, σ'_f (kPa)	80	130	200	260
Shear stress at failure, τ_f (kPa)	75	87	111	127

(a) Determine the peak strength parameters c' and ϕ'_p for each series.

(b) Normalise the results with respect to σ'_c and then determine the parameters of the single straight normalised peak strength line.

7.12 (a) Write down the general expression for the change in pore pressure in terms of the changes in isotropic and deviator stress. Explain why $B = 1$ for a saturated soil.

(b) At a given point in a mass of saturated clay, the vertical total stress is 240 kPa, the horizontal total stress is kPa and the pore pressure is 60 kPa. Assuming that coefficient $A = 0.55$, determine the pore pressure in a sample of the clay after its removal from the ground, i.e. when the total stress is reduced to zero.

(c) If a specimen cut from this sample is now subject to undrained triaxial compression at a cell pressure of 150 kPa and then fails at a deviator stress of 320 kPa, determine the pore pressure and also the effective axial and radial stresses at failure.

7.13 The following parameters are known for a peaty clayey silt:

$c' = 18$ kPa $\phi'_p = 20°$ $A = 0.40$ $B = 0.80$

Determine the pore pressures at the beginning and end of each of the following consecutive stages in a triaxial test:

(a) isotropic consolidation at a cell pressure of 200 kPa,

(b) undrained axial loading at a cell pressure raised to 400 kPa.

7.14 The following results were recorded during an undrained triaxial test on a clay soil at a cell pressure of 400 kPa. At zero cell pressure the pore pressure was also zero.

Axial strain (%)	0	2	4	6	8	10	12	14
Deviator stress (kPa)	0	258	440	588	681	729	713	662
Pore pressure (kPa)	368	557	672	665	578	529	495	482

(a) Plot the curves of deviator stress and pore pressure against axial strain and deduce the maximum value of deviator stress and the corresponding pore pressure, u_f.

(b) Calculate the value of pore pressure coefficient B.

(c) Plot the curve of pore pressure coefficient A against axial strain and from it determine the value of A_f.

7.15 Prior to carrying out strain measurements in a triaxial apparatus, a sample of soil has been consolidated at a mean normal stress of 150 kPa and a deviator stress of 75 kPa. After consolidation, the pore pressure is zero and the specific volume is 2.09. The soil is known to have the following properties:

$M = 0.96$ $H = 0.66$ $\lambda = 0.18$ $\kappa = 0.046$
$\Gamma = 3.12$ $N = 3.21$

(a) Establish whether this state lies on or below the Hvorslev surface and therefore whether or not a further increase in stresses will produce elastic strains.

(b) Calculate the volumetric strain that will take place when, under drained conditions, the radial stress is raised by 5 kPa and at the same time the vertical axial stress is raised by 15 kPa.

(c) Calculate the increase in pore pressure that will result from a similar increase in stresses under undrained conditions.

Soil Mechanics Spreadsheets and Reference Assignments and Quizzes (available on the accompanying CD):

Assignments A.7   Measurement of shear strength

Quiz Q.7   Measurement of shear strength

Chapter 8

Earth pressure and retaining walls

8.1 Types of retaining wall

In many construction operations it is often necessary to alter the ground-surface profile in such a way as to produce vertical or near-vertical faces. Sometimes these new faces may be capable of self-support, but in other instances a lateral retaining structure will be required to provide support. In stability analyses, both the nature of the wall structure and that of supported material are important, as too is the way the wall may move or yield after construction.

If a wall structure is caused to move *towards* the soil being supported, the horizontal pressures in the soil will increase; these are then referred to as **passive pressures**. If the wall moves *away* from the supported soil, the horizontal soil pressures decrease and are referred to then as **active pressures**. If the wall structure is rigid and does not yield, the horizontal soil pressures are said to be **at-rest pressures**. The various types of earth-retaining structures fall into three broad groups:

(1) **Gravity walls**: these depend largely upon their own weight for stability, have wide bases and usually a rigid construction; some examples are shown in Figs 8.1 and 8.2:

  (a) *Masonry walls*: mass concrete, brickwork or stonework; the strength of the wall material is generally much greater than that of the underlying soil; the base is usually formed in mass concrete and will typically have a breadth of one-third to one-half of the wall height.

  (b) *Gabion walls*: gabions are rectangular (occasionally cylindrical) baskets made from galvanised steel mesh or woven strips, or plastic mesh (originally wickerwork) and filled with stone rubble or cobbles, to provide free-draining wall units.

  (c) *Crib walls*: formed with interlocking pre-cast concrete units (or timber sometimes for temporary works); *stretchers* run parallel to the wall face and *headers* are laid perpendicular to the wall face; the space formed by the cribs is filled with free-draining material, such as stone rubble, cobbles or gravel.

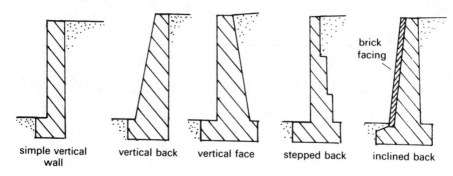

simple vertical   vertical back   vertical face   stepped back   inclined back
wall

(a) Unreinforced masonry walls

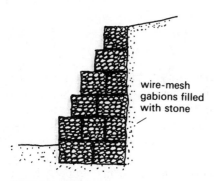

wire-mesh
gabions filled
with stone

(b) Gabion wall

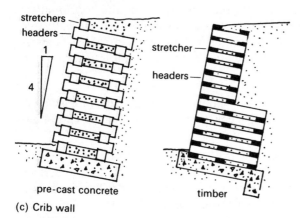

stretchers
headers

stretcher

headers

pre-cast concrete   timber

(c) Crib wall

Fig. 8.1 Types of mass gravity wall

(d) *RC walls*: reinforced concrete cantilever walls are the commonest modern form of gravity wall; either an *L-shaped* or an inverted *T-shaped*
cross-section is formed to produce a vertical cantilever slab; simple
cantilevers, some utilising the weight of backfill on the heel portion of
the slab, are suitable for walls up to 6 m in height; for greater heights,
counterfort walls or *buttressed walls* can be used; to improve resistance to
sliding a downward-projecting *key* is often incorporated into the base.

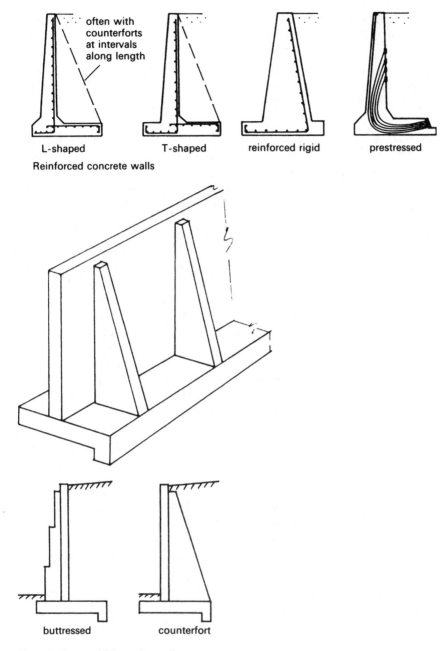

often with
counterforts
at intervals
along length

L-shaped   T-shaped   reinforced rigid   prestressed

Reinforced concrete walls

buttressed   counterfort

Fig. 8.2 Types of RC gravity wall

(e) *Counterfort walls*: basically RC gravity walls with tension stiffeners (*counterforts*) connecting the *back* of the wall slab and the base, so that the bending and shearing stresses are reduced; used for high walls or where high lateral pressures will occur, e.g. where the backfill is heavily surcharged.

(f) *Buttressed walls*: are similar except that the stiffeners are placed in *front* of the wall and act as compression braces; used for very tall walls, but are not as common.

(2) **Embedded walls**: these consist of vertical driven or placed sheets or piles that may be anchored, tied or propped, or may act as simple cantilevers; their own weight does not feature in stability analyses; some examples are shown in Fig. 8.3:

(a) *Driven sheet-pile walls*: flexible structures used particularly for temporary works, in harbour structures and in poor ground; different material types include: *timber*: suitable for temporary works and braced sheets; cantilever walls up to 3 m high; *pre-cast concrete*: used for permanent structures, but quite heavy, watertight structures formed from 'keyed' or grouted sections; *steel*: widely used, especially for cantilever and tied-back walls; variety of cross-sections, strengths and combinations; strong buckling capacity; reusable in temporary works.

Cantilever walls are economical for heights up to 4 m; used mainly for temporary works. *Anchored* or *tied-back walls* are used for a wide range of applications in different soils, for heights up to 20 m; anchor forces reduce the lateral deflection and bending moment in the pile, and also the depth of embedment required (compared with cantilever piling).

(b) *Braced or propped walls*: props, braces, shores or struts are placed in front of the wall; lateral deflection and bending moment are reduced, and embedment may not be required; in trenches, struts and wales are used; in wide excavations framed shores or raking shores are used.

(c) *Contiguous bored-pile walls*: formed from a single or double row of piles installed in contact (or in very close proximity) with each other; alternate piles are first drilled and placed; a casing guide is then used to drill alternate pile holes.

(d) *Secant bored-pile walls*: the wall is formed from bored piles, usually about 1 m diameter: piles are drilled in a row at closer spacing than the pile diameter and the concrete placed; while the concrete is still quite weak (after 2/3 days), the intermediate holes are drilled along a parallel, but slightly offset, line so that these holes cut into the first piles.

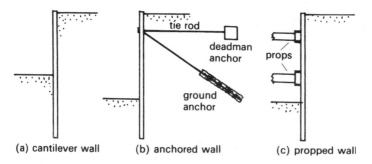

(a) cantilever wall   (b) anchored wall   (c) propped wall

Fig. 8.3 Types of embedded wall

(e)   *Diaphragm walls*: constructed in a narrow (wall width) excavated trench which is temporarily supported by a bentonite slurry; the reinforcement cage is lowered into the trench and the concrete placed through a trémie, thus displacing the slurry; used in difficult ground, where driving sheet piles would be troublesome, where unfavourable groundwater levels occur, or on restricted sites.

(3)   **Reinforced and anchored earth**: here the wall material may be the *in situ* rock or soil into which reinforcement is inserted (*soil nails* or *ground anchors*), or reinforcement may be laid between rolled layers as the fill soil is placed and compacted (*reinforced earth*); some examples are shown in Fig. 8.4:

(a)   *Reinforced earth walls*: consist of frictional backfill laid and rolled in layers, between which is placed either reinforcing strips or geotextile mesh; used in retaining walls, sea walls, dock walls, bridge abutments, earth dams and in numerous temporary works.

(b)   *Soil nailing*: an *in situ* soil reinforcement method in which steel bars or angles or other metallic elements are driven in or grouted in drilled holes (but not pre-stressed); they are typically 3–5 m long and at spacings of 0.5–2 m; the facing is often simply a layer of *shotcrete*; primarily used for temporary works, but suitable for some permanent structures with non-corrodible nails.

(c)   *Ground anchors*: tendons formed of rods, crimped bars or expanding strip elements are pressure-grouted into drilled holes; used as tie-backs for sheet-pile walls, rock faces, tunnels and mine works.

In the analysis and design of earth-retaining structures of all types it is necessary first to establish the distribution of vertical and horizontal earth pressures acting on the structure. The next sections of this chapter will examine the theories and methods of calculation in current use.

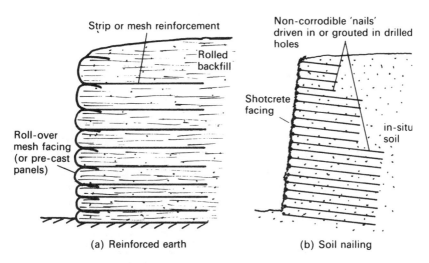

(a) Reinforced earth   (b) Soil nailing

Fig. 8.4   Reinforced earth and soil nailing

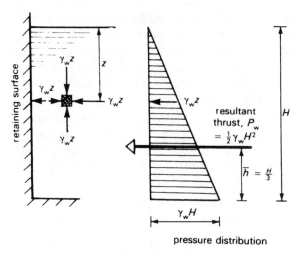

pressure distribution

Fig. 8.5 Horizontal pressure due to a liquid

8.2 Hydrostatic pressure and lateral thrust

The pressure at a point below the surface of a liquid that is in a state of hydrostatic equilibrium (i.e. no flow is occurring) is equal in all directions and increases linearly with depth. Figure 8.5 shows the distribution of horizontal hydrostatic pressure against a retaining surface. The resultant horizontal thrust on a unit length of the surface is equal to the area of the triangular pressure diagram.

Horizontal pressure, $\sigma'_h = \gamma_w z$

Horizontal thrust, $P_w = \frac{1}{2}\gamma_w H^2$ [8.1]

where γ_w = unit weight of liquid.

The line of action of P_w passes through the centre of area of the pressure diagram,

i.e. $\bar{h} = \frac{1}{3}H$.

In soil, grain, coal and other granular materials, internal frictional resistance is developed between adjacent grains. Thus, the horizontal pressure is not usually equal to the vertical pressure at the same point, although one is still a function of the other. The magnitude of lateral (horizontal) *earth pressure* is dependent on the shear strength of the soil, the lateral strain conditions, the vertical effective stress (and therefore the porewater pressure) and the state of equilibrium of the soil.

A body is said to be in a state of *elastic equilibrium* when a small change (increase or decrease) in stress acting upon it produces a corresponding and reversible change in strain. *Irreversible* strains are caused if the stress is increased beyond the *yield point* (see also Chapter 6). In a state of *plastic equilibrium* irreversible strain is taking place at constant stress; the *Mohr–Coulomb failure criterion* (Section 6.4) is one way of representing this state. It may be assumed that a mass of soil under increasing stress will remain in a state of elastic equilibrium until the plastic yield (failure or limit state) condition is reached.

The strain states relating to earth pressure calculations fall into three categories:

(a) **At-rest state**: elastic equilibrium with no lateral strain taking place.
(b) **Active state**: plastic equilibrium with lateral *expansion* taking place.
(c) **Passive state**: plastic equilibrium with lateral *compression* taking place.

8.3 Earth pressure at rest

If the stress state in a soil mass is still below the Mohr–Coulomb failure envelope (or critical state line), the soil is still in elastic equilibrium (Fig. 8.6). Under natural conditions of deposition there is a negligible amount of horizontal strain, although some lateral contraction may occur upon unloading (e.g. due to erosion or excavation). The soil mass is said to be in an *at-rest* state (or K_o state), and the horizontal effective stress is:

$$\sigma'_h = K_o \sigma'_v \qquad\qquad [8.2]$$

where K_o = coefficient of earth pressure at rest
σ'_v = vertical effective stress

The value of K_o depends on the angle of friction (ϕ') and the loading/unloading history of the soil. For normally consolidated soils Jaky (1944) proposed the following relationship, which seems to correlate well with observed values (Bishop, 1958; Brooker and Ireland, 1965):

$$K_o = 1 - \sin \phi'_c \qquad\qquad [8.3(a)]$$

where ϕ'_c = the critical angle of friction

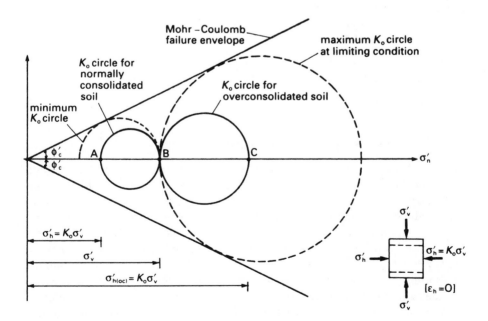

Fig. 8.6 Earth pressure at rest

K_o increases to about 1.0 for lightly overconsolidated soils, and to higher values with increasing overconsolidation ratio (R_o):

$$K_o = (1 - \sin \phi_c')\sqrt{R_o} \qquad [8.3(b)]$$

The maximum value of K_o corresponds to a state on the failure envelope (Fig. 8.6), or the critical state line (Section 6.6), so that $K_{o(max)} = \tan^2(45° + \phi_c'/2)$. If lateral expansion occurs, the value of σ_h' decreases until the active state of plastic yielding occurs, so that $K_{o(min)} = \tan^2(45° - \phi_c'/2)$.

Measured values for K_o may be obtained from one-dimensional compression tests; for example from a q'/p' plot:

$$\frac{dq'}{dp'} = \frac{3(\sigma_1' - \sigma_3')}{\sigma_1' + 2\sigma_3'} = \frac{3(1 - K_o)}{1 + 2K_o}$$

giving $K_o = \dfrac{3 - dq'/dp'}{3 + 2dq'/dp'}$ \qquad\qquad\qquad [8.3(c)]

Also, from eqn [6.19]: $K_o = \dfrac{v'}{1 - v'}$ \qquad\qquad\qquad [8.3(d)]

Using a self-boring pressuremeter (Mair and Wood, 1987), the *in situ* horizontal total stress and porewater pressure can be measured and thus an estimate found for K_o. Typical values for K_o are given in Table 8.1.

Table 8.1   Range of values for K_o

Type of soil	K_o
Loose sand	0.45–0.6
Dense sand	0.3–0.5
NC clay	0.5–0.7
OC clay	1.0–4.0
Compacted clay	0.7–2.0

8.4   Rankine's theory of granular earth pressure

The K_o circles shown in Fig. 8.6 represent states of elastic equilibrium. With lateral yielding σ_h' will either increase or decrease, with a corresponding change in the diameter of the Mohr circle (AB or BC). With lateral *expansion*, a state of plastic equilibrium (failure) is reached at the *minimum* value of σ_h'; with lateral *compression*, the limit state occurs at a *maximum* value of σ_h'. In both cases, the shearing resistance will have to be fully mobilized. Rankine's theory (Rankine, 1857) presents a solution for a mass of cohesionless soil in a state of limiting equilibrium. The magnitude of σ_h' depends only on the vertical effective stress and the shear strength of the soil, and thus the problem is statically determinate. No account is taken of displacement and so this is a *lower bound solution* (see Chapter 6).

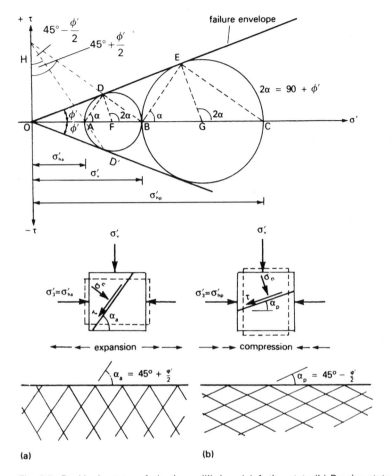

Fig. 8.7 Rankine's states of plastic equilibrium (a) Active state (b) Passive state

Active earth pressure
Consider a mass of homogeneous soil of semi-infinite extent that has an angle of friction of ϕ'. Suppose the soil mass is brought to a state of plastic equilibrium and at a given point the vertical effective stress is σ'_v. Figure 8.7 shows the Mohr circles corresponding to the two possible limit states, both touching the failure envelope. Points A and C represent the conjugate limiting stresses for the **active** and **passive** states respectively.

$OA = \sigma'_{ha}$ = active horizontal earth pressure

$OC = \sigma'_{hp}$ = passive horizontal earth pressure

Lateral *expansion* will lead to $\sigma'_{ha} < \sigma'_v$

Lateral *compression* will lead to $\sigma'_{ha} > \sigma'_v$

From the geometry of the Mohr circles, the corresponding orientations of the plane of failure developing in the soil mass are given by angles α_a and α_p:

$$\alpha_a = \tfrac{1}{2}(2\alpha) = \tfrac{1}{2}(90 + \phi') = 45° + \phi'/2 \qquad [8.4(a)]$$

$$\alpha_p = \tfrac{1}{2}(180 - 2\alpha) = \tfrac{1}{2}[180 - (90 + \phi')] = 45° - \phi'/2 \qquad [8.4(b)]$$

The limiting lateral earth pressures are defined in terms of the effective vertical stress:

The horizontal **active** earth pressure, $\sigma'_{ha} = K_a \sigma'_v$ $\qquad$ [8.5(a)]

The horizontal **passive** earth pressure, $\sigma'_{hp} = K_a \sigma_v$ $\qquad$ [8.5(b)]

where K_a = the coefficient of active earth pressure
K_p = the coefficient of passive earth pressure

By considering the trigonometrical relationships, K_a and K_p may be expressed in terms of the angle of friction ϕ':

$$K_a = \frac{\sigma'_{ha}}{\sigma'_v} = \frac{\overline{OA}}{\overline{OB}} = \frac{\overline{OF} - \overline{AF}}{\overline{OF} - \overline{FB}} = \frac{1 - \overline{AF}/\overline{OF}}{1 + \overline{FB}/\overline{OF}}$$

But $\overline{AF} = \overline{FB} = \overline{FD}$ and $\overline{FD}/\overline{OF} = \sin \phi'$

Then $K_a = \dfrac{1 - \sin \phi'}{1 + \sin \phi'}$

Also, $\overline{OH} = \dfrac{\overline{OA}}{\tan(45° - \phi'/2)} = \dfrac{\overline{OB}}{\tan(45° + \phi'/2)}$

$\therefore$ $K_a = \dfrac{\overline{OA}}{\overline{OB}} = \tan^2(45° - \phi'/2)$

Hence, $K_a = \dfrac{1 - \sin \phi'}{1 + \sin \phi'} = \tan^2(45° - \phi'/2)$ $\qquad$ [8.6(a)]

Similarly, $K_p = \dfrac{1 + \sin \phi'}{1 - \sin \phi'} = \tan^2(45° + \phi'/2)$ $\qquad$ [8.6(b)]

8.5   Lateral pressure in cohesionless soils

The simplest set of active earth pressure conditions occurs against a wall having a smooth vertical back, retaining a soil with a horizontal unloaded surface, and with no standing water behind the wall (Fig. 8.8). From Rankine's theory:

At depth z, $\sigma'_{ha} = K_a \sigma'_v = K_a \gamma z$

where γ = bulk unit weight of the soil

At the base of the wall, $z = H$ and $\sigma'_{ha} = K_a \gamma H$

The *resultant active thrust* (P_A) acting normal to the wall surface is given by:

$$P_A = \int_0^H \sigma'_v \, dz = \int_0^H K_a \gamma z \, dz \qquad [8.7(a)]$$

i.e. the area of the pressure distribution diagram.

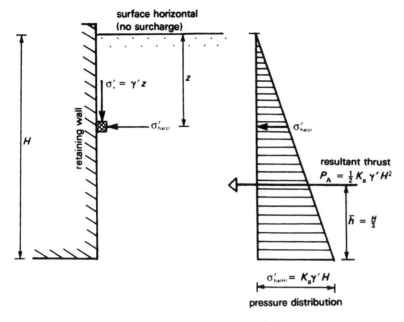

Fig. 8.8 Active pressure distribution in cohesionless soil

For the distribution shown in Fig. 8.8:

$$P_A = \tfrac{1}{2} K_a \gamma H^2 \qquad\qquad\qquad [8.7(b)]$$

The line of action of P_A passes through the centre of area at a height of $\tfrac{1}{3}H$ above the base.

Worked example 8.1 *Calculate the resultant active thrust on a vertical smooth retaining wall of height 5.4 m. The water table is well below the base of the wall.*

Soil properties: $\phi' = 30°$ $c' = 0$ $\gamma = 20 \ kN/m^3$

First, calculate $\qquad\qquad K_A = \dfrac{1 - \sin 30°}{1 + \sin 30°} = 0.333$

At a depth of 5.4 m: $\sigma'_{ha} = 0.333 \times 20 \times 5.4 = 36.0$ kPa

Resultant active thrust, $P_A = \tfrac{1}{2} \times 36.0 \times 5.4 = \underline{97.2 \ kN/m}$

This will act at a height of H/3 above the base, i.e. <u>1.8 m.</u>

Effect of groundwater level

Retaining walls should be provided with adequate back-face drainage. When the retained soil itself is not free-draining, a drainage blanket of coarse gravel and/ or geotextile layers may be laid, often with a collecting drain behind the heel of the wall. Wherever practicable, the groundwater level behind the wall should not be allowed to rise above the base. If, as a result of poor drainage, drain blockage or accidental flooding, the water level rises behind the wall, the vertical effective stresses will be reduced, but the horizontal total stresses will be increased. If the retained soil becomes fully waterlogged (i.e. the groundwater level is at the surface) the horizontal thrust acting on the wall will be *approximately doubled*. The following two worked examples illustrate the effects of a rising water level.

Worked example 8.2   *A retaining wall with a smooth vertical back of height 7.0 m retains soil having an unsurcharged horizontal surface. The soil properties are:*

$$c' = 0 \quad \phi' = 32° \quad \gamma_{sat} = 20 \; kN/m^3 \quad \text{and above} \quad WT, \gamma = 18 \; kN/m^3$$

Determine the distribution of horizontal stresses on the wall and also the magnitude and position of the resultant horizontal thrust when the groundwater level is standing (without seepage flow) at 3.0 m below the surface.

$$K_a = \frac{1 - \sin 32°}{1 + \sin 32°} = 0.3073$$

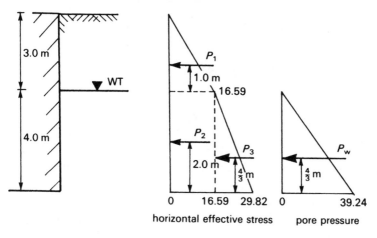

Fig. 8.9

Figure 8.9 shows the distribution of horizontal stresses; these are calculated as follows. At $z = 3.0$ m (i.e. at WT level):

Vertical total stress,	$\sigma_v = \gamma z = 18 \times 3 = 54.0$ kPa
Porewater pressure,	$u = 0$
Vertical effective stress,	$\sigma'_v = \sigma_v - u = 54.0$ kPa
Horizontal effective stress,	$\sigma'_h = K_a \sigma'_v = 0.3073 \times 54.0 = 16.59$ kPa
Horizontal total stress,	$\sigma_h = \sigma'_h + u = 16.59$ kPa

At $z = 7.0$ m (i.e. at base of wall):

Vertical total stress,	$\sigma_v = 54.0 + 20 \times 4 = 134.0$ kPa
Porewater pressure,	$u = \gamma(z - z_w) = 9.81(7.0 - 3.0) = 39.24$ kPa
Vertical effective stress,	$\sigma'_v = \sigma_v - u = 94.76$ kPa

Horizontal effective stress, $\sigma'_h = K_a\sigma'_v = 0.3073 \times 94.76 = 29.12$ kPa

Horizontal total stress, $\sigma_h = \sigma'_h + u = 68.36$ kPa

The **resultant horizontal thrust** is obtained by summing the areas of the distribution diagram:

$$P_h = P_1 + P_2 + P_3 + P_w$$

$$= \tfrac{1}{2} \times 16.59 \times 3 + 16.59 \times 4.0 + \tfrac{1}{2}(29.12 - 16.59)4.0 + \tfrac{1}{2} \times 9.81 \times 4.0^2$$

$$= 24.9 + 66.4 + 25.1 + 78.5 = \underline{194.8 \text{ kN/m}}$$

The line of action of P_h is found by equating moments about the base:

$$P_h\bar{h} = P_1 \times 5.0 + P_2 \times 2.0 + P_3 \times 4/3 + P_w \times 4/3$$

giving $$\bar{h} = \frac{24.9 \times 5.0 + 66.4 \times 2.0 + (25.1 + 78.5)4/3}{194.8} = \underline{2.03 \text{ m}}$$

Worked example 8.3   *Consider again the retaining wall and soil given in worked example 8.2. Calculate the magnitude of the total horizontal thrust for the following conditions:*

(a)   *the water table is well below the base of the wall;*
(b)   *the water table is standing at the surface (no seepage flow).*

(a)   In this case, the pore pressure will be zero at all depths, so that:

$$\sigma'_h = K_a\sigma'_v = K_a\gamma z = 0.307 \times 18z$$

Therefore $P_h = P_A = \tfrac{1}{2}K_a\gamma H^2 = \tfrac{1}{2} \times 0.307 \times 18 \times 7.0^2 = \underline{135.4 \text{ kN/m}}$

(b)   In this case the soil is fully waterlogged and $\gamma = \gamma_{sat} = 20$ kN/m^3.

At $z = 7.0$ m: $\sigma'_v = 20 \times 7.0 - 9.81 \times 7.0 = 71.33$ kPa

$$\sigma'_h = K_a\sigma'_v = 0.307 \times 71.33 = 21.9 \text{ kPa}$$

Therefore $P_h = P_A + P_w = \tfrac{1}{2} \times 21.9 \times 7.0 + \tfrac{1}{2} \times 9.81 \times 7.0^2 = \underline{317.0 \text{ kN/m}}$

Thus, it can be seen that when the groundwater rises from below the base to the ground surface the total horizontal thrust on the wall is more than doubled.

Effect of a surface surcharge load

A **uniform surcharge load** (q) applied over the *whole* surface, i.e. very wide and long, and right up to the wall, may be assumed to cause an equal increase in vertical effective stress at all depths:

At depth $= z$:

Vertical effective stress, $\sigma'_v = \gamma z + q - u$ [8.8]

Active pressure, $\sigma'_{ha} = K_a(\gamma z + q - u)$ [8.8(a)]

Passive pressure, $\sigma'_{hp} = K_p(\gamma z + q - u)$ [8.8(b)]

For surcharge loading of limited extent, e.g. **point loads, line loads, strip loads** and **area loads**, a more rigorous analysis is required. Methods have been developed using elasticity theory, plasticity theory, finite elements, etc.; some of the more common of these are given in Chapter 6. A minimum obligatory surcharge of 10 kN/m^2 is recommended in BS 8002:1994.

Worked example 8.4   *Calculate the resultant active thrust on a vertical smooth retaining wall of height 5.4 m. The water table is well below the base of the wall. An extensive uniform surcharge of 48 kPa is placed on the surface.*

Soil properties: $\phi' = 30°$ $c' = 0$ $\gamma = 20 \ kN/m^3$

First, calculate $K_A = \dfrac{1 - \sin 30°}{1 + \sin 30°} = 0.333$

At a depth of 5.4 m:

Due to the weight of soil: $\sigma'_{ha} = K_a \gamma z = 0.333 \times 20 \times 5.4 = 36.0$ kPa

Due to the surcharge: $\sigma'_{ha} = K_a q = 0.333 \times 48 = 16.0$ kPa

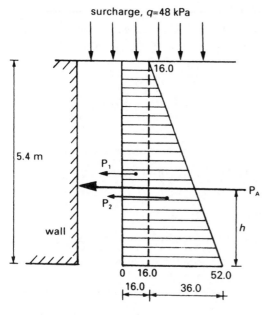

Fig. 8.10

Figure 8.10 shows the pressure distribution:

$P_1 = 16 \times 5.4 \qquad = 86.4$ kN/m

$P_2 = \frac{1}{2} \times 36.0 \times 5.4 = 97.2$ kN/m

Total active thrust, $P_A = P_1 + P_2 = \underline{183.6 \ kN/m}$

Position above the base, $\bar{h} = \dfrac{86.4 \times 2.7 + 97.2 \times 1.8}{183.6} = \underline{2.22 \ m}$

Effect of stratified soil

Where the soil behind a retaining wall consists of two or more layers, the lateral pressure distribution is determined within each layer and a composite (approximate)

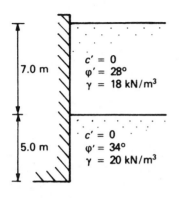
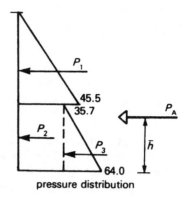

pressure distribution

Fig. 8.11

diagram drawn. Suppose at an interface between two layers (R overlying S) the soil properties are ϕ'_R, γ_R and ϕ'_S, γ_S respectively, and that the vertical effective stress is σ'_v. Then, immediately above the interface $\sigma'_{ha} = K_{aR}\sigma'_v$ and immediately below it $\sigma'_{ha} = K_{aS}\sigma'_v$. The pressure distribution (see worked example 8.5 and Fig. 8.11) shows a sudden jump in lateral pressure because of the different values of ϕ' in the two layers. In reality this does not occur, since the horizontal shear stress at the interface has been ignored. This approximation is acceptable for practical purposes, the consequent errors being quite small.

Worked example 8.5   *A retaining wall having a smooth vertical back retains soil for a depth of 12 m. The soil mass consists of two horizontal layers:*

Upper layer: c' = 0   φ' = 28°   γ = 18 kN/m³   thickness = 7.0 m
Lower layer: c' = 0   φ' = 34°   γ = 20 kN/m³

The water table is below the base of the wall. Determine the magnitude and position of the resultant active thrust.

Upper $K_a = \dfrac{1 - \sin 28°}{1 + \sin 28°} = 0.361$ Lower $K_a = \dfrac{1 - \sin 34°}{1 + \sin 34°} = 0.283$

Upper layer pressures
At $z = 0$: $\sigma'_{ha} = 0$
At $z = 7.0$ m: $\sigma'_{ha} = 0.361 \times 18 \times 7 = 45.5$ kPa

Lower layer pressures
At $z = 7.0$ m: $\sigma'_v = 18 \times 7 = 126.0$ kPa
         $\therefore$ $\sigma'_{ha} = 0.283 \times 126 = 35.7$ kPa
At $z = 12.0$ m: $\sigma'_v = 126.0 + 20 \times 5 = 226.0$ kPa
         $\therefore$ $\sigma'_{ha} = 0.283 \times 226.0 = 64.0$ kPa

The pressure distribution is shown in Fig. 8.11, the total area of which gives the resultant thrust per metre.

$P_A = P_1 + P_2 + P_3$

$= \frac{1}{2} \times 45.5 \times 7 + 35.7 \times 5 + \frac{1}{2}(64.0 - 35.7) \times 5 = \underline{409 \text{ kN/m}}$

To find $\bar{h}$, take moments about the base ($P_A\bar{h} = \Sigma(Ph)$):

$$\bar{h} = \frac{P_1(5 + \frac{7}{3}) + P_2 \times \frac{5}{2} + P_3 \times \frac{5}{3}}{P_1 + P_2 + P_3}$$

$$= \frac{159.3 \times 7.333 + 178.5 \times 2.5 + 70.8 \times 1.667}{409} = \underline{4.24 \text{ m}}$$

Effect of sloping ground surface

Where the ground surface is sloping, the vertical stress at a given depth will have a value of $\sigma'_v = (\gamma z - u) \cos \beta$ for an unloaded surface (Fig. 8.12). The lateral earth pressure against a smooth vertical wall is assumed to act parallel to the ground surface. The relationship between the lateral and vertical stresses can again be obtained analytically using the Mohr circle. Although the construction and analysis are somewhat more involved, the expressions remain essentially the same.

Active pressure, $\sigma'_{ha} = K_a\sigma'_v \cos \beta$ [8.9(a)]

Passive pressure, $\sigma'_{hp} = K_p\sigma'_v \cos \beta$ [8.9(b)]

where $$K_a = \frac{\cos \beta - \sqrt{(\cos^2 \beta - \cos^2 \phi')}}{\cos \beta + \sqrt{(\cos^2 \beta - \cos^2 \phi')}}$$ [8.10(a)]

and $$K_p = \frac{\cos \beta + \sqrt{(\cos^2 \beta - \cos^2 \phi')}}{\cos \beta - \sqrt{(\cos^2 \beta - \cos^2 \phi')}} = \frac{1}{K_a}$$ [8.10(b)]

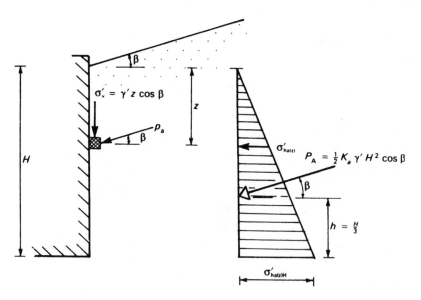

Fig. 8.12 Active pressure distribution under a sloping surface

If the angle of surface slope is equal to the angle of friction, i.e. $\beta = \phi'$, the active and passive states coincide and $\sigma'_{ha} = \sigma'_{hp} = \sigma'_v \cos \beta$. The soil will then be in a state of limiting plastic equilibrium with one of the sets of failure planes parallel to the surface. In a purely cohesionless soil ($c' = 0$), the angle of surface slope cannot exceed the angle of repose, which is the angle of friction of the soil in its loosest state. Also, when $\beta \rightarrow \phi'$, the use of Rankine's theory is not advisable (see Section 8.7).

Worked example 8.6 *Rework worked example 8.1 for the case of a supported soil mass whose surface is a planar slope of 1 vertical in 3 horizontal upward from the top of the wall.*

Angle of slope, $\beta = \arctan(\frac{1}{3}) = 18.44°$

Cos $\beta = 0.9487$ and cos $\phi' = 0.8660$

Then $K_a \cos \beta = 0.9487 \dfrac{0.9487 - \sqrt{(0.9487^2 - 0.8660^2)}}{0.9487 + \sqrt{(0.9487^2 - 0.8660^2)}} = 0.399$

Then, at a depth of 5.4 m: $\sigma'_{ha} = 0.399 \times 20.0 \times 5.4 = 43.09$ kPa

and the resultant active thrust: $P_A = \frac{1}{2} \times 42.66 \times 5.4 = \underline{116 \text{ kPa}}$

This will act at a height above the base of $5.4/3 = \underline{1.8 \text{ m}}$ inclined at 18.44° to horizontal.

8.6 Bell's solution for cohesive soils

Rankine's theory deals with earth pressure in cohesionless granular materials, i.e. assuming that $c' = 0$. However, in undrained cohesive soils, and in overconsolidated clays where $c' > 0$, the shear strength is expressed wholly or partly in terms of the apparent cohesion:

Undrained shear strength in saturated silts and clays: $\tau_f = c_u$ (or $\tau_f = s_u$)
Drained shear strength in overconsolidated clays: $\tau_f = c' + \sigma'_n \tan \phi'_p$

In 1915, Bell published an extension of Rankine's theory which included the cohesion parameter.

Drained conditions in cohesive soils

Horizontal active pressure: $\sigma'_{ha} = K_a\sigma'_v - 2c'\sqrt{K_a}$ [8.11(a)]

Horizontal passive pressure: $\sigma'_{hp} = K_p\sigma'_v + 2c'\sqrt{K_a}$ [8.11(b)]

Undrained conditions in cohesive soils (in terms of TOTAL stresses)

Horizontal active pressure: $\sigma_{ha} = \sigma_v - 2c_u$ ($= \sigma_v - 2s_u$) [8.12(a)]

Horizontal passive pressure: $\sigma_{hp} = \sigma_v + 2c_u$ ($= \sigma_v + 2s_u$) [8.12(b)]

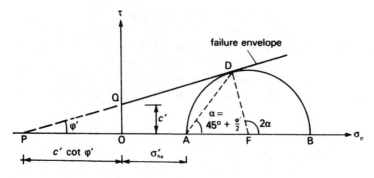

Fig. 8.13 Mohr's circle for active pressure in a cohesive soil

Figure 8.13 shows a Mohr circle for the active limiting condition in an overconsolidated soil. Starting with the ratio of effective stresses:

$$\frac{PA}{PB} = \frac{c' \cot \phi' + \sigma'_{ha}}{c' \cot \phi' + \sigma'_v}$$

But also:
$$\frac{PA}{PB} = \frac{PF - AF}{PF + BF} = \frac{1 - FD/PF}{1 + FD/BF} = \frac{1 - \sin \phi'}{1 + \sin \phi'} = K_a$$

So
$$c' \cot \phi' + \sigma'_{ha} = K_a(c' \cot \phi' + \sigma'_v)$$

Rearranging
$$\sigma'_{ha} = K_a \sigma'_v + (K_a - 1)c' \cot \phi'$$

Now
$$(K_a - 1)c' \cot \phi' = \frac{1 - \sin \phi' - (1 + \sin \phi')}{1 + \sin \phi'} \frac{\cos \phi'}{\sin \phi'}$$

$$= -2c' \frac{\cos \phi'}{1 + \sin \phi'}$$

$$= -2c' \frac{\sqrt{(1 - \sin^2 \phi')}}{1 + \sin \phi'}$$

$$= -2c' \sqrt{\left(\frac{1 - \sin \phi'}{1 + \sin \phi'}\right)} = -2c'\sqrt{K_a}$$

Hence, horizontal **active** pressure: $\sigma'_{ha} = K_a \sigma'_v - 2c'\sqrt{K_a}$

Similarly, horizontal **passive** pressure: $\sigma'_{hp} = K_p \sigma'_v + 2c'\sqrt{K_p}$

Tension cracks in cohesive soils

The lateral earth pressures under undrained conditions in cohesive soils are:

Horizontal active pressure: $\sigma_{ha} = \sigma_v - 2c$　$(= \sigma_v - 2s_u)$

Horizontal passive pressure: $\sigma_{hp} = \sigma_v + 2c$　$(= \sigma_v + 2s_u)$

Near the surface, where $\sigma_v < 2c_u$, the active pressure will therefore have a negative or tension value. This of course is an internal tension only and is not transmitted to the supporting wall surface. In fact, if the soil dries slightly tension

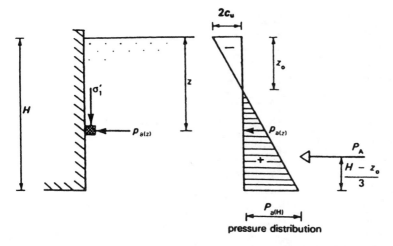

Fig. 8.14 Distribution of active pressure in an undrained cohesive soil

cracks will open from the surface downward to a depth where the active earth is zero, i.e. $\sigma_v = 2c_u$.

Figure 8.14 shows the distribution of active earth pressure against a smooth vertical wall.

At the ground surface, $z = 0$ and $\sigma_{ha} = -2c_u$

The tension zone extends down to $z = z_0$; here $\sigma_{ha} = 0 = \sigma_v - 2c_u = \gamma z_0 - 2c_u$

Thus the tension crack depth is: $\quad z_0 = \dfrac{2c_u}{\gamma}$ [8.13]

In drained conditions where $\phi' > 0$, the following expression may be used:

Since $\quad \sigma'_{ha} = 0 = K_a\gamma z_0 - 2c'\sqrt{K_a}: \quad z_0 = \dfrac{2c'\sqrt{K_a}}{K_a\gamma} = \dfrac{2c'}{\gamma\sqrt{K_a}}$ [8.14]

8.7 Active and passive thrust in undrained conditions

Active thrust

Since the negative earth pressure ('tension') in the soil cannot be transmitted to the supporting wall, it is ignored in the calculation of the active lateral thrust. The value of P_A is therefore taken as the area of the (positive) shaded portion shown in Fig. 8.14.

Resultant active thrust, $\quad P_A = \frac{1}{2}\sigma'_{ha}H(H - z_0)$ [8.15(a)]

If the surface is unloaded: $\quad P_A = \frac{1}{2}\gamma(H - z_0)^2$ [8.15(b)]

where $\quad z_0 = \dfrac{2c_u}{\gamma}$

The line of action of P_A acts through the centre of positive pressure area:
$\bar{h} = \frac{1}{3}(H - z_0)$

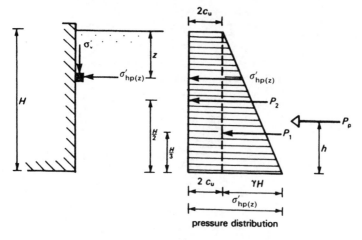

Fig. 8.15 Distribution of passive pressure in an undrained cohesive soil

Passive resistance in undrained clay

The distribution of passive pressure in an undrained clay is shown in Fig. 8.15. The total *passive resistance* provided by the soil, with the surface unloaded, is given by the shaded area:

Total passive resistance force, $P_P = P_1 + P_2 = \frac{1}{2}\gamma H^2 + 2c_u H$ [8.16]

The line of action of P_p can be obtained by taking moments about the base of the wall:

$$\bar{h} = \frac{P_1 \times \frac{1}{3}H + P_2 \times \frac{1}{2}H}{P_1 + P_2}$$

Worked example 8.7   *A retaining wall having a smooth vertical back retains a soil mass with an unloaded surface for a depth of 8 m. Calculate the active thrust acting on the wall if the soil has the following properties:*

$c_u = 32\ kPa$ $\phi_u = 0$ $\gamma = 18\ kN/m^3$

Depth of tension zone, $z_0 = 2c_u/\gamma = 2 \times 32/18 = 3.56$ m

Resultant active thrust, $P_A = \frac{1}{2}\gamma(H - z_0)^2 = \frac{1}{2} \times 18(8.0 - 3.56)^2 = \underline{178\ kN/m}$

acting at a height of $\bar{h} = \frac{1}{3}(H - z_0) = \underline{1.5\ m}$ above the base of the wall.

Effect of surface surcharge (undrained)

If an extensive uniform surcharge (q) is applied to the surface the vertical and horizontal stresses are:

Vertical total stress, $\sigma_v = \gamma z + q$ [8.17]

Horizontal active stress, $\sigma_v = \gamma z + q - 2c_u$ [8.18(a)]

Horizontal passive stress, $\sigma_v = \gamma z + q + 2c_u$ [8.18(b)]

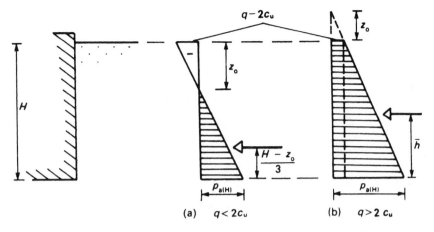

Fig. 8.16 Effect of surface surcharge on undrained active pressure distribution

In the active case, the effect of the surcharge is to reduce the depth of the tension zone. Two forms of active pressure distribution are possible, depending on whether $q < 2c_u$ or $q > 2c_u$ (Fig. 8.16). These may be distinguished by evaluating the tension depth z_0: a *positive* value results when $q < 2c_u$ (Fig. 8.14(a)) and *negative* value when $q > 2c_u$ (Fig. 8.14(b)). To obtain z_0, put $\sigma_{ha} = 0$ and $z = z_0$.

Then $\sigma_{ha} = 0 = \gamma z_0 + q - 2c_u$

giving $z_0 = \dfrac{2c_u - q}{\gamma}$ [8.19]

In either case, the resultant thrust is given by the positive area of the diagram.

For $q < 2c_u$: $P_A = \frac{1}{2}\gamma(H - z_0)^2$ [8.20(a)]

For $q > 2c_u$: $P_A = \frac{1}{2}\gamma(H - z_0)^2 - \frac{1}{2}\gamma z_0^2$

$\qquad = \frac{1}{2}\gamma(H - 2z_0)H$ [8.20(b)]

Worked example 8.8 *A retaining wall having a smooth vertical back retains soil for a depth of 8.0 m. The soil has the following properties:*

$c_u = 30\ kPa \quad \phi_u = 0 \quad \gamma = 18.5\ kN/m^3$

Calculate the magnitude of the resultant active thrust when: (a) the surface is unloaded, (b) there is a surface surcharge of 25 kPa, and (c) there is a surface surcharge of 75 kPa.

(a) *Surface unloaded*

$$z_0 = \frac{2c_u}{\gamma} = \frac{2 \times 30}{18.5} = 3.24\ m$$

Resultant active thrust, $P_A = \frac{1}{2}\gamma(H - z_0)^2$

$$= \frac{1}{2} \times 18.5(8.00 - 3.24)^2 = \underline{210\ kN/m}$$

(b) *Surcharge $q = 25$ kPa*

$$z_o = \frac{2c_u - q}{\gamma} = \frac{2 \times 30 - 25}{18.5} = 1.89 \text{ m (i.e. + ve)}$$

Resultant active thrust, $P_A = \frac{1}{2}\gamma(H - z_o)^2$

$$= \frac{1}{2} \times 18.5(8.00 - 1.89)^2 = \underline{345 \text{ kN/m}}$$

(c) *Surcharge, $q = 75$ kPa*

$$z_o = \frac{2c_u - q}{\gamma} = \frac{2 \times 30 - 75}{18.5} = -0.814 \text{ m (i.e. – ve)}$$

Resultant active thrust, $P_A = \frac{1}{2}\gamma(H - 2z_o)H$

$$= \frac{1}{2} \times 18.5(8.00 + 2 \times 0.814)8.0 = \underline{712 \text{ kN/m}}$$

Alternatively,

At $z = 0$: $\sigma_{ha} = \gamma z + q - 2c_u = 0 + 75 - 2 \times 30 = 15.0$ kPa

At $z = 8.0$ m: $\sigma_{ha} = 18.5 \times 8 + 75 - 2 \times 30$ $= 163.0$ kPa

Resultant active thrust, $P_A = \frac{1}{2} \times 8.0(15.0 + 163.0) = \underline{712 \text{ kN/m}}$

8.8 Coulomb's theory and rough walls

Rankine's theory provides a convenient form of analysis that lends itself easily to simple calculations, but it has its limitations. Since neither the nature nor the orientation of the wall is considered, calculations can only be done (strictly) in respect of a *smooth vertical* wall. The slope of the wall back and the development of friction between the soil and the wall are ignored. Thus, calculated values are somewhat pessimistic. Rankine's theory is classed as a *lower bound solution*: the yielding of the whole structure is assumed to coincide with the yielding of the first element.

Some 80 years before Rankine published his theory, Coulomb (1776) had suggested a solution based on a wedge of soil actively moving towards the wall and sliding down a planar slip surface. Thus, the limiting condition is the yielding of the whole wedge: this represents an *upper bound solution*. Although Coulomb presented a solution only in terms of total stress, he described the shear strength of the soil in terms of both internal friction and cohesion. (It was nearly 140 years later, in 1915, before Bell published his analysis for cohesive soil.) The solution for cohesionless ($c' = 0$) soil will be examined first.

Consider the wedge of soil shown in Fig. 8.17. If the wall were to yield *away*, bringing the soil to the limiting active state, the wedge ABC will slide down the slip surface BC and towards the wall. Plastic equilibrium is being maintained by three forces acting on the wedge:

P_A = active thrust supported by the wall

W = weight of soil in the wedge

R = soil reaction force on the slip plane (value not required)

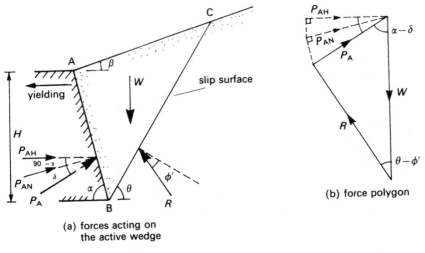

(a) forces acting on
the active wedge

(b) force polygon

Fig. 8.17 Coulomb's wedge theory

The geometry of the wedge is defined by the wall height (H) and the angles α and β, which will be known, and the angle of the slip surface θ, which is unknown. The polygon of forces (Fig. 8.17(b)) also contains two unknowns: P_A and R. The problem is therefore statically indeterminate and requires either an iterative or a calculus method of solution. If a trial value is assumed for θ, and the force polygon drawn, a value for P_A can be found; this may be repeated until the maximum value of P_A is located and evaluated. First, the weight of the wedge is required and then the solution may be obtained analytically or graphically.

Weight of wedge ABC, $W = \frac{1}{2}\gamma \cdot \sin(\alpha + \beta) \cdot \overline{AB} \cdot \overline{AC}$

where γ = unit weight of the soil

$$\overline{AC} = \text{length AC} = \overline{AB}\,\frac{\sin(\alpha + \theta)}{\sin(\theta - \beta)}$$

$\overline{AB} = \text{length AB} = H/\sin \alpha$

Worked example 8.9   *Using Coulomb's method, calculate the active thrust acting on a vertical wall of height 6.0 m due to a mass of homogeneous soil having an unsurcharged horizontal surface and the following properties.*

$c' = 0$ $\phi' = 30°$ $\delta = 15°$ $\gamma = 19\ kN/m^3$

Trial angles of the slip surface (θ) will be chosen of 56°, 58°, 60°, 62°, 64°, 66°. An analytical solution is relatively straightforward here; the model polygon of forces is shown in Fig. 8.18.

Resolving vertically: $W - R\cos(\theta - \phi') - P_A\cos(\alpha - \delta) = 0$

Resolving horizontally: $R\sin(\theta - \phi') - P_A\sin(\alpha - \delta) = 0$

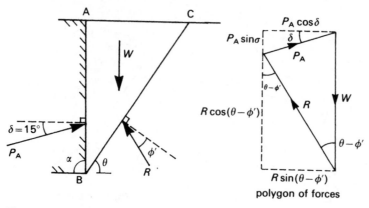

Fig. 8.18 Worked example 8.9. Polygon of forces

From which: $P_A = \dfrac{W}{\cos(\alpha - \delta) + \dfrac{\sin(\alpha - \delta)}{\tan(\theta - \phi')}}$

Now, $\alpha = 90°$ and $\sin(\alpha - \beta) = \sin 90° = 1$; $AB = H/\sin \alpha = 6.0$ m; $\cos(\alpha - \delta) = \cos(90° - 15°) = 0.2588$ and $\sin(\alpha - \delta) = 0.9659$.

$$AC = AB \sin(\alpha + \theta)/\sin(\theta - \beta) = 6.0 \times \sin(90° + \theta)/1.0$$

Weight, $W = \frac{1}{2}\gamma \cdot \sin(\alpha + \beta) \cdot \overline{AB} \cdot \overline{AC} = 9.5 \times 1.0 \times 6.0 \times AC$

and

$P_A = W/[\cos(\alpha - \delta) + \sin(\alpha - \delta)/\tan(\theta - \phi')] = W/[0.2588 + 0.9659/\tan(\theta - \phi')]$

Tabulating the results:

Trial angle θ (deg)	$\sin(\alpha + \theta)$	AC (m)	Weight W (kN/m)	$\tan(\theta - \phi')$	P_A (kN/m)
56	0.5592	3.355	191.2	0.4877	85.4
58	0.5299	3.180	181.2	0.5317	87.3
60	0.5000	3.000	171.0	0.5774	88.5
62	0.4695	2.817	160.6	0.6249	**89.0**
64	0.4384	2.630	149.9	0.6745	88.7
66	0.4067	2.440	139.1	0.7265	87.6
68	0.3746	2.248	128.1	0.7813	85.7

The variation in the magnitude of P_A with the trial angle is shown plotted in Fig. 8.19; the maximum value is the critical value: $P_A = \underline{89.0 \text{ kN/m and critical angle } \theta = 62°}$.

Angle of wall friction

The *angle of wall friction* (δ) is related to both the angle of friction (ϕ') of the soil and the roughness of the wall. Recommended arbitrary values for concrete walls are: $0.667\phi'$ on the *active* side and $0.5\phi'$ on the *passive* side; with maximum values of 20° and 15° respectively. The effect of wall friction near the base of the

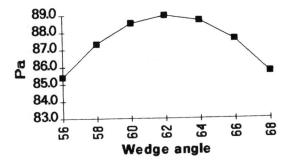

Fig. 8.19

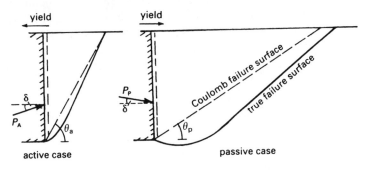

Fig. 8.20 Curvature of failure surface due to wall friction

wall is to produce a curved failure surface in both the active and passive states (Fig. 8.20). However, in Coulomb's analysis a *plane* failure surface is assumed. The consequent error is negligible in the case of active pressures, but for passive pressures involving values of $\phi' > 3$, a significant underestimation will result (see Section 8.10).

8.9 Active pressure on rough walls

(a) Drained conditions $(c' = 0)$
Referring again to Fig. 8.17, a general solution can be obtained for the maximum active thrust in a form using an earth pressure coefficient (K_a). From the geometry of the force polygon:

$$P_A = \frac{W \sin(\theta - \phi')}{\sin[(\alpha - \delta) + (\theta - \phi')]}$$

and

$$W = \tfrac{1}{2} \gamma \sin(\alpha + \beta) \cdot \overline{AB} \cdot \overline{AC} = \tfrac{1}{2} \gamma H^2 \times F\{\alpha, \beta, \theta\}$$

After substituting for W, differentiating and putting $\partial P/\partial \theta = 0$, the maximum value can be written:

$$P_A = \tfrac{1}{2}\gamma H^2 \cdot \frac{K_a}{\sin \alpha \cdot \cos \delta} \qquad [8.21]$$

$$\text{where} \quad K_a = \frac{\sin^2(\alpha + \phi') \cdot \cos \delta}{\sin \alpha \cdot \sin(\alpha - \delta)\left[1 + \sqrt{\left(\dfrac{\sin(\phi' + \delta) \cdot \sin(\phi' - \beta)}{\sin(\alpha - \delta) \cdot \sin(\alpha + \beta)}\right)}\right]^2} \qquad [8.22]$$

Thrust component **normal** to the wall, $P_{AN} = \tfrac{1}{2}\gamma H^2 \dfrac{K_a}{\sin \alpha} = P_A \cdot \cos \delta$ [8.23]

Horizontal thrust component, $P_{AH} = \tfrac{1}{2}\gamma H^2 K_a = P_A \cdot \sin \alpha \cdot \cos \delta$ [8.24]

For problems where the soil and wall surfaces are simple planes, the point of application of P_A may be taken as occurring at $H/3$ above the base – but see also Section 8.8. Values of K_a can be computed using a computer spreadsheet or interpolated from the values given in Table 8.2.

Table 8.2 Earth pressure coefficients: c' = 0

	δ*	ϕ'				
		25°	30°	35°	40°	45°
K_a	0	0.41	0.33	0.27	0.22	0.17
	10°	0.37	0.31	0.25	0.20	0.16
	20°	0.34	0.28	0.23	0.19	0.15
	30°	–	0.26	0.21	0.17	0.14
K_p	0	2.5	3.0	3.7	4.6	–
	10°	3.1	4.0	4.8	6.5	–
	20°	3.7	4.9	6.0	8.8	
	30°	–	5.8	7.3	11.4	–

* Recommended maximum values for δ:
active, $\tfrac{2}{3}\phi'$; passive $\tfrac{1}{2}\phi'$. Reduce these by 25%
for uncoated steel piling.

For a smooth vertical wall and a horizontal unsurcharged ground surface, i.e. $\alpha = 90°$, $\beta = 0$ and $\delta = 0$, the solution obtained for K_a will be the same as that given by the Rankine expression, $(1 - \sin \phi')/(1 + \sin \phi')$.

(b)   Undrained conditions ($\tau_f = c_u$)
Undrained conditions can be assumed to occur behind walls supporting saturated clays when the construction period is short and little or no dissipation of excess pore pressure may be expected for some time following. In such cases, short-term stability should be considered in terms of total stresses and the undrained shear strength of the soil, i.e. $\tau_f = c_u$. Consider the case of a vertical rough wall and a horizontal soil surface (Fig. 8.21(a)). A plane failure surface is assumed (BT) which terminates at the bottom of a tension crack of depth z_o. At the limiting active state, the equilibrium of the wedge ABTC is maintained by the following forces.

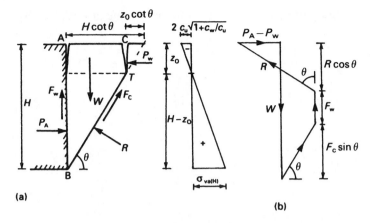

Fig. 8.21 Coulomb's theory for undrained conditions

W = weight of wedge ABTC = $\frac{1}{2}\gamma(H^2 - z_o^2)\cot\theta$

R = normal reaction on the failure plane (value not required)

P_A = active thrust acting on the wall

F_c = shear resistance force along the failure plane BT = $c_u(H - z_o)\operatorname{cosec}\theta$

F_w = shear resistance force along the wall face = $c_w(H - z_o)$

P_w = horizontal thrust due to water in tension crack = $\frac{1}{2}\gamma_w z_o^2$

In which: c_u = undrained shear strength of supported soil

c_w = undrained adhesion between soil and wall

Wall adhesion
The magnitude of *wall adhesion* (c_w) varies between $0.3c_u$ for stiff clays to c_u for soft clays, but the extent of its development along the wall surface is problematical. In the absence of empirical evidence, it is usual to adopt a value of $0.5c_u$. In BS 8002:1994 a design value for c_w is recommended of $0.75 \times$ design c_u; since design c_u = representative $c_u/1.5$, this is equivalent to $c_w = 0.5 \times$ representative c_u.

The polygon of forces acting on the wedge is shown in Fig. 8.21(b), from which:

$$R\cos\theta = W - F_c\sin\theta - F_w \quad \therefore \quad R = W\sec\theta - F_c\tan\theta - F_w\sec\theta$$

and

$$P_A - P_w = R\sin\theta - F_c\cos\theta = W\tan\theta - F_c(\sin\theta \cdot \tan\theta + \cos\theta) - F_w\tan\theta$$

After substituting for W, F_c and F_w,

$$P_A - P_w = \frac{1}{2}\gamma(H^2 - z_o^2) - c_u(H - z_o)\left[\left(1 + \frac{c_w}{c_u}\right)\tan\theta + \cot\theta\right]$$

After differentiating and putting $\partial P/\partial \theta = 0$, the optimum value of θ is given by:

$$\cot \theta_a = \sqrt{\left(1 + \frac{c_w}{c_u}\right)}$$

Substituting $P_A = \frac{1}{2}\gamma(H^2 - z_o^2) - 2c_u(H - z_o)\sqrt{\left(1 + \frac{c_w}{c_u}\right)} + P_w$ [8.25]

or putting $K_{ac} = 2\sqrt{\left(1 + \frac{c_w}{c_u}\right)}$

$$P_A = \frac{1}{2}\gamma(H^2 - z_o^2) - c_u(H - z_o)K_{ac} + P_w \qquad [8.26]$$

At a given depth the lateral pressure is:

$$\sigma_{ha} = \gamma z - c_u K_{ac} = \gamma z - 2c_u\sqrt{(1 + c_w/c_w)} \qquad [8.27]$$

At the bottom of the tension zone, $z = z_o$ and $\sigma_{ha} = 0$.

Therefore, $z_o = \dfrac{2c_u\sqrt{(1 + (c_w/c_u))}}{\gamma}$ [8.28(a)]

or, with a uniform surface surcharge q:

$$z_o = \frac{2c_u\sqrt{(1 + (c_w/c_u))} + q}{\gamma} \qquad [8.28(b)]$$

It is possible to extend this analytical solution to include a non-vertical wall face and a non-horizontal soil surface. However, the mathematical process is lengthy and results in complex expressions. In such cases, it is preferable to use a computer spreadsheet or other machine method.

(c)   Drained overconsolidated clay ($c' > 0$)
In examining the stability of a mass of heavily overconsolidated clay it is necessary to consider the potential amount of strain associated with the onset of failure. The *peak strength* ($\tau_f = c' + \sigma_n' \tan \phi_p'$) of the soil should be invoked only when it is known that very small strains precede failure. In the majority of cases involving retaining walls, the pre-yield strains are likely to be high enough to produce significant expansion. Thus, failure will occur at the *critical state*: $\tau_f = \sigma_n' \tan \phi_c'$, i.e. $c' = 0$.

As a general rule, therefore, the active thrust should be calculated using critical state parameters ($c' = 0$, $\phi' = \phi_c'$) and eqns [8.21] to [8.24]. In those cases where it is thought reasonable to suppose failure would occur at peak stress, the parameters are $c' > 0$ (as determined in a triaxial test) and $\phi' = \phi_p'$ (the peak angle of friction). A not entirely rational compromise suggested by some is to use $\phi' = \phi_c'$, together with $c' > 0$. The active thrust on the wall is then given by:

$$P_A = \frac{1}{2}\gamma H^2 K_a - c'HK_{ac} \qquad [8.29]$$

where K_a and K_{ac} are earth pressure coefficients as before, values for which may be obtained from Table 8.3.

Table 8.3 Earth pressure coefficients: c > 0

δ*		$\frac{c_w}{c}$	φ′					
			0	5°	10°	15°	20°	25°
K_a	0	All	1.00	0.85	0.70	0.59	0.48	0.40
	φ′	values	1.00	0.85	0.64	0.50	0.40	0.32
K_{ac}	0	0.0	2.00	1.83	1.68	1.54	1.40	1.29
	0	1.0	2.83	2.60	2.38	2.16	1.96	1.76
	φ′	0.5	2.45	2.10	1.82	1.55	1.32	1.15
	φ′	1.0	2.83	2.47	2.13	1.85	1.59	1.41
K_p	0	All	1.00	1.20	1.40	1.70	2.10	2.50
	φ′	values	1.00	1.30	1.60	2.20	2.90	3.90
K_{pc}	0	0.0	2.00	2.20	2.40	2.60	2.80	3.10
	0	0.5	2.40	2.60	2.90	3.20	3.50	3.80
	0	1.0	2.60	2.90	3.20	3.60	4.00	4.40
	φ′	0.5	2.40	2.80	3.30	3.80	4.50	5.50
	φ′	1.0	2.60	2.90	3.40	3.90	4.70	5.70

* See Table 8.2
Recommended maximum values for c_w

Active: $c_w = 0.5\,c_u$, but ≤50 kPa
Passive: $c_w = 0.5\,c_u$, but ≤25 kPa

8.10 Passive pressure on rough walls

(a) Drained conditions ($c' = 0$)
Coulomb's theory may be used to establish passive pressures and thrust on a rough wall that has an inclined back and where the ground surface is a regular inclined plane. The analysis follows along the same lines as that for active pressures described in Section 8.8(a). The following expressions are obtained:

$$P_P = \tfrac{1}{2}\gamma H^2 \cdot \frac{K_P}{\sin \alpha \cdot \cos \delta} \qquad [8.30]$$

$$\text{where} \quad K_P = \frac{\sin^2(\alpha - \phi) \cdot \cos \delta}{\sin \alpha \cdot \sin(\alpha - \delta)\left[1 - \sqrt{\left(\frac{\sin(\phi' + \delta)\cdot\sin(\phi' + \beta)}{\sin(\alpha + \delta)\cdot\sin(\alpha + \beta)}\right)}\right]^2} \qquad [8.31]$$

Thrust component **normal** to the wall, $P_{PN} = \tfrac{1}{2}\gamma H^2 \dfrac{K_p}{\sin \alpha} = P_p \cdot \cos \delta$ [8.32]

Horizontal thrust component, $P_{PH} = \tfrac{1}{2}\gamma H^2 K_p = P_A \cdot \sin \alpha \cdot \cos \delta$ [8.33]

Definitions of angles, etc. are given in Fig. 8.22. For problems where the soil and wall surfaces are simple planes, the point of application of P_p may be taken

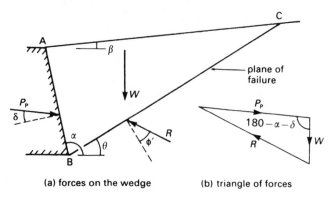

(a) forces on the wedge   (b) triangle of forces

Fig. 8.22 Coulomb's theory for passive pressure ($c' = 0$)

as occurring at H/3 above the base – but see also Section 8.8. Values of K_p can be computed using a computer spreadsheet (see *Soil Mechanics Spreadsheets and Reference*, on the accompanying CD-ROM) or interpolated from the values given in Table 8.2.

For a smooth vertical wall and a horizontal unsurcharged ground surface, i.e. $\alpha = 90°$, $\beta = 0$ and $\delta = 0$, the solution obtained for K_p will be the same as that given by the Rankine expression, i.e. $(1 + \sin \phi')/(1 - \sin \phi')$.

Effect of curved failure surface
When the value of δ is greater than about $\frac{1}{3}\phi'$, the failure surface is significantly curved at the lower end near the base of the wall. The use of Coulomb's theory in such cases, since it is based on a *planar* failure surface, leads to a marked overestimation of the passive resistance. A number of alternative methods have been proposed wherein the curved part is assumed variously to be a circular arc, an ellipse or a logarithmic spiral.

Considerable variations are apparent between values obtained by the different methods and those obtained using Coulomb's method (Table 8.5), but variations between the alternative methods themselves are less marked. Values attributed to Packshaw (1946) were given in CP2:1951, but active and passive coefficients based on the work of Caquot and Kerisel (1948) have become widely used and in an annex to BS 8002:1994 some charts published by Kerisel and Absi (1990) are given. See Tables 8.3, 8.4 and 8.5.

(b)   Undrained conditions ($\tau_f = c_u$)
Undrained conditions may be assumed in saturated clays when the construction period is short and dissipation of excess pore pressure is expected to be very slow. The analysis is carried out in terms of *total stress*. As in the drained analysis, the effect of frictional resistance tangential to the wall face produces curvature in the failure surface near the base of the wall. However, the errors incurred by assuming a planar failure surface in this (undrained) case are less significant. Figure 8.23 shows the arrangement of forces.

Table 8.4 Coefficients of passive earth pressure (K_p). After Caquot and Kerisel (1948)

(a) Factor K_β

α (deg)	β/ϕ'	Angle of friction, ϕ' (deg)								
		10	15	20	25	30	35	40	45	50
70	+1.0	1.70	2.38	3.51	5.57	9.77	19.0	42.2	111	373
	+0.8	1.68	2.33	3.40	5.29	8.89	16.4	34.0	82.3	248
	+0.6	1.62	2.18	3.09	4.60	7.33	12.7	24.3	53.6	144
	+0.4	1.54	2.03	2.76	3.94	5.95	9.66	17.2	34.4	81.9
	+0.2	1.46	1.86	2.44	3.34	4.78	7.27	11.9	21.6	45.3
	0.0	1.36	1.68	2.13	2.78	3.78	5.36	8.07	13.2	24.1
	−0.2	1.24	1.49	1.82	2.27	2.90	3.84	5.30	7.75	13.3
	−0.4	1.12	1.29	1.50	1.78	2.15	2.64	3.33	4.35	6.81
	−0.6	0.97	1.08	1.19	1.33	1.50	1.70	1.94	2.25	2.67
	−0.8	0.83	0.87	0.93	0.96	0.99	1.01	1.03	1.03	1.02
80	+1.0	1.83	2.66	4.09	6.81	12.6	26.0	61.6	176	652
	+0.8	1.81	2.60	3.97	6.45	11.4	22.3	49.5	130	432
	+0.6	1.75	2.45	3.60	5.62	9.42	17.3	35.4	84.7	251
	+0.4	1.68	2.29	3.24	4.84	7.69	13.2	25.1	54.5	143
	+0.2	1.60	2.12	2.90	4.14	6.23	10.0	17.6	34.5	79.6
	0.0	1.52	1.95	2.57	3.50	4.98	7.47	12.0	21.2	42.5
	−0.2	1.42	1.77	2.24	2.91	3.89	5.43	7.97	12.6	21.9
	−0.4	1.31	1.57	1.90	2.34	2.94	3.79	5.09	7.16	18.2
	−0.6	1.18	1.35	1.55	1.79	2.09	2.48	3.00	3.74	4.83
	−0.8	1.05	1.13	1.21	1.30	1.39	1.48	1.59	1.70	1.83
90	+1.0	1.93	2.91	4.66	8.16	15.9	34.9	88.7	275	1130
	+0.8	1.90	2.84	4.52	7.71	14.4	29.9	71.0	203	747
	+0.6	1.84	2.68	4.11	6.72	11.9	23.1	50.9	132	433
	+0.4	1.78	2.51	3.71	5.81	9.74	17.8	36.2	85.4	248
	+0.2	1.71	2.35	3.35	5.02	7.95	13.6	25.5	54.3	138
	0.0	1.64	2.19	3.01	4.29	6.42	10.2	17.5	33.5	74.3
	−0.2	1.57	2.02	2.67	3.62	5.09	7.50	11.8	20.1	38.6
	−0.4	1.48	1.83	2.31	2.96	3.90	5.32	7.61	11.6	20.2
	−0.6	1.37	1.61	1.92	2.31	2.82	3.53	4.54	6.10	8.67
	−0.8	1.24	1.37	1.52	1.69	1.89	2.11	2.39	2.76	3.25
	−1.0	0.98	0.97	0.94	0.91	0.87	0.82	0.77	0.71	0.64
100	+1.0	1.98	3.12	5.23	9.67	19.9	46.6	127	431	
	+0.8	1.96	3.05	5.07	9.11	18.0	39.8	102	316	
	+0.6	1.90	2.88	4.62	7.95	14.9	30.9	73.0	206	
	+0.4	1.84	2.70	4.19	6.90	12.2	23.8	52.0	134	429
	+0.2	1.78	2.55	3.81	5.99	10.0	18.2	36.8	85.2	240
	0.0	1.73	2.40	3.45	5.17	8.17	13.8	25.5	52.9	130
	−0.2	1.68	2.25	3.10	4.41	6.54	10.2	17.2	31.9	67.7
	−0.4	1.61	2.07	2.72	3.66	5.08	7.35	11.2	18.5	35.2
	−0.6	1.51	1.86	2.30	2.90	3.72	4.92	6.77	9.83	15.4
	−0.8	1.38	1.59	1.84	2.13	2.49	2.95	3.57	4.43	5.75
	−1.0	1.08	1.10	1.11	1.12	1.12	1.11	1.11	1.10	1.10
110	+1.0	2.01	3.30	5.83	11.4	25.0	62.5	185	680	
	+0.8	1.98	3.23	5.66	10.7	22.6	53.4	147	449	
	+0.6	1.92	3.05	5.16	9.39	18.7	41.4	106	326	

Table 8.4   Continued

α (deg)	β/φ′	Angle of friction, φ′ (deg)								
		10	15	20	25	30	35	40	45	50
	+0.4	1.87	2.87	4.69	8.17	15.4	32.0	75.4	211	
	+0.2	1.82	2.72	4.28	7.12	12.7	24.6	53.4	135	421
	0.0	1.78	2.58	3.90	6.18	10.4	18.7	37.2	84.0	228
	−0.2	1.74	2.44	3.53	5.30	8.35	14.0	25.3	51.0	119
	−0.4	1.68	2.27	3.14	4.45	6.54	10.1	16.6	29.7	60.1
	−0.6	1.60	2.06	2.68	3.55	4.83	6.80	10.0	15.8	27.3
	−0.8	1.47	1.78	2.15	2.62	3.24	4.08	5.28	7.12	10.2

(b)   Factor K_δ (use when $\delta < \phi'$)

δ/φ′	Angle of friction, φ′ (deg)								
	10	15	20	25	30	35	40	45	50
1.0	1.00	1.00	1.00	1.00	1.00	1.00	1.00	1.00	1.00
0.9	0.99	0.99	0.98	0.98	0.98	0.98	0.98	0.98	0.98
0.8	0.99	0.98	0.97	0.95	0.94	0.92	0.89	0.85	0.80
0.7	0.98	0.96	0.94	0.91	0.88	0.84	0.78	0.72	0.64
0.6	0.96	0.93	0.90	0.86	0.81	0.75	0.68	0.60	0.51
0.5	0.95	0.91	0.86	0.81	0.75	0.67	0.59	0.50	0.40
0.4	0.93	0.88	0.82	0.76	0.69	0.60	0.51	0.41	0.31
0.3	0.91	0.85	0.79	0.71	0.63	0.52	0.44	0.34	0.24
0.2	0.90	0.83	0.75	0.67	0.57	0.48	0.38	0.28	0.19
0.1	0.88	0.80	0.72	0.62	0.52	0.42	0.32	0.22	0.14
0.0	0.86	0.78	0.68	0.57	0.47	0.36	0.26	0.17	0.10

When $\phi' = 0$ or $\delta = \phi'$, $K_\delta = 1.0$ Coefficient of passive earth pressure, $K_p = K_\beta \times K_\delta$

Table 8.5   Comparisons of solutions for K_p

Method of analysis	$\phi' = 40°$			
	δ = 0	10°	20°	30°
Caquot and Kerisel (1948)	4.6	7.1	10.4	14.6
Coulomb (eqn [8.41])	4.6	6.9	11.1	21.5
CP2 (friction circle method)	4.6	6.5	8.8	11.4
Rowe and Peaker (1965)	4.6	5.8	7.2	8.8
Shields and Tolunay (1973)	4.6	6.3	7.8	9.2
Sokolovski (1965)	4.6	6.5	9.1	11.6

Planar failure surface (Fig. 8.23(a))
The analysis here is similar to that given in Section 8.9(b), except that no tension crack will occur. Thus, the angle θ is given by $\cot \theta = \sqrt{(1 + c_w/c_u)}$ and

$$P_{PU} = \tfrac{1}{2}\gamma H^2 + c_u H K_{pc}$$ [8.34]

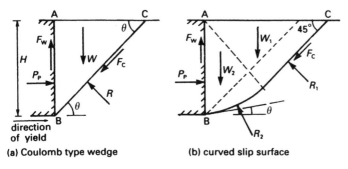

(a) Coulomb type wedge   (b) curved slip surface

Fig. 8.23 Passive pressure in undrained conditions

where $\qquad K_{pc} = 2\sqrt{\left(1 + \dfrac{c_w}{c_u}\right)}$ [8.35]

e.g. when $\qquad c_w = \frac{1}{2}c_u$, $K_{pc} = 2.449$.

Curved failure surface (Fig. 8.23(b))
The adhesion force acting tangentially to the wall produces a failure surface
angle θ, given by $\sin(2\theta) = c_w/c_u$. Thus, at a typical value of $c_w = \frac{1}{2}c_u$, $\theta = 15°$; i.e.
the angle of the slip surface *immediately* adjacent to the wall is 15°. Further
away from the wall the shear strength will be c_u and θ becomes 45°. An analysis
based on this geometry produces values for K_{pc} such as those given in Table 8.3;
used in eqn [8.34].
    When $c_w = \frac{1}{2}c_u$, $K_{pc} = 2.40$, which is only 2 per cent less than the 'straight slip
surface' result of 2.449; if $c_w = c_u$ is assumed, the variation is over 8 per cent. In
the majority of design cases, it is therefore not unreasonable to use eqn [8.35],
although interpolating in Table 8.3 is relatively easy.

(c)   Drained overconsolidated clay ($c' > 0$)
Providing that the strains preceding the onset of failure are considered to be
low (<2 per cent), the *peak strength* ($\tau_f = c' + \sigma'_n \tan \phi'_p$) may be invoked as the
shear stress at the moment of yielding (see also Section 8.9(c)). In cases where
the pre-yield strains are likely to be significantly large, the *critical state strength*
($\tau_c = \sigma'_n \tan \phi'_c$) should be used, i.e. putting $c' = 0$.
    The passive thrust at *low* lateral strain is given by

$$P_P = \tfrac{1}{2}\gamma H^2 K_p + c'HK_{pc}$$ [8.36]

Where K_p and K_{pc} are earth pressure coefficients as previously described, values
of which can be obtained from Table 8.3 based on peak strength parameters
(c' and ϕ'_p).

8.11   Compaction pressures on backfilled walls

During the process of compacting backfill behind a retaining wall the weight of
the roller causes a temporary increase in vertical stresses ($\Delta\sigma'_{vc}$) in the soil layers

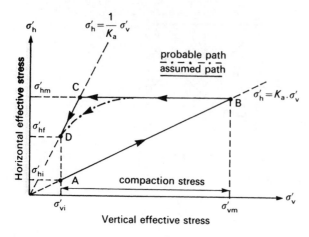

Fig. 8.24 Stress path during compaction behind a retaining wall

below. If there is sufficient lateral yielding at a given point the corresponding increase in horizontal stress would be $\Delta\sigma'_h = K_a\Delta\sigma'_{vc}$. As the roller moves on, the vertical stress decreases to the original values, but if the soil has yielded some of the increase in horizontal stress will remain. Broms (1971) presented a simplified stress path for the process (Fig. 8.24):

Point A represents the initial stresses at a point $(\sigma'_{hi}, \sigma'_{vi})$
A → B compaction stresses being applied
Point B maximum stresses induced during compaction $(\sigma'_{hm}, \sigma'_{vm})$
B → D probable unloading curve as roller moves away
B → C → D assumed simplified unloading curve
Point D final stresses $(\sigma'_{hf}, \sigma'_{vf})$ – but note that $\sigma'_{vf} = \sigma'_{vi}$

The increase in horizontal stress caused by the roller load diminishes with depth – this is shown by the curved line in Fig. 8.25(a). At very shallow depths, as the roller moves away, the horizontal stress will decrease to the limiting value $K_p\gamma z$. A critical depth (z_c) therefore exists below which the increase in horizontal stress due to compaction remains, i.e. point R in Fig. 8.25(a). When the next layer is placed and rolled the horizontal stresses below R will remain constant. The locus of point R traces a vertical path as successive layers are compacted (Fig. 8.25(b)).

The critical depth (z_c) may be calculated by considering the residual horizontal stress at point R.

$$\sigma'_{hm} = K_a\sigma'_{vm} = K_p\gamma z_c \tag{8.37}$$

where $\sigma'_{vm} = \gamma z_c + \Delta\sigma'_{vc}$
$\Delta\sigma'_{vc}$ = increase in vertical stress due to roller load

An approximate value for $\Delta\sigma'_{vc}$ can be obtained using elastic theory and considering the roller load as a line load of finite length:

$$\Delta\sigma'_{vc} \approx \frac{2p}{\pi z_c} \tag{8.38}$$

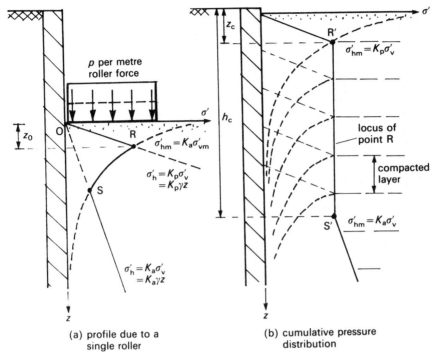

(a) profile due to a
single roller

(b) cumulative pressure
distribution

Fig. 8.25 Suggested development of horizontal stress due to compaction (after Broms (1971)
and Ingold (1979))

where p = roller line load per metre

$$z_c = \frac{K_a^2}{\gamma}\sigma'_{vm} = \frac{K_a^2}{\gamma}(\gamma z_c + \Delta\sigma'_{vc}) = \frac{K_a^2}{\gamma}\left(\gamma z_c + \frac{2p}{\pi z_c}\right)$$

From which, $$z_c = \sqrt{\left(\frac{K_a^2}{1 - K_a^2}\right)} \cdot \sqrt{\left(\frac{2p}{\pi\gamma}\right)} \qquad [8.39(a)]$$

or approximately and more simply, $z_c = K_a\sqrt{\left(\frac{2p}{\pi\gamma}\right)} \qquad [8.39(b)]$

The error incurred with eqn [8.39(b)] is small for most backfill material, e.g. 2.4
per cent when $\phi' = 40°$.

The magnitude of σ'_{hm} can now be obtained by substituting the value of z_c
into eqn [8.37]:

$$\sigma'_{hm} = \sqrt{\left(\frac{2p\gamma}{\pi}\right)} \qquad [8.40]$$

Behind high walls, a critical height (h_c) may be reached below which the
active (Rankine) pressure is greater than the residual stress due to compaction –
point S' in Fig. 8.25(b). At this critical height

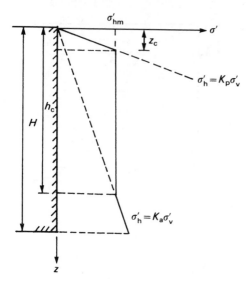

Fig. 8.26   Simplified distribution of horizontal compaction stress for design

$$\sigma'_{hm} = \sqrt{\left(\frac{2p\gamma}{\pi}\right)} = K_a\gamma h_c$$

Rearranging $$h_c = \frac{1}{K_a}\sqrt{\left(\frac{2p}{\pi\gamma}\right)} \qquad\qquad [8.41]$$

The final horizontal stress distribution recommended for design purpose is shown in Fig. 8.26. Ingold (1979) and Broms (1971) present similar approximate horizontal stress distributions. The main differences are in the choice of earth pressure coefficients: Ingold uses K_a and K_p, whereas Broms adopted K_o and $1/K_o$ respectively. The suitability of these choices may be debated, and in any case depends on the amount of lateral yielding that may be reasonably assumed to occur. If the wall is propped and rigid, K_o and $1/K_o$ would seem preferable, but where lateral yielding is likely *before* placing and rolling, K_a and K_p will be more appropriate. See worked example 8.10 for details of calculations.

Compaction pressures in cohesive soils
The analyses and equations described above include the assumption that there are no time-dependent changes in stress and cannot therefore be used in the case of clay fills. Several complications are apparent. Horizontal increases in stress tend to be minimal in clay backfill until the air-void content has been reduced to below 15 per cent. Also, compacted clays may either consolidate or swell after rolling, depending on whether the pore pressure subsequently decreases or increases. Clay fills placed 'dry' will tend to swell. To avoid this the soil should be 'wet' (i.e. have excess *positive* pore pressures after rolling), but then the shear strength may be low. Clayton, Symons and Heidra Cobo (1991) suggest that as guidelines low-plasticity clays ($I_p < 30$) are unlikely to swell, when $I_p > 50$ swelling is likely after placing fill with conventional plant.

Clayton and Symons (1992) concluded from pilot studies with cohesive fills that the horizontal total stresses after compaction are a function of the finished undrained shear strength. With clays of intermediate and high plasticity average horizontal total pressures were observed of about 20 and 40 per cent, respectively. These figures relate to rigid walls and compaction with heavy plant. The *rate* of filling may also be significant because of the amount of stress relaxation taking place between rollings. Although more research is recommended, it appears that long-term residual total stresses may be close to *at-rest* values in clays of intermediate plasticity and to *passive* values in the case of clay of high plasticity.

8.12 Stability conditions of gravity walls

In the design of retaining walls a number of possible modes of failure must be considered that are either ultimate or serviceability *limit states* (see also p. 552–3); calculations must show that adequate provision is made against the occurrence of adverse limit states pertaining to a particular type of wall. For the purpose of checking *external* stability, a gravity wall is treated as a rigid monolith (i.e. no internal yielding or distortion). In some types of construction checks are also required on *internal* stability, e.g. reinforced concrete walls, reinforced earth walls. Gravity walls depend essentially upon their own weight for external stability and the following limit states should be examined (see Fig. 8.28):

(a) Overturning   [recommended factor of safety = 2.0]
(b) Forward sliding   [recommended factor of safety = 2.0]
(c) Bearing pressure failure under base [recommended factor of safety = 3.0]
(d) Occurrence of tension in lateral joints, e.g. masonry/brickwork walls
(e) Overall slip failure, e.g. under sloping ground surface

The use of a *factor of safety* as indicated above is the traditional approach to design. The limit state approach recommended in BS 8002:1994 requires the application of a *mobilisation factor* (M) to the representative strength value in order to obtain a *design* value, e.g. design c_u = representative c_u/M. The value for M used against undrained strength should not be less than 1.5 if the wall displacements are required to be less than 0.5 per cent of the wall height, for bearing capacity calculations M should be in the range 2.0–3.0; for effective stress parameters M should not be less than 1.2. See also Appendix B. Structural forces and moments should be determined using design earth pressures and water pressures.

For convenience, the forces acting on a retaining wall are calculated in terms of vertical and horizontal·components. Vertical forces include the weight of the wall and any vertical loads it may have to carry, together with the weight of soil over the heel of L-shaped and T-shaped walls (Fig. 8.27). There may also be a vertical component of the earth pressure resultant when the support face is not vertical, the ground surface is sloping or where wall friction is included. Horizontal forces will include the resultants of earth pressure distributions and any horizontal loads. Earth pressure resultants must include (where appropriate) the effects of surface surcharge loads, compaction pressures and any horizontal loads.

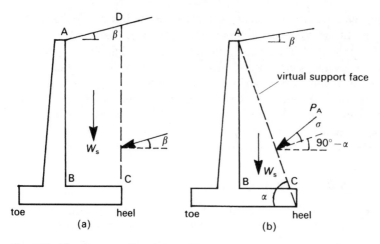

Fig. 8.27   Virtual support face behind RC walls

Virtual support face

When considering L-shaped and T-shaped reinforced concrete walls the soil resting on the base between the heel and the stem can be included as part of the wall. A *virtual support face* can be taken as either a vertical surface extending upward from the heel (CD in Fig. 8.27(a)) or an inclined surface drawn between the corners of the crest and heel (AC in Fig. 8.27(b)). The advantage of assuming CD as the virtual support face is that Rankine's K_a can be used; in this case the weight of soil within ABCD is taken as part of the wall. Assuming AC to be the virtual support face is considered to be more realistic, calculating P_A using the Müller–Breslau value of K_a with angle BCA = α; the weight of soil in the triangular section ABC is taken as part of the wall. For calculation details, see worked example 9.9.

Overturning

It is usual to assume an overturning mode about the toe (Fig. 8.28(a)), although some designers prefer to consider the edge of the middle third of the base nearest to the toe. The factor of safety is defined as the ratio of the sum of stabilising moment (M_s) to the sum of overturning moment (M_o). The factor of safety should not be less than 2.0, although where the possibility of a rising water level exists behind the wall a minimum value of 2.5 should be used.

$$F_{OT} = \frac{\Sigma(\text{stabilising moments, e.g. of } W, Q, \text{etc.})}{\Sigma(\text{overturning moments, e.g. of } P_A, \text{etc.})} \qquad [8.42]$$

Alternatively, use *design earth pressures* and establish limit state equilibrium.

Forward sliding

The factor of safety against sliding (Fig. 8.28(b)) is defined as the ratio of the sum of horizontal resistance forces to the sum of horizontal disturbing forces. The sliding resistance R_s that may be mobilised between the wall base and the soil may be taken as:

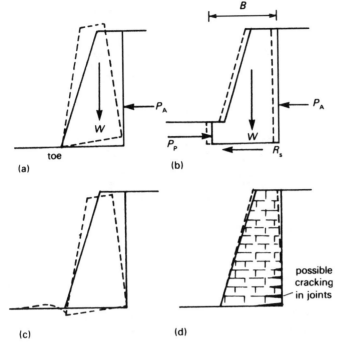

Fig. 8.28 Modes of failure of retaining walls
(a) Overturning $(F \geq 2.0)$ (b) Sliding $(F \geq 2.0)$ (c) Bearing failure in soil beneath base
(d) Tension in joints

Effective stress: $R_s = V' \tan \delta_b$ Undrained: $R_s = c_b B$

For cast *in situ* concrete, the wall friction and adhesion values may be taken as $\tan \delta_b = \tan \phi'$ and $c_b = c_u$, respectively. In calculating the passive resistance mobilised in front of the wall, account should be taken of the effects of weathering, disturbance by traffic and construction operations, the presence of excavations, service ducts, tree roots, etc. Also, the amount of yield required to mobilise passive resistance must be considered. The top 0.5–1.0 m depth of soil is usually ignored and a factor (α) applied (typically 0.5) to pressures calculated using K_p.

$$F_s = \frac{R_s + \alpha P_P}{P_A} = \frac{V \tan \phi' + \alpha P_P}{P_A} \qquad [8.43]$$

Alternatively, use *design earth pressures* and establish limit state equilibrium.

Bearing pressure beneath base
The foundation of the wall must be designed to satisfy the ultimate limit state in terms of bearing. The methods and factors described in Chapter 11 should be used to establish the *allowable bearing capacity*, incorporating a factor of safety (equivalent to M) of 2.5–3.0. The distribution of ground bearing pressure beneath the base may be either trapezoidal or triangular (Fig. 8.29); the maximum pressure usually occurs beneath the toe. This value maximum must not exceed

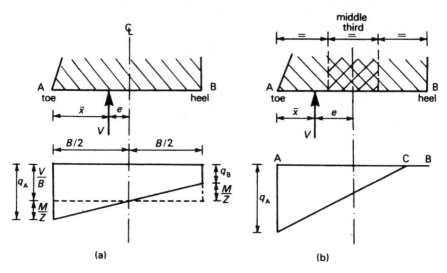

Fig. 8.29 Distribution of ground bearing pressure
(a) Trapezoidal (b) Triangular

the allowable bearing capacity of the soil, otherwise settlement or vertical yielding may occur. In the first stage of the calculation, the position ($\bar{x}$) of the ground
reaction force (V) is determined and thence the eccentricity (e) of the force
about the centre-line.

(a) Trapezoidal distribution
If the ground reaction falls within the middle third of the base (when $e < B/6$),
the pressure distribution will be trapezoidal, with compressive pressure extending between the toe and heel. The ground bearing pressure at A and B are then
given by

$$q_A = \frac{V}{B} + \frac{M}{Z} \quad \text{and} \quad q_B = \frac{V}{B} - \frac{M}{Z}$$

but $M = Ve$ and $Z = \dfrac{1 \times B^2}{6}$

Then $q_A = \dfrac{V}{B} + \dfrac{6Ve}{B^2}$ [8.44(a)]

and $q_B = \dfrac{V}{B} - \dfrac{6Ve}{B^2}$ [8.44(b)]

(b) Triangular distribution
When the ground reaction falls outside the middle third (when $e \geq B/6$), the
pressure diagram will be triangular and the maximum ground bearing pressure
will be:

$$q_A = \frac{2V}{3\bar{x}}$$ [8.45]

Between C and B the bearing pressure is zero since tension cannot be developed between the soil and the base. The distance BC can be described as the *tension crack width*.

Tension crack width $= B - 3\bar{x}$

Tension failure in joints
Although tension cannot develop between the soil and the base, it could develop in lateral joints in the masonry. A calculation based on eqn [8.44(b)] could be carried out at each joint level to check crack potential. However, in masonry, brickwork, crib and gabion walls it is preferable to eliminate tension in all lateral joints. This limit state can be examined by noting the position of the ground reaction (V). If V falls within the *middle third* of the base, the pressure distribution will be trapezoidal and no tension will occur. This is referred as the *middle-third rule*.

Rotational slip failure
In cohesive soils, especially where the soil surface is inclined upward above the wall, there may be a potential failure mode in the form of a slip along a cylindrical surface which passes beneath the wall base (Fig. 8.30). This may be brought about if the shear strength of the soil is reduced due to weathering or when the water levels behind the wall rise. The analysis of this type of failure is dealt with in Chapter 9.

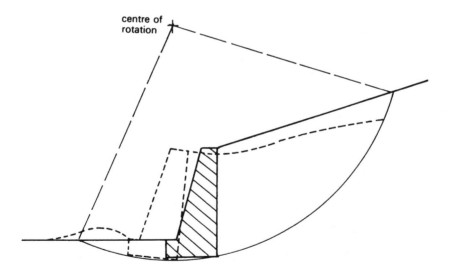

centre of
rotation

Fig. 8.30 Rotational slip failure

Worked example 8.10 *Check the stability of the reinforced concrete retaining wall shown in Fig. 8.31. The soil is a silty gravel sand with a horizontal surface and no surcharge. The water table lies well below the base. The wall was propped while the backfill was placed using only light compaction.*

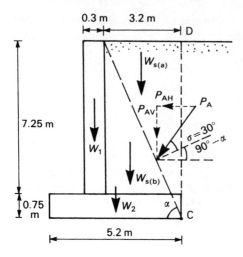

Fig. 8.31

Soil properties: $c' = 0$ $\phi' = 40°$ $\gamma = 20 \; kN/m^3$ Concrete: $\gamma = 24 \; kN/m^3$

Two example calculations will be done: considering first a virtual support face that is vertical and then an inclined face.

(a)   *Virtual support face vertical*

Rankine's $K_a = \dfrac{1 - \sin 40°}{1 + \sin 40°} = 0.217$

Since no yielding along CD will occur, $\delta = 0$.
Horizontal active resultant force, $P_A = \frac{1}{2} \times 20 \times 8.0^2 \times 0.217 = 138.9$ kN/m

Vertical forces: $W_1 = 24 \times 0.3 \times 7.25 = 52.2$ kN/m

$W_2 = 24 \times 5.2 \times 0.75 = 93.6$ kN/m

$W_s = 20 \times 3.2 \times 7.25 = 464.0$ kN/m

Total, $V = 609.8$ kN/m

Sliding: $F_s = \dfrac{609.8 \tan 40°}{138.9} = \underline{3.68}$ [>2.0, $\therefore$ adequate]

Overturning: Overturning moment, $M_o = 138.9 \times 8/3 = 370$ kNm/m

Stabilising moment, $M_s = 52.2 \times 1.85 + 93.6 \times 2.6 + 464(2.0 + 3.2/2)$

$= 2010$ kNm/m

$F_{OT} = \dfrac{2010}{370} = \underline{5.43}$ [>2.0, $\therefore$ adequate]

Bearing pressure: $\bar{x} = \dfrac{2010 - 370}{609.8} = 2.689$ m

Eccentricity, $e = \frac{1}{2} \times 5.2 - 2.689 = -0.089$ (i.e. on the heel side)

Maximum bearing pressure, $q_{max} = \dfrac{609.8}{5.2} + \dfrac{6 \times 609.8 \times 0.089}{5.2^2} = \underline{129 \text{ kPa}}$

[under the heel]

(b) *Now considering an inclined virtual support face*

The virtual support face AC is at angle α (Fig. 8.31):

$\alpha = \arctan \dfrac{8.0}{3.2} = 66.20°$

The section of soil ABC may be assumed to move with the wall, therefore assume $\delta = 30°$. From eqn [8.22], $K_a = 0.349$

Active resultant force, $\quad P_A = \dfrac{\frac{1}{2} \times 20 \times 8.0^2 \times 0.349}{\sin 66.2° \times \cos 30°} = 281.9 \text{ kN/m}$

Horizontal component, $\quad P_{AH} = 281.9 \cos(90 - \alpha + \delta) = 166.5 \text{ kN/m}$

Vertical component, $\quad P_{AV} = 281.9 \sin(90 - \alpha + \delta) = 227.5 \text{ kN/m}$

Vertical forces: $\quad W_1 = 24 \times 0.3 \times 7.25 = 52.2 \text{ kN/m}$

$\quad\quad\quad\quad\quad\quad\quad W_2 = 24 \times 5.2 \times 0.75 = 93.6 \text{ kN/m}$

$\quad\quad\quad\quad\quad\quad\quad W_s = \frac{1}{2} \times 20 \times 3.2 \times 7.25 = 232.0 \text{ kN/m}$

Total, $\quad V = 52.2 + 93.6 + 230.0 + 227.5 = 605.3 \text{ kN/m}.$

Sliding: $\quad F_s = \dfrac{605.3 \tan 40°}{166.5} = \underline{3.05} \quad [>2.0, \therefore \text{ adequate}]$

Overturning: Overturning moment, $M_o = 166.5 \times 8/3 = 444 \text{ kNm/m}$

$\quad\quad\quad\quad\quad$ Stabilizing moment, $M_s = 52.2 \times 1.85 + 93.6 \times 2.6 + 233(2.0 + 3.2/3)$
$\quad\quad\quad\quad\quad\quad\quad\quad\quad\quad\quad\quad\quad\quad\quad + 227.5(5.2 - 1.18)$

$\quad\quad\quad\quad\quad\quad\quad\quad\quad\quad\quad\quad\quad\quad = 1966 \text{ kNm/m}$

$F_{OT} = \dfrac{1966}{444} = \underline{4.43} \quad [>2.0, \therefore \text{ adequate}]$

Bearing pressure: $\quad\quad\quad\quad\quad\quad \bar{x} = \dfrac{1966 - 444}{605.3} = 2.514 \text{ m}$

Eccentricity, $\quad\quad\quad\quad\quad\quad\quad e = \frac{1}{2} \times 5.2 - 2.514 = 0.086$ (i.e. on the heel side)

Maximum bearing pressure, $q_{max} = \dfrac{605.3}{5.2} + \dfrac{6 \times 605.3 \times 0.086}{5.2^2} = \underline{128 \text{ kPa}}$

The ultimate bearing capacity for a strip footing for $\phi' = 40°$ (see Table 11.2) and ignoring depth of embedment will be

$q'_f = \frac{1}{2} B \gamma N \gamma = \frac{1}{2} \times 4.0 \times 20 \times 95.5 = 3820.0 \text{ kN/m per metre}$

Factor of safety, $\quad F_q = \dfrac{3820}{128} = 29.8 \quad [\text{adequate}]$

The factors of safety calculated here are all well above the recommended minimum values; there may therefore be some argument for reducing the size of the base. This would be considered after a preliminary check on bending moments in the wall.

Worked example 8.11   *Check the stability of a backfilled masonry wall of height 9.0 m, crest width 1.2 m and base width 4.0 m; the back is vertical and the finished soil surface horizontal and level with the crest. The soil will be compacted using a roller which will trasnmit an equivalent line load of 60 kN/m. Use the following properties:*

Soil: $c' = 0$ $\phi' = 40°$ $\gamma = 20 \ kN/m^3$ Masonry: $\gamma = 24 \ kN/m^3$ $\delta = 20°$

Vertical forces

$W_1 = 24 \times 9.0 \times 1.2 = 259.2$ kN/m

$W_2 = 24 \times (4.0 - 1.2)/2 \times 9 = 302.4$ kN/m

Total vertical force, $V = 561.6$ kN/m.

Horizontal forces due to compaction

Rankine
$$K_a = \frac{1 - \sin 40°}{1 + \sin 40°} = 0.217$$

From eqn [8.40],
$$\sigma'_{hm} = \sqrt{\left(\frac{2p\gamma}{\pi}\right)} = \sqrt{\left(\frac{2 \times 60 \times 20}{\pi}\right)} = 27.64 \text{ kPa}$$

From eqn [8.39(b)], critical depth, $z_c = K_a \sqrt{\left(\frac{2p}{\pi\gamma}\right)} = 0.217 \sqrt{\left(\frac{2 \times 60}{20\pi}\right)} = 0.300 \text{ m}$

From eqn [8.41], critical height, $h_c = \frac{1}{K_a} \sqrt{\left(\frac{2p}{\pi\gamma}\right)} = \frac{1}{0.217} \sqrt{\left(\frac{2 \times 60}{20\pi}\right)} = 6.37 \text{ m}$

Horizontal pressure at base of wall, $\sigma'_{hH} = K_a \gamma H = 0.217 \times 20 \times 9.0 = 39.06$ kPa

The pressure distribution is shown in Fig. 8.32, from which the resultant horizontal force due to compaction is:

$P_c = 27.64 \times 9.0 - \frac{1}{2} \times 27.64 \times 0.300 + \frac{1}{2} \times (39.06 - 27.64) \times (9.00 - 6.37)$

$= 248.76 - 4.15 + 15.02 = 259.6$ kN/m

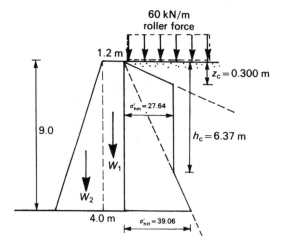

Fig. 8.32

Check sliding

Factor of safety, $F_s = \dfrac{561.6 \times \tan 40°}{259.6} = \underline{1.82}$ [<2.0 and $\therefore$ inadequate]

Check overturning

Overturning moment, $M_o = 248.76 \times 9.0/2 - 4.15(9.00 - 0.300/3 + 15.02(9.00 - 6.37)/3$

$$= 1095.6 \text{ kN/m}$$

Stabilising moment about the toe, $M_s = 259.6(4.0 - 0.6) + 302.4(4.0 - 1.2)2/3$

$$= 1447.1 \text{ kN/m}$$

Factor of safety, $F_{OT} = \dfrac{1447.1}{1095.6} = \underline{1.32}$ [<2.0, $\therefore$ inadequate]

Check maximum ground bearing pressure

Position of ground resultant, $\bar{x} = \dfrac{M_s - M_o}{V} = \dfrac{1447.1 - 1095.6}{561.6}$

$$= 0.626 \text{ m from the toe [i.e. outside the middle third]}$$

Maximum ground bearing pressure, $q_{max} = \dfrac{2V}{3x} = \dfrac{2 \times 561.6}{3 \times 0.626} = \underline{598 \text{ kPa}}$

The ultimate bearing capacity for a strip footing for $\phi' = 40°$ (see Table 11.2) and ignoring depth of embedment will be

$q_f' = \frac{1}{2}B\gamma N\gamma = \frac{1}{2} \times 4.0 \times 20 \times 95.5 = 3820.0$ kN/m per metre

Factor of safety, $F_q = \dfrac{3820}{598} = 6.39$ [adequate]

8.13 Reinforced earth walls

The use of reinforced earth was first introduced by Henri Vidal, a French architect and engineer (Vidal, 1966, 1969). The construction consists of frictional backfill rolled in layers with strip or mesh reinforcement placed between. Reinforced earth is now used in retaining walls, sea walls, dock walls, bridge abutments, earth dams and in numerous temporary works. The use of prefabricated reinforcing strips and facing panels, together with the inherent flexibility of the material, allows fairly rapid construction. A reinforced earth wall weighs much less than a similar sized concrete structure and so lower ground bearing pressures are imposed, giving an advantage on poor or soft ground. Reinforced earth can absorb considerable strains without sustaining structural damage. For design recommendations refer to BS8006 (1995), Jones (1996).

The main components of reinforced earth are:

(a) *Backfill.* Carefully selected frictional backfill is required: ≯15 per cent weight < 80 μm, ≯25 per cent weight > 150 mm, a maximum particle size of 300 mm and $\phi' \geq 25°$. *Cohesive* frictional fill may be used provided $\phi' \geq 20°$, clay content ≤ 10 per cent and $w_L \leq 45$ per cent.

(b) *Reinforcement.* The properties required are adequate strength, moderate to high stiffness, low creep with time, good bond with soil and good resistance to mechanical and chemical damage. Common high-stiffness materials used are galvanised steel, stainless steel and plastic-coated metals. Geosynthetic meshes and grids are widely used; these have lower deformation moduli, but are lighter, cheaper, have better durability and very good frictional properties (Murray, 1980). Reinforcement in the form of strips and bars is only suitable in highly frictional soils; mesh or grid reinforcement provides better bond in cohesive fills.

(c) *Facing skin.* The main role of the facing skin is to provide containment and protection against corrosion; no contribution is made to the mechanical stability of the wall. In the case of rigid facings, allowance must be made for vertical compression in the wall. Four main types of facings are used: (i) *pre-cast*: concrete or pre-formed metal panels, often interlocking to give a continuous face; (ii) *wrap-around*: mesh or grid reinforcement rolled upward and into the layer interface above; (iii) *gabions*: stone-filled gabions or concrete-filled bags; (iv) *sprayed face*: shotcrete (concrete grout) or bitumen sprayed on after construction, also vegetation.

Design principles

The reinforcement is provided to enhance the shear strength of the soil. Compressive and tensile strains are induced under shear loading; reinforcement is therefore placed in the direction of tensile strain, which is horizontal. The compressive strains are mainly vertical and result in some settlement. The design process therefore seeks to establish the reinforcement forces that are *required* to provide equilibrium of the soil mass, and to match these with the forces *available* in the reinforcement. The *required* forces are the same for each system of reinforced or anchored walls, but the *available* force depends on the type, form and orientation of the reinforcement. Both *external* and *internal* stability must be checked.

External stability

The external stability conditions govern the *length* of reinforcement required. These conditions are based on the same limit states that apply to gravity walls in gravity (see Section 8.12), i.e. overturning, sliding, bearing pressure, overall slip failure. The vertical and horizontal forces acting on the wall must first be calculated (Fig. 8.33).

Vertical forces

W = weight of reinforced soil block = $\gamma_R H L$
W_s = surcharge force on surface of reinforced soil block = $q_R L$
V = ground reaction force = total downward force = $W + W_s = (\gamma_R H + q_R)L$

Horizontal forces

P_1 = active thrust due to weight of unreinforced soil = $\frac{1}{2}K_a\gamma H^2$
P_s = active thrust due to surcharge on unreinforced soil = $K_a q_s H$
Total horizontal force = $P_1 + P_s = \frac{1}{2}K_a(\gamma H + 2q_s)H$

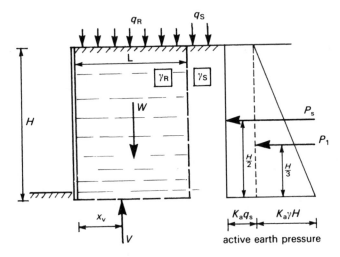

Fig. 8.33 Forces on a reinforced earth wall

where γ_R = unit weight of reinforced soil

γ = unit weight of unreinforced soil

Overturning moment about the base

$$M_o = P_1 H/3 + P_s H/2 = K_a(\gamma H + 3q_s)H^2/6$$

Position of ground reaction

$$\bar{x} = \frac{L}{2} - \frac{M_o}{V}$$

Determine length of reinforcement
Either a trial reinforcement length (L) is chosen and the sliding, overturning and bearing pressure limit states checked, or limit state factors are set and the corresponding minimum values of L calculated.

Forward sliding (Fig. 8.34(a))

Factor of safety, $$F_s = \frac{\mu V}{P_A} = \frac{2\mu(\gamma_R H + q_R)L}{K_a(\gamma H + 2q_s)H} \qquad [8.46(a)]$$

For $F_s = 2.0$, $$L_{min} = \frac{2P_A}{\mu V} = \frac{K_a(\gamma H + 2q_s)H}{\mu(\gamma_R H + q_R)} \qquad [8.46(b)]$$

Overturning (Fig 8.34(b))

Factor of safety, $$F_{OT} = \frac{\frac{1}{2}VL}{M_o} = \frac{\frac{1}{2}(\gamma_R H + q_R)L^2}{\frac{1}{6}K_a(\gamma H + 3q_s)H^2} \qquad [8.47(a)]$$

For $F_{OT} = 2.0$, $$L_{min} = \sqrt{\left(\frac{4M_o}{\gamma_R H + q_R}\right)} = H\sqrt{\left(\frac{2K_a(\gamma H + 3q_s)}{3(\gamma_R H + q_R)}\right)} \qquad [8.47(b)]$$

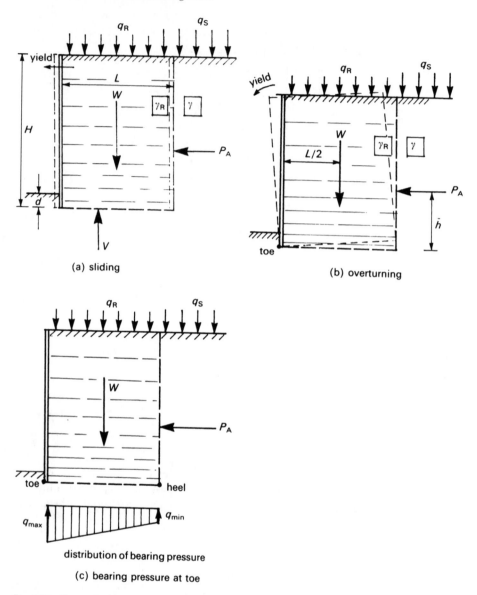

(a) sliding

(b) overturning

distribution of bearing pressure

(c) bearing pressure at toe

Fig. 8.34 External stability of a reinforced earth wall

Ground bearing pressure (Fig. 8.34(c))

The maximum bearing pressure will occur beneath the toe; the maximum ground bearing pressure is therefore:

$$\sigma'_{v(max)} = \frac{V}{L} + \frac{6M_o}{L^2} = (\gamma_R H + q_R) + \frac{K_a(\gamma H + 3q)H^2}{L^2} \qquad [8.48(a)]$$

This value must not exceed the allowable bearing capacity, q_a.

$$\text{for } \sigma'_{v(\max)} = q_a, \ L_{\min} = \sqrt{\left(\frac{6M_o}{q_a - (\gamma_R H + q_R)}\right)} = H\sqrt{\left(\frac{K_a(\gamma H + 3q_s)}{q_a - (\gamma_R H + q_R)}\right)}$$

$$[8.48(b)]$$

Internal stability

The internal stability conditions govern the *vertical spacing* of the reinforcement. Within the block of reinforced soil, the net horizontal thrust is transferred to the reinforcement; the magnitude of this thrust varies at each reinforcement level. An internal stability check is carried out in two stages: (i) *tension failure* – the stability is considered of an internal wedge of soil above each reinforcement level; (ii) *pullout failure* – the pullout capacity is checked by considering the bond length of reinforcement required beyond each wedge of soil above that layer. The tension forces thus required for equilibrium are then compared with the design strength of the reinforcement.

Tension failure and vertical spacing of reinforcement

A *tie-back wedge analysis* (Fig. 8.35). Consider a reinforcement element at depth z in wall of overall depth H. Part of the element lies within the active wedge as defined in Rankine's or Coulomb's theory. The distribution of vertical effective stress at this level along the element may be taken as trapezoidal. It can be shown that the maximum value (adjacent to the face) is given by:

Maximum vertical effective stress, $\sigma'_{vz} = (\gamma_R z + q_R) + K_a(\gamma z + 3q_s)\dfrac{z^2}{L^2}$

The tensile force required per metre run in the reinforcement will be:

$$T_r = K_r \sigma'_{vz} S_v \qquad\qquad [8.49]$$

where S_v = vertical spacing of reinforcement elements
           K_r = an earth pressure coefficient ($K_a \leq K_r \leq K_o$)

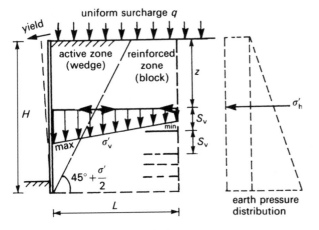

Fig. 8.35 Tie-back wedge analysis

The magnitude of T_r must be not greater than the *design strength* of the reinforcement which is determined from the characteristic strength (obtained from manufacturer's literature or pullout tests) and partial factors of safety.

$$\text{Design strength, } P_d = \frac{\text{Characteristic strength}}{f_m f_d} \qquad [8.50]$$

The partial factors of safety are typically: $f_m = 1.3\text{--}1.5$, $f_d = 1.35$. Thus the *vertical spacing* can be determined:

$$S_v = \frac{P_d}{K_r \sigma'_{vz}} = \frac{P_d}{K_r \left[(\gamma_R z + q_R) + K_a(\gamma z + 3q_s)\dfrac{z^2}{L^2} \right]} \qquad [8.51]$$

For practicability, the actual spacing at a given level should be a multiple of the rolled layer thickness.

Pullout failure
Both *local bond* and *wedge stability* must be checked.

Local bond
The pullout resistance per metre width of reinforcement is given by:

$$R_p = 2L_b \mu \sigma'_n \qquad [8.52]$$

where L_b = bond length behind wedge = $L - (H - z) \tan \theta$
   μ = coefficient of friction = $\alpha_b \tan \phi'$
   α_b = bond coefficient depending on type of reinforcement $(0 < \alpha_b \le 1.0)$
     (e.g. smooth strips = 0.5–0.6, ribbed strips and meshes = 0.6–1.0)
   σ'_n = average effective normal (vertical) stress over bond length L_b
     $\approx (\gamma_R z + q_R)$ [i.e. neglecting the effect of active pressure]

Wedge stability (or overall bond stability) (Fig. 8.36)
The stability of wedges of soil above each reinforcement level must be considered and the total tensile force required to hold each in equilibrium calculated

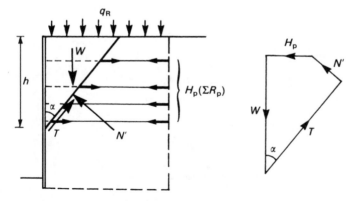

Fig. 8.36 Wedge stability analysis

and compared with the sum of the resistances provided by the reinforcement layers in the wedge. A typical trial wedge is shown in Fig. 8.36. From the polygon of forces, the tensile force required to maintain equilibrium is found to be:

Horizontal tension required for trial wedge, $\quad T_H = \dfrac{h \tan \theta(\frac{1}{2}\gamma_R h + q_R)}{\tan(\theta + \phi')}$ [8.53]

Factor of safety, $\qquad\qquad\qquad\qquad\qquad\qquad F_T = \dfrac{\Sigma R_p}{T_H} \geq 2.0$

Worked example 8.12 *A reinforced earth wall is to have a finished height of 6.6 m. The surface will be required to carry an extensive uniform surcharge of 15 kPa. Assuming that the water table will remain below the base of the wall, and given the following data, determine the breadth of wall required and check the external and internal stability.*

Backfill (after compacting in 225 m layers): $\quad \gamma = 18 \ kN/m^3 \quad c' = 0 \quad \phi' = 35°$

Foundation soil: allowable bearing capacity, $\quad q_a = 190 \ kN/m^2$

Geotextile reinforcement (non-woven):

Characteristic strength, $P_k = 33 \ kN/m \quad$ *Bond coefficient,* $\alpha_b = 0.8$

Factors of safety:

Partial factor, $f_m = 1.3 \quad$ *Design strength factor,* $f_d = 1.35$

Against sliding, overturning, pullout, F = 2.0

Allowing 10 per cent for embedment, actual height, $\quad H = 6.6 \times 1.1 = 7.26$ m. <u>Say 7.25 m.</u>

For the finished backfill, $\quad K_a = \dfrac{1 - \sin 35°}{1 + \sin 35°} = 0.271$

Vertical forces

Weight of reinforced soil, $W_1 = \gamma_R HL = 18 \times 7.25 \ L = 130.5L$ kN/m

Surcharge force on surface of reinforced soil block, $W_s = q_R L = 15.0L$ kN/m

$V =$ ground reaction force = total downward force = $W + W_s = 145.5L$ kN/m

Horizontal forces

Active thrust due to unreinforced soil, $\quad P_1 = \frac{1}{2}K_a \gamma H^2$

$\qquad\qquad\qquad\qquad\qquad\qquad = \frac{1}{2} \times 0.271 \times 18.0 \times 7.25^2 \ = 128.2$ kN/m

Active thrust due to surcharge, $\qquad P_s = K_a q_s H = 0.271 \times 15 \times 7.25 = 29.5$ kN/m

Overturning moment about the base

$M_o = 128.2 \times 7.25/3 + 29.5 \times 7.25/2 = 416.8$ kNm/m run

External stability – to determine L_{min}

(a) *To prevent sliding:*

Coefficient of friction, $\quad \mu = \alpha_b \tan \phi' = 0.8 \times \tan 35° = 0.560$

From eqn [8.46(b)], $\quad L_{min} = \dfrac{2(128.2 + 29.5)}{0.560 \times 145.5} = 3.87$ m

(b) *To prevent overturning:*

From eqn [8.47(b)], $L_{min} = \sqrt{\left(\dfrac{4.0 \times 416.8}{145.5}\right)} = 3.39$ m

(c) *To prevent bearing overstress beneath toe:*

From eqn [8.48(b)], $L_{min} = \sqrt{\left(\dfrac{6.0 \times 416.8}{190.0 - 145.5}\right)} = 7.50$ m

Thus the design value for $L = \underline{7.50\ m}$.

Design strength of reinforcement

Design strength, $P_d = \dfrac{\text{Characteristic strength}}{f_m f_d} = \dfrac{33}{1.3 \times 1.35} = 18.8$ kN/m

Internal stability – to determine vertical spacing
The vertical spacing can be determined from eqn [8.51]:

$$S_v = \frac{P_d}{K_r \sigma'_{vz}} = \frac{P_d}{K_r \left[(\gamma_R z + q_R) + K_a(\gamma z + 3q_s)\dfrac{z^2}{L^2} \right]}$$

where $K_r = K_a$ when using geotextiles, i.e. 0.271

Thus, $S_v = \dfrac{18.8}{0.271\left[(18.0z + 15.0) + 0.271(18.0z + 45.0)\dfrac{z^2}{7.5^2} \right]}$

$$= \frac{69.36}{18.0z + 15.0 + 0.00482(18.0z + 45.0)z^2}$$

Values of S_v have been calculated using a spreadsheet (Fig. 8.37), corresponding to depths in multiples of 225 mm, i.e. the compacted layer thickness.

Internal wedge stability
For an active wedge of depth h and angle θ, the horizontal tension (T_H) required for stability is given by eqn [8.53]. Then, putting $\theta = 45° - \phi/2 = 27.5°$:

$$T_H = \frac{h \tan \theta(\tfrac{1}{2}\gamma_R h + q_R)}{\tan(\theta + \phi')} = \frac{h \tan 27.5°(\tfrac{1}{2}18.0h + 15.0)}{\tan(27.5° + 35°)} = 2.439h^2 + 4.065h$$

For each reinforcement depth (z), the available tensile resistance is the *lesser* of the pullout resistance (R_p/F_p) and the design strength (P_d). From eqn [8.52], $R_p = 2L_b\mu\sigma'_n$ in which

$L_b = L - (h - z) \tan \theta = 7.5 - 0.521(h - z)$ and $\sigma'_n = 18.0z + 15.0$

Now, putting $F_p = 2.0$, values of R_p/F_p can be calculated. Trial wedges are selected, in this case at depths of 2, 4, 5.5 and 7.25 m, and for each of these is calculated the required tensile force T_H and R_p/F_p. The *available* tensile force will be the *least* of R_p/F_p and P_d. These calculations have been carried using a spreadsheet (see, *Soil Mechanics Spreadsheets and Reference* on the accompanying CD-ROM), an extract of the results is shown in

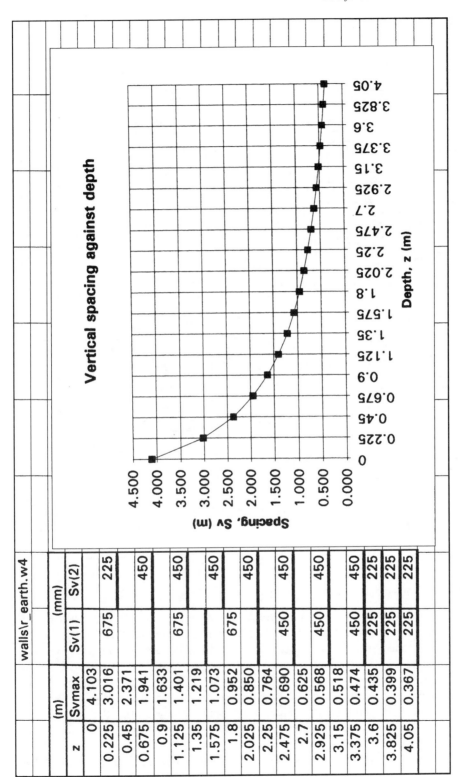

z (m)	Svmax	Sv(1)	Sv(2)
0	4.103		
0.225	3.016	675	225
0.45	2.371		
0.675	1.941		450
0.9	1.633		
1.125	1.401	675	450
1.35	1.219		
1.575	1.073	675	450
1.8	0.952		
2.025	0.850		450
2.25	0.764	450	
2.475	0.690		450
2.7	0.625	450	
2.925	0.568		450
3.15	0.518	450	
3.375	0.474		450
3.6	0.435	225	225
3.825	0.399	225	225
4.05	0.367	225	225

walls\r_earth.w4

Fig. 8.37

re2e811		Reinforced earth wall: Internal stability - wedge stability check								
		BSM3 - WE8.11								
Trial wedge angle,	u =	27.5	deg	Friction angle	φ =	35	Base, L =	7.5	m	
Coefficient of friction,	m =	0.560		Bond coef	α =	0.8	Unit weight =	18	kN/m^3	
Design strength of element, Pd =		18.8	kN/m		Fp =	2	Surcharge =	15	kN/m^2	

Trial wedge depth h(m)	Depth BS z(m)	Bond length Lb	Available resistance force (either Rp/Fs or Pd) = Ra							Required force per wedge	
			Rp/Fs q=0	Rp/Fs with q	Pd	Re = least q=0	Re = least with q	ΣRa q=0	ΣRa with q	Th(q=0)	Th(q)
2	1.125	7.045	79.91	139.10	18.8	18.80	18.80	18.80	18.80	9.76	17.89
										ok!	ok!
4	1.125	6.003	68.10	118.54	18.8	18.80	18.80				
4	2.025	6.472	132.14	186.52	18.8	18.80	18.80				
4	2.925	6.940	204.69	263.01	18.8	18.80	18.80				
4	3.6	7.292	264.68	325.95	18.8	18.80	18.80	56.40	56.40	39.02	55.28
										ok!	ok!
5.5	1.125	5.223	59.24	103.12	18.8	18.80	18.80				
5.5	2.025	5.691	116.20	164.02	18.8	18.80	18.80				
5.5	2.925	6.160	181.66	233.42	18.8	18.80	18.80				
5.5	3.6	6.511	236.34	291.05	18.8	18.80	18.80				
5.5	4.275	6.862	295.80	253.46	18.8	18.80	18.80				
5.5	4.725	7.097	338.10	397.72	18.8	18.80	18.80				
5.5	5.175	7.331	382.52	444.12	18.8	18.80	18.80	131.60	131.60	73.78	96.13
										ok!	ok!
7.25	1.125	4.312	48.91	85.13	18.8	18.80	18.80				
7.25	2.025	4.780	97.60	137.76	18.8	18.80	18.80				
7.25	2.925	5.249	154.79	198.89	18.8	18.80	18.80				
7.25	3.6	5.600	203.27	250.32	18.8	18.80	18.80				
7.25	4.275	5.951	256.53	306.54	18.8	18.80	18.80				
7.25	4.725	6.186	294.69	346.67	18.8	18.80	18.80				
7.25	5.175	6.420	334.98	388.93	18.8	18.80	18.80				
7.25	5.625	6.654	377.40	433.31	18.8	18.80	18.80				
7.25	6.075	6.888	421.94	479.82	18.8	18.80	18.80				
7.25	6.3	7.005	445.01	503.87	18.8	18.80	18.80				
7.25	6.525	7.123	468.61	528.45	18.8	18.80	18.80				
7.25	6.75	7.240	492.74	553.57	18.8	18.80	18.80				
7.25	6.975	7.357	517.40	579.21	18.8	18.80	18.80				
7.25	7.2	7.474	542.59	605.39	18.8	18.80	18.80	263.20	263.20	128.20	157.67
										ok!	ok!

Fig. 8.38

Fig. 8.38. It can be seen that, in this case, for all the trial wedges the available resistance exceeds the required tensile force, and that the design strength of the reinforcement governs each one.

8.14 Embedded walls

This section deals with retaining walls that are usually installed in undisturbed ground, after which excavation takes place on one side. Structurally, such walls can be formed from driven sheet-piles, secant piles, contiguous piles or as a diaphragm wall. They may act as simple cantilevers or be provided with single or multiple anchors or props. Embedded walls differ from gravity walls in several ways: (i) they are flexural and are designed structurally as simple or propped cantilevers, (ii) their weight is ignored in stability analyses, (iii) they depend for support upon passive resistance mobilised in the soil, (iv) excavation usually follows installation, rather than backfilling.

The modes of failure that need to be examined in design are shown in Fig. 8.39:

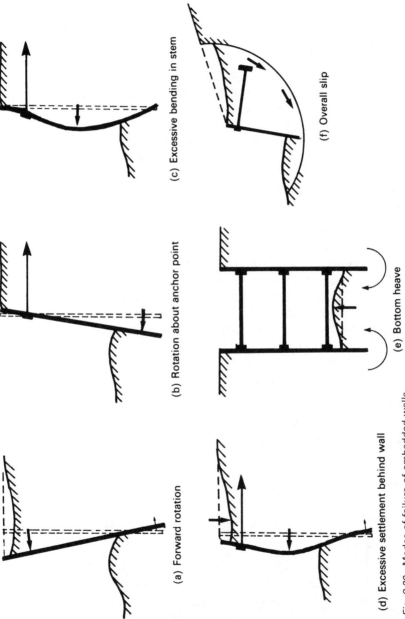

(a) Forward rotation

(b) Rotation about anchor point

(c) Excessive bending in stem

(d) Excessive settlement behind wall

(e) Bottom heave

(f) Overall slip

Fig. 8.39 Modes of failure of embedded walls

(a) Forward rotation due to inadequate passive resistance (applies to canti-
     lever walls).
(b) Rotation about anchor/prop point due to inadequate passive resistance.
(c) Failure of anchor or prop system.
(d) Failure of wall stem in bending or excessive deflection.
(e) Excessive settlement behind wall due to forward movement.
(f) Failure in excavation due to bottom heave.
(g) Overall slip failure.

The first stage in stability calculations for embedded walls is usually to determine the distribution of earth pressures, the classical theories can be used for
this purpose. For most permanent structures effective stress analyses are required in the calculation of both active and passive pressures, and, where a
water flow is anticipated, seepage pressures must be determined (see p. 333). As
with gravity walls the choice of shear strength parameters depends largely upon
the anticipated movement of the wall, with peak values only being justified when
movement will be small. In the case of embedded walls in clays analyses based
on undrained strength may be required, especially for temporary structures.

Cantilever pile walls $(c' = 0)$
Cantilever sheet-pile walls are mainly used as temporary structures in cohesionless
soils and depend for their stability on the passive resistance mobilised on the
embedded portion. It is usual to assume that the wall (Fig. 8.40(a)) will fail by
rotating about a point C just above the toe D, in which case passive pressure will
be developed along BC in front of the wall and along CD behind it (Fig. 8.40(b)).
To simplify the analysis, the assumption may be made that the passive resistance
along CD acts as a point load at point C, and the assumed driving depth is
$d = BC$ (Fig. 8.40(c)). Thus, three forces maintain equilibrium at the limit state:

Active thrust: $P_A = \frac{1}{2}K_a\gamma(H + d)^2$

Passive thrust: $P_P = \frac{1}{2}K_p\gamma d^2$

Passive resistance: $R = P_P - P_A$ (value not required)

For equilibrium, $\Sigma M_c = 0 = P_P \times d/3 - P_A \times (H + d)/3$

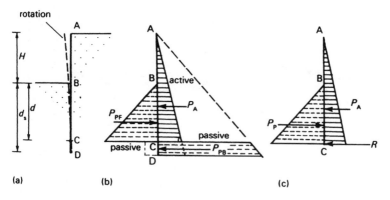

Fig. 8.40 (a) Cantilever sheet pile wall (b) Theoretical pressure distribution
           (c) Simplified pressure distribution

Substituting for P_P and P_A, and ignoring wall friction (Corbett and Stroud, 1975):

$$0 = \tfrac{1}{6}K_p\gamma d^3 - \tfrac{1}{6}K_a\gamma(H + r)^3$$

Then $\qquad K_p d^3 = K_a(H + d)^3$

Solving, $\qquad d = \dfrac{H}{(K_p/K_a)^{1/3} - 1}$ [8.54(a)]

If $\qquad K_a = 1/K_p, \quad d = \dfrac{H}{K_p^{2/3} - 1}$ [8.54(b)]

The required driving depth is $d_s = \overline{CD}$ (Fig. 8.40); this is usually obtained by multiplying d by a *factor of safety on embedment* (F_d): $d_s = dF_d$. For permanent works, the recommended range for F_d is 1.2–1.6. The adequacy of the design depth d_s can be checked by calculating the net lateral thrust developed over the length CD; for equilibrium this should be $\geq P_P - P_A$ (calculated between A and C). An analysis along these lines produces the following expression:

$$F_{d(min)} = \frac{\sqrt{((K_pH)^2 + (K_p - K_a)d\,[2H(K_p - K_a) + d(2K_p - 3K_a) - K_aH^2/d]) - K_pH}}{(K_p - K_a)d}$$
[8.55]

The maximum bending moment occurs at the point of zero shear force: z_s below B (Fig. 8.41).

Equating forces: $\quad 0 = \tfrac{1}{2}K_p\gamma z_s^2 - \tfrac{1}{2}K_a\gamma(H + z_s)^2$

From which: $\qquad z_s = \dfrac{H}{\sqrt{(K_p/K_a)} - 1} = \dfrac{H}{K_p - 1}$ [8.56]

The maximum bending moment will then be:

$$M_{max} = \tfrac{1}{6}\gamma[K_a(H + z_s)^3 - K_p z_s^3]$$ [8.57]

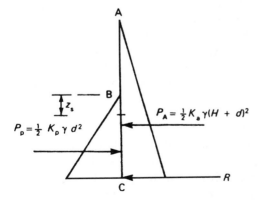

Fig. 8.41

Worked example 8.13   *A cantilever sheet-pile wall is to support the side of an excavation 3.0 m deep. Using a factor of safety on shear strength of 1.4 and a factor of safety on embedment of 1.2, determine the safe driving depth required and then the maximum bending moment induced in the piling. The soil properties are: c′ = 0, φ′ = 30°, γ = 20 kN/m³.*

The factored angle of friction, ϕ'_{mob} = arctan(tan 30°/1.4) = 22.41°.

K_p = tan²(45° + 22.41°/2) = 2.232   and K_a = 1/K_p = 0.448

From eqn [8.54(a)], $d = \dfrac{3.0}{2.232^{2/3} - 1}$ = 4.24 m

Then safe driving depth, d_s = 4.24 × 1.2 = 5.08 m

For horizontal equilibrium when

d = 4.24 (eqn [8.55]), $F_{d(min)}$ = 1.07 [<1.2 OK]

From eqn [8.56], the point of zero shear force (Fig. 8.41):

$z_s = \dfrac{3.0}{2.232 - 1}$ = 2.435 m

From eqn [8.57], the maximum bending moment will be:

$M_{max} = \frac{1}{6} \times$ 20[0.448(3.0 + 2.435)³ − 2.232 × 2.435³] = 132 kNm/m

Anchored sheet piles (c' = 0)

Cantilever walls are best suited for temporary works and for walls of low height. With the provision of one or more anchors (ties or props) the required embedment depth is reduced, as is the lateral deflection and bending moment in the pile. Two methods of analysis are available for anchored or propped walls, the difference between them being the assumption made with regard to the restraint provided by the soil at the bottom of the embedded length. In the *free-earth support* method it is assumed that the piling is very stiff compared to the soil and there is insufficient embedment to prevent rotation of the toe of the wall, although the wall is still in equilibrium. In the *fixed-earth support* method the piling is assumed to be flexible, the embedment sufficient and the soil stiff enough to prevent rotation of the toe. Fixed-earth support conditions clearly apply in the case of a cantilever wall, but in current practice in the UK this method is considered to produce excessive embedment depths and piling weights. Following the work of Rowe (1952), a modification is applied to the free-earth support method by which the bending moment can be reduced by a factor dependent on the relative flexibilities of the piling and soil. It is this method that is widely used in the UK and the USA.

Free-earth support method

It is assumed that the piling is very stiff compared with the soil, and that rotation occurs about the anchor point. Sufficient yielding is also assumed and therefore the development of active and passive pressures, with *critical state* conditions (c' = 0) in cohesive soils. For homogeneous soil conditions, the distribution of earth pressures is shown in Fig. 8.42. The forces maintaining equilibrium are:

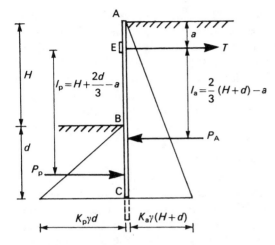

Fig. 8.42 Anchored sheet piling: free-earth support method – assumed pressure distribution (CIRIA 104, p. 124)

T = force in the anchor tie (or prop)

P_A = active thrust force on the back of the wall = $\frac{1}{2}K_a\gamma(H + d)^2$

P_P = passive thrust force on the front of the wall = $\frac{1}{2}K_p\gamma d^2$

in which K_a and K_p are determined with respect to appropriate values of δ: e.g. $\delta_a = \frac{2}{3}\phi'$ and $\delta_p = \frac{1}{2}\phi'$ (Padfield and Mair, 1987) – also see Appendix B.

Where hydrostatic and/or steady-state seepage conditions occur, the porewater pressure distribution must be evaluated and the resultant forces included in the equilibrium calculations.

For equilibrium, the net moment about the anchor point E is zero.

$$0 = \Sigma M_E = P_P \times l_p - P_A \times l_a \qquad [8.58]$$

This results in a cubic equation in d^3, the solution of which can be achieved by the repeated substitution of values for d. The *design embedment depth* (d_d) is obtained by the following methods:

Method 1. The shear strength is not factored (i.e. $F_s = 1.0$), but the resulting value of d is multiplied by a *factor of safety on embedment* (F_d): $d_s = dF_d$. For permanent works, the recommended range for F_d is 1.2–1.6 (see also Appendix B).

Method 2. A *factor on shear strength* is applied (tan ϕ'_{mob} = tan ϕ'/F_s), from which values are obtained for δ_a and δ_p, and then values for K_a (eqn [8.6(a)]) and K_p (Table 8.4). The value of d is then taken as the *design embedment depth*: $d_s = d$. For permanent works, the recommended range for F_s is 1.5–2.0 (see also Appendix B).

The anchor or prop force required can now be obtained by equating horizontal forces:

$$0 = \Sigma H = T + P_P - P_A \qquad [8.59]$$

from which a value is obtained per metre run of wall. Anchors are usually spaced at 2–3 m intervals and secured to stiffening wales. The limiting equilibrium method should be used to calculate the prop force ($F_s = 1.0$) and the resulting value increased by 25 per cent to allow for flexibility in the piling and arching in the soil. A factor of safety (F_T) of at least 2 should be applied to obtain the *design anchor force* when determining tie-rod or prop sizes; an allowance should be made for corrosion if conditions are expected to be aggressive.

Design shear force and bending moment
Method 1 ($F_s = 1.0$) should be used to establish the distribution of shear force and bending moment on the wall, assuming constant wall flexibility. The maximum shear force will generally occur at the anchor point. After the position of zero shear force is established, the maximum bending moment can be calculated. Some engineers prefer to use only Method 2 (factored shear strength), and thus only one pressure distribution; Padfield and Mair (1987), do not recommend this method. Rowe (1952) has proposed an empirical moment reduction method which can be applied when the bending moment distribution which has been obtained using factored shear strength or factored earth pressure coefficients.

The point of zero shear force below the top of the wall (z_s) is found by equating forces:

$$0 = T - \tfrac{1}{2}K_a\gamma z_s^2 \quad \text{(provided there are no porewater pressures)}$$

giving,
$$z_s = \sqrt{\left(\frac{2T}{K_a\gamma}\right)} \qquad [8.60]$$

The maximum bending moment will be:

$$M_{max} = T(z_s - a) - \tfrac{1}{6}K_a\gamma z_s^3 \qquad [8.61]$$

Worked example 8.14 *An anchored sheet-pile wall is required to support an excavation of depth 9 m. The anchor will be attached at a point 1.5 m below the top of the wall (i.e. below the ground surface). No groundwater conditions need to be considered. Determine the design embedment depth and the design anchor force per metre run. The factors of safety will be:*

$F_s = 1.5$, $F_d = 1.2$, $F_T = 2.0$
$c' = 0$ $\phi' = 28°$ $\gamma = 20 \ kN/m^3$

(a) *Using Method 1: $F_s = 1.0$*
  Wall friction: for $\phi' = 28°$, $\delta_a = \tfrac{2}{3}\phi' = 18.67°$ and $\delta_p = \tfrac{1}{2}\phi' = 14.0°$
  From eqn [8.22], $K_a = 0.304$ and from Table 8.4, $K_p = 4.292$.

$$P_A = \tfrac{1}{2}K_a\gamma(H + d)^2 = \tfrac{1}{2} \times 0.304 \times 20(9.0 + d)^2 = 30.4(9.0 + d)^2$$

$$P_P = \tfrac{1}{2}K_p\gamma d^2 = \tfrac{1}{2} \times 4.29 \times 20d^2 = 42.9d^2$$

Lever arm $l_a = \tfrac{2}{3}(9.0 + d) - 1.5 = 4.5 + \tfrac{2}{3}d$ and lever arm $l_p = 9.0 - 1.5 + \tfrac{2}{3}d = 7.5 + \tfrac{2}{3}d$

Taking moments about the anchor point E:

$0 = \Sigma M_E = P_A \times l_a - P_P \times l_p$ [a spreadsheet (see *Soil Mechanics Spreadsheets and Reference* on the CD-ROM) is used for evaluation]

Embedment depth required for equilibrium, $d = 2.51$ m
Design embedment depth, $d_s = dF_d = 2.51 \times 1.2 = \underline{3.01 \text{ m}}$

(b) *Using Method 2: $F_s = 1.5$*
Mobilised angle of friction, $\phi'_{mob} = \arctan(\tan 28°/1.5) = 19.52°$.
From eqn [8.22], $K_a = 0.435$ and from Table 8.4, $K_p = 2.539$, from which the embedment depth, $d = 4.88$ m and $d_s = \underline{4.88 \text{ m}}$.

(c) *Anchor force*: using Method 1, $T = P_p - P_a = 131.8$ kN.
Design anchor capacity per metre run, $T_d = T \times F_d = 131.8 \times 2.0 = \underline{263.6 \text{ kN}}$

(d) *Maximum shear force (occurs at anchor point)*: using Method 1

$$Q_s = T - \tfrac{1}{2} \times 0.304 \times 20 \times 1.5^2 = 125.0 \text{ kN/m}$$

(e) *Maximum bending moment*: using Method 1

From eqn [8.60], $\qquad z_s = \sqrt{\left(\dfrac{2 \times 131.8}{0.304 \times 20}\right)} = 6.58$ m

Maximum bending moment, $\quad M_{max} = 131.8(6.58 - 1.5) - \tfrac{1}{2} \times 0.304 \times 20 \times 6.58^3$

$$= \underline{381 \text{ kN/m}}$$

Porewater pressure and seepage
If the water levels are different in front of and behind the wall, a differential porewater pressure distribution must be included in the calculation of forces. Water levels may vary due to pumping, rainfall or tidal conditions. Where piling is embedded in impermeable soils, such as clays, little or no seepage will take place, but the distribution of hydrostatic porewater pressure must be considered. In permeable soils, seepage pressures also must be included.

Piles embedded in clays
Figure 8.43(a) shows the distribution of hydrostatic pressure. Below the level of the water table on the excavation side the hydrostatic pressures on either side cancel out, so that the net hydrostatic forces on the active side are:

$P_{w1} = \tfrac{1}{2}\gamma h_1^2$ and $P_{w2} = \tfrac{1}{2}\gamma h_1 h_2$

Piles embedded in permeable soils
When steady-stage seepage is taking place, the distribution of porewater pressure may be determined using a flow net. However, this can be time-consuming, so that a simplified method is usually employed in which the head difference (h) is assumed to be dissipated uniformly over a flow path of FCE or FCB (Fig. 8.43(b)), depending on whether the water table on the excavation side is below or above excavation level.

The assumed hydraulic gradient is:

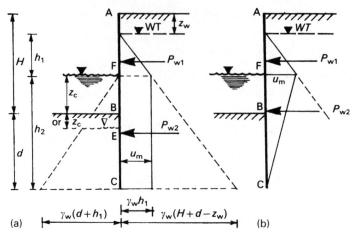

Fig. 8.43 Porewater pressures behind embedded walls
(a) In clays (b) In sands

$$i = \frac{H - z_w + z_e}{H + 2d - z_w - z_c}$$

The maximum porewater pressure occurs on the active side at the level of the water table on the excavation side (point F), and will have a value of:

$$u_m = (2d - z_e)\gamma_w = \frac{(2d - z_e)(H - z_w + z_e)\gamma_w}{H + 2d - z_w - z_e} \qquad [8.62]$$

and the water pressure forces will be:

$$P_{w1} = \tfrac{1}{2}u_m h_1 \quad \text{and} \quad P_{w2} = \tfrac{1}{2}u_m h_2$$

Temporary works and total stress design
The design approach for temporary works is similar to that for permanent works, except for the following differences:

(a) Initially the soil has a greater shear strength because of negative excess pore pressures.
(b) During construction, loading conditions are different and variable.
(c) Temporary walls rely more on the soil for support, while more permanent aspects of support are being constructed.
(d) Lower factors of safety are used, resulting in a greater mobilisation of soil strength, but ground movements can have the same consequences as for permanent works.

There are three design methods available:

(1) *Effective stress design.* This method is the same as that described above, except that a lower factor of safety will be used (see Appendix B) and, if worst credible strength parameters ($c' = 0$ and ϕ') are used, only an equilibrium check with $F = 1$ is required.

(2) *Modified total stress design.* This applies only to walls embedded in cohesive soils. The undrained shear strength is used, but must be carefully estimated, bearing in mind the usual wide scatter of test results. An allowance must be made on the passive side for the softening that takes place during the period of construction: a commonly adopted rule for overconsolidated clays is to assume a profile of $c_{us} = 0$ at passive side surface down to $c_{us} = (0.7 - 0.8)c_u$ at a depth of 1 m. Caution should be exercised here, since much experience is required in arriving at reliable estimates of softened strengths. Behind the wall the active pressure used should be either the conventionally calculated (positive) value ($\gamma z - 2c_u$) or the *minimum equivalent fluid pressure* (MEFP) of $5z$ kPa, whichever is the greater. If free water at the surface is expected, the hydrostatic pressure developed over the full depth of tension crack ($z_o = 2c_u/\gamma$) must be included. See worked example 8.14 for more details and calculations.

(3) *Mixed total and effective stress design.* An effective stress analysis is used with full water pressures on the active side of the wall, at least down to the tension crack depth ($z_o = 2c_u/\gamma$); if z_o is near to the total depth of wall ($H + d$), an effective stress design is more appropriate. Below the tension crack, if this is not too deep, a total stress analysis is used, invoking the minimum equivalent fluid pressure rule. A total stress analysis is used on the passive side, making due allowance for softening. This method is a clear choice when piling is embedded in a layer of clay that is overlain with a permeable soil.

Worked example 8.15 *A temporary anchored sheet-pile wall is required to support an excavation in clay of depth 9 m. The anchor will be attached at a point 1.5 m below the top of the wall (i.e. below the ground surface). The surface may become flooded. Using a mixed effective stress and total stress analysis, determine the design embedment depth required for an adequate factor of safety on moments, and also the design anchor force per metre run. Soil properties:*

$c' = 0 \quad \phi' = 24° \quad \gamma = 20 \ kN/m^3$

Undrained strength (allowing for softening), $c_u = 65 \ kPa$

The earth pressure distribution down to the tension crack depth will be obtained in terms of *effective stress*; below the tension crack on the active side, *total* stresses will be calculated. On the passive side, *total* stresses will be calculated, but with an allowance for full softening at the surface down to a depth of 1.0 m.

Earth pressure coefficients
Effective stress: from eqn [8.22], $K_a = 0.359$.
Total stress, assuming $c_w = \frac{1}{2}c_u$, $K_{ac} = K_{pc} = 2\sqrt{(1 + 0.5)} = 2.449$
Tension crack depth, $z_o = 2c_u/\gamma = 2 \times 65/20 = 6.5$ m
The distribution of horizontal pressures is shown in Fig. 8.44.

On the active side, the minimum fluid pressure rule must be applied: at any depth below the tension crack depth, the minimum total horizontal stress must be the greater of the active pressure or $5z$. At the critical depth (z_{MEFP}) where these are equal:

$5.0z_{MEFP} = \gamma z_{MEFP} - K_{ac}c_u \qquad \therefore \quad z_{MEFP} = K_{ac}c_u/(\gamma - 5.0) = 2.449 \times 65/15 = 10.61$ m

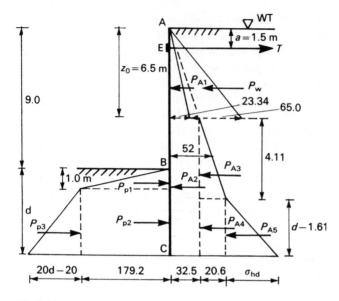

Fig. 8.44

Horizontal pressures on the active side
At $z = 6.5$ m: $\sigma'_{ha} = K_a(\gamma z_o - \gamma_w z_o) = 0.359(20 - 10)6.5 = 23.34$ kPa
$\qquad u = 10 \times 6.5 = 65.0$ kPa

Just below $z = 6.5$ m, $\sigma_h = 5 \times 6.5 = 32.5$ kPa
At $\qquad z = z_{MEFP} = 10.61, \sigma_h = 5 \times 10.61 = 53.05$ kPa
At $\qquad z = H + d, \sigma_{ha} = 20(9.0 + d) - 2.449 \times 65 = 20.82 + 20d$

Horizontal pressures on the passive side
At $z = 0$, $\sigma_{hp} = 0$ (allowing for full softening)
At $z = 1.0$ m, $\sigma_{hp} = 20 \times 1.0 + 2.449 \times 65 = 179.2$ kPa
At $z = d$, $\sigma_{hp} = 20d + 2.449 \times 65 = 20d + 159.2$ kPa

The horizontal forces (P_{A1}, P_w, P_{A2}, P_{A3}, P_{A4}, P_{P1}, P_{P2}, P_{P3}) can now be calculated. To determine the design embedment depth, the moments of these forces about the anchor point E are equated, with a *factor of safety on moments* (F_m) being applied to passive moments. Thus

$$F_m = \Sigma(P_P L_P) = \Sigma(P_A L_A)$$

where L_P and L_A = lever arms about the anchor point (E)

An optimum value for d is obtained at which F_m reaches a maximum value. Increasing the embedment depth beyond the optimum reduces the factor of safety, until (if d were large enough) it would equal 1.0. The only advantage of increasing d beyond the optimum would be in the anchor force (T) required; this is calculated from:

$$T = (\Sigma P_A - \Sigma P_P) - \frac{(\Sigma P_A L_A - \Sigma P_P L_P)}{H + d - a}$$

For a value of 1.0 for F_m, the second term would of course be zero. The calculations are quite lengthy when done by hand, but are easily set out on a spreadsheet. The results below show how a range of values can be extracted, allowing the designer a number of alternative choices.

d (m)	6.0	6.2	6.4	6.6	6.9
F_m	1.927	1.929	1.930	1.929	1.926
T (kN/m)	135.2	127.9	120.4	112.8	101.1

The usual minimum value for F_m is 2.0; therefore in this case the designer would have to decide whether or not the *maximum achievable* value of 1.93 is adequate.

Anchors for sheet piling
The anchorage system is a significant part of sheet piling, indeed the commonest failure modes include forward movement of an anchor block and failure of the tie rod. Several methods of providing anchorage are available: some typical arrangements are shown in Fig. 8.45. A parallel row of sheet piles (Fig. 8.45(a)) may be used to advantage where space is limited or for temporary work. In fill or soft ground, to avoid bending, pairs of raking piles (Fig. 8.45(b)) can be used. Ground anchors (Fig. 8.45(c), see also Section 8.16) consisting of tensioned rods or cables embedded in cement grout are particularly useful when surface access is impracticable. The commonest form of anchor used in firm ground is the 'deadman' type, where a mass concrete wall or series of blocks is cast into the ground (Fig. 8.45(d)). It is important to locate deadman anchors so that the soil failure zones of the wall and the anchor do not overlap. A simple rule for ensuring this is shown (Fig. 8.45(d)): the lower end of the active wedge against the wall is assumed to occur at the actual toe (design depth) where there is free-earth support, or at a point a quarter of the way up the pile if fixed-earth support has been assumed.
The allowable pull for a deadman anchor may be determined as follows:

Continuous deadman: $T_a = \frac{1}{2}\gamma d^2 S(K_p - K_a)$ [8.63(a)]

Separate block: $T_a = \frac{1}{2}\gamma d^2[L(K_p - K_a) + K_o(d_2 - d_1)]\tan\phi'_{mob}$ [8.63(b)]

where
γ = unit weight of soil
$d = d_2$ (if $d_1 \le \frac{1}{2}d_2$), otherwise $= d_2 - d_1$
S = horizontal spacing of anchor ties
L = length of separate deadman
K_a, K_p and K_o = earth pressure coefficients based on ϕ'_{mob}

8.15   Excavation support

Almost all construction projects involve excavation to form basements, trenches, etc. which will require support. In some cases cantilever sheet piling can provide temporary support, but most excavation support systems will take the form of propped sheeting. There are four main considerations in the design of such systems: (a) determination of forces induced in props or struts, (b) a check on the possibility of base heave, (c) a check on the possibility of ground movements

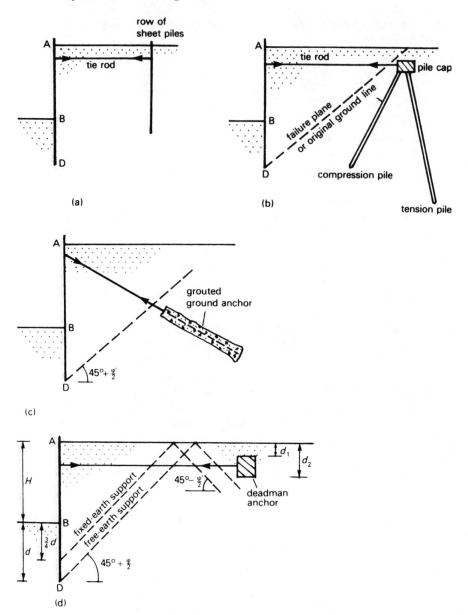

Fig. 8.45 Types of anchor for sheet piling
(a) Parallel row of piles (b) Pairs of raking piles (c) Ground anchor (d) Deadman

in adjacent soil masses or structures and (d) an assessment of the influence of the construction process or excavation sequence. The majority of excavation support systems are installed on a temporary basis and so factors of safety are chosen accordingly (see Appendix B). Nevertheless, careful design is necessary to ensure structurally viable and safe working conditions.

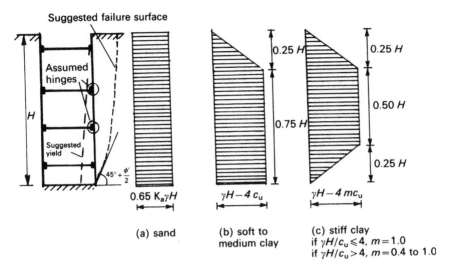

Fig. 8.46 Pressures on trench sheeting (after Terzaghi and Peck (1967) and Peck (1969))

Internally braced trenches

When the lateral struts or braces are inserted in trenches, an element of compression is induced by wedging or jacking. Under these conditions the soil is unlikely to be in an active state. Furthermore, the pressure distribution will change during the process, the distribution depending largely on the arrangement and order of installation. It is common to use empirical or approximation solutions, rather than employ rigorous methods such as finite elements (except for very large excavations or where threat to adjacent buildings exists). Although a number of methods have been suggested, the commonest in current use was proposed by Terzaghi and Peck (1948, 1967, 1969) and modified by Peck, following observations of actual loads and model tests (Fig. 8.46). The main features are as follows:

(a)   *In sands and gravels*
   Assumed drained with a rectangular distribution: $\sigma'_{h(max)} = 0.65K_a\gamma H$.

   Total thrust, $P_H = 0.65K_a\gamma H^2$ [8.64(a)]

(b)   *In soft to medium clays* (i.e. $N = \gamma H/c_u > 5-6$)
   Assumed undrained with a trapezoidal/rectangular distribution:

   $\sigma_{h(max)} = \gamma H(1 - 4m/N)$ [where $m = 1.0$ or 0.4 for deep NC clay]

   Total thrust, $P_H = 0.875\gamma H^2(1 - 4m/N)$ [8.64(b)]

(c)   *In stiff clays* (i.e. $N = \gamma H/c_u < 5-6$)
   Assumed undrained with a trapezoidal distribution:
   $\sigma_{h(max)} = m\gamma H$ [where $m = 0.2$ when $N < 4$, and $m = 0.4$ when $4 < N < 6$]

   Total thrust, $P_H = 0.75m\gamma H^2$ [8.64(c)]

   For stiff fissured clays

   Total thrust, $P_H = 0.875m\gamma H^2$ [8.64(d)]

It is recommended (Ward, 1955) that the first prop be installed before the excavation depth has reached $2c_u/\gamma$. The prop forces can be evaluated by dividing the pressure diagram half-way between each prop. It must be stressed that this method is approximate, and may overestimate the prop forces. However, the effects of over-jacking may increase them temporarily and produce variations in sheeting pressures.

Worked example 8.16   *A trench in a sand of depth 6.5 m is to be supported by timbering with horizontal struts at 1 m, 3 m and 5 m below ground level and spaced at 2 m intervals. Determine the estimated strut forces induced.*

Soil properties: $\phi' = 35°$ $\gamma = 20 \ kN/m^3$

$$K_a = \frac{1 - \sin 35°}{1 + \sin 35°} = 0.271$$

Assuming a rectangular distribution of horizontal pressure (Fig. 8.46(a))

$\sigma'_h = 0.65 K_a \gamma H = 0.65 \times 0.271 \times 20 \times 6.5 = 22.9$ kPa

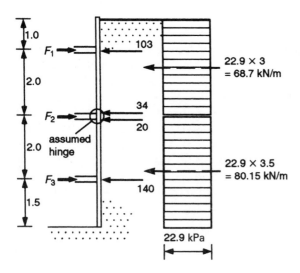

Fig. 8.47

Figure 8.47 shows the pressure distribution and the forces carried by each strut.

$F_1 = 68.7 \times (1.5/2) \times 2 = 103$ kN

$F_3 = 80.15 \times (1.75/2) \times 2 = 140$ kN

$F_2 = 68.7 \times 2 - 103 + 80.15 \times 2 - 140 = 34 + 20 = 54$ kN

Base heave in trenches

The upward movement of soil in the base of excavations in clay is a common form of failure. Two main mechanisms are possible: (a) a general base heave

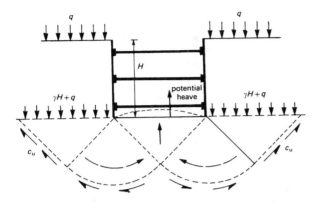

Fig. 8.48 Base heave in an excavation

involving the whole of the soil below and adjacent to the excavation, and (b) a local failure close to the sheeting due to softening or excessive inward movement. In the latter case, the probability of failure can be minimised by carefully controlling the installation process and attention to adequate drainage. A factor of safety against general base heave can be obtained by applying bearing capacity theory.

The removal of vertical stress by excavation, especially in soft cohesive soils, produces an inverted bearing pressure effect. The soil adjacent to the excavation provides a 'hydrostatic head' which tends to drive the soil in the base upward. This has been described as *passive stress relief* (Burland and Fourie, 1985). A simple approach is to consider the balance of forces on a horizontal plane passing through the base of the excavation. If the vertical pressure on the excavation surface is zero, the plastic equilibrium diagram may be assumed to be that shown in Fig. 8.48. The factor of safety is then:

$$F_b = \frac{N_c c_u}{\gamma H + q} \quad [\geq 2.5\text{--}3.0] \tag{8.65}$$

where N_c = Skempton's bearing capacity factor (Fig. 11.6).

Trench support using bentonite mud

In the construction of diaphragm walls for basements, underpasses, coffer dams, etc. the trench walls may be supported temporarily by bentonite mud or slurry. The trench is kept filled with mud while excavation is being carried out using a mechanical grab. Concrete is then placed in the trench using a tremie pipe, displacing the mud from the bottom upward.

In *clays*, there is an initial absorption of water from the mud into the trench sides. This quickly results in the formation of a filter cake which is almost impervious, and thus provides a seal. Since this support is only temporary, undrained conditions are assumed and the failure plane is at 45° (Fig. 8.49(a)). An approximate factor of safety can be evaluated by considering the forces on the wedge ABC.

The mobilised value of force C required for stability,

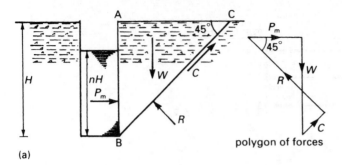

(a)

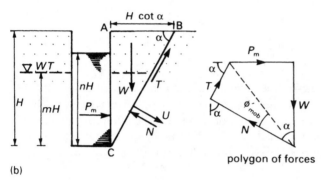

(b)

Fig. 8.49 Pressures on a mud-filled trench
(a) Cohesive soil (b) Cohesionless soil

$C_{mob} = (W - P_m)/\sqrt{2}$

where $W = \frac{1}{2}\gamma H^2$ (γ = soil unit weight)

$P_m = \frac{1}{2}\gamma_m(nH)^2$ (γ_m = mud unit weight)

If the shear strength of the soil = c_u, then $C = c_u H \sqrt{2}$. Therefore, the factor of safety is

$$F = \frac{C}{C_{mob}} = \frac{4c_u}{H(\gamma - \gamma_m n^2)} \qquad [8.66]$$

In *sands*, although some penetration and loss into the pores takes place, the mud soon forms a thixotropic gel, thus making a seal. The stability of the trench is very sensitive to the level of the water table, as well as to the depth of the slurry. Figure 8.49(b) shows the forces acting on the supported wedge of soil.

Resolving forces hoizontally and vertically:

$$P_m = N \tan \alpha - T \cos \alpha \quad \text{and} \quad W = N \cos \alpha + T \sin \alpha$$

But

$$N = N' + U \quad \text{and} \quad T = N' \tan \phi'_{mob}$$

Then

$$P_m = N'(\sin \alpha - \tan \phi'_{mob} \cos \alpha) + U \sin \alpha$$

$$= N' \cos \alpha(\tan \alpha - \tan \phi'_{mob}) + U \sin \alpha$$

and

$$W = N' \cos \alpha(1 - \tan \alpha \tan \phi'_{mob}) - U \cos \alpha$$

Equating: $\quad N' \cos \alpha = \dfrac{P_{\mathrm{m}} - U \sin \alpha}{\tan \alpha - \tan \phi'_{\mathrm{mob}}} = \dfrac{W - U \cos \alpha}{1 + \tan \alpha \tan \phi'_{\mathrm{mob}}}$

From which: $\quad \tan(\alpha - \phi'_{\mathrm{mob}}) = \dfrac{P - U \sin \alpha}{W - U \cos \alpha}$

Substituting for the three forces and the slip surface angle:

$P_{\mathrm{m}} = \frac{1}{2}\gamma_{\mathrm{m}}(nH)^2, \quad W = \frac{1}{2}\gamma H^2 \cot \alpha \quad U = \frac{1}{2}\gamma_{\mathrm{w}}(mH)^2 \operatorname{cosec} \alpha$

and $\quad \alpha = 45° + \phi'_{\mathrm{mob}}/2$

$$\tan\left(45° - \frac{\phi'_{\mathrm{mob}}}{2}\right) = \frac{\gamma_{\mathrm{m}} n^2 - \gamma_{\mathrm{w}} m^2}{\gamma \cot \alpha - \gamma_{\mathrm{w}} m^2 \cot \alpha}$$

$$\therefore \quad \tan^2\left(45° - \frac{\phi'_{\mathrm{mob}}}{2}\right) = \frac{\gamma_{\mathrm{m}} n^2 - \gamma_{\mathrm{w}} m^2}{\gamma - \gamma_{\mathrm{w}} m^2} = K_{\mathrm{m}} \qquad [8.67]$$

From this the required slurry weight can be determined:

$$\gamma_{\mathrm{m}} = \frac{K_{\mathrm{m}}(\gamma - \gamma_{\mathrm{w}} m^2) + \gamma_{\mathrm{w}} m^2}{n^2} \qquad [8.68]$$

Also, from eqn [8.67], it can be shown that

$$\tan \phi'_{\mathrm{mob}} = \frac{1 - K_{\mathrm{m}}}{2\sqrt{K_{\mathrm{m}}}}$$

Then the factor of safety is independent of the depth of the trench and is given by:

$$F = \frac{\tan \phi'}{\tan \phi'_{\mathrm{mob}}} = \frac{2\sqrt{(K_{\mathrm{m}})}\tan \phi'}{1 - K_{\mathrm{m}}} \qquad [8.69]$$

The true factor of safety may be higher than this since the shear strength of the mud is ignored, as are the thixotropic properties and the effects of caking. However, the method provides a simple and reasonably reliable design guide. The effect of the water table level is illustrated in the following worked example.

Worked example 8.17 *A narrow trench is to be excavated to a depth of 6 m and will be supported up to the surface with bentonite mud. The surrounding soil is sand having an angle of friction of 36° and a saturated unit weight of 19 kN/m³.*

(a) *Calculate the factor of safety against shear failure when using a mud of unit weight 12 kN/m³ and when the water table is expected to vary between the surface and a depth of 2 m.*

(b) *Calculate the unit weight required for the mud if the minimum factor of safety is 1.5 when the water table is at the surface.*

(a) Since the mud fills the trench, $n = 1$, and so, using eqn [8.69]:

$$\text{Factor of safety,} \quad F = \frac{2\sqrt{(K_{\mathrm{m}})}\tan 36°}{1 - K_{\mathrm{m}}}$$

where $K_m = \dfrac{12 - 9.81m^2}{19 - 9.81m^2}$

WT 2 m below surface: $m = 0.667$; $K_m = 0.522$ and $F = \underline{2.20}$
WT 1 m below surface: $m = 0.833$; $K_m = 0.426$ and $F = \underline{1.65}$
WT at the surface: $m = 1.000$; $K_m = 0.238$ and $F = \underline{0.93}$
Thus, failure is predicted if the water table rises to the surface.

(b)   Required $\sigma'_{mob} = \arctan (\tan 36°/1.5) = 25.84°$

$K_m = \tan^2(45° - \phi'_{mob}/2) = 0.3928$

Using eqn [8.68], the required unit weight for the mud is:

$\gamma_m = 0.3928(19.0 - 9.81 \times 1.0) + 9.81 \times 1.0 = \underline{13.42 \text{ kN/m}^3}$

8.16   Ground anchors

A ground anchor is a device for transferring a tensile load to a secure load-bearing rock or soil mass. There are essentially three parts: (a) the stressing head, (b) the tendon and (c) the fixed anchor.

High-tensile steel ($f_y \geq 800$ MPa) is usually used for tendons, which may be in the form of rods or multi-stranded cables. The three main types of ground anchor are shown in Fig. 8.50. The pullout resistance will be developed partly by side-shear and partly by end-bearing on the grout plug, depending on its shape and type. In clays and intact rocks, low-pressure cement grouting is used. In coarse soils and fissured or broken rocks high-pressure grout can be made to penetrate pores, cracks, etc. and so enhance pullout resistance. The type of anchor is generally chosen to suit the soil/rock conditions.

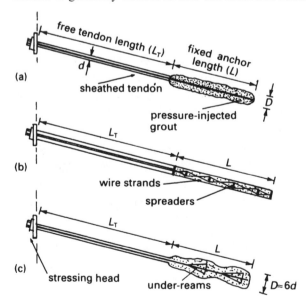

Fig. 8.50   Ground anchors
            (a) Pressure grouted (b) Multiple stranded (c) Multiple under-reamed

Intact rocks and very stiff clays

Low-pressure grouting with crimped rods or expanding multi-strand tendons are used in straight bores. Pullout resistance is entirely due to side-shear:

$$T_R = \tau_b \pi d L \qquad\qquad [8.70]$$

where τ_b = maximum bond strength between the grout plug and sides of bore
d = bore diameter, L = fixed anchor length

For rocks: τ_b is often determined from pullout tests, or may be estimated from SPT N-values, or from the unconfined compressive strength (UCS) of the intact rock (e.g. 10 per cent of UCS).
For very stiff clays:

$$\tau_b = \alpha \bar{c}_u.$$

where α = adhesion factor in the range 0.6–1.0 and
$\bar{c}_u$ = average undrained strength

Firm clays and marls

Since grout penetration is poor, an under-reamed type is used. Pullout resistance is developed by a combination of side-shear and end-bearing. Usually, the under-ream diameter will be about six times that of the bore. Soils for which $\bar{c}_u < 90$ kPa are unsuitable for anchor embedment.

$$T_R = \bar{c}_u = \left[\alpha_u \pi D_u L_u + N_c \frac{\pi}{4}(D_u^2 - d^2) + \alpha \pi d L_n \right] \qquad [8.71]$$

where $\bar{c}_u$ = average undrained shear strength
α_u = adhesion factor associated with the under-reamed length (0.7–0.9)
α = adhesion factor associated with the non-under-reamed length (0.3–0.4)
D_u = diameter of the under-reamed portion (typically = $6d$)
d = bore diameter
N_c = bearing capacity factor (usually taken as 9.0)
L_u = length of under-reamed portion
L_n = length of non-under-reamed portion = $L - L_u$

Coarse soils and weak or fissured rocks

Low-pressure grouting, or two-stage low/high pressure grouting, is used in lined bores (the lining being withdrawn as the bore fills). Grout penetration into pores and cracks enhances the pullout resistance, which can be estimated in either of two ways.

(1) *Using an empirical expression*

$$T_R = Lm \tan \phi' \qquad\qquad [8.72(a)]$$

where m = an overall strength factor dependent on the type and relative density of the soil, and also on the type of anchor and grout pressure; the range is 150–600 depending on particle size

Littlejohn (1977) has recommended the expression log $m = 2.52 + 0.34$ log D_{50}, where D_{50} is the 50 per cent grading characteristic.

(2)   *Using a theoretical method*

$$T_R = K_f \sigma'_v \tan \phi' \pi DL + N_f \sigma'_v \frac{\pi}{4}(D^2 - d^2)$$ [8.72(b)]

where K_f = an earth pressure coefficient (range 1.0–30)
   N_f = bearing capacity coefficient, for which the suggested value
    $= N_q/4$
   N_q = bearing capacity coefficient given by Berezantsev *et al.* in
    Fig. 11.20 corresponding to the ratio L/D and ϕ'.

Fine sands, silts and silty sands

In soils having permeabilities $<1 \times 10^{-4}$ m/s, grout penetration may be poor. Expanding multi-stranded tendons are often used, inserted in lined bores and embedded in low- or high-pressure grout. The expanded diameter of the grout plug may be difficult to estimate, therefore $D = d$ is often assumed. The pullout resistance may be estimated using eqns [8.72(a)] or [8.72(b)].

Factors of safety

The pullout resistances obtained from the equations above must be considered as *ultimate* values and must be divided by an appropriate factor to obtain design values. The factors recommended in BS 8081:1989 are:

Temporary works <6 months $F \geq 2.0$
Temporary works ≥6 months $F \geq 2.5$
Permanent works $F \geq 3.0$

Overall stability of anchored walls

For walls tied with a single row of anchors, the methods described above are usually adequate; however, where a sloping or irregular surface is involved a

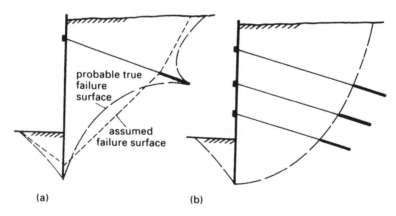

(a)   (b)

Fig. 8.51   Failure surfaces for anchored walls
    (a) Single row of anchors (b) Multiple anchors

check should be made on overall stability. The failure surface may be approximated as shown in Fig. 8.51(a) and the limiting equilibrium of the resulting soil wedges considered. Multiple anchorages will produce a curved failure surface (Fig. 8.51(b)), for which an analysis using a method of slices should be used, together with a safety check on each anchor.

Exercises

Exercises 8.1–8.11 inclusive are designed to illustrate the variations that will result in the magnitude and position of the resultant active thrust due to different soil conditions behind a retaining wall. In each case, the wall has a smooth vertical back and supports a 12 m depth of soil.

For each case, calculate the magnitude and position (above the wall base) of the resultant active thrust (P_A), assuming that the groundwater level is well below the base of the wall.

c' = apparent cohesion
ϕ' = angle of shearing resistance $\Big\}$ of the supported soil
γ = unit height

8.1   Surface horizontal: no surcharge: single layer:

$c' = 0$ $\phi' = 30°$ $\gamma = 18.0$ kN/m³

8.2   Surface horizontal: uniform surcharge of 24 kPa: single layer:

$c' = 0$ $\phi' = 32°$ $\gamma = 19.5$ kN/m³

8.3   Surface horizontal: no surcharge: two layers:

0–5 m below surface: $c' = 0$ $\phi' = 30°$ $\gamma = 18.0$ kN/m³
below 5 m: $c' = 0$ $\phi' = 36°$ $\gamma = 20.0$ kN/m³

8.4   Surface horizontal: no surcharge: two layers:

0–4 m below surface: $c' = 0$ $\phi' = 34°$ $\gamma = 19.6$ kN/m³
below 4 m: $c' = 0$ $\phi' = 28°$ $\gamma = 18.4$ kN/m³

8.5   Surface inclined at 15° to horizontal from the top of the wall: no surcharge: single layer:

$c' = 0$ $\phi' = 30°$ $\gamma = 19.0$ kN/m³

8.6   Surface horizontal: no surcharge: single layer:

$c_u = 45$ kPa $\phi_u = 0$ $\gamma = 18.0$ kN/m³

8.7   Surface horizontal: no surcharge: single layer:

$c' = 15$ kPa $\phi' = 20°$ $\gamma = 18.0$ kN/m³

8.8   Surface horizontal: uniform surcharge of 24 kPa: single layer:

$c' = 15$ kPa $\phi' = 20°$ $\gamma = 18.0$ kN/m³

8.9   Surface horizontal: uniform surcharge of 60 kPa: single layer:

$c' = 15$ kPa $\phi' = 20°$ $\gamma = 18.0$ kN/m³

8.10   Surface horizontal: no surcharge: two layers:

0–5 m below surface: c_u = 12 kPa γ = 17.0 kN/m³
below 5 m: c_u = 35 kPa γ = 18.0 kN/m³

8.11   Surface horizontal: no surcharge: two layers:

0–4 m below surface: c' = 0 ϕ' = 30° γ = 19.6 kN/m³
below 4 m: c' = 25 kPa ϕ' = 15° γ = 18.2 kN/m³

8.12   A concrete retaining wall has a smooth vertical back and supports an 8 m depth of soil having the following properties:

c' = 0 ϕ = 32° γ = 19 kN/m³ $\gamma_{sat.}$ = 21 kN/m³

There is no surface surcharge.

   Calculate the magnitude of the resultant active thrust acting on the wall for the following conditions: (a) fully drained and free to yield, (b) fully drained, but with the top of the wall restrained against yielding, and (c) free to yield, but with the groundwater level standing at the surface.

8.13   A concrete retaining wall with a smooth vertical back supports a 12 m depth of soil, comprising two layers:

0–7 m below surface: c' = 0 ϕ' = 28°   drained γ = 19.2 kN/m³
                                        saturated γ = 20.6 kN/m³
below 7 m: c_u = 25 kPa γ = 18.1 kN/m³

Calculate the magnitude of the resultant active thrust acting on the wall for the following groundwater level conditions: (a) groundwater level well below the base of the wall, (b) groundwater level 7 m below the surface, and (c) groundwater level at the surface.

8.14   A concrete retaining wall with a vertical back supports a 5 m depth of soil having the following properties:

c' = 0 ϕ' = 32°   drained γ = 18.8 kN/m³

A vertical line-load surcharge of 80 kN/m is applied parallel to and 3 m away from the back of the wall.

(a)   Calculate the total lateral thrust acting on the back of the wall, assuming the soil to be fully drained and: (i) the wall to be smooth, (ii) an angle of wall friction between wall and soil of 25°.
(b)   If the mass of the wall is 222 kN/m, calculate the factor of safety against sliding (assuming a smooth vertical support face).

8.15   The back of a retaining wall is inclined at 80° to the horizontal and has a vertical height of 9 m. The surface of the supported soil slopes upward from the top of the wall at an angle of 20°. The soil properties are:

c' = 0 ϕ' = 30° γ = 20 kN/m³ δ = 24°

Determine the lateral active thrust imposed on the wall due to trial wedges of soil having failure planes at 75°, 60° and 45° to the horizontal.

8.16   The retaining wall shown in Fig. 8.52 retains soil having the following properties:

c' = 0 ϕ' = 33° γ = 19 kN/m³ δ = 20°

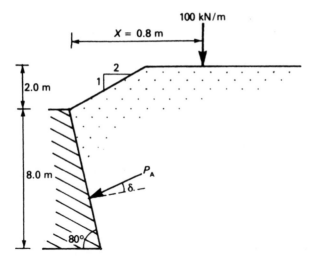

Fig. 8.52

(a) Determine the magnitude of the active thrust P_A acting on the wall: (i) when the surface carries no surcharge, and (ii) when a surface surcharge line-load of 100 kN/m is applied as shown.

(b) Determine the minimum distance (x) behind the wall that this surcharge could be located so as not to cause an increase in the earth pressure on the wall.

8.17 Using the earth pressure coefficients based on the wedge theory, calculate: (a) the active thrust acting on a vertical wall face of 16 m, and (b) the passive resistance offered to a vertical face of 4 m, in the case of a soil having the following properties:

$c' = 25$ kPa $\quad c_w = 12$ kPa
$\phi' = 15°$ $\qquad \delta = 10°$ $\quad \gamma = 19$ kN/m³

8.18 The cross-section of a proposed concrete retaining wall is shown in Fig. 8.53, the unit weight of the concrete being 24 kN/m³. Determine the factor of safety against

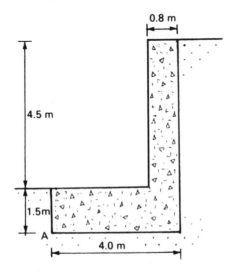

Fig. 8.53

overturning about A, the factor of safety against sliding (ignoring passive resistance) and the vertical bearing stress imposed on the soil beneath A, for the following conditions:

(a) Supporting a cohesionless soil fully drained:

$$c' = 0 \quad \phi' = 34° \quad \gamma = 20 \text{ kN/m}^3$$

(b) Supporting a cohesive soil, assuming that cracks will develop in the tension
zone which may fill with water:

$$c_u = 22 \text{ kPa} \quad \gamma = 18 \text{ kN/m}^3$$

8.19　Figure 8.54 shows the cross-section of a concrete crib wall constructed with vertical faces. The retained soil is used to fill the cribs and has the following properties:

$$c' = 0 \quad \phi' = 32° \quad \text{drained } \gamma = 19 \text{ kN/m}^3$$
$$\delta = 20° \quad \text{saturated } \gamma = 21 \text{ kN/m}^2$$

Average unit weight of crib and soil = 22 kN/m³.

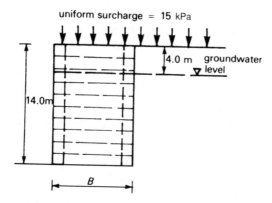

Fig. 8.54

Neglecting passive resistance, but allowing for the surcharge, determine the width
of the crib base required to satisfy the following design criteria: (a) factor of safety
against overturning ≮2.0, (b) factor of safety against sliding ≮1.5, and (c) bearing
pressure imposed below base ≯600 kPa.

8.20　A long narrow trench of depth 8 m is to be supported using a bentonite mud
having a unit weight of 12 kN/m³. The water table and trench mud levels stand at
1.8 m and 0.5 m respectively below the ground surface. Determine the factor of
safety against side collapse in each of the following soils:

(a) A sand: $c' = 0$ $\phi' = 32°$ $\gamma = 20 \text{ kN/m}^3$
(b) A clay: $c_u = 40 \text{ kPa}$ $\gamma = 19 \text{ kN/m}^3$

8.21　A cantilevered sheet-pile wall is required to support 5 m depth of soil having the
following properties:

$$c' = 0 \quad \phi' = 35° \quad \gamma = 20.8 \text{ kN/m}^3$$

Determine the length of pile required if a safety factor of 1.5 is applied to the
strength parameter and allowance made for an approximate analysis.

8.22 An anchored sheet-pile wall supports 6 m of fully-drained soil having the following properties:

$c' = 0$ $\phi' = 38°$ $\gamma = 19$ kN/m^3

Horizontal anchors are attached at a depth of 1.4 m and at 2.5 m centres. Using the free-earth support method and using a factor of safety of 1.4, determine: (a) minimum safe driving depth required, and (b) the force transmitted by each anchor rod.

8.23 An anchor sheet-pile wall retains a 5.9 m depth of drained soil having the following properties:

$c' = 0$ $\phi' = 34°$ $\gamma = 20$ kN/m^3

The piles have a total length of 10 m and are anchored at a point 1.3 m below the upper ground surface with horizontal tie rods at 3.0 m centres. Using the free-earth support method and neglecting friction, determine: (a) what proportion of the possible passive resistance is mobilised on the embedded length of piles, and (b) the force transmitted by each anchor rod.

Soil Mechanics Spreadsheets and Reference Assignments and Quizzes (available on the accompanying CD):

Assignments A.8   Earth pressure and retaining walls
Quiz Q.8   Earth pressure and retaining walls

Chapter 9

Stability of slopes

9.1 Movement in natural and artificial slopes

Soil or rock masses with sloping surfaces may be the result of natural agencies or they may be man-made; some examples are given in Fig. 9.1. In all slopes there exists an inherent tendency to degrade to a more stable form – ultimately towards horizontal – and in this context *instability* is construed as the tendency to move, and *failure* as an actual mass movement. The forces which cause instability are mainly those associated with gravity and seepage, while resistance to failure is derived mainly from a combination of slope geometry and the shear strength of the rock or soil mass itself.

Mass movement may take place as the result of a shear failure along an internal surface or when a general decrease in effective stress between particles causes full or partial liquefaction. A wide variety of types of movement (failure) have been observed; it is convenient here, however, to consider three classes.

Falls. These are characterised by movement *away* from existing discontinuities, such as joints, fissures, steeply-inclined bedding planes, fault planes, etc. and within which the failure condition may be assisted or precipitated by the effects of water or ice pressure (Fig. 9.2(a)).

Slides. In this form of movement the mass remains essentially intact while sliding along a definite failure surface. Two structural sub-divisions are apparent:

(a) *Translational slides* which may involve linear movement of rock blocks along bedding planes or of a soil layer lying near to the (sloping) surface. Such movements are normally fairly shallow and parallel to the surface (Fig. 9.2(b)).

(b) *Rotational slips* occur characteristically in homogeneous soft rocks or cohesive soils; the movement taking place along a curved shear surface in such a way that the slipping mass slumps down near the top of the slope and bulges up near the toe (Fig. 9.2(c)).

Flows. Here the slipping mass is *internally disrupted* and moves partially or wholly as a fluid. Flows often occur in weak saturated soils when the pore pressure has increased sufficiently to produce a general loss of shear strength; true shear surface development may be intermittent or mostly absent.

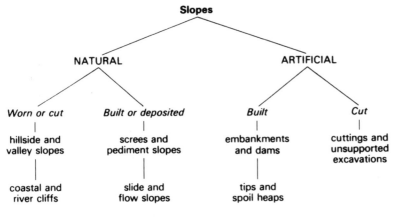

Fig. 9.1   Natural and artificial slopes

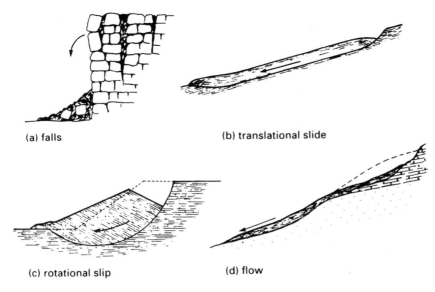

Fig. 9.2   Types of mass movement

Slope failures are usually precipitated by a variation in conditions, such as a change in rainfall, drainage, loading or surface stability (e.g. removal of vegetation). Such changes may occur immediately after construction, or they may develop slowly over a number of years, or they may be imposed suddenly at any time. In the analysis of both cut and built slopes it is necessary to consider both immediate and long-term stability conditions. It is also necessary to consider whether failure is likely along a newly created slip surface, or along a pre-existing one, since the difference between the peak and residual shear strength in some soils may be considerable.

Some slopes may exist for years in a state of *incipient failure*, i.e. on the brink of further movement. This is particularly evident in natural slopes and the slopes of spoil tips, and it should be borne in mind that many hill slopes have been naturally degraded by weathering and may also be close to a failure state. In these situations, man-made interference, such as the removal of trees or other vegetation, or cutting into the toe of the slope, may precipitate movement.

The work of this chapter is essentially concerned with the stability of cut and built soil slopes. For the sake of clarity in the introduction of fundamental ideas, only simple cases of drained and undrained loading will be examined. However, the methods and analyses dealt with here may equally be applied to a wider spectrum of problems involving, for example, natural slopes, spoil tips and soils of complex origin and composition. For details of *falls and flows* the reader should consult an appropriate text on engineering geology.

9.2   Choice of shear strength parameters for slope design

As discussed in Chapters 6 and 7, the parameters used to quantify shear strength depend fundamentally on both the stress history of the soil and the operational drainage conditions. In the case of slope stability problems, a further consideration is whether the failure movement will take place on a newly created slip surface or on an existing surface where slipping has occurred previously.

The following recommendations are made for general purpose design, but it should be remembered that many soil engineering problems are individualistic. *Guidelines* for design are helpful, but *general rules* can often lead the unwary into difficulties. Problems falling into *geotechnical risk category 3* (see Appendix A) require consideration from geotechnical engineers with a sound background of experience.

Slides along pre-existing slip surfaces
These failure states occur only where large displacements have already occurred; perhaps of several metres. *Residual strength* parameters should be used:

Undrained: $\tau_r = c_{ur}$
Drained: $\tau'_r = \sigma'_n \tan \phi'_r$

Slides creating new slip surfaces
A *new* slip surface will occur either when the critical strength or the peak strength of the soil is reached consistently within the soil mass. The choice of parameter should be made only after careful examination of the stress history and the predicted drainage state(s).

Normally consolidated soils
Both normally consolidated and *lightly* overconsolidated soils, i.e. soils that are *wetter* (or less dense) than the critical stage, are included.

For *undrained* conditions, the *peak* strength will be lower than the *critical* strength, so use:

$\tau_f = c_u$

For *drained* conditions, the *critical* strength should be used:

$\tau'_f = \sigma'_n \tan \phi'_c$

Heavily overconsolidated soils
These soils will be *drier* (or more dense) than the critical state and so, at small strains, the *peak* strength is higher than the *critical* strength. Designers will be well advised to be cautious in cases where the strains are unknown or difficult to predict.

Small strains —undrained: $\tau_f = c_u$
　　　　　　　 drained: $\tau'_f = c' + \sigma'_n \tan \phi'_f$ (Based on peak strength)
Otherwise —undrained: $\tau_f = c_u$
　　　　　　　 drained: $\tau'_f = \sigma'_n \tan \phi'_c$ (Based on critical strength)

For *very large* strains in heavily overconsolidated clays the fall-off in strength after the peak strain can be considerable. Under such circumstances it would be more appropriate to use (perhaps with a reduced factor of safety):

$\tau_f = c_{ur}$ and $\tau'_f = \sigma'_n \tan \phi'_r$

Sands

Denser than critical state: $\tau'_f = \tan \phi'_f$ (Peak strength)
Looser than critical state: $\tau'_f = \tan \phi'_c$ (Critical strength)

Where maximum and minimum density tests have been carried out, an interpolated value of ϕ' can be obtained corresponding to the *in situ* density.

Designers should ensure that they understand the true nature of test results and the parameters given in site and laboratory reports. The quick undrained triaxial test is still widely used, although it can only produce *total stress* parameters, which are often of little value in slope design. From a consolidated-undrained triaxial test, with pore pressures measured, the appropriate effective stress parameters can be evaluated. Estimates of residual strength parameters should be obtained from ring-shear or reversing shear box tests (see Chapter 7).

9.3 Translational slide on an infinite slope

The term *infinite slope slide* is commonly used to describe a *plane* translational movement at a shallow depth parallel to a long slope. Often the presence of an underlying harder stratum will constrain the failure surface to a plane. The effects of curvature at the extreme top and bottom of the slope and at the side are usually ignored; this leads to a conservative result for laterally straight or concave slopes, but may overestimate stability along convex curves and corners. Failure is often precipitated by a sudden increase in pore pressure, especially in partially desiccated soil where the surface layer binds together and moves as a thin flat slab; the term *flake slide* is sometimes used.

Undrained infinite slope
Consider a section in an infinite slope which is expected to fail along a slip plane parallel to the surface (Fig. 9.3). The undrained stability of a prismatic element depends on the following forces:

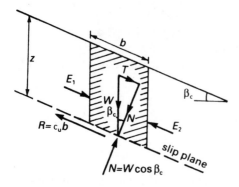

Fig. 9.3 Forces on an element in an undrained slope

Weight of the element,	$W = \gamma z b \cos \beta_c$
Normal reaction on the slip plane,	$N = W \cos \beta_c$
Tangential force down the slope,	$T = W \sin \beta_c$
Shear resistance force up the slope,	$R = \tau b$

where β_c = critical angle of the slope
τ_f = undrained shear strength of the soil = c_u

The inter-slice forces E_1 and E_2 may be considered equal and opposite and so cancel out.

For limiting equilibrium: $R - T = 0$

Therefore, $c_u b = W \sin \beta_c = \gamma z b \cos \beta_c \sin \beta$

giving $\dfrac{c_u}{\gamma z} = \sin \beta \cos \beta = \frac{1}{2} \sin 2\beta_c$

or $\sin 2\beta_c = \dfrac{2c_u}{\gamma z}$ [9.1]

Clearly, this stability condition is only valid for $0° < \beta_c < 45°$,

i.e. $\dfrac{2c_u}{\gamma z} \leq 1$

It is more realistic to restate eqn [9.1] in terms of *critical depth*, i.e. the depth at which the slip surface may be expected to develop.

Critical depth, $z_c = \dfrac{2c_u}{\gamma \sin 2\beta}$ [9.2]

It should be remembered, however, that in normally consolidated clays c_u increases with depth, except in the partially desiccated zone near to the surface. Also, this equilibrium condition presumes a *shallow* failure. A non-shallow z_c value calculated from eqn [9.2] does not necessarily indicate a safe slope, but may mean a deep-seated rotational slip is possible.

Worked example 9.1 *The soil in a long slope has an undrained shear strength of 50 kPa and a unit weight of 18 kN/m³. Using the infinite slope method, estimate the depth at which a shear slip may develop when the slope angle is 22°.*

From eqn [9.2]: critical depth, $z_c = \dfrac{2 \times 50}{18 \times \sin 44°} = \underline{8.0\ m}$

For a depth of 8 m to be 'shallow', the slope would have to be very long! It is likely here that a deep-seated rotational failure could be a problem, and this should be investigated.

Worked example 9.2 *Calculate the factor of safety relating to the undrained stability of a long slope of 1 vertical: 1.5 horizontal if at a depth of 1.8 m a weak layer of cohesive soil occurs for which:* $c_u = 24\ kPa$ *and for the overburden* $\gamma = 18.5\ kN/m³$.

From eqn [9.1]: $\sin 2\beta_c = \dfrac{2 \times 24}{18.5 \times 1.8} = 1.44$

Actual slope angle, $\beta = \arctan \dfrac{1}{1.5} = 33.69°$

Factor of safety $F_s = \dfrac{1.44}{\sin(2 \times 33.69°)} = \underline{1.56}$

Drained infinite slope

Under *drained* conditions the shear strength of the soil is given by:

$$\tau_f' = \sigma_n' \tan \phi'$$

or $\tau_f' = c' + \sigma_n' \tan \phi'$

with values for c' and ϕ' selected as appropriate (see Section 9.2).

Consider a prismatic element in a drained infinite slope (Fig. 9.4). Initially, the general case will be examined in which the groundwater level is lying parallel to the ground surface, so that the pore pressure varies with depth, but is constant along the slip plane, i.e. a condition of steady seepage taking place parallel to the ground surface. The forces on the element in equilibrium are:

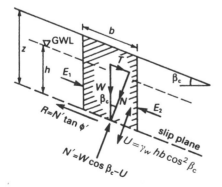

Fig. 9.4 Forces on an element in a drained slope

Weight of the element, $\qquad W = \gamma zb \cos \beta_c$
Normal reaction on the slip plane, $\qquad N = W \cos \beta_c$
Tangential force down the slope, $\qquad T = W \sin \beta_c$
Pore pressure force on the slip plane, $\qquad U = \gamma_w hb \cos^2 \beta_c$
Shear resistance force up the slope, $\qquad R = \tau'b$

where $\quad \beta_c$ = critical angle of the slope
$\qquad \tau'_f$ = drained shear strength of the soil

General expression when c' = 0

For limiting equilibrium: $R = T$

In which $\quad R = N' \tan \phi' = (W \cos \beta_c - \gamma_w hb \cos^2 \beta_c) \tan \phi'$

$$= (\gamma z - \gamma_w h)b \cos^2 \beta_c \tan \phi'$$

and $\quad T = W \sin \beta_c = \gamma zb \cos \beta_c \sin \beta_c$

Then $\quad (\gamma z - \gamma_w h)b \cos^2 \beta_c \tan \phi' = \gamma zb \cos \beta_c \sin \beta_c$

Therefore $\quad \dfrac{\gamma z - \gamma_w h}{\gamma z} = \dfrac{\tan \beta_c}{\tan \phi'}$

So that $\quad \tan \beta_c = \left(1 - \dfrac{\gamma_w h}{\gamma z}\right) \tan \phi'$ $\qquad\qquad$ [9.3]

and factor of safety, $\quad F = \left(1 - \dfrac{\gamma_w h}{\gamma z}\right) \dfrac{\tan \phi'}{\tan \beta}$ $\qquad\qquad$ [9.4]

where $\quad \beta$ = actual slope angle

Alternatively, putting pore pressure, $u = \gamma_w h \cos^2 \beta$:

$$F = \left(1 - \dfrac{u}{\gamma z \cos^2 \beta}\right) \dfrac{\tan \phi'}{\tan \beta} \qquad\qquad [9.5]$$

Having established these general statements for stability, it is worth examining several particular cases in which the drainage conditions have unique or special interest.

(a)   Dry sand or gravel
Consider a cohesionless soil ($c' = 0$) that is either dry or is sufficiently coarse-grained not to allow a significant level of capillary suction.

In this case: $\quad h = 0$

Thus $\quad \tan \beta_c = \tan \phi'$ $\qquad\qquad$ [9.6]

(b)   Groundwater level coincident with the slip plane
In a coarse-grained soil, $c' = 0$ and $h = 0$.

Thus $\quad \tan \beta_c = \tan \phi'$ (as for dry sand)

(c)   Groundwater level below the slip plane
In fine sands and silts, negative pore pressure will develop due to capillary attraction, hence, the effective stress at the slip plane will be increased by suction.
If h_s = distance of GWL below the slip plane

$$\text{Then}\quad \tan \beta_c = \left(1 + \frac{\gamma_w h_s}{\gamma z}\right) \tan \phi' \qquad [9.7]$$

So that β_c may be very steep, even vertical, as clearly demonstrated by a seaside sandcastle.

(d)   Waterlogged slope with steady parallel seepage
With seepage taking place parallel to the slope, the upper flow line is coincident with the ground surface and $h = z$.

$$\text{Then}\quad \tan \beta_c = \left(1 - \frac{\gamma_w}{\gamma}\right) \tan \phi' \qquad [9.8(a)]$$

$$\text{or}\quad \tan \beta_c = \frac{\gamma'}{\gamma} \tan \phi' \qquad [9.8(b)]$$

(e)   Waterlogged slope with steady vertical seepage
Where groundwater is draining vertically downward into a filter or other permeable layer below the slip plane (Fig. 9.5), the pore pressure throughout the soil will be zero, provided that no parallel flow occurs above slip plane.
Then $h = 0$ and $\tan \beta_c = \tan \phi'_p$

(f)   Slope in c′ > 0 conditions
Under certain circumstances, e.g. in slightly cemented or heavily overconsolidated soils, the peak value of c' may be taken as greater than zero. The shear strength of the soil is then:

$$\tau' = c' + \sigma'_n \tan \phi'_p$$

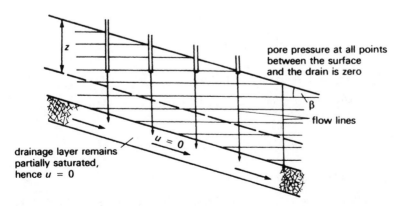

pore pressure at all points between the surface and the drain is zero

flow lines

drainage layer remains partially saturated, hence $u = 0$

Fig. 9.5 Vertical seepage to a drainage layer

and the expression for the factor of safety becomes:

$$F = \frac{c'b + N' \tan \phi'_p}{T}$$

$$= \frac{c' + (\gamma z - \gamma_w h) \cos^2 \beta \tan \phi'_p}{\gamma z \sin \beta \cos \beta} \qquad [9.9]$$

(g)   Using a pore pressure ratio
It is sometimes convenient to express the pore pressure in terms of either the slip plane depth or the overburden pressure using a pore pressure ratio (or coefficient).

(i)   Let $m = h/z$

Then $\tan \beta_c = (1 - m\gamma_w/\gamma) \tan \phi'$ [9.10]

(ii)   Let $r_u = \dfrac{u}{\gamma z} = \dfrac{\gamma_w h \cos^2 \beta}{\gamma z}$

So that $\gamma_w h = r_u \gamma z \sec^2 \beta$

Then $\tan \beta = (1 - r_u \sec^2 \beta) \tan \phi'$

Hence $F = (1 - r_u \sec^2 \beta) \dfrac{\tan \phi'}{\tan \beta}$ [9.11]

Worked example 9.3   *Adopting a factor of safety of 1.5, determine the maximum permissible angle for a slope in the following sandy soil: (a) when dry, and (b) when just waterlogged with steady seepage parallel to the surface:*

$\phi' = 36°$ $c' = 0$ $\gamma = 20 \ kN/m^3$

(c)   Calculate the factor of safety of the waterlogged slope against failure along a slip plane parallel to the surface at a depth of 4 m, if at this depth there is a thin layer of cohesive soil with the following properties:

$c' = 12 \ kPa$ $\phi'_p = 24°$ $\gamma = 20 \ kN/m^3$

(a)   When the soil is dry, $h = 0$, so that from eqn [9.6]:

$$F = \frac{\tan \phi'}{\tan \beta} \qquad \therefore \quad \tan \beta = \frac{\tan 36°}{1.5} = 0.4844$$

Hence, $\beta_{max.} = \underline{26°}$ (or a slope ratio of 1v:2h).

(b)   When the soil is waterlogged and the groundwater level is at the ground surface, $h = z$, so that from eqn [9.8]:

$$F = \frac{\gamma' \tan \phi'}{\gamma \tan \beta} \qquad \therefore \quad \tan \beta = \frac{(20.0 - 9.81) \tan 36°}{20.0 \times 1.5} = 0.2470$$

Hence, $\beta_{max.} = \underline{14°}$ (or a slope ratio of 1v:4h).

(c)   If the slope is therefore inclined at 14°, and considering a cohesive layer at $z = 4$ m, then from eqn [9.9]:

$$F = \frac{c' + (\gamma z - \gamma_w h) \cos^2 \beta \tan \phi_p'}{\gamma z \sin \beta \cos \beta}$$

$$= \frac{12 + (20 \times 4 - 9.81 \times 4) \cos^2 14° \times \tan 24°}{20 \times 4 \times \sin 14° \times \cos 14°} = \underline{1.55}$$

9.4 Slope failure mechanisms in cohesive soils

The most usual methods of providing an analysis of stability in slopes in cohesive soils are based on a consideration of *limiting plastic equilibrium*. Fundamentally, a condition of limiting plastic equilibrium exists from the moment that a shear slip movement commences and strain continues at constant stress. It is first necessary to define the geometry of the slip surface; the mass of soil about to move over this surface is then considered as a free body in equilibrium. The forces or moments acting on this free body are evaluated and those shear forces acting along the slip surface compared with the available shear resistance offered by the soil.

Several forms of slip surface may be considered for cohesive soils as shown in Fig. 9.6. The simplest of these, suggested by Cullmann in 1866, consists of an infinitely long plane passing through the toe of the slope. Although the analysis of the free body equilibrium is simple in this case, the method yields factors of safety which grossly overestimate the true stability conditions. On the other hand, while the choice of a more complex surface, such as a log-spiral or an irregular shape, may produce results near to the actual value, the analysis tends to be long and tedious. For most purposes, a *cylindrical* surface, i.e. *circular* in cross-section, will yield satisfactorily accurate results without involving analytical routines of any great complexity.

The stability of a cut or built slope depends very largely on changes in the pore pressure regime. During the construction of *embankments* pore pressures

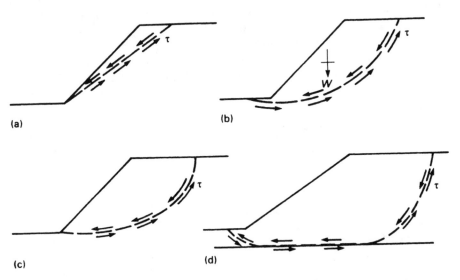

(a)

(b)

(c)

(d)

Fig. 9.6 Types of failure surface

will rise and, after construction, they will gradually fall. In cuttings, however, excavation causes an initial fall in pore pressures, but as seepage develops they gradually rise. Effective stresses and therefore shear strengths are generally inversely related to pore pressures. The most critical (lowest) factor of safety may therefore be expected to occur immediately after or during the construction of an embankment; after this the soil will gradually get stronger. In contrast, the shear strength in a cutting diminishes with time and so does the factor of safety.

Thus, it is necessary to consider both *short-term* (end-of-construction) and *long-term* stability. In this context it is convenient to think of short-term conditions as being completely *undrained*, in which the shear strength is given by $\tau = c_u$. In the next section, the first case to be considered will be that of undrained stability of a slope in a saturated clay; this type of analysis is often referred to as a *total stress*, method.

For long-term problems, and problems where changes in conditions may occur long after the end of construction (such as the sudden draw-down of level in a reservoir), a form of *effective stress* analysis is required. These methods may take the form of either a force- or moment-equilibrium analysis, involving plane, circular or irregular slip surfaces. For complex problems, stress path and slip line field methods are used. The shear strength parameters must be chosen with care (see Section 9.2).

9.5   Undrained stability – total stress analyses

A total stress analysis may be applied to the case of a newly cut or newly constructed slope in a fully saturated clay, the undrained shear strength being $\tau = c_u$. It is assumed that the failure surface will take the cross-sectional form of a circular arc, usually referred to as the *slip circle*. The centre of the critical slip circle will be somewhere above the top of the slope. The critical (failure) slip circle is one of an infinite number of potential circles that may be drawn having different radii and centres (Fig. 9.7). Some circles will pass through the toe of the slope and some will cut the ground surface in front of the toe.

The *critical circle* is the one along which failure is most likely to occur and for which the factor of safety is the lowest. A number of trial circles are chosen and the analysis repeated for each until the lowest factor of safety is obtained.

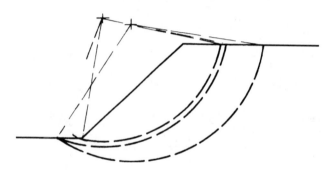

Fig. 9.7 Slip circles at different radii and centres

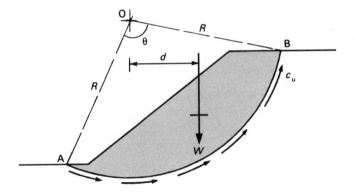

Fig. 9.8 Total stress (ϕ_u) analysis

Figure 9.8 shows the cross-section of a slope together with a trial slip circle of radius R and centre O. Instability tends to be caused due to the moment of the body weight W of the portion above the slip circle.

Disturbing moment = Wd

The tendency to move is resisted by the moment of the mobilised shear strength acting along the circular arc AB.

Length of arc $\qquad\qquad$ AB = $R\theta$

Shear resistance force along $\quad$ AB = $c_u R\theta$

Shear resistance moment $\qquad$ = $c_u R^2 \theta$

Then factor of safety, $\qquad F = \dfrac{\text{shear resistance moment}}{\text{disturbing moment}}$

$$= \frac{c_u R^2 \theta}{Wd} \qquad\qquad\qquad [9.12]$$

The values of W and d are obtained by dividing the shaded area into slices or triangular/rectangular segments and then taking area-moments about a vertical axis passing through the toe, or other convenient point.

Tension cracks
In cohesive soils, a tension crack tends to form near the top of the slope as the condition of limiting equilibrium (and failure) develops. From Chapter 8 [eqn 8.13] it will be seen that the tension crack depth may be taken as

$$z_o = \frac{2c_u}{\gamma}$$

The development of the slip circle is terminated at the tension crack depth and so its arc length is really AC as shown in Fig. 9.9. The free body weight W of the slipping mass is the shaded area bounded by the ground surface, the

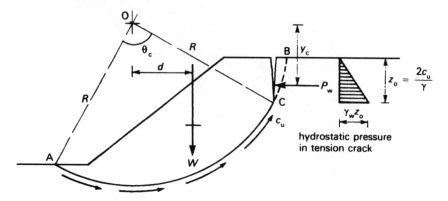

Fig. 9.9 Effect of tension crack in total stress analysis

slip circle arc and the tension crack. No shear strength can be developed in the tension crack, but, if it can fill with water, allowance must be made for the hydrostatic force P_w, which acts horizontally adding to the disturbing moment:

$$P_w = \tfrac{1}{2}\gamma_w z_o^2$$

Taking this into account, together with the fact that the slip circle arc is reduced, the factor of safety expression becomes:

$$F = \frac{c_u R^2 \theta_c}{Wd + P_w y_c} \tag{9.13}$$

Worked example 9.4   *A cutting in a saturated clay is inclined at a slope of 1 vertical:1.5 horizontal and has a vertical height of 10.0 m. The bulk unit weight of the soil is 18.5 kN/m³ and its undrained cohesion is 40 kPa. Determine the factors of safety against immediate shear failure along the slip circle shown in Fig. 9.10: (a) ignoring the tension crack, (b) allowing for the tension crack empty of water, and (c) allowing for the tension crack when full of water.*

The factors of safety against immediate shear failure may be obtained using the total stress method of analysis. Firstly, it is necessary to establish the geometry and area of the slip mass.

(a)   In the case ignoring the tension crack, the slip mass is bounded by the ground surface and the circular arc AB, for which the following may be calculated.

>   Radius, $R = OA = \sqrt{(5^2 + 16.7^2)} = 17.43$ m
>   Sector angle, $\theta = 84.06°$
>   Area of slip mass, $A = 102.1$ m²
>   Centroid distance from O, $d = 6.54$ m

>   Then from eqn [9.12]: $F = \dfrac{c_u R^2 \theta}{Wd}$

$$= \frac{40 \times 17.43^2 \times 84.06 \times \pi}{102.1 \times 18.5 \times 6.54 \times 180} = \underline{1.44}$$

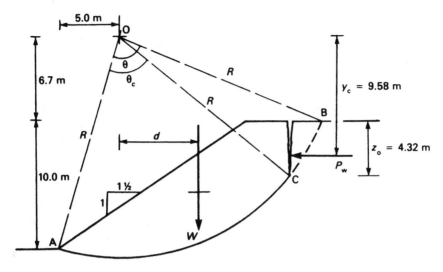

Fig. 9.10

(b) The effect of the tension crack is to reduce the arc length from AB to AC.

Depth of tension crack,	$z_0 = 2 \times 40/18.5 = 4.32$ m
Sector angle,	$\theta_c = 67.44°$
Area of slip mass,	$A = 71.64$ m^2
Centroid distance from O,	$d = 5.86$ m^2
In this case,	$P_w = 0$

Then from eqn [9.13]:

$$F = \frac{c_u R^2 \theta_c}{Wd}$$

$$= \frac{40 \times 17.43^2 \times 67.44 \times \pi}{71.64 \times 18.5 \times 5.86 \times 180}$$

$$= \frac{14\ 304}{7766} = \underline{1.84}$$

(c) When the tension crack is full of water, a horizontal force P_w will be exerted on the slip mass.

$$P_w = \tfrac{1}{2}\gamma_w z_0^2 = \tfrac{1}{2} \times 9.81 \times 4.32^2 = 91.54 \text{ kN/m}$$

The lever arm of P_w about O, $y_c = 6.7 + 2 \times 4.32/3 = 9.58$ m

Then from eqn [9.13]:

$$F = \frac{c_u R^2 \theta_c}{Wd + P_w y_c}$$

$$= \frac{14\ 304}{7766 + 91.54 \times 9.58} = \underline{1.65}$$

9.6   Undrained non-homogeneous slopes

A truly homogeneous soil would exhibit the same undrained shear strength (c_u) at all points throughout its mass. Although it is obvious that such a condition can only exist in theoretical terms, for certain cases the assumption of constant c_u can provide a perfectly reasonable estimate of the factors of safety. However, the non-homogeneous nature of the soil cannot be ignored when: (a) two or more layers of distinctly different soils are present, (b) in a single layer there is a significant variation in the undrained shear strength with depth.

The multi-layer problem
A common type of multi-layered slope occurs when one soil is either tipped or built-up to form an embankment on the pre-existing surface of another soil (Fig. 9.11(a)). A multi-layered system may also be exposed when a cutting is driven through stratified deposits (Fig. 9.11(b)). The treatment of this type of

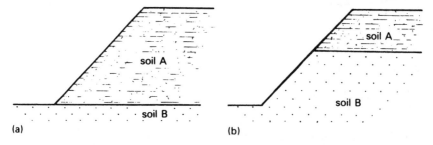

(a)   (b)

Fig. 9.11   Multi-layered slopes
              (a) Embankment or tip (b) Cutting in stratified deposits

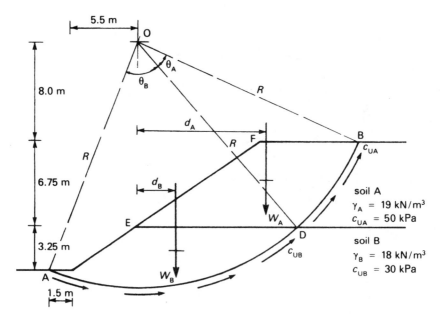

Fig. 9.12

problem depends mainly on the comparative values of c_u in the soils on either side of an interface. Where the soils have reasonably similar shear strengths, the total stress analysis described in the previous section can be used.

The slip mass corresponding to a given trial circle is divided into zones (Fig. 9.12) and the area, centroid position and sector angle determined for each. Equation [9.12] may be modified as follows:

$$F = \frac{R^2(c_{uA}\theta_A + c_{uB}\theta_B + \dots)}{(W_A d_A + W_B d_B + \dots)} \tag{9.14}$$

Worked example 9.5 *The slope of a cutting is 1 vertical:1.5 horizontal and the vertical height is 10 m. The soil mass comprises two saturated clay layers as shown in Fig. 9.12. Using the total stress method, determine the factor of safety against shear failure along the trial slip circle shown.*

The slip mass is first considered as two separate zones for which the common radius R is found to be 19.31 m.

For zone A (FBDE):
Sector angle, $\theta_A = 25.32°$
Area, $A_A = 65.943$ m^2
Centroid distance from O, $d_A = 9.79$ m

For zone B (EDA):
Sector angle, $\theta_B = 57.20°$
Area, $A_B = 53.74$ m^2
Centroid distance from O, $d_B = 2.85$ m

Then from eqn [9.14]:

$$F = \frac{19.31^2(50 \times 25.32 + 30 \times 57.20)\pi/180}{(65.94 \times 19.0 \times 9.79 + 53.74 \times 18.0 \times 2.85)}$$

$$= \frac{19\,407}{15\,022} = \underline{1.29}$$

Effect of a hard layer

When the underlying layer has a much greater shear strength, the critical slip circle is constrained to develop only in the weaker layer above. In the case of a built-up slope on a hard existing soil, all the trial circles should be taken through or just above the toe. In multi-layered soils, the stability of the soft upper layer must be checked on its own as well as that of the whole slope (Fig. 9.13).

9.7 Submerged slopes

In the case of water-retaining embankments, such as earth dams, canal banks, lagoon banks and the like, a part or, at times, all of the slope may be submerged. Figure 9.14 shows a partially submerged slope from which it may be deduced that the moment about O of the mass of water in the half-segment EFH exactly balances that in FGH. Thus the net water pressure moment is zero, providing

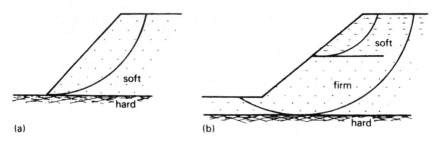

Fig. 9.13 Effect of an underlying hard layer

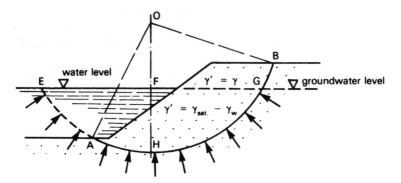

Fig. 9.14 Effect of a submerged slope

the soil is saturated. This being the case, the weight of that portion of the slip mass below EFG (the external water level) is calculated on the basis of the submerged unit weight ($\gamma_{sat.} - \gamma_w$). The bulk unit weight (γ) is still used for the portion above EFG.

In effect, the water resting on the submerged part of the slope provides an additional component of resistance moment, so that the factor of safety increases as the water level rises and decreases as it falls.

Worked example 9.6   *The slope of a water-retaining embankment is 1 vertical:2 horizontal and the vertical height is 10 m. The soil is fully saturated and has an undrained cohesion of 30 kPa and a unit weight of 18 kN/m³. Determine the factor of safety against shear failure along the trial circle shown in Fig. 9.15: (a) when the water level is at the toe of the slope, and (b) when the water level is 6 m above the toe.*

The slip mass is first considered as two separate zones for which the common radius R is found to be 22.83 m.

For zone A (FBDE):
Area, $A_A = 41.92 \text{ m}^2$
Centroid distance from O, $d_A = 13.00 \text{ m}$

For zone B (AED):
Area, $A_B = 144.11 \text{ m}^2$
Centroid distance from O, $d_B = 4.44 \text{ m}$

Tension crack depth, $z_0 = 2 \times 30/18 = 3.33 \text{ m}$
Sector angle, $\theta = 83.07°$

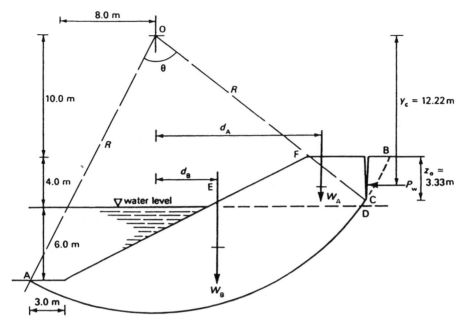

Fig. 9.15

Hydraulic thrust in tension crack, $P_w = \frac{1}{2}\gamma_z z_o^2$

$$= \frac{1}{2} \times 9.81 \times 3.33^2 = 54.4 \text{ kN/m}$$

Lever arm of hydraulic thrust, $y_c = 10 + 2 \times 3.33/3 = 12.22$ m

In each case, the hydraulic thrust in the tension crack will be included.

(a) Water level at toe:

$$F = \frac{30 \times 22.83^2 \times 83.07 \times \dfrac{\pi}{180}}{41.92 \times 18 \times 13.00 + 144.11 \times 18 \times 4.44 + 54.4 \times 12.22}$$

$$= \frac{22\,670}{9809 + 11\,517 + 665} = \underline{1.03}$$

i.e. shear failure will occur.

(b) Water level 6 m vertically above toe:

$$F = \frac{22\,670}{9809 + 144.11(18.0 - 9.81)4.44 + 665} = \underline{1.44}$$

9.8 Location of the most critical circle

The most critical circle is the one for which the calculated factor of safety has
the lowest value. The minimum factor of safety is clearly the criterion required
for design. The problem of locating the most critical circle may be approached
in one of two ways:

(a) By a process of trial and error, using a reasonable number of 'trial' circles and a thoughtful search pattern.
(b) By employing an empirical rule to prescribe an assumed critical circle and setting the limiting factor of safety high enough to allow for imperfections in the rule.

In the trial and error approach, the method has to allow for variation in three of the geometric parameters: the position of the centre, the radius and the intercept distance in front of the toe. For acceptable reliability, a very large number of trials may have to be made. The use of computers has made this method much more feasible and also more reliable.

Even when a large number of trials is to be the method employed, it is still useful to produce a good estimate for the first or 'seed' value. Slip circle behaviour is not completely random; on the contrary, some definite patterns may be observed. For example, when the angle of friction is greater than 3°, the critical circle will almost always pass through the toe of the slope. This is also the case when, irrespective of the value of ϕ, the slope angle exceeds 53° (Fig. 9.16).

Figure 9.17 shows a chart from which a first trial centre may be obtained for homogeneous undrained conditions. Values of Y_c/H and X_c/H are read off corresponding to the slope angle β, where:

X_c = horizontal distance from the toe to the circle centre
Y_c = vertical distance from the toe to the circle centre

The first trial centre can then be made the centre of the first group of nine trial centres. After evaluation of the factors of safety, new centres are chosen tactically according to the trend of the lowest values. Variations in radius may also be incorporated in the calculations. Simple grid plotting and contouring techniques are usually employed to 'home-in' on the most critical circle.

Where a harder or stiffer layer of rock or soil lies beneath the slope, the depth will be limited (Fig. 9.18). The most critical circle may in fact touch the lower harder surface.

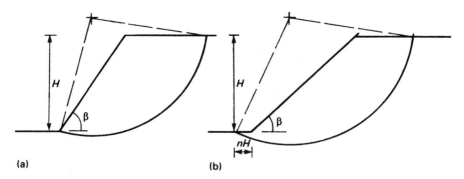

(a)   (b)

Fig. 9.16 Factors affecting the location of the critical circle (Through toe if $\phi > 3°$ or $\beta > 53°$
(b) In front of toe if $\phi \geq 3°$ or $\beta \geq 53°$

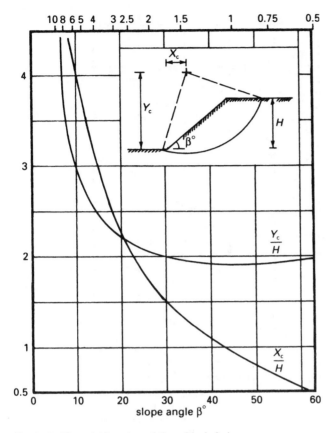

Fig. 9.17 First trial location of the critical circle

9.9 Taylor's stability number method

In 1948, D. W. Taylor proposed a simple method of determining the minimum factor of safety for a slope in a homogeneous soil. Using a total stress analysis and ignoring the possibility of tension cracks, he produced a series of curves which relate a *stability number* (N) to the slope angle β.

Consider the basic expression used in a total stress analysis

$$F = \frac{c_u RL}{Wd} \quad \text{(from eqn [9.12])}$$

It will be seen that $L \propto H$ and $W \propto \gamma H^2$, i.e. $L = K_1 H$, $W = K_2 \gamma H^2$

Then $\quad F = \dfrac{c_u R H K_1}{\gamma H^2 K_2 d}$

The *stability number* is dependent on the geometry of the slip circle and may be defined as:

$$N = \frac{K_2 d}{K_1 R} = \frac{c_u}{F \gamma H}$$

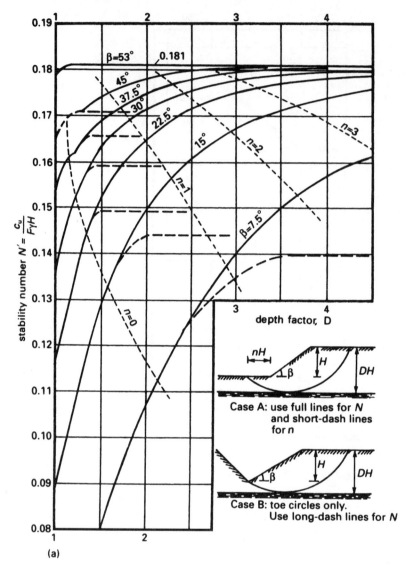

Fig. 9.18   Stability numbers for total stress analyses
(a) In terms of depth factor D (b) Toe circles in terms of slope angle β

Hence, $F = \dfrac{c_u}{N\gamma H}$ [9.15]

or since $c_{mob.} = c_u/F$, required $c_{mob.} = N\gamma H$

Values of N related to the slope angle β, the angle of shearing resistance ϕ_u and the depth factor D are given in the charts shown in Figs 9.18(a) and (b). For slope angles greater than 53°, the critical circle passes through the toe of the slope and the chart shown in Fig. 9.18(b) is used. For slope angles less than 53°,

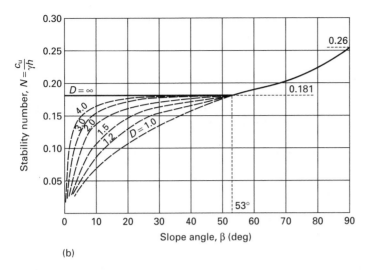

(b)

Fig. 9.18 Continued

the critical circle may pass in front of the toe and the chart shown in Fig. 9.18(a) is used. When the critical circle will be restricted to passing through the toe, the heavy broken lines on the chart must be used. The value of n, giving the break-out point of the critical circle in front of the toe, can be obtained from the light broken lines.

Worked example 9.7   *A cutting in a saturated clay has a depth of 10 m. At a depth of 6 m below the floor of the cutting there is a layer of hard rock. The clay has an undrained cohesion of 34 kPa and a bulk unit weight of 19 kN/m³. Calculate the maximum safe slope that will provide a factor of safety of 1.25 against short-term shear failure.*

Refer to Fig. 9.18(a).

$H = 10$ m and $DH = 16$ m $\therefore$ $D = 1.5$

Required stability number, $N = \dfrac{c_u}{1.25\gamma H} = \dfrac{34}{1.25 \times 19 \times 10} = 0.143$

The point on the chart located by $D = 1.5$ and $N = 0.143$ gives a slope angle of $\beta = 18°$. Also, from the chart, $n = 0.2$. Hence the circle will break out 2.0 m in front of the toe.

Worked example 9.8   *A cutting in a cohesive soil has a slope angle of 35° and a vertical height of 8 m. Using Taylor's stability method, determine the factor of safety against shear failure for the following cases:*

(a) $c_u = 40$ kPa $\gamma = 18$ kN/m³   D is large
(b) $c_u = 40$ kPa $\gamma = 18$ kN/m³   D = 1.5

(a)   When D is large, with $\beta < 53°$ then $N = 0.181$

Factor of safety, $F = \dfrac{40}{0.181 \times 18 \times 8} = \underline{1.53}$

(b)   When $D = 1.5$, $\beta = 35°$ and $\phi_u = 0$, then from Fig. 9.18(b):

$N = 0.168$ and $n = 0.6$

Factor of safety, $F = \dfrac{40}{0.168 \times 18 \times 8} = \underline{1.65}$

Thus, the presence of the harder layer constrains the failure mode to a smaller critical circle and so the factor of safety is larger, and the break-out point (nH) of this circle will be $0.6 \times 8 = 4.8$ m.

9.10   Drained stability – effective stress analyses

Stability analyses should be carried out in terms of effective stresses in problems where changes in pore pressure take place, such as existing embankments and spoil tips; also to estimate the long-term stability of slopes and in the case of overconsolidated clays for both immediate and long term conditions. Because of the variations in the stresses along a trial slip surface, the slip mass is considered as a series of slices. A trial slip circle is selected having a centre O and a radius R (Fig. 9.19), and the horizontal distance between the two ends A and B divided into slices of equal breadth b.

The forces acting on a slice of length 1 m will be as follows:

W = the body weight of the slice = γhb
N' = the effective normal reacting force at the base of the slice
T = the shearing force induced along the base
  = $W \sin \alpha$
R_1 and R_2 = forces imposed on the sides from adjacent slices – which may be resolved into:

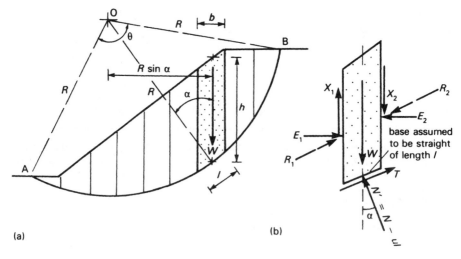

(a)   (b)

Fig. 9.19   Method of slices
  (a) Division of slip mass (b) Forces on a slice

E_1 and E_2 = normal interslice forces
X_1 and X_2 = tangential interslice forces

The effects of any surcharge on the surface must be included in the computation of the body weight and other forces.

At the point of limiting equilibrium, the total disturbing moment will be exactly balanced by the moment of the total mobilised shear force along AB.

$$\sum \tau_m l R = \sum \frac{\tau_f}{F} l R = \sum W \sin \alpha R$$

giving

$$F = \frac{\sum \tau_f l}{\sum W \sin \alpha}$$

Now in terms of effective stress,

$$\tau_f = c' + \sigma_n' \tan \phi'$$

and

$$\tau_f l = c' l + N' \tan \phi'$$

So that

$$F = \frac{\sum c' l + \sum N' \tan \phi'}{\sum W \sin \alpha}$$

or if the soil is homogeneous

$$F = \frac{c' L_{AB} + \tan \phi' \sum N'}{\sum W \sin \alpha} \qquad [9.16]$$

where L_{AB} = arc length AB = θR

A lot depends on how the values of N' are obtained. A number of methods have been suggested, some relatively simple and some which are quite rigorous. The most accurate estimates may be expected from rigorous methods, but may only be possible if a computer routine can be employed. A compromise may be arrived at by combining a simpler method of analysis with an increased factor of safety.

Fellenius' method

In this method, it is assumed that the interslice forces are equal and opposite and cancel each other out, i.e. $E_1 = E_2$ and $X_1 = X_2$. It is now only necessary to resolve the forces acting on the base of the slice, so that:

$N' = W \cos \alpha - ul$
$\quad = \gamma h b \cos \alpha - ub \sec \alpha \quad (l = b \sec \alpha)$

or putting $u = r_u \gamma h$

$N' = \gamma h (\cos \alpha - r_u \sec \alpha) b$

or $\quad \sum N' = \gamma b \sum h (\cos \alpha - r_u \sec \alpha)$

Then substituting in eqn [9.13]:

$$F = \frac{c' L_{AB} + \gamma b \tan \phi' \sum h (\cos \alpha - r_u \sec \alpha)}{\sum W \sin \alpha} \qquad [9.17]$$

The number of slices taken should not be less than five, and obviously a larger number would yield a better estimate of F. Even so, this method tends to give a value for F which may be as much as 50 per cent on the low side. Errors may also arise when r_u is high and the circle is deep-seated or has a relatively short radius. In such cases, Bishop's method is preferable.

Bishop's simplified method

In reasonably uniform conditions and when also r_u is nearly constant, it may be assumed that the tangential interslice forces are equal and opposite, i.e. $X_1 = X_2$ but that $E_1 \neq E_2$ (Fig. 9.20).

For equilibrium along the base of the slice:

$$0 = W \sin \alpha - \frac{\tau_f}{F}l = W \sin \alpha - \frac{c'L + N' \tan \phi'}{F}$$

So that $$F = \frac{\Sigma(c'l + N' \tan \phi')}{\Sigma W \sin \alpha}$$

For equilibrium in a vertical direction:

$$0 = W - N' \cos \alpha - ul \cos \alpha - \frac{\tau_f}{F}l \sin \alpha$$

$$= W - N' \cos \alpha - ul \cos \alpha - \frac{c'}{F}l \sin \alpha - \frac{N' \tan \phi'}{F} \sin \alpha$$

Then $$N' = \frac{W - (c'/F)l \sin \alpha - ul \cos \alpha}{\cos \alpha + (\tan \phi'/F) \sin \alpha}$$

After substituting for $l = b \sec \alpha$ and N' rearranging

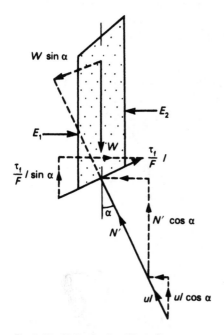

Fig. 9.20 Bishop's simplified slice

$$F = \frac{1}{\Sigma W \sin \alpha} \Sigma \frac{[c'b + (W - ub) \tan \phi'] \sec \alpha}{1 + \tan \alpha \tan \phi'/F}$$ [9.18]

The procedure is commenced by assuming a trial value for the F on the right-hand side and then, using an iterative process, to converge on the true value of F for a given trial circle. This is the routine procedure commonly used in programs designed for use on computers. Many of the program packages now available offer also the inclusion of multi-layer conditions, surcharge loads, variable pore pressure distribution and even the provision of berms and drains. In most problems, it is acceptable to adopt a constant average value for r_u. The factors of safety computed by this method may be slight underestimates, but with errors not usually exceeding 3 per cent, except in occasional unusual cases with deep base failure circles and F less than unity. More accurate lower factors of safety are claimed for methods which account for the variation in seepage forces on and in the slice (King, 1989). However, such refinements depend even more on good estimates of pore pressures. Morrison and Greenwood (1989) have further examined the assumptions made in this method.

Worked example 9.9 *Determine the factor of safety in terms of effective stress for the slope in Fig. 9.21 in respect of the trial circle shown. The soil properties are as follows:*

$c' = 10 \ kPa \quad \phi' = 28° \quad \gamma = 18 \ kN/m^3$

The pore pressure distribution along the trial circle is obtained by sketching equipotentials at each slice centre.

Figure 9.21 shows how a partially graphical solution may be obtained. The slip mass has been divided into slices of width 3 m and the section drawn to an appropriate scale. The average height of each slice (h) is scaled off the diagram and its weight calculated.

$W = \gamma hb = 18 \times h \times 3 = 54.0h \ kN/m$

The length of the chord at the base of each slice is scaled off and the pore pressure force calculated.

$ul = h_w \times 9.81 \times l$

A triangle of forces is drawn at the base of each slice to obtain values of N and T.
    The calculations are tabulated below:

Slice no.	h (m)	W (kN/m)	h_w (m)	l (m)	N	ul	N'	T
1	0.6	32	0.6	3.2	31	19	12	–8
2	2.4	130	1.7	3.1	128	52	76	–20
3	4.6	248	3.6	3.0	248	106	142	0
4	6.4	346	4.6	3.1	341	140	201	57
5	7.4	400	5.3	3.2	378	166	212	130
6	8.3	448	5.1	3.5	381	175	206	218
7	7.2	389	4.1	3.9	298	157	141	251
8*	3.7	246	1.3	6.9	150	88	62	186
							Σ 1052	814

* Width = 3.7 m

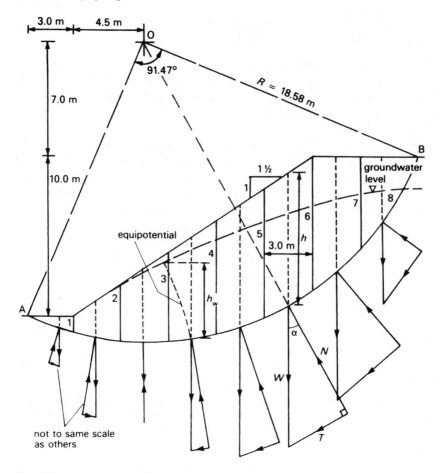

Fig. 9.21

$$L_{AB} = \theta R = 91.47 \times \frac{\pi}{180} \times 18.58 = 29.7 \text{ m}$$

From eqn [6.16]

$$F = \frac{c'L_{AB} + \tan \phi' \sum N'}{\sum T}$$

$$= \frac{10 \times 29.7 + \tan 28° \times 1052}{814} = \underline{1.05}$$

A similar analysis carried out on a microcomputer gave $F = 1.074$. Another computer analysis using Bishop's simplified method gave $F = 1.22$.

Worked example 9.10   *Determine the factor of safety in terms of effective stress of the slope shown in Fig. 9.22 in respect of the trial circle shown. Assume that the pore pressure ratio $r_u = 0.3$ and that the soil properties are as follows:*

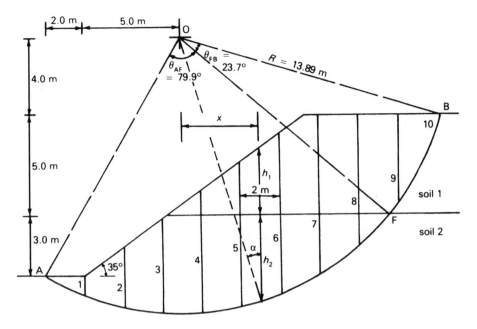

Fig. 9.22

Upper layer: $c'_1 = 25\ kPa$ $\phi'_1 = 12°$ $\gamma_1 = 18.0\ kN/m^3$
Lower layer: $c'_2 = 7\ kPa$ $\phi'_2 = 25°$ $\gamma_2 = 19.5\ kN/m^3$

The slip mass is divided into a convenient number of slices of width *b* as in worked example 9.9. The average height of each slice is measured off a scale drawing of the cross-section in two components h_1 and h_2. The weight of each slice is therefore:

$$W = (\gamma_1 h_1 + \gamma_2 h_2)b$$

As an alternative method to that used in worked example 9.9 (i.e. drawing the triangles of forces): in this example the angle α has been evaluated for each slice:

$$\alpha = \arcsin(x/R)$$

From eqn [9.15]:

$$F = \frac{\Sigma c'l + \tan \phi' \ \Sigma [W(\cos \alpha - r_u \sec \alpha)]}{\Sigma W \sin \alpha}$$

which, since there are two layers, becomes:

$$F = \frac{[c'_1 L_{FB} + c'_2 L_{AF}] + \left[\tan \phi'_1 \sum_{B}^{F} W(\cos \alpha - r_u \sec \alpha) + \tan \phi'_2 \sum_{F}^{A} W(\cos \alpha - r_u \sec \alpha) \right]}{\Sigma W \sin \alpha}$$

The results are tabulated below for a solution in which ten slices were taken:

Slice no.	h_1 (m)	h_2 (m)	W (kN/m)	x (m)	α (deg.)	cos α– r_u sec α	N'	T
1	0.00	0.6	23.4	5.67	24.10	0.584	13.7	–9.6
2	0.00	2.1	81.9	4.00	16.70	0.645	52.8	–23.5
3	0.00	3.9	152.1	2.00	8.28	0.686	104.3	–21.9
4	0.55	5.0	214.8	0.00	0.00	0.700	150.4	0.0
5	2.00	4.8	259.2	2.00	8.28	0.686	177.8	37.3
6	3.30	4.4	290.4	4.00	16.70	0.645	187.3	83.5
7	4.70	3.5	305.7	6.00	25.60	0.569	173.9	132.1
8	5.00	2.5	277.5	8.00	35.20	0.450	124.9	160.0
9	5.00	0.7	207.3	10.00	46.00	0.265	54.5	149.0
				Σ N' (A → F)			1040	
10*	2.9	0.0	125.3	11.8	58.1	–0.039	0[†]	106.4
				Σ T(A → B)				613

* Width = 2.4 m

[†] When cos $\alpha - r_u$ sec $\alpha < 0$, N' is set to 0, since N' cannot be negative

$N' = W(\cos \alpha - r_u \sec \alpha)$ $T = W \sin \alpha$

$$L_{AF} = \theta_{AF} R = 79.9 \times \frac{\pi}{180} \times 13.39 = 19.37 \text{ m} \quad L_{FB} = 23.7 \times \frac{\pi}{180} \times 13.89 = 5.75 \text{ m}$$

Then $$F = \frac{[25 \times 5.75 + 7 \times 19.39] + [\tan 12°(0) + \tan 25°(1040)]}{613}$$

$$= \frac{279 + 485}{613} = \underline{1.25}$$

A similar analysis carried out on a microcomputer gave $F = 1.27$. Another computer analysis using Bishop's simplified method gave $F = 1.42$.

9.11   *Effective stress stability coefficients*

A method involving the use of stability coefficients similar to that devised by Taylor, but in terms of effective stress, was suggested by Bishop and Morgenstern (1960). The factor of safety (F) is dependent on five problem variables:

(a)   slope angle β
(b)   depth factor D (as in Taylor's method – Fig. 9.18)
(c)   angle of shearing resistance ϕ'
(d)   a non-dimensional parameter $c'/\gamma H$
(e)   pore pressure coefficient r_u

The factor of safety varies linearly with r_u and is given by

$$F = m - nr_u \tag{9.19}$$

where m and n are coefficients related to the variables listed above. In the original work sets of charts were provided for ranges of values: $\cot \beta = 0.5$ to 5.0, $\varphi' = 20°$ to $40°$, $D = 1$ to 1.5 and $c'/\gamma H = 0$ to 0.05, and were thus suitable for relatively weak soils. O'Connor and Mitchell (1977) extended the range of charts to include $c'/\gamma H = 0.075$ and 0.100, but some stiff soils and soft rocks have higher values than this. The Author has recalculated the m and n values over a further extended range of $c'/\gamma H = 0$ to 0.150, and these are given in Table 9.1. The calculations were carried out using Bishop's simplified method as a basis and statistical search and fitting methods to establish critical values for m and n. Similar results have since been published by Chandler and Peiris (1989) which are found to correlate well with the values given here.

Table 9.1   Effective stress stability coefficients

c'/γH = 0

Slope cot β:		0.5:1		1:1		2:1		3:1		4:1		5:1	
D	φ'	m	n	m	n	m	n	m	n	m	n	m	n
ALL	20	0.18	0.90	0.36	0.72	0.73	0.90	1.08	1.21	1.45	1.54	1.81	1.88
	25	0.23	1.16	0.47	0.92	0.92	1.16	1.40	1.55	1.86	1.97	2.32	2.41
	30	0.29	1.43	0.58	1.15	1.15	1.43	1.72	1.91	2.30	2.44	2.88	2.98
	35	0.35	1.74	0.70	1.39	1.39	1.74	2.10	2.32	2.79	2.97	3.48	3.62
	40	0.42	2.09	0.83	1.67	1.67	2.09	2.51	2.79	3.34	3.55	4.18	4.34

c'/γH = 0.025

Slope cot β:		0.5:1		1:1		2:1		3:1		4:1		5:1	
D	φ'	m	n	m	n	m	n	m	n	m	n	m	n
1.00	20	0.52	0.72	0.70	0.76	1.11	1.01	1.53	1.34	1.95	1.69	2.37	2.04
	25	0.59	0.79	0.83	0.96	1.35	1.27	1.87	1.69	2.39	2.13	2.91	2.59
	30	0.67	0.88	0.97	1.19	1.60	1.56	2.23	2.07	2.86	2.61	3.41	3.17
	35	0.76	1.00	1.13	1.44	1.87	1.88	2.63	2.50	3.38	3.15	4.14	3.83
	40	0.86	1.17	1.30	1.72	2.18	2.24	3.07	2.98	3.95	3.76	4.86	4.56
1.25	20	1.00	0.93	1.07	1.02	1.29	1.20	1.60	1.45	1.93	1.76	2.30	2.06
	25	1.22	1.18	1.31	1.30	1.60	1.53	1.97	1.87	2.42	2.25	2.87	2.65
	30	1.46	1.47	1.59	1.62	1.95	1.91	2.41	2.33	2.93	2.80	3.49	3.28
	35	1.74	1.76	1.90	1.96	2.32	2.31	2.89	2.83	3.50	3.38	4.17	3.98
	40	2.04	2.11	2.23	2.35	2.74	2.75	3.43	3.39	4.14	4.04	4.93	4.75

c'/γH = 0.050

Slope cot β:		0.5:1		1:1		2:1		3:1		4:1		5:1	
D	φ'	m	n	m	n	m	n	m	n	m	n	m	n
1.00	20	0.69	0.78	0.90	0.83	1.37	1.06	1.83	1.38	2.32	1.77	2.77	2.08
	25	0.80	0.98	1.05	1.03	1.61	1.33	2.18	1.75	2.77	2.20	3.33	2.64
	30	0.91	1.21	1.21	1.24	1.88	1.62	2.56	2.15	3.24	2.68	3.91	3.24
	35	1.02	1.40	1.37	1.46	2.17	1.95	2.99	3.78	2.58	3.25	4.57	3.96
	40	1.14	1.61	1.55	1.71	2.50	2.32	3.44	3.06	4.40	3.91	5.30	4.64

Table 9.1 Continued

c′/γH = *0.050*

Slope cot β:		0.5:1		1:1		2:1		3:1		4:1		5:1	
D	φ′	m	n	m	n	m	n	m	n	m	n	m	n
1.25	20	1.16	0.98	1.24	1.07	1.50	1.26	1.82	1.48	2.22	1.79	2.63	<u>2.10</u>
	25	1.40	1.23	1.50	1.35	1.81	1.59	2.21	1.89	2.70	2.28	3.19	<u>2.67</u>
	30	1.65	1.51	1.77	1.66	2.14	1.94	2.63	2.33	3.20	2.81	3.81	3.30
	35	1.93	1.82	2.08	2.00	2.53	2.33	3.10	2.84	3.78	3.39	4.48	4.01
	40	2.24	2.16	2.42	2.38	2.94	2.78	3.63	3.38	4.41	4.07	5.22	4.78
1.50	20	1.48	1.28	1.55	1.33	1.74	1.49	2.00	1.69	2.33	1.98	2.68	2.27
	25	1.82	1.63	1.90	1.70	2.13	1.89	2.46	2.17	2.85	2.52	3.28	2.88
	30	2.18	2.01	2.28	2.09	2.56	2.33	2.95	2.69	3.42	3.10	3.95	3.56
	35	2.57	2.42	2.68	2.52	3.02	2.82	3.50	3.25	4.05	3.75	4.69	4.31
	40	3.02	2.91	3.16	3.02	3.55	3.37	4.11	3.90	4.77	4.48	5.50	5.12

c′/γH = *0.075*

Slope cot β:		0.5:1		1:1		2:1		3:1		4:1		5:1	
D	φ′	m	n	m	n	m	n	m	n	m	n	m	n
1.00	20	0.85	0.80	1.09	0.84	1.61	<u>1.10</u>	2.14	<u>1.44</u>	2.66	<u>1.80</u>	3.17	<u>2.13</u>
	25	0.95	1.01	1.25	1.05	1.86	1.38	2.50	<u>1.80</u>	3.13	<u>2.26</u>	3.74	<u>2.72</u>
	30	1.06	1.24	1.42	1.30	2.14	1.69	2.88	<u>2.20</u>	3.62	<u>2.76</u>	4.36	<u>3.33</u>
	35	1.19	1.49	1.61	1.56	2.44	2.03	3.31	<u>2.66</u>	4.18	<u>3.33</u>	5.02	<u>4.00</u>
	40	1.33	1.76	1.80	1.82	2.77	2.39	3.78	<u>3.15</u>	4.79	<u>3.95</u>	5.78	<u>4.76</u>
1.25	20	1.34	1.02	1.39	1.09	1.69	1.29	2.07	1.54	2.49	<u>1.82</u>	2.95	<u>2.17</u>
	25	1.58	1.28	1.66	1.39	2.00	1.64	2.47	1.96	2.97	<u>2.32</u>	3.52	<u>2.73</u>
	30	1.83	1.56	1.94	1.70	2.35	2.01	2.89	2.39	3.50	2.86	4.15	<u>3.36</u>
	35	2.11	1.87	2.25	2.03	2.73	2.39	3.36	2.87	4.08	3.46	4.83	<u>4.04</u>
	40	2.42	2.21	2.58	2.40	3.15	2.84	3.89	3.43	4.73	4.13	5.60	<u>4.83</u>
1.50	20	1.64	1.31	1.71	1.35	1.92	1.51	2.20	1.73	2.55	1.99	2.93	2.27
	25	1.98	1.66	2.05	1.71	2.31	1.91	2.66	2.20	3.08	2.53	3.55	2.92
	30	2.34	2.04	2.43	2.10	2.74	2.36	3.16	2.71	3.66	3.13	4.22	3.59
	35	2.74	2.46	2.84	2.54	3.21	2.85	3.71	3.29	4.30	3.79	4.96	4.34
	40	3.19	2.93	3.31	3.03	3.74	3.40	4.33	3.93	5.03	4.53	5.79	5.19

c′/γH = *0.100*

Slope cot β:		0.5:1		1:1		2:1		3:1		4:1		5:1	
D	φ′	m	n	m	n	m	n	m	n	m	n	m	n
1.00	20	0.98	0.80	1.25	0.86	1.83	<u>1.13</u>	2.41	<u>1.46</u>	2.97	<u>1.83</u>	3.53	<u>2.15</u>
	25	1.10	1.02	1.41	1.07	2.09	<u>1.42</u>	2.78	<u>1.84</u>	3.36	<u>2.29</u>	4.09	<u>2.72</u>
	30	1.21	1.25	1.58	1.30	2.37	1.72	3.17	<u>2.25</u>	3.91	<u>2.80</u>	4.71	<u>3.34</u>
	35	1.34	1.50	1.77	1.57	2.68	2.08	3.59	<u>2.71</u>	4.49	<u>3.34</u>	5.39	<u>4.03</u>
	40	1.48	1.78	1.99	1.87	3.01	2.44	4.07	<u>3.21</u>	5.10	<u>3.97</u>	6.14	<u>4.80</u>
1.25	20	1.48	1.03	1.52	1.09	1.86	1.29	2.27	1.55	2.74	<u>1.83</u>	3.23	<u>2.15</u>
	25	1.72	1.29	1.79	1.38	2.19	1.63	2.67	1.96	3.21	<u>2.32</u>	3.81	<u>2.74</u>
	30	1.99	1.59	2.08	1.73	2.53	2.00	3.09	2.41	3.73	<u>2.84</u>	4.42	<u>3.35</u>
	35	2.27	1.90	2.40	2.07	2.91	2.41	3.58	2.90	4.30	3.44	5.10	<u>4.04</u>
	40	2.58	2.23	2.74	2.44	3.33	2.85	4.09	3.44	4.96	4.11	5.88	<u>4.84</u>

Table 9.1 Continued

c'/γH = 0.100

Slope cot β:		0.5:1		1:1		2:1		3:1		4:1		5:1	
D	φ'	m	n	m	n	m	n	m	n	m	n	m	n
1.50	20	1.77	1.30	1.85	1.36	2.07	1.52	2.38	1.73	2.76	2.00	3.14	2.28
	25	2.11	1.66	2.20	1.72	2.47	1.93	2.83	2.21	3.28	2.53	3.78	2.91
	30	2.48	2.05	2.58	2.11	2.90	2.38	3.33	2.72	3.86	3.12	4.44	3.59
	35	2.88	2.47	2.98	2.54	3.37	2.86	3.88	3.28	4.49	3.78	5.17	4.34
	40	3.33	2.94	3.45	3.03	3.90	3.42	4.49	3.92	5.21	4.51	5.99	5.16

c'/γH = 0.125

Slope cot β:		0.5:1		1:1		2:1		3:1		4:1		5:1	
D	φ'	m	n	m	n	m	n	m	n	m	n	m	n
1.00	20	1.13	0.81	1.43	0.88	2.04	<u>1.15</u>	2.69	<u>1.54</u>	3.26	<u>1.78</u>	3.87	<u>2.12</u>
	25	1.25	1.04	1.60	1.11	2.32	<u>1.45</u>	3.06	<u>1.91</u>	3.74	<u>2.27</u>	4.45	<u>2.72</u>
	30	1.38	1.27	1.77	1.34	2.62	<u>1.78</u>	3.46	<u>2.30</u>	4.25	<u>2.81</u>	5.07	<u>3.37</u>
	35	1.50	1.51	1.96	1.59	2.93	2.12	3.88	<u>2.71</u>	4.82	<u>3.41</u>	5.77	<u>4.05</u>
	40	1.61	1.75	2.17	1.89	3.27	2.48	4.36	<u>3.18</u>	5.46	<u>4.06</u>	6.55	<u>4.89</u>
1.25	20	1.64	1.06	1.67	1.10	2.05	1.32	2.49	<u>1.58</u>	2.98	<u>1.86</u>	3.50	<u>2.17</u>
	25	1.89	1.33	1.94	1.40	2.38	1.67	2.89	1.99	3.48	<u>2.38</u>	4.08	<u>2.75</u>
	30	2.16	1.63	2.23	1.73	2.73	2.04	3.32	2.43	4.01	<u>2.92</u>	4.71	<u>3.41</u>
	35	2.45	1.95	2.56	2.09	3.11	2.45	3.80	2.93	4.59	<u>3.50</u>	5.41	<u>4.13</u>
	40	2.77	2.30	2.92	2.49	3.54	2.91	4.33	3.49	5.24	4.16	6.21	<u>4.95</u>
1.50	20	1.92	1.32	2.02	1.39	2.23	1.55	2.57	1.75	2.96	2.00	3.40	2.29
	25	2.26	1.68	2.37	1.75	2.64	1.97	3.03	2.23	3.50	2.55	4.02	2.91
	30	2.63	2.07	2.75	2.15	3.07	2.43	3.53	2.75	4.08	3.15	4.69	3.60
	35	3.04	2.50	3.16	2.58	3.55	2.92	4.08	3.32	4.73	3.81	5.44	4.36
	40	3.50	2.98	3.63	3.07	4.09	3.49	4.71	3.98	5.46	4.57	6.28	5.23

c'/γH = 0.150

Slope cot β:		0.5:1		1:1		2:1		3:1		4:1		5:1	
D	φ'	m	n	m	n	m	n	m	n	m	n	m	n
1.00	20	1.25	0.81	1.58	0.89	2.25	<u>1.16</u>	2.89	<u>1.44</u>	3.57	<u>1.80</u>	4.21	<u>2.15</u>
	25	1.37	1.02	1.75	1.12	2.53	<u>1.45</u>	3.24	<u>1.80</u>	4.01	<u>2.27</u>	4.78	<u>2.77</u>
	30	1.50	1.25	1.93	1.36	2.83	<u>1.78</u>	3.64	<u>2.24</u>	4.54	<u>2.79</u>	5.41	<u>3.39</u>
	35	1.65	1.53	2.12	1.61	3.14	<u>2.14</u>	4.09	<u>2.71</u>	5.10	<u>3.38</u>	6.09	<u>4.09</u>
	40	1.80	1.82	2.33	1.89	3.49	<u>2.53</u>	4.57	<u>3.24</u>	5.74	<u>4.05</u>	6.86	<u>4.85</u>
1.25	20	1.79	1.07	1.80	1.10	2.22	1.32	2.69	<u>1.59</u>	3.22	<u>1.86</u>	3.77	<u>2.17</u>
	25	2.03	1.33	2.07	1.40	2.55	1.68	3.09	<u>2.01</u>	3.71	<u>2.37</u>	4.33	<u>2.76</u>
	30	2.30	1.63	2.37	1.74	2.90	2.06	3.51	2.44	4.22	2.92	4.96	<u>3.38</u>
	35	2.60	1.96	2.69	2.08	3.28	2.47	4.00	2.94	4.81	<u>3.50</u>	5.66	<u>4.10</u>
	40	2.92	2.33	3.05	2.44	3.72	2.92	4.53	3.48	5.46	<u>4.17</u>	6.44	<u>4.92</u>
1.50	20	2.05	1.33	2.15	1.39	2.38	1.54	2.74	1.75	3.15	2.01	3.63	2.30
	25	2.39	1.68	2.51	1.76	2.77	1.97	3.19	2.23	3.67	2.55	4.23	2.90
	30	2.76	2.07	2.89	2.16	3.22	2.43	3.70	2.75	4.26	3.14	4.90	3.57
	35	3.16	2.50	3.30	2.59	3.69	2.92	4.24	3.31	4.90	3.79	5.64	4.33
	40	3.62	2.98	3.76	3.07	4.23	3.48	4.87	3.95	5.63	4.54	6.47	5.19

From the soil and slope data, a value is first calculated for $c'/\gamma H$. A section of Table 9.1 is now selected for which $D = 1.00$ and $c'/\gamma H$ just *greater* than that in the problem. Using linear interpolation, values are obtained for m and n, corresponding to problem values of $\cot \beta$ and ϕ'. If n is found to be underlined, a more critical (lower F) circle may exist at a greater depth. In this case the table section for $D = 1.25$ (and, if necessary, $D = 1.50$) is used until a non-underlined value for n is obtained. This procedure is repeated for table sections corresponding to a $c'/\gamma H$ value just *less* than that in the problem. Using eqn [9.17], two factor of safety values (F_1 and F_2) are calculated for $c'/\gamma H$ values respectively above and below that in the problem. The value of F is then obtained by interpolating linearly between F_1 and F_2.

The pore pressure parameter r_u is determined by averaging within the slip mass between the surface and the slip circle (see Section 9.12). The factor of safety thus calculated tends to be overestimated, in extreme cases up about 7 per cent.

Worked example 9.11   *A cutting in a clay soil has a slope of 1 vertical:3 horizontal and vertical height of 17 m. The value of r_u is 0.2 and the soil has the following properties:*

$$c' = 12 \; kPa \quad \phi' = 22° \quad \gamma = 18 \; kN/m^3$$

Using effective stress stability coefficients, obtain a value for the factor of safety.

Interpolating linearly in Table 9.1 for $\cot \beta = 3$ and $\phi' = 22°$:

For $c'/\gamma H = 0.050$ and $D = 1.00$: $m = 1.97$, $n = \underline{1.528}$

i.e. underlined; therefore a deeper critical circle may exist.

For $c'/\gamma H = 0.050$ and $D = 1.25$: $m = 1.976$, $n = 1.644$
Therefore $F_1 = 1.976 - 1.644 \times 0.2 = 1.647$

For $c'/\gamma H = 0.025$ and $D = 1.00$: $m = 1.666$, $n = 1.480$
Therefore $F_2 = 1.666 - 1.480 \times 0.2 = 1.370$

Interpolating now between F_1 and F_2:

$$F = 1.370 + (1.647 - 1.370)\frac{0.0392 - 0.025}{0.050 - 0.025} = 1.53$$

Other analyses of the same problem yielded the following results:

(1) Using Bishop and Morgenstern's method: $F = 1.51$
(2) Using the Chandler and Peiris table: $F = 1.56$
(3) Using the Fellenius method on a computer: $F = 1.41$

9.12   Determination of pore pressures

The total stress analysis provides an estimate of the minimum factor of safety under immediate (i.e. undrained) conditions. In order to obtain an estimate for the long-term condition, an analysis based on effective stresses is required. It is therefore necessary to determine the distribution of pore pressure on each trial slip circle. The following sets of conditions may have to be considered.

(a) End of construction
Since the construction period of earth dams and large embankments is usually fairly long, some dissipation of the excess pore pressure is likely. A total stress analysis would therefore yield value of F on the low side.

At any point the pore pressure is:

$$u_0 + \Delta u$$

where u_0 = initial value

Δu = change in pore pressure due to a change in the major principal stress $(\Delta\sigma_1)$

Using the pore pressure coefficient $\bar{B}$ (see Chapter 7, eqn [7.26]):

$$u = u_0 + \bar{B}\Delta\sigma_1$$

Putting $\qquad\qquad \dfrac{u}{\gamma h} = r_u$ (pore pressure ratio)

Then $\qquad\qquad r_u = \dfrac{u_0}{\gamma h} + \bar{B}\dfrac{\Delta\sigma_1}{\gamma h}$

But generally $\Delta\sigma_1 = \gamma h$ (i.e. no surcharge)

So that $\qquad\qquad r_u = \dfrac{u_0}{\gamma h} + \bar{B}$ $\qquad\qquad$ [9.20]

The pore pressure coefficient $\bar{B}$ may be obtained using a form of triaxial test (see Chapter 6). If when the soil is placed its water content is below the optimum, the value of u_0 may be almost zero, in which case $r_u = \bar{B}$. The value of r_u may be reduced by increasing the rate of dissipation of pore pressure by the incorporation of drainage layers.

(b) Steady seepage
A condition of steady seepage will be established after a reservoir has been full for a time. The procedure for constructing a flow net for this condition is described in Chapter 5. The pore pressure at a given point on a trial slip surface is obtained from the value of the equipotential passing through the point.

In Fig. 9.23, the pressure represented by the equipotential passing through P is equal to h_w:

Then $\quad r_u = \dfrac{h_w \gamma_w}{\gamma h}$ $\qquad\qquad$ [9.21]

In homogeneous conditions the value of r_u may vary up 0.45, but the presence of internal drainage layers will produce lower values.

(c) Rapid drawdown
In the event of the water level being reduced quickly, the water within the soil will tend to flow back into the reservoir through the upstream face. In this context, even a period of several weeks may bring about a 'rapid' change in the pore pressure distribution: the flow net for the steady seepage condition before

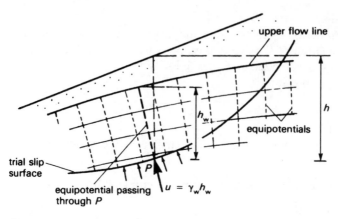

Fig. 9.23 Pore pressure at steady seepage

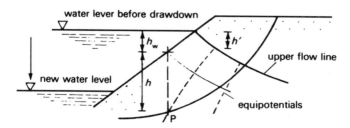

Fig. 9.24 Pore pressure after rapid drawdown

drawdown is again used. From Fig. 9.24, it will be seen that the pore pressure at point P (on a trial slip circle) before drawdown is given by:

$$u_0 = \gamma_w (h + h_w - h')$$

Again it is assumed that the total major principal stress at P is equal to the overburden pressure ($\sigma_1 = \gamma h$). When the drawdown reduces the phreatic surface to a level below h_w, the change in total major principal stress is given by:

$$\Delta\sigma_1 = -\gamma_w h_w$$

giving a corresponding change in pore pressure of:

$$\Delta u = -\bar{B}\gamma_w h_w$$

Therefore, immediately after drawdown, the pore pressure at P is:

$$
\begin{aligned}
u &= u_0 + \Delta u \\
&= \gamma_w (h + h_w - h') - \bar{B}\gamma_w h_w \\
&= \gamma_w [h + h_w (1 - \bar{B}) - h']
\end{aligned}
$$

So that $r_u = \dfrac{u}{\gamma h}$

$$= \frac{\gamma_w}{\gamma}\left[1 - \frac{h_w}{h}(1 - \bar{B}) - \frac{h'}{h}\right] \qquad [9.22]$$

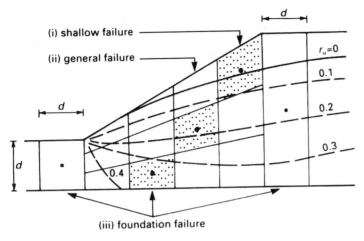

Fig. 9.25 Pore pressure ratio by averaging

Since h' is often quite small, it may be neglected and that r_u will commonly have values in the range 0.3 to 0.4.

(d)   Estimating r_u by averaging

It is important that sensible estimates are made of the average pore pressure ratio (r_u), especially when using methods such as that proposed by Bishop and Morgenstern (Section 9.11). A method suggested by Bromhead (1986) offers a relatively simple systematic solution. The soil is divided into strips of equal breadth (Fig. 9.25) and each strip divided into three equal vertical portions. After plotting isobars of r_u (or using a flow net), the average value of r_u is obtained for the appropriate zone of each slice and the overall average for the slope evaluated from:

$$r_u = \frac{\Sigma(\Delta A r_u)}{\Sigma \Delta A} \qquad [9.23]$$

where ΔA = slice zone area

The slice zones should be selected on the following basis (see also Fig. 9.25):

(a)   For shallow failure modes: use upper thirds.
(b)   For general failures: use middle thirds.
(c)   For foundation failure modes: use lower thirds, plus the toe and head slices.

Zone values for r_u may be obtained from the chart shown in Fig. 9.26.

(e)   Natural slopes

Pore pressures are determined from site measurements using observation wells or piezometers over a long enough period to include the 'worst' conditions.

(f)   Beneath fill

Data on the pore pressure response to the placing of fill may not be available, in which case use $\bar{B} = 1$. Often, seepage will develop towards the toe of the slope. In compacted fill, the immediate usual response is a suction, but further placement will bring about positive pore pressures. If in doubt, use $\bar{B} = 1$.

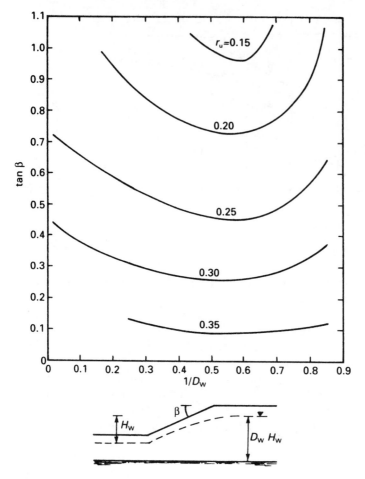

Fig. 9.26

9.13 Non-circular slips

The usefulness of Bishop's simplified method, which assumes a cylindrical slip surface, has led to several proposals along similar lines to deal with slip surfaces of any shape. Janbu (1954) was the first to publish such a method, although in an incorrect form. Morgenstern and Price (1965, 1967) suggested a similar method and Janbu (1973) finally published a corrected version of what is now accepted as the general routine method.

The form of analysis is similar to Bishop's, yielding the following expression for the factor of safety:

$$F = \frac{\sum [c'b + (W - ub + \Delta X) \tan \phi'] \dfrac{\sec^2 \alpha}{1 + \tan \alpha \tan \phi'/F}}{\sum (W \sin \alpha)} \qquad [9.24]$$

where $\Delta X = X_1 - X_2 =$ difference between vertical interslice forces

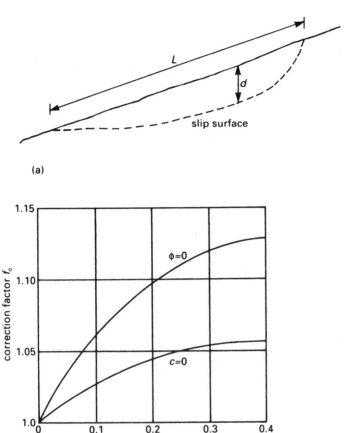

(a)

(b)

Fig. 9.27 Factor of safety correction for non-circular slips

and other symbols have the same meaning as in eqn [9.16]. As with Bishop's simplified method, a systematic iterative procedure is adopted which converges towards a final (minimum) value of F. It was further suggested by Janbu that this method tended to underestimate the factor of safety by up to 13 per cent, as the ratio of the depth to length of the slipping mass increases (Fig. 9.27(a)). Thus, using the chart shown in Fig. 9.27(b), a correction factor should be applied:

$$\text{Corrected} \quad F = f_o F \qquad\qquad [9.25]$$

In the case of very shallow slip surfaces, it would be prudent also to estimate a value for F using the *infinite slope method* (see Section 9.3):

$$\text{Hence} \quad F = (1 - \gamma_u \sec^2 \beta)\frac{\tan \phi'}{\tan \beta} \quad (\text{eqn } [9.11])$$

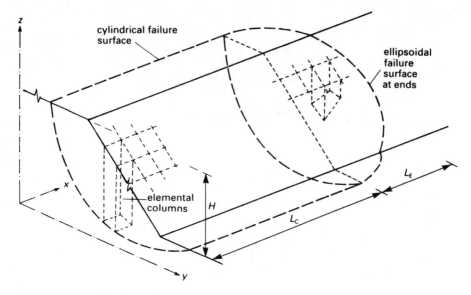

Fig. 9.28 Effect of end curvature

This latter solution is the one that the *circle* methods will tend towards, since a slip surface parallel to the ground surface is equivalent to a circle of infinitely large radius.

9.14   Effects of end curvature on narrow slides

The plane strain assumption common to routine limit equilibrium stability methods ignores the effects of lateral curvature at the ends of the slipping mass. In the majority of cases, the breadth of the central cylindrical portion greatly exceeds that of the end portions (Fig. 9.28). Also, the ratio of the central breadth to the slope height will usually be high. In all such cases the effect of the end portions is negligible.

However, in *narrow slides*, i.e. when $L_c/H < 4$ (Fig. 9.28), significant underestimates of factors of safety will result from two-dimensional analyses. These cases may arise in shafts, deep narrow excavations, sharply concave slopes and at concave corners. At convex corners, overestimates will result.

A number of three-dimensional methods have been proposed: Hovlund (1977), Humphrey and Dunne (1982), Chen and Chameau (1983), Hutchinson and Sarma (1985) and Hungr (1987). The general approach is to divide the slipping mass into a series of vertical columns for which a limiting equilibrium algorithm is developed. A summative iterative procedure similar to Bishop's method is then followed. In all cases, a computer-based analysis is used for parametric evaluations, generally involving 100–160 column elements.

While marginal disagreements are apparent between the different methods, it is clear that the three-dimensional factor of safety (F_3) is generally equal to or is higher than the equivalent plane strain value (F_2). The following typical values illustrate this point:

Ratio of F_3/F_2 for a 1 vertical: 2.5 horizontal slope

L_c/H		$c' = 0, \phi' = 40°$ $\gamma = 19 \ kN/m^3$		$c' = 30 \ kPa, \phi' = 15°$ $\gamma = 19 \ kN/m^3$	
	L_E/H	0.8	3.0	0.8	3.0
0.5		1.08	1.00	1.30	1.08
4.0		1.02	1.00	1.07	1.03

The figures suggest that only a very small underestimate occurs in $c' = 0$ cases, e.g. normally consolidated clays and sands, except for very narrow slides ($L_c/H < 0.5$). For slopes in overconsolidated clays using peak strength ($c' > 0$), some caution should be exercised in construing plane strain estimates of up to 40 per cent higher, since lower overestimates must be presumed when using residual parameters ($c' = 0$ and $\phi' = \phi'_r$). Higher factors of safety should be demanded at convex corners, since F_3 may be as little as $0.6F_2$.

9.15 Wedge methods of analysis

In slope stability problems when the configuration of the slip surface is fully or partially controlled by the presence of a layer of hard soil or rock, a form of *wedge analysis* will often be appropriate. The slope section is divided into convenient wedges over a postulated slip surface consisting of straight lines. The forces acting on each component wedge are then evaluated, starting at the top of the slope, using either an analytical or a graphical method.

Any number of triangular or quadrangular wedges may be used, but with more than five the procedures become tediously long. The precision of the factor of safety obtained depends (apart from data reliability) on the assumption concerning the direction of the interslice forces.

Figure 9.29 shows a typical wedge over a portion of slip surface AB. The forces acting on the wedge are:

W = weight of soil = $\gamma \times$ cross-sectional area
U = pore pressure force
   = average pore pressure $\times$ length AB = $\bar{u}L$
C = cohesive resistance force = $c'L/F$
R = resultant reaction due to friction along AB
N' = net normal force acting on AB
S = frictional shear resistance along AB
   = $N' \tan \phi'/F$
E_1 and E_2 = interslice forces which are inclined respectively at δ_1 and δ_2 to the interslice normal

Resolving forces normal to the slope:

$N' = W \cos \alpha - U + E_1 \sin(\delta_1 - \alpha) - E_2 \sin(\delta_2 - \alpha)$

This may be simplified by putting $\delta_2 = \alpha$. Resolving forces along the slope:

$E_2 \cos(\delta_2 - \alpha) - E_1 \cos(\delta_1 - \alpha) + N' \tan \phi'/F + c'L/F = 0$

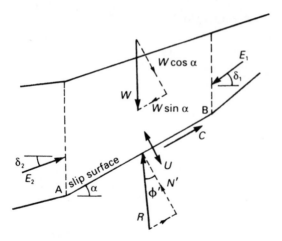

Fig. 9.29 Slope analysis using wedges

Thus, the lower interslice force may be evaluated approximately:

$$E_2 \cos(\delta_2 - \alpha) - E_2 \sin(\delta_2 - \alpha) \tan \phi'/F$$

$$= E_1 \cos(\delta_1 - \alpha) + W \sin \alpha - [W \cos \alpha - U + E_1 \sin(\delta_1 - \alpha)] \tan \phi'/F - c'L/F$$

$$[9.26]$$

If the wedges are dealt with successively from the top, the interslice forces are 'passed down' to the last wedge. Here the residual value of E_R represents the out-of-balance force and may be dealt with in one of two ways:

(a) If the slip surface is *fixed*, i.e. by a harder layer or similar, and the value of F was the limiting value, a retaining structure should be provided to support a lateral force equal to E_R.

(b) If the minimum factor of safety for a slope is being sought, the value of $\phi'_{mob.}$ may be adjusted as given below and the analysis repeated until the residual interslice force $E_R = 0$.

$$\Delta \tan \phi'_{mob.} = E_R/(\Sigma \, N' + \Sigma \, c'L)$$

For those used to the drawing board, a simple graphical method can be employed without serious error. A composite polygon of forces is drawn, starting at the top of the slope and proceeding to a scaled evaluation of the resultant E_R. After modifying $\phi'_{mob.}$ and F, a second polygon is drawn, and so on until $E_R = 0$.

Some care needs to be exercised over the choice of the interslice force angles (δ). The value of E_R is sensitive to δ; however, the effect on the final (minimum) value of F is less marked. It is usual to set δ equal to either ϕ' or α at the lower interface of each wedge. The following example illustrates the analytical procedure.

Worked example 9.12   *Figure 9.30 shows the cross-section through a slope in a cohesive soil which is underlain by a hard layer. A trial slip surface ABCD has been proposed, lying partly on the hard layer. Using a wedge analysis method, determine the factor of safety against shear slip. The properties of the upper soil are:*

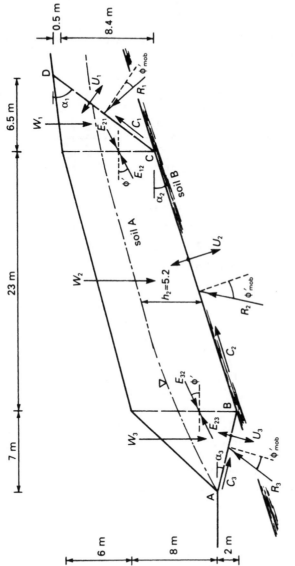

Fig. 9.30

$c' = 24 \ kPa \quad \phi' = 22° \quad \gamma = 19 \ kN/m^3$

Using the notation given in Fig. 9.30, the angles are first calculated:

$\alpha_1 = \arctan[(0.5 + 8.4)/6.5] = 53.9°$

$\alpha_2 = \arctan[(2.0 + 8.0 + 6.0 - 8.4)]/23.0 = 18.3°$

$\alpha_3 = \arctan(2/7) = -16.0°$

Let interslice inclination angle $\delta = \phi' = 22°$

Let $F = 1.5$: then $\tan \phi'_{mob.} = \tan 22°/1.5 = 0.269 \quad \therefore \quad \phi'_{mob.} = 15.06°$

The length and areas can be calculated or measured off a scaled cross-section, and then $E_2 \cos(\delta_2 - \alpha)$ evaluated using eqn [9.24].

The calculations are tabulated below:

Wedge	$E_1 \cos(\delta_1 - \alpha)$	$W \sin \alpha$	$W \cos \alpha - U + E_1 \sin(\delta_1 - \alpha)$	$c'L$	$E_2 \cos(\delta_2 - \alpha)$
1	0	444	324 − 165 + 0	264	225
2	265	1264	3817 − 1528 + 17	581	518
3	409	−183	639 − 193 + 320	175	−97
		$\Sigma = 2894 + 337$		$\Sigma = 1020$	

$\phi'_{mob.}$ correction: $\Delta \tan \phi'_{mob.} = E_R/(\Sigma \ N' + \Sigma \ c'L) = -97/4251 = -0.023$

Then let $\tan \phi'_{mob.} = 0.269 - 0.023 = 0.246$:

i.e. $\phi'_{mob.} = 13.8°$ and $F = \underline{1.64}$

Wedge	$E_1 \cos(\delta_1 - \alpha)$	$W \sin \alpha$	$W \cos \alpha - U + E_1 \sin(\delta_1 - \alpha)$	$c'L$	$E_2 \cos(\delta_2 - \alpha)$
1	0	444	324 − 165 + 0	264	244
2	286	1262	3817 − 1528 + 19	581	626
3	494	−183	639 − 193 + 386	175	0

Thus the factor of safety = 1.64, when $\delta = 22°$. The precision of the method does not really extend to the second place, so the best estimate of F is $\underline{1.6}$.

A similar analysis in which the interslice forces were assumed to be parallel to the slip planes, i.e. $\delta = \alpha$, gave a value of $F = \underline{1.72}$.

9.16 Factors in slope design

In approaching the design of a slope the engineer is required at the outset to consider a number of fundamental questions:

(a) Is it a cutting or a built embankment?

(b) What are the consequences of construction relating to pore pressures, effective stresses and volume changes?

(c) Which of these consequences are short-term and which are long-term?

(d) How many conditions change in the future?

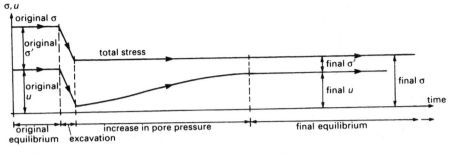

(a) cuttings

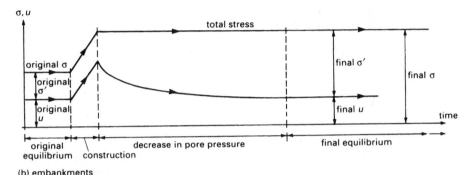

(b) embankments

Fig. 9.31 Changes in stresses during slope construction

The art and skill of engineering design is embodied in the ability to make sound decisions regarding questions such as these. It is worth while here making an examination of some of the factors that need to be considered.

Cuttings

Cuttings are excavated, whereas *embankments* are built. During excavation the removal of soil brings about a decrease in total stresses and a consequent decrease in pore pressures. After construction a seepage regime develops in the slope depending on the drainage conditions and the permeability of the soil. The result of this is an *increase* in pore pressure, while the total stress now remains constant: the effective stress therefore *decreases* (Fig. 9.31(a)). Since the shear strength of the soil is proportional to the effective normal stress, it too decreases. The long-term stability of a cutting is therefore more critical than its stability at the end of construction.

The nature of the slip surface depends on soil conditions. In homogeneous sands and sandy silts, translational slides are likely close to the surface. Desiccation cracks in the surface layer in silty or clayey soils may admit water quickly and precipitate movement; the provision of surface vegetation can prevent, or at least minimise, the risk. In homogeneous clays, failure is more likely to be deep seated along a slip surface which is circular in cross-section. In stratified deposits, several slip patterns need to be considered, especially where the presence of hard layers may tend to control the development of failure.

In heavily overconsolidated clays (e.g. London Clay, Keuper marl and some clay shales), the effect of excavation is to 'free' very high locked-in stresses. Rapid expansion results which may in turn draw in water, thus reducing the shear strength; also, considerable lateral strains may be imposed on supporting structures. In such cases, it is necessary to compute shear strength in terms of remoulded values (ϕ'_{rm}) and perhaps carry out a strain (displacement) analysis.

The provision of drains must be considered carefully, since this will have a marked effect on the allowable slope angle. If the factor of safety is 1.25, the allowable slope in a drained (or dry) slope is about $0.8\phi'$, while, for a water-logged slope, it will be about $0.4\phi'$. For permanent cuttings, stability calculations based on undrained strength should only be used as a preliminary guide, since only calculations based on effective stress can be considered to be sufficiently reliable (Chandler and Skempton, 1974). Furthermore, provision should be made for future possible adverse drainage changes.

Embankments

Embankments are built by rolling or otherwise compacting layers of selected soil in succession. Initially, the compaction process squeezes out air, but as the built-up height increases, the lower layers experience an increase in pore pressure (Fig. 9.31(b)). In coarse-grained soils, the excess pore pressures dissipate quickly.

In fine-grained soils, the excess pore pressure is slow to dissipate and consolidation may continue for several years. The installation of horizontal or vertical drainage blankets is used to speed up this process. In the course of time, the pore pressure decreases and the effective stresses and therefore the shear strength increase. Thus, the most critical stability condition for an embankment occurs at the end of construction, or sometimes during construction.

In the case of *earth dams*, a further decrease in shear strength may arise on the upstream face when the reservoir is emptied quickly and the supporting effect of the water is removed. This is known as *rapid draw-down instability* (Fig. 9.32). For earth dams *four* sets of stability conditions need to be examined:

(a)   end of construction: before filling – downstream face
(b)   end of construction: before filling – upstream face
(c)   after filling and development of steady seepage – downstream face
(d)   after rapid draw-down – upstream face

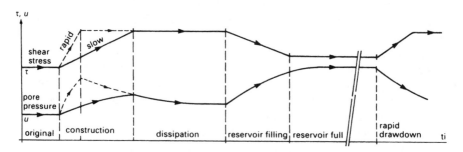

Fig. 9.32 Changes in effective stresses and pore pressure in an earth dam

A further point to consider in earth dams and those embankments expected to carry heavy seepage is the provision of toe drains (see also Section 5.13). The reduction of seepage pressures near the toe is often a pre-eminent condition of sound and economic design.

Where embankments are constructed over soil of nearly equal or lesser strength, the most critical failure mode may be a *foundation failure* (Fig. 9.6). When the foundation soil is much stronger, a *slope failure* will be more likely, with slip surface emerging at or near the toe of the slope.

9.17   Factors of safety

In deciding the minimum factor of safety for a particular problem a number of factors need to be considered:

(a) The consequences of the event that is being factored against, e.g. slip of an embankment or cutting.
(b) The numerical effect on the F value of variations in the parameters involved.
(c) The reliability of the measured or assumed values of the parameters involved.
(d) The economics of the problem.

The consequences of an embankment or cutting failure may simply be to cause some repair work to be done, or to block a road, in which case they are a matter of inconvenience and probably monetary expense. Where there is a threat to the safety of buildings and people, however, the matter is considerably more serious. In certain cases, the minimum F for a slope might be as low as 1.2, but where buildings and/or people may be at risk the minimum figure should be 2.0.

The numerical effect is apparent in soil problems since the shear strength equation has two components, one or both of which may be factored:

$$\tau_m = \frac{c'}{F} + \sigma'_n \tan \phi' \quad \text{or} \quad \tau_m = \frac{c'}{F_c} + \sigma'_n \frac{\tan \phi'}{F_\phi}$$

In some procedures the final factor of safety ensures that $F_c = F_\phi$, e.g. Bishop's effective stress method (see also worked example 9.10). The value of F may be very sensitive to changes in some parameters, but remain relatively insensitive to others. This is seen in the construction of Bishop and Morgenstern's charts, where, for example, ϕ' has a marked effect, but only three values of D need to be considered (see also Section 11.14).

It is axiomatic that the reliability of any calculation depends on the reliability of the problem parameter values themselves. Even with properly conducted and truly appropriate laboratory tests, measured values of the soil parameters will rarely be representatively accurate to better than ± 5 per cent. Where the samples are not adequately representative, or the test procedure doubtful, and particularly where *assumed* values are used for key parameters, the minimum factor of safety should be increased accordingly.

From the point of view of economics, each problem should be treated on its own merits and a minimum value of F decided on the basis of all the factors involved, but as a generalised guide the following values are suggested:

End of construction (embankments and cuttings):   1.30
Steady seepage condition: 1.25
After sudden drawdown: 1.20
Natural slope of long standing: 1.10–1.20
Spoil tip: 1.50
Problems involving buildings: 2.0

Exercises

9.1   A long slope in an overconsolidated clay is inclined at an angle of 12°. The ground-water level is lying at the slope surface, with steady seepage taking place parallel to the surface. Determine the factor of safety against plane translational shear slip on planes parallel to the surface at depths of 3 m, 4 m and 5 m respectively.

$c' = 14$ kPa $\phi' = 25°$ $\gamma = 20$ kN/m³

9.2   An extensive slope of a sandy soil is inclined at an angle of 17°; a hard imperme-able layer lies 4 m below and parallel to the slope surface. The properties of the sand are:

$c' = 0$ $\phi' = 32°$ $\gamma = 18$ kN/m³ $\gamma_{sat.} = 20$ kN/m³

Determine the factors of safety against movement and the depth of the slip surface they relate to in the following cases:

(a)   Negligible water in the slope.
(b)   Waterlogged slope, with parallel seepage.
(c)   Groundwater level parallel to surface at a depth of 2 m, with parallel seepage.

9.3   In a long-established slope inclined at 12° a slip movement has occurred at a depth of 5 m along a large plane surface parallel to the slope. Assuming that the ground was waterlogged (groundwater level at surface) and that parallel seepage was oc-curring, calculate the residual value of ϕ' (take $c' = 0$ and $\gamma = 20$ kN/m³).

9.4   Figure 9.33 shows the cross-section of a proposed cutting in a homogeneous clay soil having an undrained shear strength of 35 kPa and a bulk density of 19 kN/m³. Calculate the factor of safety against shear slip along the surface AB:

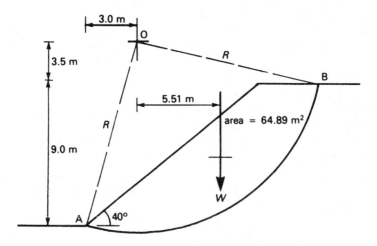

Fig. 9.33

(a) Ignoring the possibility of tension cracks.

(b) Allowing for a tension crack empty of water.

(c) Allowing for a tension crack full of water.

9.5 Figure 9.34 shows the cross-section of a cutting in a homogeneous cohesive soil in which a slip failure has taken place along the surface ACB, which may be taken as a circular arc. Assuming a unit weight for the soil of 18 kN/m³ and that $\tau_f = c_u$, and allowing for a tension crack at depth 2 m, obtain an estimate of the undrained shear strength of the soil at the time of failure.

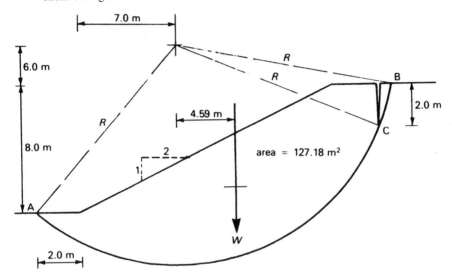

Fig. 9.34

9.6 A proposed cutting in a deep layer of a homogeneous fine soil is to be 10 m deep and inclined at 35°. Using the Fellenius–Jumikis method, obtain trial centres for the most critical slip circle passing 2 m in front of the toe of the slope. By successive trials obtain an estimate of the lowest factor of safety of this set of circles and state the location of its centre (horizontally and vertically from the toe).

$c_u = 40$ kPa $\quad \gamma = 18$ kN/m³

9.7 Figure 9.35 shows the cross-section of a proposed cutting in a fine soil. The undrained shear strengths (c_u) may be taken as 25 kPa in the upper layer and 42 kPa in the lower layer; the unit weight is 19 kN/m³ in both layers.

Calculate the factor of safety against a shear slip failure along the circular arc surface AB:

(a) Ignoring the possibility of tension cracks.

(b) Allowing for a tension crack empty of water.

(c) Allowing for a tension crack full of water.

9.8 A proposed cutting in a deep layer of homogeneous fine soil will be 9 m deep and have a slope of 25°. Using Taylor's stability numbers, determine the factor of safety against shear slip failure in respect of the following soil:

$c_u = 45$ kPa $\quad \gamma = 19$ kN/m³

9.9 On a site where a road cutting is to be made, a layer of homogeneous soil overlies a hard layer of shale. The properties of the upper layer are:

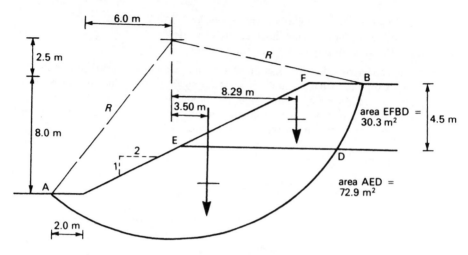

Fig. 9.35

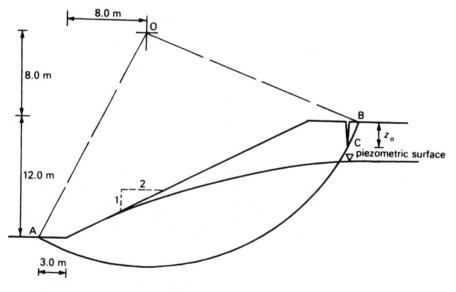

Fig. 9.36

$c_u = 30$ kPa $\gamma = 19.5$ kN/m³

Calculate the factor of safety against shear slip failure for the following sections of the cutting where the slope is 1 vertical: 1.5 horizontal:

(a) Depth to shale = 10 m   Depth of cutting = 8 m
(b) Depth to shale = 6 m   Depth of cutting = 6 m
(c) Depth to shale = 4 m   Depth of cutting = 8 m

9.10 Determine the safe depth to which a vertically-sided trench may be excavated without support in a clay soil having properties of:

$c_u = 35$ kPa $\gamma = 18$ kN/m³

Adopt a factor of safety of 2.0.

9.11 Using Fellenius' method of slices, determine the factor of safety against shear failure with respect to effective stress along the surface ABC in the slope shown in Fig. 9.36. The pore pressures along the slip surface may be estimated from the piezometric surface shown, assuming a condition of steady seepage. The soil properties are:

$c' = 12$ kPa $\phi' = 24°$ $\gamma = 19$ kN/m³

9.12 Using Bishop's simplified method of slices determine the factor of safety in terms of effective stress for the trial circle in the slope shown in Fig. 9.37. The pore pressure ratio (r_u) may be taken as 0.25 and the soil properties as:

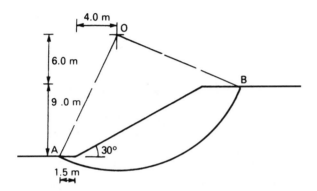

Fig. 9.37

$c' = 8$ kPa $\phi' = 25°$ $\gamma = 19$ kN/m³

9.13 A cutting of depth 15 m is to be formed in a fine soil with a slope of 1 vertical: 2.5 horizontal. The soil has a unit weight of 20 kPa and the following shear strength parameters:

$c_u = 90$ kPa

$c' = 10$ kPa $\phi' = 28°$

The average pore pressure ratio r_u may be taken as 0.40. Determine the factors of safety against: (a) immediate shear failure, and (b) long-term shear failure.

9.14 An earth embankment of height 24 m is to be constructed with soil which will have the following properties after placing and rolling:

$c' = 17$ kPa $\phi' = 26°$ $\gamma = 20$ kN/m³

(a) Determine the factor of safety of a 1 vertical:3 horizontal slope in the embankment if the average pore pressure ratio $r_u = 0.2$.
(b) Adopting a factor of safety of 1.5 and assuming a maximum r_u of 0.35, determine a suitable slope.

Soil Mechanics Spreadsheets and Reference Assignments and Quizzes (available on the accompanying CD):

Assignments A.9 Stability of slopes
Quiz Q.9   Stability of slopes

Chapter 10

Settlement and consolidation

10.1 Types of ground movement and causes of settlement

The term **settlement** refers to the vertical downward displacement at the base of a foundation or other structure due to ground movement. There are several mechanisms which may produce ground movement, and there are many types of structure, with varying potentials to withstand or to be distressed by movement. Brick and masonry buildings are brittle and may sustain cracks and even structural damage following very small foundation displacements; other structures may be constructed to sustain considerable movements without suffering real damage.

Also, soil conditions are apt to change, often considerably, from before, to during, and also after construction. Most building damage occurs when unforeseen soil conditions arise; inadequate site investigations and a lack of understanding of soil behaviour are largely to blame. Methods are available by which both the *amount* and the *rate* of foundation settlement can be estimated. These estimates will remain reasonably reliable providing that the assumed soil conditions represent the actual conditions, and are likely to persist throughout the life of the building.

Some mechanisms producing settlement can be modelled mathematically and offer theoretical approaches, e.g. elastic displacement, while others require empirical or experiential methods. It is useful to start a study of settlement by considering briefly a number of ground movement mechanisms which are potential *causes of settlement*.

Compaction

Compaction occurs when soil particles are forced into a closer state of packing with a reduction in volume and the expulsion of air (see also Chapter 3). Mechanical energy is required, which can be due to self-weight loading or a surface surcharge. Vibrations due to traffic movement, heavy machinery and certain construction operations, such as pile-driving, can also cause compaction settlement. In earthquake zones, seismic shock waves may have a similar effect. The most susceptible soils are loosely-packed sands or gravel-sands and fill material, particularly that which has been placed without adequate rolling or tamping.

Consolidation

In saturated cohesive soils the effect of loading is to squeeze out porewater; this process is called *consolidation*. A gradual reduction in volume occurs until internal pore pressure equilibrium is reached. Unloading results in swelling, providing the soil can remain saturated. A detailed study of the consolidation process and methods of assessing the resulting settlements follows in this chapter. The rate of consolidation depends on a soil's permeability and can be very slow in fine soils, so that it may take several years for the final settlement to be achieved.

The most susceptible soils are normally consolidated clays and silts and certain types of saturated fill. Peat and peaty soils can be highly compressible, resulting in changes in stratum thickness of as much as 20 per cent under quite modest loading.

Elastic or immediate settlement

In overconsolidated clays increases in effective stress which do not exceed the yield point (σ'_y) cause almost elastic compression. Beyond the yield point, non-linear (consolidation) settlement occurs. In heavily overconsolidated clays, since the yield point will be very high, settlement calculations can be based on elastic theory, using parameters referred to effective stresses (E', E'_0, v', C_s, etc.) (see also Sections 6.3 and 6.8).

Moisture movement

Some clay soils show a marked increase or decrease in volume as the water content is respectively increased or decreased; these are alternatively called *shrinkable clays* or *expansive clays*. Examples are found in parts of southern and eastern England, the main geological strata being the Lower Lias, Oxford Clay, Kimmeridge Clay, Gault Clay, London Clay and some in the Woolwich and Reading Beds (see also Section 1.5).

In the UK, the effects of seasonal variations in water content can extend down to about 0.8 m below ground surface. Annual surface movements in SE England as high as 50 mm may be expected. These clays characteristically possess high liquid limits and plasticity indices. Some so-called 'black cotton soils', occurring in poorly drained tropical conditions in parts of Australia, Central Africa and Central America, expand and shrink with water content changes to a very marked degree (Clarke 1957). The shrinkage/swelling potential of a soil is related to its clay content and plasticity (i.e. its activity – eqn [2.8]). The chart in Fig. 10.1 shows a classification of shrinkage/swelling potential based on the work of several authors, notably Holtz and Gibbs (1956), Holtz (1959), Williams and Pidgeon (1982) and Van de Merwe (1964), and also a similar classification given in BRE Digest 240 (1980). Brackley (1980) has proposed an expression by which the maximum movement due to swelling may be estimated:

$$\text{Swell potential (\%)} = 0.1(I_p - 10) \log(s/p')$$

where I_p = plasticity index (% units)
s = suction prior to construction (kPa)
p' = final bearing pressure (kPa)

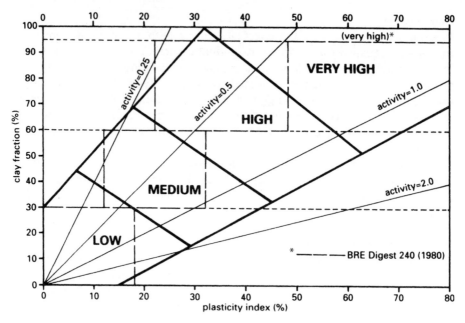

Fig. 10.1 Shrinking/swelling potential of clays

It has been suggested (BRE Digest No. 241 (1980)) that in the case of ordin-ary low-rise buildings, such as dwelling-houses, footings in these soils should be at a minimum depth of 0.9 m below the surface.

Effects of vegetation
Also associated with highly plastic clays is the draining effect of the roots of trees. The radial extent of some tree root systems is greater than the height of the tree; they may also reach depths of several metres. A movement of 100 mm was recorded in house foundations 25 m away from a row of poplars. The removal of such trees means that more water is held in the soil, and so swelling occurs. Where well-established trees and shrubs have to be removed from a site, a period of one or two winters should be allowed so that equilibrium may be achieved.

The planting of seedlings adjacent to buildings should be carefully controlled where shrinkable clays exist. A useful rule is always to site new plantings at a distance away from the building of at least 1.5 times the mature height of the tree. Pruning and pollarding can also affect the ground moisture loss and thus bring about swelling. The degree of desiccation (and therefore swelling potential) can be estimated by comparing water content and liquid limit readings taken near to existing (or recently removed) trees with readings taken in similar soil in open ground.

Effects of groundwater lowering
As water is pumped from an excavation, the water table in the surrounding ground may be lowered. Settlement can result from this change in hydrostatic conditions

due to two processes. Firstly, in some clays, a decrease in water content will result in a decrease in volume and the soil *above* the reduced groundwater level may therefore shrink. Secondly, a reduction in hydrostatic pore pressure results in an increase in the effective overburden stress; accordingly, the soil (especially in soft clays or peat) *below* the reduced groundwater level may consolidate.

Effects of temperature changes

Quite severe shrinkage can occur in clay soils as they dry out beneath foundations to furnaces, kilns, ovens and boilers. In one case, a boiler building on London Clay settled 150 mm at the centre and 75 mm at the sides in less than two years. It is usual to provide an open or rubble-filled air gap between such heat sources and the foundation soil.

In some soils, such as silts, fine sands, and chalk and chalky soils, there is a possibility of frost heave at sustained low temperatures. Water expands by approximately 9 per cent upon freezing and thus a 1 m layer of saturated soil having a porosity of 45 per cent would expand by $0.09 \times 0.45 \times 10^3 = 40$ mm. In the UK, the depth of frozen ground due to winter frost rarely exceeds 400 mm. Nevertheless, frost expansions of 30 per cent of the surface layer thickness have been observed (see also Section 4.8). Severe expansion can also take place in soils under cold storage buildings unless insulation is provided.

Effects of seepage and scouring

In certain sandy soils, such as fine dry sands and loess, the movement of water can move some of the fine particles. *Scouring* is the removal of material by surface water and streams, but this can also occur where sewers or water mains have been fractured. Where excavations are taken well below groundwater level within coffer dams and the like, the upward flow of water may cause a form of instability called *piping* (see also Section 5.15). In arid areas the same soils are liable to surface erosion due to wind action.

In certain rocks and soils the mineral cement in the matrix may be dissolved due to groundwater movement. The formation of caverns and swallets is a feature of limestone and chalk areas, and subsidence following the collapse of these is not uncommon. In Cheshire, extensive subsidence has resulted from the collapse of old rock-salt workings, brought about by the solution of the salt pillars left in as roof supports.

Loss of lateral support

Serious, even catastrophic, building failures can be associated with the excavation of deep holes alongside the foundation. Many cases have been recorded (Feld, 1967, 1968; Hammond, 1956; McKaig, 1962) in which adjacent excavation has resulted in a failure. The bearing capacity of the soil directly beneath a footing is dependent on the lateral support afforded it by the soil alongside (see also Chapter 11). If this lateral support is removed, as may occur in unsupported excavations, the likely outcome is a shear slip in the soil beneath the footing, taking the footing into excavation (Fig. 10.2(a)). Similarly, settlement might occur as a result of movement of natural earth slopes or cuttings, due to sliding or flowing (Fig. 10.2(b)).

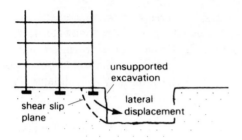

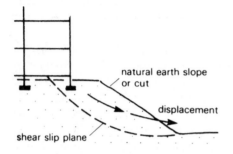

Fig. 10.2 Settlement due to loss of lateral support
(a) Resulting from an adjacent excavation (b) Resulting from slope movement

Effects of mining subsidence

The current method of mining coal is by the *longwall* method, in which the coal is continuously worked across a wide face. As the workings advance the space left is partly filled with waste material and the pit props removed. The unsupported roof then slowly drops, and with it the overburden up to the surface where settlement takes place. A settlement wave is apparent at the surface; the displacement lagging behind slightly, but following the direction of the advance of the coal face. It is possible to calculate the extent of this settlement and to predict the delay.

Protection must be afforded to buildings against the effects of subsidence according to the law and/or compensation paid. Methods of protection fall under two headings: (a) control of mining, e.g. leaving in pillars, cutting in strips, and (b) specialised construction, e.g. articulated structures, shallow raft footings.

10.2   *Undrained or immediate settlement*

The stress–strain behaviour of an undrained saturated mass of soil subject to surface loading is similar to that of an elastic solid body, providing that the strains are relatively low. This is likely in the case of foundations, where, with a factor of safety of three, actual stresses may be only one-third of ultimate values (whereas in slopes and excavations $F \approx 1.5$). Also, the highest values of stress occur immediately under the point of loading and diminish laterally and vertically.

The *immediate* or *elastic settlement* problem is modelled on a semi-infinite elastic half-space loaded at the surface. The standard expression was given in Section 6.11 (eqn [6.74]):

Immediate settlement, s_i or $\rho_i = \dfrac{qB(1-v^2)}{E_u}I_p$

For undrained conditions, $v = 0.5$; otherwise refer to Table 6.5. E_u can be obtained from undrained triaxial or oedometer tests. For homogeneous strata E_u may be considered to be constant, although it does increase with depth. For layer thicknesses over 3 m it is better to consider a series of sub-layers, with appropriate soil properties ascribed to each. Values for the influence factor I_p can be obtained from Tables 6.4 and 6.6.

Worked example 10.1 *A concrete raft foundation of length 32 m and breadth 18 m will transmit to the soil below a uniform contact pressure of 240 kPa at a depth of 2.0 m. Determine the amount of immediate settlement that is likely to occur under the centre of the foundation, assuming that it is flexible. $E_u = 45$ MPa, $v = 0.5$ and $\gamma = 20$ kN/m³*

$\dfrac{L}{B} = \dfrac{32}{18} = 1.78$

From Table 6.3 (interpolating): $I_p = 1.36 + (1.53 - 1.36)\dfrac{1.78 - 1.50}{2.00 - 1.50} = 1.46$

Net contact pressure, allowing for excavation, $q = 240 - 20 \times 2 = 200$ kPa

Immediate settlement, $s_i = \dfrac{200 \times 18(1 - 0.5^2) \times 146 \times 10^3}{45 \times 10^3} = \underline{88 \text{ mm}}$

Immediate settlement of a thin layer

The values for I_p given in Table 6.4 are based on a layer extending below the foundation to a considerable depth. However, if the layer thickness is less than twice the breadth an overestimation of settlement will result. For thin layers underlain by a hard stratum, the following expression may be used to determine the average settlement under a *flexible* foundation (Janbu *et al.*, 1956):

$$s_i = \dfrac{\mu_0\mu_1 qB(1 - v^2)}{E_u} \qquad\qquad [6.73]$$

Values of μ_0 and μ_1 are given in Fig. 6.58 and are related to the breadth and depth of the foundation and thickness of the layer below. In the case of a *softer* thin layer below the founding layer (Fig. 10.3), the immediate settlement may be

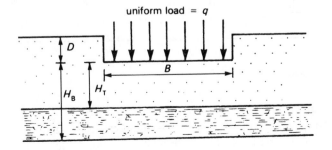

uniform load = q

Fig. 10.3

computed by obtaining first a value $\mu_{1(B)}$ corresponding to a layer thickness H_B and then a value $\mu_{1(T)}$ corresponding to a layer of thickness H_T. The immediate settlement due to the softer thin layer is then calculated using eqn [6.78], in which $\mu_1 = \mu_{1(B)} - \mu_{1(T)}$.

Worked example 10.2   *A foundation of dimensions 6 × 3 m is to transmit a uniform net contact pressure of 175 kPa at a depth of 1.5 m in a layer clay which extends to a depth of 5 m beneath the surface. Assuming $E_u = 40$ MPa and $v = 0.5$, determine the average amount of immediate settlement that is likely to occur.*

See Fig. 6.58: $\dfrac{D}{B} = \dfrac{1.5}{3.0} = 0.5$ $\dfrac{L}{B} = \dfrac{6}{3} = 2$ $\therefore \mu_0 = 0.9$

$\dfrac{H}{B} = \dfrac{3.5}{3.0} = 1.17$ $\dfrac{L}{B} = 2$ $\therefore \mu_1 = 0.55$

From eqn [6.78]: $s_i = \dfrac{0.9 \times 0.55 \times 175 \times 3 \times 10^3(1 - 0.5^2)}{40 \times 10^3} = \underline{4.9\ mm}$

Worked example 10.3   *A flexible foundation of dimensions 12 × 18 m is required to transmit a uniform contact pressure of 160 kPa at a depth of 1.7 m below the ground surface. A layer of sandy clay is located between 5 and 10 m below the ground surface for which $E_u = 35$ kPa, $\gamma = 19$ kN/m³ and $v = 0.3$. Determine the average amount of immediate settlement that is likely to occur due to the elastic compression of the sandy clay layer.*

Net contact pressure, $160 - 19 \times 1.7 = 128$ kPa

See Fig. 6.58: $\dfrac{D}{B} = \dfrac{1.7}{8.0} = 0.2$ $\dfrac{L}{B} = \dfrac{12}{8} = 1.5$ $\therefore \mu_0 = 0.97$

For layer between 0 and 10 m, $H_B = 10.0 - 1.7 = 8.3$ m

$\dfrac{H_B}{B} = \dfrac{8.3}{8.0} = 1.04$ $\dfrac{L}{B} = 1.5$ $\therefore \mu_{1(B)} = 0.49$

For layer between 0 and 5 m, $H_{(T)} = 5.0 - 1.7 = 3.3$ m

$\dfrac{H_T}{B} = \dfrac{3.3}{8.0} = 0.41$ $\dfrac{L}{B} = 1.5$ $\mu_{1(T)} = 0.27$

Then $\mu_1 = 0.49 - 0.27 = 0.22$

From eqn [6.78]: $s_i = \dfrac{0.97 \times 0.22 \times 128 \times 8 \times 10^3(1 - 0.3^2)}{35 \times 10^3} = \underline{5.7\ mm}$

10.3 Consolidation settlement

The process of consolidation

When a saturated mass of soil is loaded, say by a foundation, an immediate increase in pore pressure occurs and a hydraulic gradient is set up so that seepage flow takes place into surrounding soil. This excess pore pressure dissipates

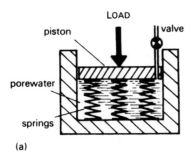

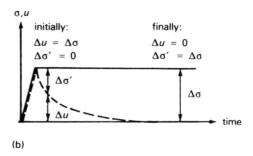

Fig. 10.4 One-dimensional consolidation
(a) Terzaghi's model (b) Stress–time curve

as water drains from the soil: very quickly in coarse soils (sands and gravels), and very slowly in fine soils (silts and clays) which have low permeabilities. As water leaves the soil a change in volume occurs, the rate gradually diminishing until steady state conditions are regained. The process is called **consolidation**.

In the natural process of deposition, fine-grained soils, such as silts and clays, consolidate slowly, as water between the particles is gradually squeezed out by the weight of the layers deposited above. After a period of time (perhaps many years) a state of equilibrium is reached and the compression ceases. A soil is said to be fully consolidated when its volume remains constant under a constant state of stress. A soil is said to be *normally consolidated* when it is currently in a state corresponding to its maximum consolidation pressure. A soil is *overconsolidated* when the present-day overburden pressure is less than the highest historic consolidation pressure; for example, soils consolidated under an ice sheet that has now retreated, or where some of the original overburden has been eroded away (see also Section 6.3).

In studying the mechanics of consolidation, it is necessary to consider the transient seepage of pore water during loading, since the change in soil volume (and settlement) is equal to the change in water volume. Terzaghi (1943) suggested the model shown in Fig. 10.4(a) to illustrate one-dimensional consolidation, with steel springs to represent the soil. It is assumed that the frictionless piston is supported by the springs and the cylinder filled with water. If a load is applied to the piston with the valve closed, the length of the springs remains unchanged since water is (assumed) incompressible (i.e. undrained conditions). If the load induces an increase in total stress of $\Delta\sigma$, then the whole of this must be taken up initially by an equal increase in porewater pressure Δu (Fig. 10.4(b)). When the valve is opened, the excess porewater pressure causes the water to flow out, the porewater pressure decreases and the piston sinks as the springs are compressed. Thus, the load is gradually transferred to the springs, causing them to shorten, until it is all carried by the springs. At the final stage, therefore, the increase in effective stress $\Delta\sigma'$ is equal to the increase in total stress, and the excess porewater pressure has been reduced to zero. The rate of compression obviously depends on the extent to which the valve is opened; this is analogous to the permeability of the soil.

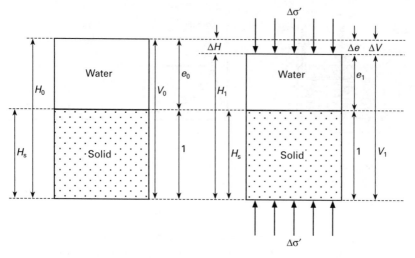

Fig. 10.5 Interpretation of compressibility using soil model
(a) Before loading (b) After loading

Amount of consolidation settlement

Consider a model soil sample subject to an increase in effective stress (Fig. 10.5). The process will be assumed to be *one-dimensional*, i.e. no change in lateral dimensions takes place, only a change in thickness. The change in volume (ΔV) which results from an increase in effective stress ($\Delta\sigma'$) may be represented by either the change in thickness (ΔH) or the change in void ratio (Δe) (see also Section 6.3). Equating volumetric strains:

$$\frac{\Delta V}{V_0} = \frac{\Delta H}{H_0} = \frac{\Delta e}{1 + e_0}$$

The change in thickness of a layer initially H_0 thick is, therefore:

$$\Delta H = \frac{\Delta e}{1 + e_0} H_0 \qquad\qquad [10.1]$$

The volumetric strain is also a function of the increase in stress, so that the *amount of consolidation settlement* may be obtained from:

$$s_c = \Delta H = m_v \Delta\sigma' H_0 \qquad\qquad [10.2]$$

where m_v = coefficient of volume compressibility, i.e. the change in unit volume per unit increase in effective stress (units will be the inverse of stress, i.e. m²/kN or m²/MN).

The one-dimensional consolidation test (oedometer test)

The compressibility characteristics of a soil relating both to the amount and rate of settlement are usually determined from the *consolidation test*, using an apparatus called an *oedometer* (Fig. 10.6). A soil specimen in the form of a disc

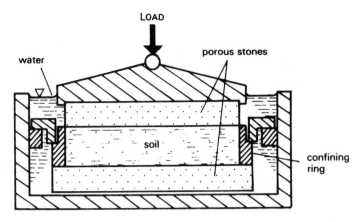

Fig. 10.6 The oedometer cell

(usually of diameter 75 mm and thickness 15–20 mm) is cut from an undisturbed sample. The specimen, enclosed in a metal ring, is sandwiched between two saturated porous stone discs, the upper one having a diameter slightly smaller than that of the metal ring, with the lower one slightly larger.

This assembly is placed in the cell and held in place by a clamping ring. A vertical static load is then applied through a lever system and the cell flooded with water. Changes in thickness of the sample are measured by means of a displacement dial gauge or transducer. Readings are continued until the specimen is fully consolidated: usually for a period of 24 or 48 hr. Further increments of load are then applied, each being double the previous increment, and the readings repeated. The number and value of the load increments will depend on the type of soil and on the range of stress anticipated on site. The pressure applied for the first stage should normally be equal to the *in situ* vertical stress at the depth from which the sample was obtained, except for soft and very soft clays, when a lower value must be used: 25, 12 or 6 kPa; or even 1 kPa for very soft organic clays and peats.

After full consolidation has been reached under the final load, the load is removed, either in one or in several stages, and the sample is allowed to swell. The swelling stage enables the specimen to stabilise before the final water content and thickness are determined; otherwise swelling might occur as the specimen is being removed from the oedometer, leading to error. If a detailed swelling curve is required, the unloading is carried out in stages and changes in thickness recorded.

First, the final void ratio after swelling is determined:

Method A

Final water content after swelling $= w_f$

Void ratio after swelling $= e_f = w_f G_s$ [10.3(a)]

(from eqn [3.11], since the soil is saturated, $S_r = 1.0$)

Method B

Dry mass after swelling $= M_d$

Final thickness after swelling period $= H_f$

Area of test sample $= A$

Equivalent thickness of solid particles $= H_s = M_d/AG_s\rho_w$

Void ratio after swelling $= e_f = \dfrac{H_f - H_s}{H_s} = \dfrac{H_f}{H_s} - 1$ [10.3(b)]

The void ratios at the start of each load increment stage may be obtained by working backwards from the final water content and thickness readings.

Void ratio at the end of a stage $= e_1$

Thicknesses at start and end of a stage $= H_0$ and H_1

Change in thickness $= \Delta H = H_0 - H_1$

From eqn [10.1]: Change in void ratio, $\Delta e = \dfrac{\Delta H}{H_1}(1 + e_1)$ [10.4]

Void ratio at start of a stage, $e_0 = e_1 + \Delta e$

Alternatively, $e_0 = \dfrac{H_0}{H_s} - 1$

Plotting compression curves
The data collected from the oedometer test enables changes in void ratio to be related to changes in effective stress. This is done by plotting a curve or graph of stage void ratios (e) against effective stresses (σ') (Fig. 10.7). The amount of consolidation settlement can then be obtained using the e/σ' curve.

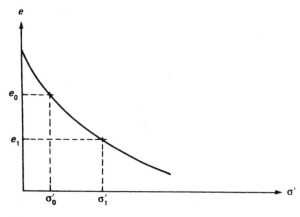

Fig. 10.7   Void ratio/effective stress curve

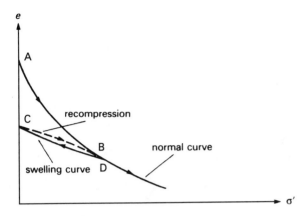

Fig. 10.8 Swelling and recompression

Suppose a stratum of clay of thickness H_0 is subject to a change in effective stress from σ'_0 to σ'_1. From the e/σ' curve the corresponding values of e_0 and e_1 are obtained and then, from eqn [10.1]:

$$\text{Consolidation settlement, } s_c = \Delta H = \frac{e_0 - e_1}{1 + e_0} H_0 \qquad [10.5]$$

The shape of the e/σ' curve depends on the consolidation history of the soil. If the soil was initially *normally consolidated* the compression path (AB) moves along the normal compression line (NCL) (Fig. 10.8). After unloading from point B it follows the swelling line (SRL) to C – reloading would cause recompression to D and further loading normal compression from D onward. (See Section 6.3 for details of compression and swelling.)

Determination of the compression index (C_c) (see also Section 6.2)
If the test data are plotted as e against $\log \sigma'$, the normal compression line is found to be mainly a straight line (Fig. 10.9). The initial curved part is representative of the preconsolidation; the greater the length of this initial curve, the greater the amount of *overconsolidation*. If a swelling and recompression sequence is carried out, the SRL should rejoin the straight NCL at the same point. The equations and parameters of the NCL and SRL are given in eqns [6.30(a) and (b)]. The *compression index* (C_c) is the slope of the NCL and is measured from the plot:

$$C_c = \frac{\Delta e}{\Delta \log \sigma'} = \frac{e_0 - e_1}{\log(\sigma'_1/\sigma'_0)} \qquad [10.6]$$

Thus, the change in void ratio for a given change in effective stress is:

$$e_0 - e_1 = C_c \log(\sigma'_1/\sigma'_0) \qquad [10.7]$$

The compression index for a given soil may be taken as a constant in settlement computations, providing the range of effective stress involved lies within the limits of the NCL: i.e. it is acting as a normally consolidated clay.

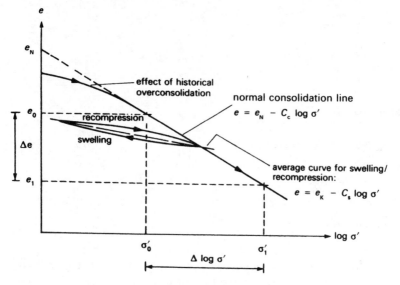

Fig. 10.9 e/log σ′ curve and compression index

Consolidation settlement, $s_s = \dfrac{C_c}{1+e_0}\log(\sigma_1'/\sigma_0')H_0$ [10.8]

Terzaghi and Peck (1967) demonstrated from laboratory results an approximate relationship between the compression index of normally consolidated clays and the liquid limit:

$C_c \approx 0.009(w_L - 10)$ [10.9]

where w_L = percentage liquid limit

The average slope of the swelling/recompression curve (SRL) is referred to as the *swelling index* (C_s); this can also be determined from the e/log σ′ plot similarly to C_c.

Determination of preconsolidation stress

The *pre-consolidation stress* (σ_{pc}') is the highest historical stress experienced by the soil. This may have been imposed during sample preparation (to simulate site conditions, etc.), or may be due to natural depositional loading with or without unloading (due to erosion, etc.). An estimate of the pre-consolidation stress can be obtained from the oedometer test results. Casagrande (1936) suggested an empirical graphical method using the e/log σ′ curve (Fig. 10.10). Firstly, point P is located at the point of maximum curvature between A and B, and then two lines are drawn passing through P: one is a tangent to the curve TPT and the other, PQ, is parallel to the stress axis. The point of intersection S of the bisector PR (of angle QPT) and the projection of the straight portion BC of the curve gives an approximate value for the preconsolidation stress (σ_{pc}').

The preconsolidation stress is not the yield stress (σ_y'), nor is it necessarily the maximum stress ever experienced by the soil, but it can be used as a useful guide

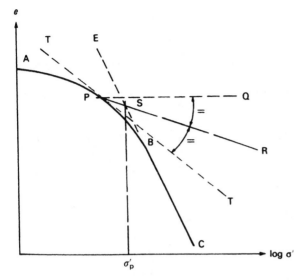

Fig. 10.10 Casagrande's method for determining preconsolidation stress

to limit settlement in overconsolidated clays, since the amount of consolidation settlement will not be great for effective stress changes less than σ'_{pc}.

Determination of the coefficient of volume compressibility
The *coefficient of volume compressibility* (m_v) has been defined in eqn [6.33] and represents the amount of change in unit volume due a unit increase in effective stress. The value of m_v is not constant for a given soil, but varies with the level of effective stress; oedometer test results can be used to obtain a range of values.

From eqn [10.2]: $m_v = \dfrac{\Delta H}{\Delta \sigma' H}$

But $\dfrac{\Delta H}{H} = \dfrac{\Delta e}{1 + e_0}$

Therefore $m_v = \dfrac{\Delta e}{\Delta \sigma'} \dfrac{1}{1 + e_0}$ [10.10]

where $\Delta e / \Delta \sigma'$ = slope of the e/σ' curve

It is quite usual to quote for a soil the value of $m_{v(100)}$, i.e. the value corresponding to $\sigma' = 100$ kPa. Also, m_v is the reciprocal of the confined modulus: $m_v = 1/E'_0$

Worked example 10.4   *The following readings were obtained from an oedometer test on a specimen of saturated clay. The load being held constant for 24 hr before the addition of the next increment.*

Applied stress (kPa)	0	25	50	100	200	400	800
Thickness (mm)	19.60	19.25	18.98	18.61	18.14	17.68	17.24

At the end of the last load period the load was removed and the sample allowed to expand for 24 hr, at the end of which time its thickness was 17.92 mm and its water content found to be 31.8 per cent. The specific gravity of the soil was 2.66.

(a) *Plot the e/σ' curve and determine the coefficient of volume compressibility (m_v) for an effective stress range of 220–360 kPa.*

(b) *Plot the $e/\log \sigma'$ curve and from it determine the compressibility index (C_c) and the preconsolidation pressure (σ'_{pc}).*

(c) *Plot a m_v/σ curve for the soil.*

(d) *Use the data obtained in (a), (b) and (c) to obtain and compare the values for consolidation settlement for 4 m thick layer of the clay when the average effective stress changes from 220–360 kPa.*

Firstly, determine the final void ratio.

Since $S_r = 1.0$ $\qquad\qquad e_1 = w_1 G_s$
$$= 0.318 \times 2.66 = 0.842$$

Change in void ratio, $\qquad \Delta e = \dfrac{\Delta h}{h_0}(1 + e_0)$

e.g. during swelling stage: $\quad \Delta e = \dfrac{0.68}{17.92}(1.842) = 0.070$

$\qquad$ during 400–800 stage: $\quad \Delta e = \dfrac{-0.44}{17.24}(1.772) = -0.045$

The rest of the results and calculations are tabulated below:

σ' (kPa)	$\Delta\sigma'$ (kPa)	h (mm)	Δh (mm)	Δe	e	$\log \sigma'$	$\dfrac{\Delta e}{\Delta\sigma'} \times 10^{-3}$	m_v (m²/MN)
0		19.60			1.014			
	25		−0.35	−0.036			1.440	0.715
25		19.25			0.978	1.40		
	25		−0.27	−0.028			1.120	0.566
50		18.98			0.950	1.70		
	50		−0.37	−0.038			1.760	0.390
100		18.61			0.912	2.00		
	100		−0.47	−0.048			0.480	0.251
200		18.14			0.864	2.30		
	200		−0.46	−0.047			0.235	0.126
400		17.68			0.817	2.60		
	400		−0.44	−0.045			0.113	0.062
800		17.24			0.772	2.90		
			0.68	0.070				
0		17.92			0.842			

(a) The e/σ' curve is shown plotted in Fig. 10.11.

$\qquad$ From the curve: $\quad$ for $\sigma'_0 = 220, \quad e = 0.858$
$\qquad\qquad\qquad\qquad\quad$ for $\sigma'_1 = 360, \quad e = 0.825$

$\qquad$ From eqn [10.8]:

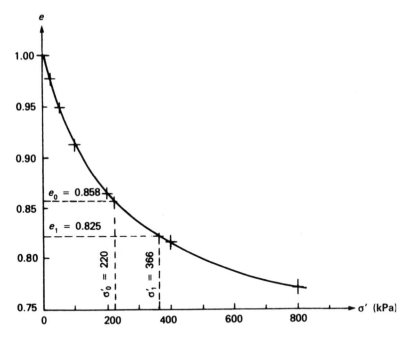

Fig. 10.11

$$m_v = \frac{\Delta e}{\Delta\sigma'}\frac{1}{1+e_0}$$

$$= \frac{(0.858 - 0.825) \times 10^3}{(360 - 220)1.858} = \underline{0.127 \text{ m}^2/\text{MN}}$$

(b) The $e/\log \sigma'$ curve is shown plotted in Fig. 10.12.

Compressibility index, C_c = slope of straight portion

$$= \frac{0.864 - 0.772}{\log 800 - \log 200} = \underline{0.153}$$

Using the Casagrande method, the preconsolidation stress $\sigma'_p = \underline{43 \text{ kPa}}$

(c) The column headed m_v is obtained from:

$$m_v = \frac{\Delta e}{\Delta\sigma'}\frac{1}{1+e_0}$$

The m_v/σ' curve is shown plotted in Fig. 10.13 with the values of m_v located at the end-point of each stress stage. (Alternatively, the mid-point values of m_v could have been calculated, corresponding to the median e for each stage.)

(d) From data (a): $s_c = m_v\Delta\sigma' H$

$$= 0.127 \times 10^{-3}(360 - 220) \times 4 \times 10^3 = \underline{71 \text{ mm}}$$

From data (b): $s_c = \dfrac{C_c}{1+e_0}\log(\sigma'_1/\sigma'_0)H_0$

$$= \frac{0.153}{1.858}\log(360/220) \times 4 \times 10^3 = \underline{70 \text{ mm}}$$

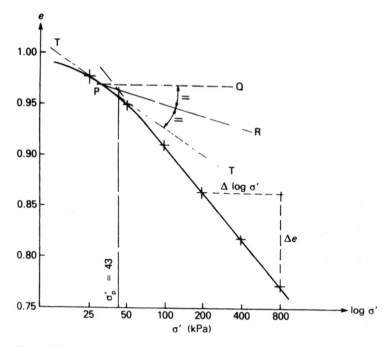

Fig. 10.12

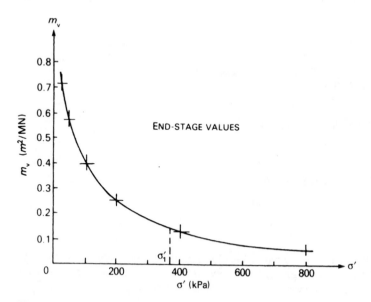

Fig. 10.13

From data (c):   Using Fig. 10.13, the value of corresponding to the end-stage stress
($\sigma'_1 = 360$ kPa) $m_v = 0.13$ m²/MN

Then $S_c = 0.13 \times 10^{-3}(360 - 220)4 \times 10^3 = \underline{73 \text{ mm}}$

10.4 *Rate of consolidation*

Immediate settlement due to elastic compression takes place instantaneously following an increase in stress. With consolidation, however, some time must elapse as water seeps from the soil and the excess porewater pressure is dissipated. The rate at which consolidation occurs depends mainly on the permeability of the soil and the length of drainage path. To a lesser extent it also depends on deformational creep in the soil particle skeleton and on the compressibility of soil constituents such as air/water vapour and organic matter. The soils which give rise to most settlement problems are normally consolidated clays and silts of recent origin often from alluvial or estuarine depositions.

The term *primary consolidation* refers to the volume change occurring up to the full dissipation of excess pore pressures, while *secondary compression* is that resulting from creep, slippage between particles, etc.

Terzaghi's theory of consolidation
In 1925, in his native Vienna, Terzaghi presented a theory, based on the model shown in Fig. 10.4, for the evaluation of *primary consolidation*. This was later incorporated in his *Theoretical Soil Mechanics* (1943). The following assumptions are made:

(a) The soil is fully saturated and homogeneous.
(b) Both the water and the soil particles are incompressible.
(c) Darcy's law of water flow applies.
(d) The change in volume is one-dimensional in the direction of the applied stress.
(e) The coefficients of permeability and volume compressibility remain constant.
(f) The change in volume corresponds to the change in void ratio and $\partial e / \partial \sigma'$ remains constant.

Figure 10.14 shows a layer of clay which is subject to a sudden increase in total vertical stress ($\Delta\sigma$), which is distributed uniformly over a semi-infinite area. At the instant of loading (time = 0), the porewater pressure in the layer will increase by Δu_0 (where $\Delta u_0 = \Delta\sigma$) uniformly over the thickness H of the layer: indicated by the diagram *abcd* in Fig. 10.14(b). After time t has elapsed, drainage into the sand layers above and below will have lowered the excess porewater pressure to be reduced to the profile shown by the *unshaded* portion of *abcd*. Note that the excess pore pressure at the drainage faces is now zero, rising to a maximum at the centre of the layer.

Now consider a soil element within the clay stratum of thickness dz and lateral dimension dx and dy. At time t, the fall in porewater pressure across the element is dh (shown by the hypothetical standpipes in Fig. 10.14(a)). The prismatic element is shown enlarged in Fig. 10.15 with the one-dimensional flow-in and flow-out conditions.

The hydraulic gradient, $\quad i_z = \dfrac{\partial h}{\partial z} = \dfrac{1}{\gamma_w} \dfrac{\partial \bar{u}}{\partial z}$

where $\qquad\qquad\qquad \bar{u}$ = excess pore pressure

and $\qquad\qquad\qquad A$ = area of flow path = dxdy

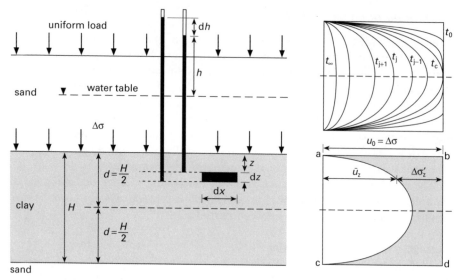

(a) Sectional elevation

(b) Excess pore pressure distribution

Fig. 10.14 Distribution of excess pore pressure in a clay layer subject to a uniform increase
in loading

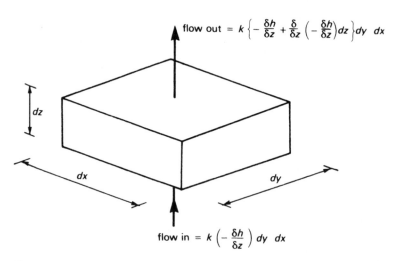

Fig. 10.15 One-dimensional flow through a prismatic element

Then, according to Darcy's law:

Flow in: $q_{z(\text{in})} = kiA = -\dfrac{k}{\gamma_w} \dfrac{\partial \bar{u}}{\partial z} \mathrm{d}x \mathrm{d}y$

Flow out: $q_{z(\text{out})} = -\dfrac{k}{\gamma_w} \dfrac{\partial \bar{u}}{\partial z} \mathrm{d}x \mathrm{d}y + \dfrac{k}{\gamma_w} \dfrac{\partial}{\partial z}\left(\dfrac{\partial \bar{u}}{\partial z}\right) \mathrm{d}x \mathrm{d}y \mathrm{d}z$

The net rate of flow out of the element is therefore:

$$-\frac{k}{\gamma_w}\frac{\partial}{\partial z}\left(\frac{\partial \bar{u}}{\partial z}\right)dxdydz \qquad [10.11(a)]$$

For continuity, this must equal the rate of change of volume (V), i.e. dV/dt. Also, volume change is related to effective stress: $\Delta V = m_v\,\Delta\sigma'\,dydxdz$.

As the excess pore pressure dissipates the effective stress on the soil fabric increases, so that $\partial\sigma' = \partial u$.

Then
$$\frac{dV}{dt} = -m_v\frac{\partial \bar{u}}{\partial t}dxdydz \qquad [10.11(b)]$$

Equating eqns [10.11(a)] and (b)]: $m_v\dfrac{\partial \bar{u}}{\partial t} = \dfrac{k}{\gamma_w}\dfrac{\partial^2 \bar{u}}{\partial z^2}$

and defining the *coefficient of consolidation, c_v* $= \dfrac{k}{m_v\gamma_w}$ $\qquad [10.12]$

The units of c_v will be length^2/time, e.g. m^2/year, mm^2/min, etc.

Thus the differential equation for one-dimensional consolidation is:

$$c_v\frac{\partial^2 \bar{u}}{\partial z^2} = \frac{\partial \bar{u}}{\partial t} \qquad [10.13]$$

Consolidation equation solution and isochrones

Solutions to eqn [10.13] give a set of curves called *isochrones* (from the Greek: *isos* = equal and *khronos* = time) representing the distribution of excess pore pressure ($\bar{u}$) with depth (z) and time (t) (Fig. 10.14(b)). In order to obtain a solution, the following non-dimensional factors are substituted:

Degree of consolidation, $\quad U_z = \dfrac{e_0 - e_t}{e_0 - e_1} = \dfrac{\bar{u}_0 - \bar{u}_t}{\bar{u}_0} = \dfrac{s_t}{s_c}$ $\qquad [10.14]$

Time factor, $\qquad T_v = \dfrac{c_v t}{d^2}$ $\qquad [10.15]$

Drainage path ratio, $\qquad Z = \dfrac{z}{d}$

where e_0 = initial void ratio, e_1 = final void ratio, e_t = void ratio after time t
$\bar{u}_0$ = initial excess pore pressure, $\bar{u}_t$ = excess pore pressure after time t
s_c = final consolidation settlement, s_t = consolidation settlement after time t
d = length of drainage path

Substituting, eqn [10.13] now becomes:

$$\frac{\partial^2 U_z}{\partial Z^2} = \frac{\partial U_z}{\partial T_v} \qquad [10.16]$$

The differential equation will now yield curves of degree of consolidation (U_z) against time factor (T_v). Two analytical approaches are commonly used: (a) Taylor's Fourier series method and (b) the method of parabolic isochrones.

(a)   Taylor's Fourier series method

The precise shape of the isochrones depends on the initial distribution of excess pore pressure and the drainage conditions at the upper and lower boundaries of the layer. The most common set of conditions is shown in Fig. 10.14, i.e. with the two-way one-dimensional drainage and the distribution of initial excess pore pressure uniform across the layer; and with the total stress remaining constant with time. The same conditions apply in the case of the oedometer test. The boundary conditions will be:

When $T_v = 0$ ($t = 0$), $\bar{u}_z = \bar{u}_0 = \Delta\sigma$,

When $T_v = 0$ to ∞ and $z = d$, $\bar{u} = 0$

When $T_v = 0$ to ∞ and $z = 0$, $0 < \bar{u}_t < \bar{u}_0$

When $T_v = \infty$, $\bar{u} = 0$

The degree of consolidation is then given by

$$U_z = 1 - \sum_{m=0}^{m=\infty} \frac{2}{M}\sin(MZ)\,\exp^{-2M^2T_v} \qquad [10.17]$$

where $M = \dfrac{\pi}{2}(2m + 1)$

and $m = 0, 1, 2, \ldots$ etc.

This is the degree of consolidation corresponding to the particular depth z. The *average* degree of consolidation in the layer is given by:

$$\bar{U} = 1 - \sum_{m=0}^{m=\infty} \frac{2}{M^2}\exp^{M^2T_v} \qquad [10.18]$$

where $\bar{U} = \dfrac{S_t}{S_c}$

S_t = consolidation settlement at time t corresponding to a particular value of T_v

S_c = final consolidation settlement

For values of $\bar{U} < 0.6$, $\bar{U} = \sqrt{\left(\dfrac{4T_v}{\pi}\right)}$ $\qquad [10.19]$

Effect of drainage and initial stress conditions

Equations [10.17, 10.18 and 10.19] apply to a layer that may drain freely from both its upper and lower surfaces; this is said to be an *open layer* and the length of the drainage path d is equal to half the thickness (Fig. 10.16). If free drainage can only take place at one boundary, the layer is said to be *half-closed*, and the length of the drainage path equals the thickness.

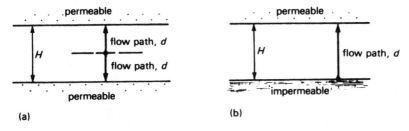

Fig. 10.16   Open and half-closed layers
  (a) Open layer: $d = H/2$ (b) Half-closed layer: $d = H$

An impermeable boundary may occur in the form of a concrete foundation (although some concrete may be permeable) or another layer of clay or rock with very low permeability. Permeable boundaries occur in the form of cohesionless soils, pervious rocks (e.g. sandstone), foundations underlain by hardcore, layers of peat, etc.

A number of initial stress distributions may arise in practice, according to the type of construction situation. Those that give (or approximate to) simple linear distributions are considered as basic types. Figure 10.17 shows the basic types in both open and half-closed layers. More complex distributions will occur, but may be treated as intermediate situations between the basic types. For example, the distribution within a new embankment carrying a surface surcharge will be approximately trapezoidal, i.e. between uniform and increasing with depth. In order to cover most practical situations, only three solution cases need to be considered for eqn [10.18]; these are:

Case 0   Applies to all linear distributions in open layers and to uniform distributions in half-closed layers as shown in Figs. 10.17(a), (b), (c) and (d).

$$\bar{U}_0 = 1 - \frac{8}{\pi^2}\left(\exp^{-(\pi^2/4)T_v} + \frac{1}{9}\exp^{-(9\pi^2/4)T_v} + \frac{1}{25}\exp^{-(25\pi^2/4)T_v} + \ldots\right) \qquad [10.20]$$

Case 1   Applies to a stress distribution in a half-closed layer which increases linearly with depth as shown in Fig. 10.17(e).

$$\bar{U}_1 = 1 - \frac{32}{\pi^3}\left(\exp^{-(\pi^2/4)T_v} + \frac{1}{27}\exp^{-(9\pi^2/4)T_v} + \frac{1}{125}\exp^{-(25\pi^2/4)T_v} + \ldots\right) \qquad [10.21]$$

Case 2   Applies to a stress distribution in a half-closed layer which decreases linearly with depth as shown in Fig. 10.17(e).

$$\bar{U}_2 = 1 - \frac{16}{\pi^3}\left((\pi - 2)\exp^{-(\pi^2/4)T_v} + \frac{1}{27}(3\pi - 2)\exp^{-(9\pi^2/4)T_v}\right.$$

$$\left. + \frac{1}{125}(5\pi - 2)\exp^{-(25\pi^2/4)T_v} + \ldots\right) \qquad [10.22]$$

It can also be shown that

$$\bar{U}_2 = 2\bar{U}_0 - \bar{U}_1 \qquad [10.23]$$

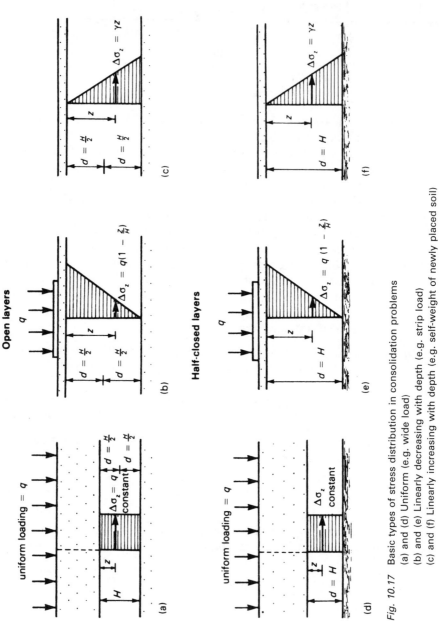

Fig. 10.17 Basic types of stress distribution in consolidation problems
(a) and (d) Uniform (e.g. wide load)
(b) and (e) Linearly decreasing with depth (e.g. strip load)
(c) and (f) Linearly increasing with depth (e.g. self-weight of newly placed soil)

Table 10.1   Values of time factor T$_v$

Average degree of consolidation	Time factor T$_v$		
	Case 0	Case 1	Case 2
$\bar{U} = \dfrac{s_t}{s_f}$	and all linear distributions in *open* layers		
0.1	0.008	0.047	0.003
0.2	0.031	0.100	0.009
0.3	0.071	0.158	0.024
0.4	0.126	0.221	0.048
0.5	0.197	0.294	0.092
0.6	0.287	0.383	0.160
0.7	0.403	0.500	0.271
0.8	0.567	0.665	0.440
0.9	0.848	0.940	0.720
1.0	∞	∞	∞
When $T_v = 2.0$, $\bar{U} = 0.994$		0.993	0.996

To enable calculations to be done simply, values of T_v corresponding to a range of values of $\bar{U}$ have been calculated for these three cases, and these are given in Table 10.1.

Trapezoidal cases
Values of T_v for trapezoidal cases can be obtained from the following expressions:

Cases 0–1: $T_v = T_{v(0)} + (T_{v(1)} - T_{v(0)})I_{01}$ [10.24]

Cases 0–2: $T_v = T_{v(0)} + (T_{v(2)} - T_{v(0)})I_{02}$ [10.25]

where the values of $T_{v(0)}$, $T_{v(1)}$ and $T_{v(2)}$ correspond to cases 0, 1 and 2, respectively; and values of coefficients I_{01} and I_{02} are obtained from Table 10.2.

(b)   The method of parabolic isochrones
In Taylor's method the isochrones are modelled as exponential curves; an alternative is to model them as parabolic curves. This provides a simple solution, accurate enough for most purposes, for cases where the initial distribution of excess pore pressure is uniform throughout the layer. Two expressions must be evaluated, one each for elapsed times on either side of time t_c, which is the time at which the excess pore pressure *first* begins to fall at the centre of the layer with two-way drainage or at the impermeable layer boundary with one-way drainage (Fig. 10.18).

Table 10.2 Coefficients to find T_v in trapezoidal cases

Cases 0–1		Cases 0–2	
$\dfrac{\Delta\sigma_o}{\Delta\sigma_d}$	I_{01}	$\dfrac{\Delta\sigma_o}{\Delta\sigma_d}$	I_{02}
0	1.0	1.0	1.0
0.1	0.84	1.5	0.83
0.2	0.69	2.0	0.71
0.3	0.56	2.5	0.62
0.4	0.46	3.0	0.55
0.5	0.36	3.5	0.50
0.6	0.27	4	0.45
0.7	0.19	5	0.39
0.8	0.12	7	0.30
0.9	0.06	10	0.23
1.0	0	20	0.13

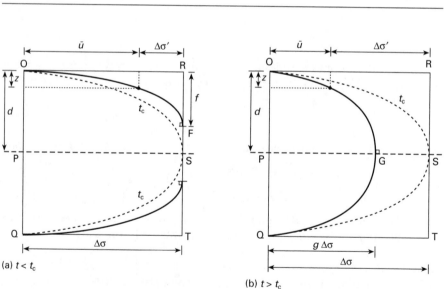

Fig. 10.18 Parabolic isochrones

When $t < t_c$

Figure 10.18(a) shows a parabolic isochrone for time $t < t_c$. Considering the half-layer area OPRS: at the instant of loading the excess pore pressure is uniform throughout ($\bar{u} = \Delta\sigma$), but as seepage commences it immediately drops to zero at the drainage interface OR. After a given time t, the amount of settlement is given by $m_v \times$ the area swept so far (i.e. ORF).

Therefore the settlement at $t < t_c$, $s_t = \frac{1}{3} m_v f \Delta\sigma$ [10.26]

Differentiating with respect to t, the rate of settlement is:

$$\frac{ds_t}{dt} = \frac{1}{3} m_v \Delta\sigma \frac{df}{dt} \quad\quad\quad [10.27]$$

The equations describing the parabolic isochrone are;

$$\bar{u} = az^2 + bz \quad \text{and} \quad \frac{d\bar{u}}{dz} = 2az + b$$

but when $z = f$, $\dfrac{d\bar{u}}{dz} = 0$, so $b = -2af$

Also when $z = f$, $\bar{u} = \Delta\sigma = af^2 + bf = af^2 - 2af^2$, so $a = -\Delta\sigma/f^2$

giving $\bar{u} = \left(2\dfrac{z}{f} - \dfrac{z^2}{f^2}\right)\Delta\sigma$ [10.28]

and $\dfrac{d\bar{u}}{dz} = 2az - 2af = \dfrac{2\Delta\sigma}{f^2}(f - z)$ [10.29]

The gradient of the isochrone at O ($z = 0$) is: $\dfrac{d\bar{u}}{dz} = \dfrac{2\Delta\sigma}{f}$ [10.30]

Now the rate of settlement is equal to the rate of seepage and (invoking Darcy's law) $v = ki$. Then

$$\frac{ds_t}{dt} = ki_t = \frac{1}{\gamma_w} \frac{d\bar{u}_t}{dz}$$

Substituting $\dfrac{ds_t}{dt} = \dfrac{k}{\gamma_w} \dfrac{2\Delta\sigma}{f}$ [10.31]

Equating [10.27] and [10.31]:

$$\frac{1}{3} m_v \Delta\sigma \frac{df}{dt} \gamma = \frac{k}{\gamma_w} \frac{2\Delta\sigma}{f}$$

giving $f\dfrac{df}{dt} = 6\dfrac{k}{m_v \gamma_w} = 6c_v$

Integrating ($f = 0$ and $t = 0$): $f^2 = 12c_v t$

or $f = \sqrt{(12c_v t)} = \sqrt{(12c_v d^2)}$ [10.32]

Then the settlement at time $t(<t_c)$, $s_t = \frac{1}{3} m_v \Delta\sigma \sqrt{(12c_v t)}$ [10.33]

However, the final settlement of the half-layer ($z = d$) is given by $s_c = m_v\Delta\sigma d$. Therefore, the degree of consolidation is

$$U_t = \frac{s_t}{s_c} = \frac{\sqrt{(12c_v t)}}{3d} = \sqrt{\left(\frac{4c_v t}{3d^2}\right)} = \sqrt{\left(\frac{4T_v}{3}\right)} \qquad [10.34]$$

The boundaries for this expression are $0 < t < t_c$.

At $t = t_c$, $f = d = \sqrt{(12c_v t)} = \sqrt{(12T_v d^2)}$

and therefore $T_v = \frac{1}{12}$ and $U_t = \frac{1}{3}$

When $t > t_c$

In Figure 10.18(b) an isochrone is shown for $t > t_c$. The excess pore pressure at the impermeable interface has now fallen below the original $\Delta\sigma$ and has a maximum value of $g\Delta\sigma$. Following similar reasoning to that given above for $t < t_c$:

Settlement at time t, $s_t = m_v \times$ the swept area ORSG

$$= m_v(\Delta\sigma - \tfrac{2}{3}g\Delta\sigma)d$$

$$= m_v(1 - \tfrac{2}{3}g)\Delta\sigma d \qquad [10.35]$$

Rate of settlement, $\dfrac{ds_t}{dt} = -\dfrac{2}{3}m_v\Delta\sigma d\dfrac{dg}{dt}$ $\qquad [10.36]$

Also, $\dfrac{ds_t}{dt} = \dfrac{k}{\gamma_w}\dfrac{2g\Delta\sigma}{d}$ $\qquad [10.37]$

Equating equations [10.36] and [10.37]:

$$\frac{1}{g}\frac{dg}{dt} = -\frac{3k}{m_v\gamma_w d^2} = -\frac{3c_v}{d^2} = -\frac{3T_v}{t} \qquad [10.38]$$

Integrating between $g = 1$ ($T_v = \frac{1}{12}$) and $g = 0$ ($T_v = \infty$) gives

$$g = \exp(0.25 - 3T_v) = \exp\left(0.25 - \frac{3c_v t}{d^2}\right) \qquad [10.39]$$

The settlement at time t ($>t_c$), $s_t = m_v[1 - \tfrac{2}{3}\exp(0.25 - 3T_v)]\Delta\sigma d$ $\qquad [10.40]$

and the degree of consolidation, $U_t = \dfrac{s_t}{s_c} = 1 - \tfrac{2}{3}\exp(0.25 - 3T_v)$ $\qquad [10.41]$

Since d is the drainage path length the (parabolic isochrone) expressions given above for both $t < t_c$ and $t > t_c$ can be applied to open layers ($d = H/2$) and half-closed layers ($d = H$), but are not recommended for non-uniform initial excess pore pressure distributions.

Calculation of pore pressures

From the foregoing analyses expressions can be derived from which excess pore pressures at given depths at given times can be calculated:

When $t < t_c$ (i.e. when $0 < T_v < \frac{1}{12}$ or $0 < U_t < \frac{1}{3}$)

$$\bar{u}_t = \left(2\frac{z}{f} - \frac{z^2}{f^2}\right)\Delta\sigma \qquad\qquad [10.42]$$

where $f = \sqrt{(12c_v t)} = \sqrt{(12T_v d^2)}$

When $t > t_c$ (i.e. when $\frac{1}{12} < T_v < \infty$ or $\frac{1}{3} < U_t < 1$)

$$\bar{u}_t = \left(2\frac{z}{d} - \frac{z^2}{d^2}\right)g\Delta\sigma \qquad\qquad [10.43]$$

where $g = \exp(0.25 - 3T_v) = \exp\left(0.25 - \frac{3c_v t}{d^2}\right)$

10.5 Determination of c_v from oedometer test

The results of a single stage of the oedometer test can be used to obtain a value for the coefficient of consolidation c_v. The test must be continued to almost complete consolidation ($U_t > 0.99$) so that degrees of consolidation and time factors can be related. Then the curve of laboratory results can be fitted against the theoretical curve of U_t/T_v. Curve fitting can be accomplished using either a square-root-of-time method or a log-time method.

Square root of time method (Taylor's method)
Equations [10.18] and [10.19] can be used as the theoretical basis, or alternatively the parabolic isochrone equations [10.34] and [10.41] can be used. If U_t is plotted against $\sqrt{T_v}$, or settlement (s_t) against $\sqrt{\text{time}(t)}$, the first portion ($0 < U_t < 0.6$) of the each curve is approximately a straight line. The theoretical and laboratory curves can be fitted to each other providing two sets of congruent coordinates are known – the first set is obviously (0,0).

Figure 10.19 shows the dimensionless theoretical curve based on eqn [10.19]:

$$\bar{U} = \sqrt{\left(\frac{4T_v}{\pi}\right)} \quad \text{alternatively, eqn [10.34] could be used: } U_t = \sqrt{\left(\frac{4T_v}{3}\right)}$$

If the laboratory curve flattens sufficiently so that final consolidation ($U_t = 1$) is achieved the curves can be matched. However, in most tests this flattening is not achieved and the settlement corresponding to $U_t = 1$ remains unknown, so another point must be used. Points can be located easily up to $U_t \approx 0.5$, but for greater accuracy the point representing $U_t = 0.9$ is commonly used. Point C on the curve corresponds to $U_t = 0.9$ and is located as follows:

In Fig. 10.19: $AB = 0.9 \times \sqrt{\left(\frac{\pi}{4}\right)} = 0.7976$

From Table 10.1, $AC = \sqrt{0.848} = 0.9209$

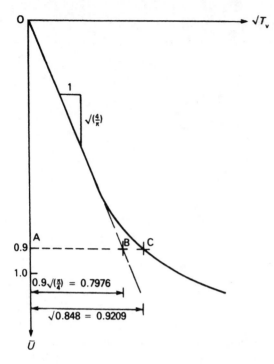

Fig. 10.19 Theoretical primary consolidation curve

Line OC therefore has abscissae greater than line OB by

$$\frac{0.9209}{0.7976} = 1.155$$

The laboratory readings of thickness are plotted against $\sqrt{\text{time}(t)}$ (Fig. 10.20) and the best straight line drawn through the first part of the curve. Line OC is drawn with abscissae 1.155 greater, point C being on the laboratory curve at $U_t = 0.9$ and giving an intercept on the time axis equal to $\sqrt{t_{90}}$, which corresponds to $\sqrt{T_{v90}}$ on the theoretical curve.

Now, by definition, $T_v = \dfrac{c_v t}{d^2}$ and so $T_{v90} = \dfrac{c_v t_{90}}{d^2}$

giving $c_v = \dfrac{T_{v90} d^2}{t_{90}}$ [10.44]

Compression ratios

The laboratory curve consists of three parts:

(a) *Initial compression.* Indicated by a short curved line, which mainly results from the compression of any air that is present, or slight particle reorientation; it is assumed that air, and not water, is expelled during this stage. Correct for zero after this stage.

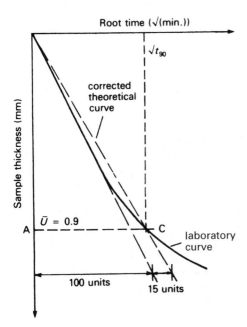

Fig. 10.20   Square root of time method

(b) *Primary compression.* Commences as soon as seepage begins and is represented by the portion of the curve from $\bar{U} = 0$ to approximately $\bar{U} = 1.0$ (although some secondary compression is occurring here as well).

(c) *Secondary compression.* Indicated by the final curved portion of the curve and is the compression that continues to take place after the excess porewater pressure has reached zero; is thought to be due to delayed reorientation of the particles due to the high viscosity of the adsorbed water layers.

The true zero for the primary compression is obtained by projecting the straight portion of the curve back to the thickness axis at zero time, i.e. point F (Fig. 10.21); this point represents $\bar{U} = 0$. The following ratios are sometimes used to show the relative amounts of initial, primary and secondary compression:

On the thickness axis: OF = amount of initial compression
FD = amount of primary compression = FA/0.9
DG = amount of secondary compression
OG = total amount of compression

Initial compression ratio, $r_i = \dfrac{OF}{OG}$ [10.45]

Primary compression ratio, $r_p = \dfrac{FD}{OG}$ [10.46]

Secondary compression ratio, $r_s = \dfrac{DG}{OG} = 1 - r_i - r_p$ [10.47]

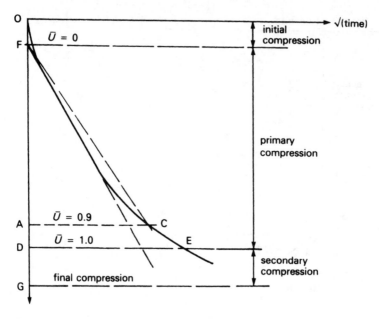

Fig. 10.21   Initial, primary and secondary compression

Worked example 10.5   *During a loading stage in a oedometer test the stress was increased by 100 kPa and the following changes in thickness were recorded:*

Time (min.)	0.00	0.04	0.25	0.50	1.00	2.25	4.00	6.25	9.00
Change in thickness (mm)	0.00	0.121	0.233	0.302	0.390	0.551	0.706	0.859	0.970

Time (min.)	12.25	16.00	25.00	36.00	64.00	100	360	1440
Change in thickness (mm)	1.065	1.127	1.205	1.251	1.300	1.327	1.401	1.482

At the end of the last stage (1440 min.), the thickness of the specimen was 17.53 mm and the void ratio 0.667.

Using the root-time method, determine: (a) the coefficient of consolidation (c_v), (b) the initial and primary compression ratios, and (c) the coefficient of volume compressibility (mv) and permeability (k).

Firstly, choosing appropriate scales, set up the root-time and change-in-thickness axes, and plot the points of the laboratory readings curve (Fig. 10.22).

Now, draw the best straight line through the points in the first 50–60 per cent of the plot. The intersection of this straight line with the thickness axis (point F) locates $\bar{U} = 0$, which corresponds to a change in thickness of $\Delta H_0 = 0.078$ mm.

Next, draw a straight line with abscissae 1.155 times those of the first straight line. This line is assumed to intersect the laboratory curve at $\bar{U} = 0.90$ (point C).

(a)   From the plot, $\sqrt{t_{90}} = 3.79$; $t_{90} = 14.36$ min.

   Using Taylor's Fourier series method

   From Table 10.1, $T_{v90} = 0.848$.

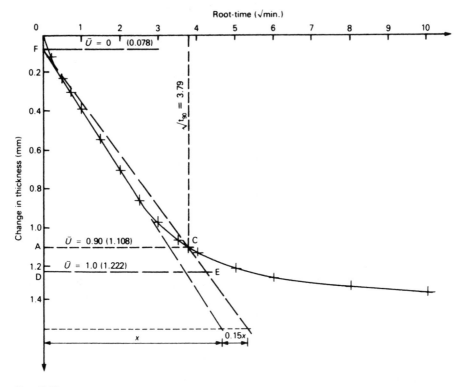

Fig. 10.22

Average thickness of the specimen during the stage = 17.53 + 1.482/2 = 18.27 mm

∴ length of drainage path, d = 18.27/2 = 9.14 mm

So from eqn [10.44]: $c_v = \dfrac{T_{v90}d^2}{t_{90}} = \dfrac{0.848 \times 9.14^2}{14.36} = \underline{4.93 \text{ mm}^2/\text{min}}$

Using parabolic isochrones

From eqn [10.41]: $T_v = \dfrac{0.25 - \ln(1-\bar{U})}{3}$ ∴ when $\bar{U} = 0.90$, $T_v = 0.8509$

giving $c_v = \underline{4.95 \text{ mm}^2/\text{min}}$

(b) From the plot, ΔH_{90} = 1.108 mm and ΔH_0 = 0.078 mm.

Then $\Delta H_{100} = \dfrac{1.108 - 0.078}{0.90} = 1.222 \text{ mm}$

Total $\Delta H = 1.482 \text{ mm}$

Hence, initial compression ratio, $r_i = \dfrac{0.078}{1.482} = \underline{0.053}$

and primary compression ratio, $r_p = \dfrac{1.222 - 0.078}{1.482} = \underline{0.772}$

(c)   Final void ratio, $e_f = 0.667$

Initial thickness, $H_0 = 17.53 + 1.472 = 19.00$ mm

Change in void ratio, $\Delta e = \dfrac{\Delta H}{H_0}(1 + e_0) = \dfrac{1.482 \times 1.667}{19.00} = 0.130$

Initial void ratio, $e_0 = 0.667 + 0.130 = 0.797$

From eqn [10.10]: $m_v = \dfrac{\Delta e}{\Delta\sigma(1 + e_0)}$

$$= \dfrac{0.130 \times 10^3}{100 \times 1.797} = \underline{0.723 \text{ m}^2/\text{kN}}$$

From eqn [10.9]: $k = c_v m_v \gamma_w = 4.93 \times 10^{-6} \times 0.723 \times 10^{-3} \times 9.81$

$$= \underline{3.5 \times 10^{-8} \text{ m/min}}$$

Log-time method (Casagrande's method)

An alternative method of finding c_v suggested by Casagrande, involves a laboratory curve in which thickness is plotted against the logarithm of time. This theoretical curve comprises three parts: an approximately parabolic initial curve, a middle straight-line portion and a final curve to which the time axis is an asymptote (Fig. 10.23).

After plotting the laboratory results, point F (corresponding to $\bar{U} = 0$) must be located. Since the first part of the curve is approximately parabolic, this is done by selecting two points P and Q on the curve for which the values of t are in a ratio of 1:4, e.g. 1 min. and 4 min., 2 min. and 8 min., etc. The vertical distances

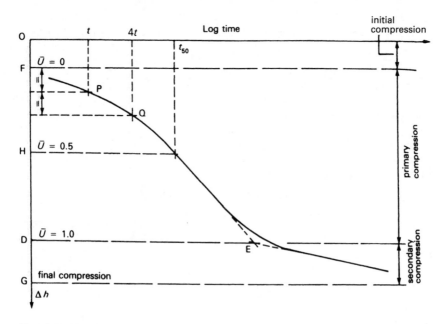

Fig. 10.23 Log-time method

(changes in thickness) PQ and FP are equal, enabling F to be located. OF therefore equals the amount of *initial compression*.

The final part of the laboratory curve should be approximately straight, but not horizontal. Point E, representing $\bar{U} = 1.0$, is located at the intersection of the projections of this final straight portion and the middle straight portion. Then FD equals the amount of *primary compression*. The compression ratios may be calculated using eqns [10.45], [10.46] and [10.47].

The point corresponding to $\bar{U} = 0.5$ is found by projecting point H on to the curve (FH = HD = $\frac{1}{2}$FD). The value of t_{50} is now obtained on the time axis and the value of c_v calculated from

$$c_v = \frac{T_{v50}d^2}{t_{50}} \qquad [10.48]$$

Calculation of settlement times

The coefficient of consolidation is not actually a constant for a given soil, since its value depends on the coefficient of permeability (k) and the coefficient of volume compressibility (m_v), both of which vary: k with changes in void ratio and m_v with changes in effective stress. However, the variation in c_v will be relatively small over limited stress ranges. Once c_v has been determined the time required for a given amount of consolidation to occur can be calculated:

From eqn [10.12], $t = \dfrac{T_v d^2}{c_v}$ $\qquad [10.49]$

Worked example 10.6   *Rework worked example 10.5 using the log-time method.*

First, choosing appropriate scales, set up the log-time and change-in-thickness axes and plot the points of the laboratory curve (Fig. 10.24).

(a) To locate $\bar{U} = 0$, select two points on the curve, P and Q, so that $t_Q = 4t_P$. Since the curve is approximately parabolic then parallel to the thickness axis: FP = PQ.

From the plot: at $t_P = 0.25$ min., $\Delta H_P = 0.233$ mm
at $t_Q = 1.00$ min., $\Delta H_Q = 0.390$ mm

Then $\Delta H_F = 0.233 - (0.390 - 0.233) = 0.076$ mm

Now, draw two straight lines: one through the final part of the laboratory curve and the other through its middle portion. Point E, at the intersection of these two lines, represents $\bar{U} = 1.0$.

From the plot, $\Delta H_{100} = 1.224$ mm

Then $\Delta H_{50} = \frac{1}{2}(1.224 - 0.076) + 0.076 = 0.650$ mm

After locating ΔH_{50} on the plot, log t_{50} is found to be 0.525, giving $t_{50} = 3.35$ min.

From Table 10.1, $T_{v50} = 0.197$

As in worked example 10.5, $d = 9.14$ mm

From eqn [10.27]: $c_v = \dfrac{T_{v50}d^2}{t_{50}} = \dfrac{0.197 \times 9.14^2}{3.35} = \underline{4.91 \text{ mm}^2/\text{min}}$

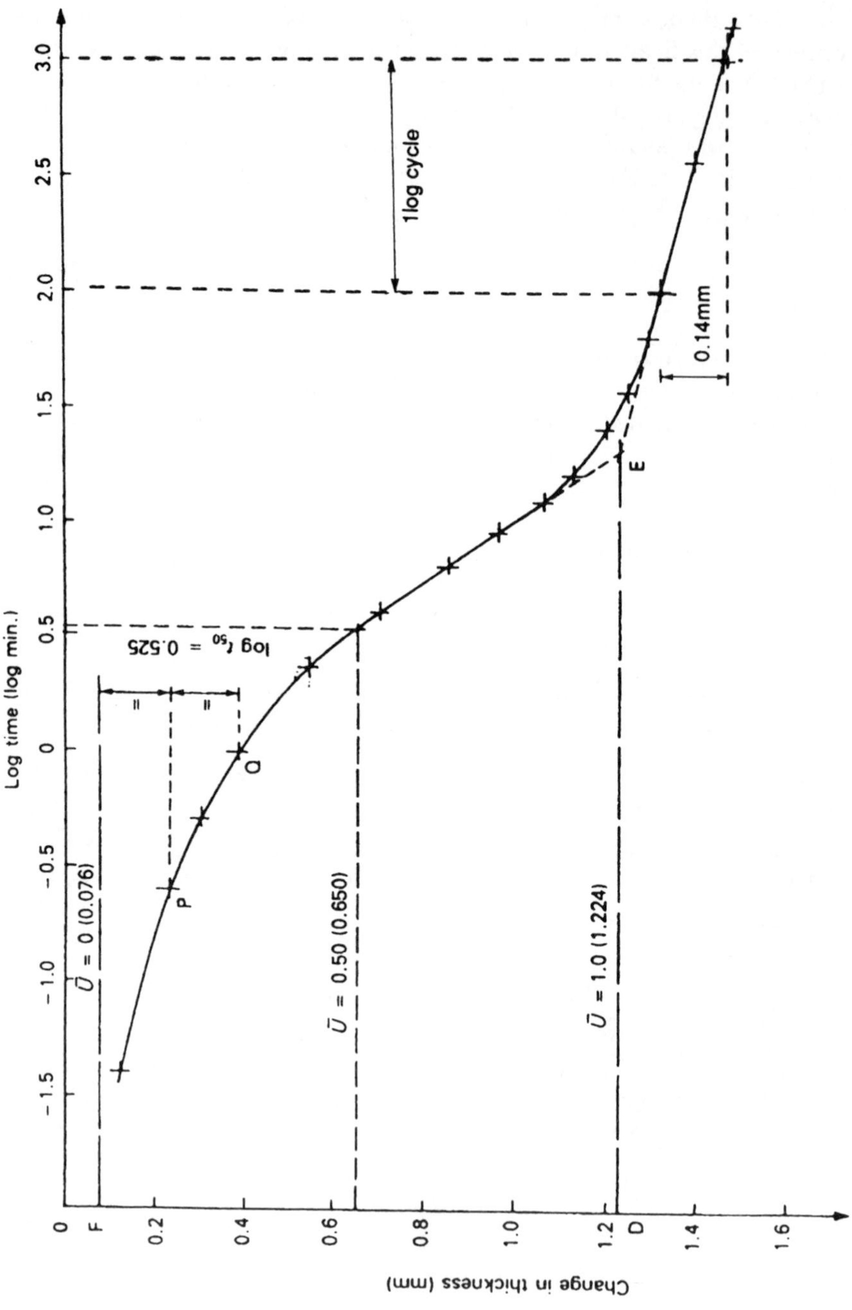

Fig. 10.24

(b) From the plot: $\Delta H_{100} = 1.224$ mm, $\Delta H_0 = 0.076$ mm

and Total $\Delta H = 1.482$ mm.

Hence, initial compression ratio, $r_i = \dfrac{0.076}{1.482} = \underline{0.051}$

and primary compression ratio, $r_p = \dfrac{1.222 - 0.076}{1.482} = \underline{0.775}$

(c) This part of the solution is identical to that given for part (c) of worked example 10.5.

Worked example 10.7 *The coefficient of consolidation* (c_v) *for a clay was found to be 0.955 mm²/min. The final consolidation settlement estimated for a 5 m thick layer of this clay was calculated at 280 mm. Assuming there is a permeable layer both above and below, and a uniform initial excess porewater pressure distribution, calculate the settlement time for (a) 90 per cent primary consolidation and (b) a settlement of 100 mm.*

Since this is an 'open' layer, the drainage path, $d = \frac{1}{2} \times 5.0 \times 10^3 = 2500$ mm and since the initial $\bar{u}$ distribution was uniform, $m = 1$.

(a) From Table 10.1, for $\bar{U} = 0.90$, $T_{v90} = 0.848$.
Then the time required for 90 per cent settlement is:

$$t_{90} = \frac{T_{v90}d^2}{c_v} = \frac{0.848 \times 2500^2}{0.955} = 5.55 \times 10^6 \text{ min} = \underline{10.55 \text{ yr}}$$

(b) For 100 mm settlement, $\bar{U} = \dfrac{100}{285} = 0.357$

From Table 10.1, $T_{v35.7} = 0.102$ (interpolating linearly). Then the time for 100 mm settlements:

$$t_{35.7} = \frac{0.102 \times 2500^2}{0.955} = 0.668 \times 10^6 \text{ min} = \underline{1.25 \text{ yr}}$$

The drainage/time relationship

If two samples of the same clay, subject to the same increase in effective stress, reach the same degree of consolidation, then theoretically at that point the T_v/c_v ratio must be the same in each. The following relationship therefore exists:

$$\frac{T_v}{c_v} = \frac{t_A}{d_A^2} = \frac{t_B}{d_B^2} \qquad [10.50]$$

where t_A = settlement time in sample/layer A
t_B = settlement time in sample/layer B
d_A = drainage path in sample/layer A
d_B = drainage path in sample/layer B

This relationship provides an easy method for computing site settlement times from laboratory results, while avoiding the necessity of evaluating c_v.

Worked example 10.8   *On a particular site a layer of clay occurs of 6.0 m thickness. In a laboratory oedometer test on a 19.0 mm thick specimen of the clay, 50 per cent consolidation was reached after 12.0 min. Determine the site settlement time for 50 per cent consolidation when the clay layer is (a) fully drained top and bottom and (b) drained from one surface only.*

(a)   Laboratory:

$t_A = 12.0$ min. $d_A = \frac{1}{2} \times 19.0 = 9.5$ mm

Site:

$t_B = ?$ $d_B = \frac{1}{2} \times 6.0 \times 10^3 = 3000$ mm

From eqn [10.50], $t_B = \dfrac{12.0 \times 3000^2}{9.5^2} = 1.197 \times 10^6$ min $= \underline{2.28 \text{ yr}}$

(b)   Laboratory:

$t_A = 12.0$ min. $d_A = \frac{1}{2} \times 19.0 = 9.5$ mm

From eqn [10.50], $t_B = \dfrac{12.0 \times 6000^2}{9.5^2} = 4.787 \times 10^6$ min $= \underline{9.10 \text{ yr}}$

Worked example 10.9   *Using the data from worked example 10.8, determine both the laboratory and the site times for 90 per cent consolidation.*

The time taken for 50 per cent consolidation is known, so that T_{v50} may be obtained either from Table 10.1 or, since $\bar{U} < 0.6$, from eqn [10.19], i.e. $\bar{U}_t = \sqrt{(4T_v/\pi)}$.

$T_{v50} = 0.5^2 \times \pi/4 = 0.196$

From Table 10.1, $T_{90} = 0.848$

So with d constant, using eqn [10.50]:

Laboratory $t_{90} = t_{50} \times \dfrac{T_{v90}}{T_{v50}} = \dfrac{12.0 \times 0.848}{0.196} = \underline{51.9 \text{ min.}}$

For an open layer: site $t_{90} = 51.9 \times 3000^2/9.5^2 = 5.175 \times 10^6$ min $= \underline{9.84 \text{ yr}}$

and for a half-closed layer: site $t_{90} = 51.9 \times 6000^2/9.5^2 = 20.7 \times 10^2$ min $= \underline{39.4 \text{ yr}}$

10.6   Secondary compression or creep

If Terzaghi's theory were a perfect model of consolidation, no further compression would occur after the excess pore pressure had been fully dissipated. An examination of typical $e/\log t$ curves shows this not to be the case. The latter part of the $e/\log t$ curve is usually found to be sloping and approximately linear. The stage is referred to as *secondary compression*, or sometimes *creep*, and is generally held to be the result of some kind of internal particle repositioning mechanism, similar to compaction but occurring slowly.

A number of factors are thought to influence the amount and rate of secondary compression: the principal stress ratio (σ_1'/σ_3'), the rate of load increase, the ambient temperature, the stress history, the layer thickness. Also, secondary compression seems to be greater in organic soils, at stresses below the pre-consolidation stress and when the loading is one-dimensional as opposed to isotropic loading. At present, no theory of confirmed reliability has emerged to provide a model for this phenomenon (or phenomena). For design purposes, it is acceptable to use the following parameters:

Coefficient of secondary compression, $C_\alpha = \dfrac{\Delta e}{\Delta \log t}$ 　　　　[10.51]

i.e. the change in void ratio (or unit thickness) per log cycle of time after primary consolidation ($\bar{U} = 1.0$) has been exceeded.

Secondary compression index, $C_{\alpha\varepsilon} = \dfrac{C_\alpha}{1 + e_\alpha}$ 　　　　[10.52]

also called the *rate of secondary compression*, where e_α = void ratio at the start of the linear portion of the $e/\log t$ curve (at approximately $\bar{U} = 1.0$).

The *amount* of secondary compression (s_s) is the change in thickness caused by the sustained application of load after the full dissipation of excess pore pressures. For practical purposes, C_α may be measured from the slope of the $e/\log t$ curve between $t = 100$ min and $t = 1000$ min, or between $t = 1000$ min and $t = 10\,000$ min. Values of C_α range approximately as follows:

Overconsolidated clays:　　　　$C_\alpha = {<}0.005$

Normally consolidated clays:　　$C_\alpha = 0.005 - 0.05$

Organic soils:　　　　　　　　$C_\alpha = 0.05 - 0.5$

The ratio C_α/C_c will usually be in the range 0.025–0.10 for natural soils, with higher values applying to organic soils (Mesri and Godiewski, 1977).

Worked example 10.10　*Using the data given in worked example 10.6 and plotted in Fig. 10.24, determine the coefficient of secondary consolidation for the soil.*

In Fig. 10.24, the thicknesses over the log cycle from $t = 100$ min to $t = 1000$ min are:

Thickness at $t = 100$ min = $17.53 + (1.482 - 1.327) = 17.69$ mm
Thickness at $t = 1000$ min = $17.69 - 0.14$ 　　　　　= 17.55 mm

Average thickness during this log cycle = 17.62 mm

Therefore, $C_\alpha = (17.69 - 17.55)/17.62 = \underline{0.0079}$

The effect of secondary consolidation on primary consolidation measurement in subsequent stages is often questioned. A number of studies (Olson, 1986) show that the void ratio should be calculated using the *total* compression for a given stage and that no correction for secondary consolidation should be made. However, the shape of the primary root-time curve may be affected, since some downward concavity has been observed in load stages that follow considerable secondary consolidation.

10.7　Continuous loading tests

Incremental loading procedures, such as that employed in the standard oedometer test, are widely used and provide simple methods for routine tests, but they do have some serious disadvantages. These disadvantages are either eliminated or greatly reduced with continuous loading tests:

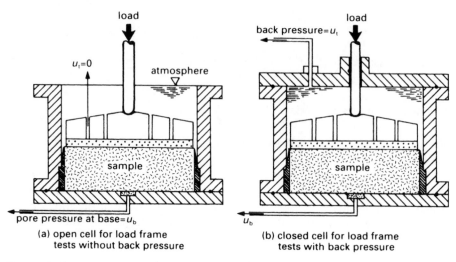

Fig. 10.25 Compression cells for continuous loading tests

(a)   The length of time to complete the test can be reduced, releasing apparatus and increasing job turnover.
(b)   The loading procedure can be automated and is thus less labour intensive.
(c)   The rate of testing can be selected by the operator, so that soils with higher permeabilities may be tested more quickly, e.g. approximately two hours for Kaolin compared with 48 hours for London Clay.
(d)   Continuous stress/strain/time data is collected, providing better data definition and reliability.
(e)   A variety of loading procedures and criteria can be adopted, e.g. constant rate of strain, constant rate of stress increase, constant pore pressure ratio, etc. Automated data logging and control techniques are easily applied.
(g)   Data processing is easily facilitated, especially with the use of computer-driven plotters and printers.

Several types of continuous loading procedure have been suggested and described in the literature. The apparatus required mainly features a hydraulic compression cell, e.g. Rowe cell, Oxford University cell or an adaptation of an oedometer-type cell, e.g. Bristol cell (Fig. 10.25). Computer-aided monitoring and control is an essential feature of most types of test. A continuous e/σ' or $e/\log \sigma'$ curve can be produced, with an abundance of plot points, and thus having high data reliability. Some procedures are also cheaper to run than incremental loading tests. Their main disadvantages relate to the apparatus required: it is complex and expensive, and requires well-trained technicians to operate it. The main types of continuous loading tests are described below.

Constant rate of strain (CRS) test
The constant rate of strain test appears to have been developed in the late 1950s (Hamilton and Crawford, 1959) in an attempt to gather more (e, σ') plot points when testing sensitive soils. The rate of vertical displacement is kept constant by using a mechanical screw drive with a cell such as that shown in Fig. 10.25(a).

The applied stress, sample thickness and pore pressure are monitored continuously against time. The rate of displacement must be slow enough to yield pore pressures less than about 10 per cent of the applied total stress.

Constant rate of loading (CRL) test
The applied stress is increased at a constant rate, i.e. $d\sigma'/dt$ = constant, using a hydraulic cell (Fig. 10.25(b)). Equations [10.54]–[10.57] are valid except for the initial stages. The test can be completed in 3–72 hours, using loading rates of 6–300 kPa per hour. The best loading rate must be determined by trial; from a practical point of view it seems sensible to adopt a loading rate which will allow completion in 1, 2 or 3 working days.

Constant gradient (CG) test
The loading rate is controlled so that the pore pressure (u_b) at the impermeable base of the specimen is maintained at a constant value. The pore pressure at the exit face is equal to the back pressure, so, if this remains constant, the pore pressure gradient across the sample thickness is also constant. For $\bar{U} > 0.8$ eqns [10.54]–[10.57] are valid.

Constant pore pressure ratio (CPR) test
The ratio of the base pore pressure to the applied total stress (u_b/σ) is maintained at a constant value. This procedure is described by Janbu *et al.* (1981) as a 'continuous loading test', but the more specific description seems necessary in the light of other procedures.

Simplified analysis of continuous loading tests
The essential feature of the data collected is the abundance of plot points in the stress/strain/time continuum. Analyses are based on the slopes of the stress–strain and strain (or displacement)–time curves. A number of detailed analytical procedures have been suggested, some of which involve empirical corrections, e.g. Wissa *et al.* (1971), Head (1985), Janbu *et al.* (1981), and others. However, for most purposes, the following simplified analysis should prove adequate.
   For a given amount of settlement ΔH, the void ratio of the specimen is given by:

$$e = \frac{H_0 - \Delta H}{H_s} \tag{10.53}$$

where H_0 = initial height (thickness) of test specimen
  H_s = notional height of the solid phase of the specimen (which remains constant throughout the test)
   = $M_s/AG_s\,\rho_w$ in which M_s = mass of solids
      A = area of specimen
      $G_s\,\rho_w$ = density of solids

Now, the one-dimensional or confined modulus is:

$$E_0' = \frac{1}{m_v} = \frac{\delta\sigma_v'}{\delta\varepsilon} = \frac{\delta\sigma_v'}{\delta H}H_0 \tag{10.54}$$

Thus, E' (or mv) may be obtained from the stress/displacement curve.

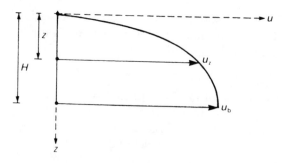

Fig. 10.26   Continuous loading tests: parabolic isochrones

In a Rowe or similar cell, the base is impermeable and the drainage takes place upward through the thickness of the specimen. The average effective stress within the specimen is therefore:

$$\sigma' = \sigma - \alpha u_b \qquad\qquad [10.55]$$

where u_b = pore pressure at the impermeable base
α = a pore pressure coefficient

The value of α in the standard oedometer test is 1.0 at the start, falling to about $\frac{2}{3}$. When either the stress–time or stress–strain curve is linear, $\alpha \approx \frac{2}{3}$, and if the pore pressure distribution through the specimen is parabolic α is exactly $\frac{2}{3}$ (Fig. 10.26). Parabolic isochrones were dealt with in Section 10.4 and can be evaluated as follows:

$$u_z = \frac{u_b}{H_0 z}(2H_0 z - z^2) \qquad\qquad [10.56]$$

Differentiating, $\dfrac{d^2 u}{dz^2} = \dfrac{2u_b}{H_0 z}$ $\qquad\qquad [10.57]$

The rate of change of volume is given by the rate of change in the slope of the isochrone (Fig. 10.26):

$$\frac{\delta\varepsilon}{\delta t} = \frac{\delta H}{\delta t}\frac{1}{H_0} = \frac{k}{\gamma_w}\frac{d^2 u}{dt^2} \qquad\qquad [10.58]$$

Giving $k = \dfrac{\gamma_w H_0}{2u_b}\dfrac{\delta H}{\delta t}$ $\qquad\qquad [10.59]$

and therefore $c_v = \dfrac{k}{m_v \gamma_w}$

Alternatively, c_v can be evaluated from the stress–time curve:

$$c_v = \frac{H_0^2}{2u_b}\frac{\delta\sigma_v'}{\delta t} \qquad\qquad [10.60]$$

Comments on continuous loading tests

The main advantage of continuous loading tests is the ability to record a more or less continuous definition of the parameters e, ε, k, m_v and c_v as functions of effective stress. The main disadvantage (apart from equipment costs) is the masking of secondary compression effects. This problem can be partially overcome by holding the applied stress constant for a short time at a number of tactically placed intervals along the strain/stress curve.

Strain rates need to be chosen with care: less than 5 per cent per hour for plastic clays and not greater than 10 per cent per hour for stiff clays. The $e/\log \sigma'$ curve is affected by the strain rate and a correction is required to provide normalisation. The following empirical correction is suggested here (after Lerouriel *et al.*, 1985) which is sufficiently valid in the range $10^{-8} < d\varepsilon/dt < 10^{-4}$:

$$\alpha_\varepsilon = 2.0 + 0.14\Delta\varepsilon_s \qquad [10.61]$$

where $\Delta\varepsilon_s$ = strain increase per second

Then a given void ratio recorded at a test stress of σ'_T should be plotted as:

$$\sigma' = \sigma'_T/\alpha_\varepsilon \qquad [10.62]$$

After this correction, values obtained for m_v appear to correlate well with field values. Measurements of c_v appear to differ dramatically from those obtained from standard (oedometer) tests: up to 100 times higher in some recorded cases, although they are nearer to values obtained from field back-analyses. The two main problems seem to be secondary consolidation and incomplete transient flow, both of which are affected by the strain rate.

In order to minimise these effects, and in the interests of good laboratory management, it would seem that a loading period of at least 24 hours is advisable for most soils. Under these circumstances, the results of continuous loading tests may be considered to be at least as reliable as those obtained from traditional incremental loading procedures. The higher c_v values, although closer to field values, are still not in good agreement in many cases. Field values will always be higher because of factors such as horizontal drainage, soil structure distortion, non-model stress distribution, etc. Laboratory results should not be considered in isolation, but should be compared with observations of site settlements and pore pressure measurements.

10.8 *Validity and reliability of consolidation tests*

A *valid* test measures the properties it is claimed to measure, and a *reliable* test, when repeated several times, should produce identical quantitative results. In consolidation testing, measured stress–strain relationships are generally acceptable, with minor corrections in some cases. Measurements of strain–time relationships, however, cannot be treated with a very high level of confidence. The main sources of error in standard tests are related to the assumption that k and $\delta\varepsilon/\delta\sigma$ remain constant. However, some observed rates of settlement (Taylor, 1948; Davis and Raymond, 1965) for normally consolidated clays are similar to those predicted by Terzaghi's method.

The majority of site observations indicate that actual rates of settlement are generally greater than those predicted by any of the laboratory tests, although continuous loading test values appear to be closer to field values than those recorded in oedometer tests. These discrepancies may be partly due to the effect of variations in the soil fabric and variations from assumed stress distributions. Anisotropic drainage conditions result when features occur such as sand or silt layers (as in varved and other post-glacial clays), laminations and fissures, and the presence of organic matter or small voids (e.g. root holes). Where such conditions exist, larger specimens can be tested in a hydraulic cell, such as that suggested by Rowe and Barden (1966).

The effect of secondary compression during primary consolidation also distorts the measurement of c_v, since the proportional amount of secondary compression (r_s) increases as the specimen thickness decreases, and hence is more apparent in the test than in the field. Also, r_s increases as the $\Delta\sigma'/\sigma'$ ratio decreases; for this reason the incremental increase in stress should always be equal to the current stress, i.e. the applied stress doubled each time, thus keeping $\Delta\sigma'/\sigma'$ constant. The amount of secondary compression is greater in normally consolidated clays than in overconsolidated clays, although the ratio of secondary to primary may be less. Secondary compression is greatest in organic soils.

It is possible to identify a number of error-prone areas in sampling and testing procedures.

Ring friction

The effective stress actually applied to the soil is reduced due to friction between the consolidation ring and the sides of the soil specimen. In estimating the amount of this reduction, the type of material and whether or not the ring is effectively greased is important. The following guide figures are recommended for the loading of normally consolidated clays:

Friction corrected average applied stress, $\sigma_F = \alpha_F \, \sigma$ [10.63]

where α_F = ring friction coefficient

e.g. greased smooth metal: $\alpha_F = 0.85$–0.95
ungreased smooth metal: $\alpha_F = 0.70$–0.90
greased plastic: $\alpha_F = 0.90$–0.98
ungreased plastic: $\alpha_F = 0.80$–0.95

This friction effect is greater during unloading.

Boundary flow impedance

The porous stones above and below the specimen must be sufficiently fine-grained to prevent clogging by soil particles, otherwise the flow of expelled pore-water will be impeded. Clogging may also be prevented by using a disc of filter paper (Whatman No. 54) between the soil and each porous stone.

Sample disturbance

Good consolidation test samples should be *undisturbed*, but this is a relative term, in fact. Sample disturbance can result from a combination of a number of factors in varying degrees.

Sampler effects
(a) Stress relief due to removal of overburden stress.
(b) Shear strain in the soil beneath the tube or piston.
(c) Change in moisture content during sampling: decrease due to pressure on driving and/or increase due to suction on withdrawal.
(d) Internal vertical shear in soil near to inner face of tube.
(e) Smearing along sides of sample.
(f) Changes in density: increase due to driving pressure or decreases due to lateral expansion at clearance diameter.

Transport and storage effects
(a) Ineffective sealing, producing water content change.
(b) Mechanical, vibratory or shock damage.
(c) Lateral and vertical moisture movement.
(d) Oxidation and ion transfer from steel tubes.
(e) Crystallisation of salts or other groundwater solubles.

Sample preparation
(a) Density and moisture content changes due to extrusion from sample tube.
(b) Smearing or plucking damage to drainage faces affecting boundary flow; an estimated 2 mm of sample thickness is often affected.
(c) Premature shrinkage or swelling.

One of the main effects of sample disturbance is to produce curvature in the recorded $e/\log \sigma'$ plot, so that for a given σ' the recorded value of e is reduced. Also, the location of σ'_y can be ambiguous (Fig. 10.27).

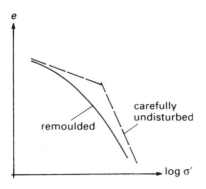

Fig. 10.27 Effect of sample disturbance

The size of the test specimen has an apparently small effect on the $e/\log \sigma'$ curve, but quite a dramatic effect on c_v. Rowe (1971) found that the ratio of c_v values measured using 250 mm × 90 mm specimens to those measured using conventional 76 mm × 20 mm specimens ranged from 2 to 100. Whatever size of specimen is used, correlation with field conditions continues to be the main problem. Observations of site settlements and pore pressure measurements should be used as a form of control.

Rapid loading tests

In order to speed up the conventional incremental loading procedure, a number of attempts have been made to justify adding the next load increment before the ultimate stage settlement has been reached. It can be shown that adding a further increment of loading at $\bar{U} = 0.9$ produces an error in t_{50} of 7 per cent, and adding an increment at $\bar{U} = 0.50$ produces an error of 14 per cent. It may be argued, therefore, that adding an increment at or after $\bar{U} = 100$ per cent will produce only a minimal error in m_v and a c_v value that is lower, but not too erroneous. However, this reasoning only applies if the effects of secondary consolidation are negligible. Leonards and Ramiah (1960) found that, with a daily reloading cycle, measured values of c_v were higher than when using either hourly or weekly cycles. Another important practical point is that for load incrementation to take place at t_{100} or sooner, constant monitoring is required and technician-manned loading must be carried out regardless of the time of day. Such procedures will inevitably be awkward to plan and costly to run.

Comparison of root-time and log-time methods

Both of these methods are attempts to fit a model of behaviour to actual performance. Neither produces a comprehensively perfect fit, but each is acceptable in certain respects and both are simple to apply. The sensible approach for many problems is to use both, and to be selective in the choice of alternative interpretations.

(a) The root-time method is preferred for the location of $\bar{U}_0$, since it is easier and less ambiguous to fit the $\bar{U} = \sqrt{(\pi T_v/4)}$ theoretical curve than to use the trial fitting procedure on the log-time curve.
(b) For the location of $\bar{U}_{100}$ both methods are equally valid when no secondary compression is occurring. The root-time method can be used if secondary effects up $\bar{U}_{90}$ are negligible, otherwise it is better to use the log-time curve. However, the log-time curve also becomes ambiguous when the secondary compression is more than marginally non-linear.
(c) Values of c_v measured entirely by the log-time method are almost always higher than those measured entirely by the root-time method.
(d) Curve fitting using a computer is more easily automated using the root-time method, but printers and plotters can produce hard copy output of either.

Design considerations

The quality of data used in design depends partly on eliminating or minimising sampling and testing errors, and partly on choosing the most appropriate methods of test. For projects involving large predicted settlements, e.g. embankments on soft clay, acceptably reliable results can be obtained from conventional oedometer tests. Sampling should be carried out in soft clays using a thin-walled piston sampler, and then specimens prepared to a smaller diameter than the field sample so as to exclude the outer 'damaged' zone.

In incremental loading tests, the loading increase should be $\Delta\sigma = \sigma$, which is acceptable for almost all purposes. For practical reasons, and to avoid transient and secondary compression problems, a 24 or 48 hour loading cycle is recommended. In the analysis of results, the root-time method is preferable to establish $\bar{U}_0$, $\bar{U}_{90}$ and $\bar{U}_{100}$; and hence t_{50}, t_{90}, etc. The log-time method should be used to measure secondary consolidation (C_α).

Finally, it should be remembered that, in the words of Karl Terzaghi: 'the results of soil tests . . . gradually close up the gaps in knowledge, and if necessary [*the designer should*] modify the design during construction'. The 'gaps' being closed by test results are usually quite small in comparison with the total span of knowledge required as a whole for design. Remember, too, that there is a very wide variation in soils and a vast range of natural properties, e.g.

Gault clay: $\qquad e \approx 0.75$, $\sigma'_y \approx 7000$ kPa (Samuels, 1975)

Mexico City clay: $\quad e \approx 14.0$, $\sigma'_y \approx 90$ kPa $\quad$ (Mesri *et al.*, 1975)

10.9 Estimate of total settlement by the Skempton–Bjerrum method

One source of error inherent in estimates of consolidation settlement stems from the fact that conditions in the oedometer test are truly one-dimensional, whereas actual site conditions are not. Under true one-dimensional conditions, the lateral strain is zero and the pore pressure coefficient A is equal to unity, i.e. the initial increase in pore pressure is equal to the increase in total stress.

Under site conditions, the coefficient A will only be close to unity when the extent of the loaded area is large compared with the thickness of the layer. In layers of significant thickness the lateral strain will have a measurable effect and the initial excess pore pressures will be less than $\Delta\sigma$. In such cases, the total settlement (s) will comprise two components:

(a) *Immediate settlement* (s_i) under undrained conditions, which may be estimated using eqns [6.74] or [6.80], using the undrained modulus E_u and $v = 0.5$ (see Sections 10.2 and 6.11).

(b) *Consolidation settlement* (s_c) due to drainage as the excess pore pressure is dissipated.

Total settlement, $s = s_i + s_c$

Skempton and Bjerrum (1957) have suggested that the effects of lateral expansion (due to lateral strain) and recompression (due to increase in lateral effective stress as the pore pressure dissipates) may be ignored. Although it is admitted that this could produce errors of up to 20 per cent, and errors of 15 per cent have been observed, the simplification is still recommended since it allows the use of the results of the oedometer test. Lambe (1967) has proposed a rigorous method which involves establishing the exact stress-path plot for the consolidation processes and then running laboratory triaxial tests along this plot while measuring the strains. The settlement is then computed from the strain measurements. Although this latter method may be shown to be reliable in principle, it requires elaborate and lengthy procedures and is therefore inappropriate for routine work.

However, the oedometer value (s_{oed}) can be modified in order to make allowance for the variation in excess pore pressure due to the lateral strain. Under one-dimensional conditions (i.e. zero lateral strain):

$$s_{oed} = m_v \Delta\sigma_1 H = m_v \Delta u H$$

or integrating, $s_{oed} = \int_0^H m_v \Delta\sigma_1 dz = \int_0^H m_v \Delta u dz$

where H = thickness of layer

Now from eqn [4.10] (where $B = 1$ for a saturated soil):

$$\Delta u = \Delta\sigma_3 + A(\Delta\sigma_1 - \Delta\sigma_3)$$

$$= \Delta\sigma_1 \left(A + \frac{\Delta\sigma_3}{\Delta\sigma_1}(1 - A) \right)$$

So that when $A \neq 1.0$:

$$s_c = \int_0^H m_v \Delta\sigma_1 \left(A + \frac{\Delta\sigma_3}{\Delta\sigma_1}(1 - A) \right) dz$$

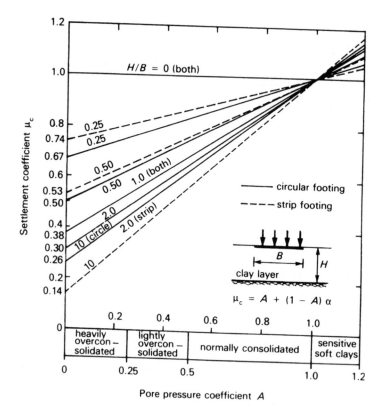

Fig. 10.28   Settlement coefficient μ_c (after Skempton and Bjerrum, 1957)

Thus, a *settlement coefficient* can be defined as:

$$\mu_c = \frac{S_c}{S_{oed}} = \frac{\int_0^H m_v \Delta\sigma_1 \left(A + \frac{\Delta\sigma_3}{\Delta\sigma_1}(1 - A) \right) dz}{\int_0^H m_v \Delta\sigma_1 dz} \qquad [10.64]$$

It is convenient in practice to assume that in a layer or sub-layer the values of m_v and A remain constant, so that the coefficient μ_c may be expressed as:

$$\mu_c = A + (1 - A)\alpha_c \qquad [10.65]$$

$$\alpha_c = \frac{\int_0^H \Delta\sigma_3 dz}{\int_0^H \Delta\sigma_1 dz}$$

Values of α_c may be obtained using elastic theory, taking the undrained value $v = 0.5$. Values μ_c corresponding to various layer-thickness/breadth ratios are given in Fig. 10.28.

10.10 Correction for gradual application of load

In the development of consolidation theory in the preceding pages, it was assumed that the load was applied instantaneously. In practice, however, most consolidation problems are connected with construction processes extending over several months, or even years. During the excavation stage of construction there may even be a significant reduction in applied load and therefore swelling, in which case new consolidation will not commence until the reduction has been replaced by new loading. A settlement–time curve plotted on the basis of instantaneous loading would give an overestimate settlement at given time intervals if, in fact, the load had been gradually applied.

Terzaghi has suggested a method of correcting the instantaneous curve. Figure 10.29 shows a graph representing the averaged total stress–time relationship during the construction period. During the excavation period the level of stress decreases and then increases, so that after a time the level of stress is again zero. At this point new consolidation (and hence settlement) commences. The total stress continues to increase to the end of the construction period during a further time t_1, after which it remains constant at a value of σ_1.

The instantaneous settlement–time curve is plotted from point O, where consolidation actually commences. To obtain the correction, it is assumed that the amount of consolidation settlement at time t_1 is the same as that given by the instantaneous curve at time $\frac{1}{2}t_1$. Thus point P is located on the instantaneous curve at time $\frac{1}{2}t_1$, and point Q found by projecting PQ parallel to the time axis to meet the ordinate at t_1. From points P and Q onwards the abscissae of the corrected curve will be $\frac{1}{2}t_1$ greater than on the instantaneous curve. To obtain the corrected curve during the construction period itself, point R is located on the instantaneous curve at time $\frac{1}{2}t$ and point S found by projecting horizontally

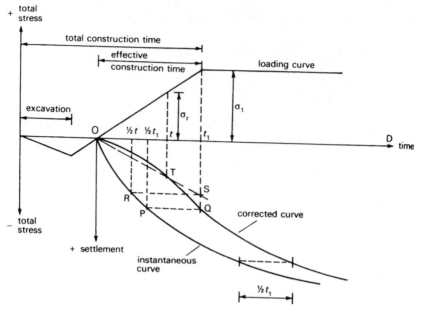

Fig. 10.29 Correction for gradual load application during construction

on to the ordinate at time t_1. Point T (on the corrected curve) is then found at the intersection of OS and the ordinate at time t. This process is repeated until there are sufficient points to draw the corrected curve.

10.11 Compressibility of granular soils

In the majority of cases, the design of foundations on granular soils will be governed by a settlement criterion and not by shear strength. Because of the high permeability of sands and gravels, most of the settlement will take place during the construction process, and will be largely complete at the end of the construction period. The effects of creep are likely to be negligible, except in the case of very wide foundations on variable soils, or where sand–silt mixtures occur. Other post-construction settlement problems may be related to vibration-induced compaction, rapid changes in water table or other rapid loading changes, and earthquake effects.

Since granular soils are generally much less homogeneous than clays, the ratio of differential settlement to total settlement will be greater: $\Delta s/s = 0.5$–1.0. Also, it is virtually impossible to obtain good undisturbed samples of granular soils, and recompacted specimens do not usually mimic, with any reliability, field conditions and properties. The principal methods to determine the compressibility characteristics of granular soil are therefore site tests:

(a) Plate bearing test
(b) Standard penetration test
(c) Cone penetration test
(d) Pressuremeter test

Since design methods for foundations on granular soils are mainly centred on the interpretation of the site test results, descriptions of the methods of analysis are dealt with in Chapter 11, and some explanation of associated field techniques given in Chapter 12.

Exercises

10.1 A rectangular foundation 6 × 4 m is to transmit a uniform contact pressure of 180 kPa at a depth of 1.2 m in a deep layer of saturated clay. If the undrained modulus E_u for the clay is 42 MPa, determine the average immediate settlement expected to occur under the foundations. Unit weight of clay = 19 kN/m³.

10.2 The following data were obtained from an undrained triaxial compression test on a specimen of clay:

Axial strain (%)	0	1	2	3	4	5	6	7	8
Deviator stress (kPa)	0	61	108	143	161	170	175	176	172

Obtain an estimate of E_u and then determine the immediate settlement likely to occur under a 3 m square footing that transmits a uniform load of 2.2 MN at a depth of 1.4 m in a 9 m thick stratum of the clay. Assume the clay to be underlain by a hard layer. Assume unit weight of clay = 18.5 kN/m³ and $v = 0.5$. Consider: (a) a corner, and (b) the centre of the foundation.

10.3 It has been found from field tests that the undrained modulus of a saturated clay varies with depth, approximately $E_u = 15 + 2.2z$ (z = depth below surface). Also the unit weight of the clay is 20 kN/m³ and $v = 0.5$. A flexible foundation 8 m long × 3 m wide is founded at a depth of 3 m in this clay and will transmit a uniform contact pressure of 280 kPa. The total depth of the clay layer is 12 m down to a hard rock stratum. By dividing the clay layer into four equal sub-layers, obtain an estimate of the average immediate settlement that may be anticipated.

10.4 During an oedometer consolidation test on a specimen of clay the thickness decreased from 19.36 m to 19.08 mm after the applied stress had been increased from 200 to 400 kPa and maintained for 24 hr. The stress was then removed and the sample allowed to swell for 24 hr, at the end of which time the water content was found to be 28.9 per cent. Assuming $G_s = 2.7$, calculate: (a) the void ratio at the beginning and end of this stress–increment stage, and (b) the coefficient of volume compressibility for this stress range.

10.5 In consolidation test on a specimen of clay each pressure stage was maintained for 24 hr. At the end of the last (unloading) stage the water content of the sample was found to be 28.6 per cent. The specific gravity is 2.70 and the changes in thickness were recorded as follows:

Stress (kPa)	0	50	100	200	400	800	0
Thickness (mm) (after 24 hr)		19.80	19.39	19.24	18.97	18.68	18.45	19.02

(a) Calculate the void ratio corresponding to the end of each stage and plot the void ratio/log effective-stress curve.
(b) Determine the ultimate settlement that may be expected due to the consolidation of a 4 m thick layer of this clay when the average effective stress in the layer changes:
  (i) From 136 kPa to 320 kPa.
  (ii) From 85 kPa to 240 kPa.

10.6   On an extensive reclamation site, where the water table is at the ground surface, a layer of coarse sand 4 m thick overlies a layer of soft clay 5 m thick. A 3 m thick layer of fill is to be laid over the site. The following data have been determined:

Unit weights: fill = 21 kN/m³   Coefficient of volume compressibility of the
sand = 20 kN/m³   clay = 0.22 m²/MN
clay = 18 kN/m³

(a)   Calculate the vertical effective stress at the centre of the clay layer, both before and after the placing of the fill.

(b)   Calculate the ultimate amount of settlement that may be expected due to the consolidation of the clay.

10.7   The foundation of a large building will be at a depth of 2.5 m in a stratum of dense sand. The sand is 5.5 m thick from the surface down to a layer of clay which is 6 m thick down to a layer of hard shale. The groundwater level is at a depth of 3.6 m. It has been calculated that the foundation loading will cause an increase in vertical effective stress of 140 kPa at the top of the clay and 75 kPa at the bottom of the clay. The results of an oedometer and other tests are given below. Calculate the ultimate settlement that may be expected due to the consolidation of the clay.

Effective stress (kPa)	25	50	100	200	400	800
Void ratio	0.892	0.884	0.866	0.834	0.800	0.766

Unit weights: sand = 21.2 kN/m³ (saturated)
and 19.6 kN/m³ (drained)
clay = 19.5 kN/m³

10.8   The construction of a foundation above a 3 m thick stratum of normally consolidated clay is expected to cause an average increase in effective stress from an initial value of 160 kPa to a final value of 340 kPa. If the liquid limit of the clay is known to be 52 per cent and a specimen of the clay was found to have a void ratio of 0.712 when subject to an effective stress of 200 kPa, estimate the expected ultimate settlement due to consolidation of the clay.

10.9   A clay layer of thickness 4.4 m is subject to a uniform increase in effective stress of 180 kPa.

(a)   Given that the coefficient of volume compressibility (m_v) is 0.25 m²/MN, calculate the ultimate consolidation settlement that may be expected to take place.

(b)   Given that the coefficient of permeability (k) of the soil is 5 mm/yr and the time factor (T_v) for full consolidation is 2.0, calculate the estimated time required for the ultimate settlement to take place (assume double drainage).

10.10   A clay layer of thickness 5.8 m is underlain by an impermeable layer of shale and overlain by a moderately permeable sand. The loading will be such that the effective stress in the clay will be increased uniformly over its whole thickness over a wide area. In a laboratory oedometer test a specimen of the clay of thickness 20 mm, subject to the same level of effective stress, showed a change in void ratio from 0.827 to 0.806. It was also observed that, after 30 min. had elapsed, 65 per cent of the consolidation had taken place.

(a)   Calculate the expected ultimate amount of consolidation settlement.

(b)   Calculate the time required for: (i) half, and (ii) three-quarters of the ultimate settlement to take place. (Refer to Table 10.1 for values of T_v.)

10.11 Observations of a building have shown that over the first two years after construction an average settlement of 65 mm has taken place. Laboratory tests on specimens from a clay layer beneath the building indicate an ultimate consolidation settlement of 285 mm. Assume the 65 mm to be all due to consolidation under two-way drainage and give an estimate of:

(a) The time required for 50 per cent of the ultimate settlement to take place.
(b) The amount of consolidation settlement to be expected after a period of 15 years (from construction) has elapsed.

10.12 The following results were recorded during an oedometer test when the applied stress was increased from 100 kPa to 200 kPa.

Elapsed time (min)	0	0.04	0.25	0.5	1.0	2.25
Thickness of specimen (mm)	18.98	18.91	18.81	18.75	18.67	18.52

Elapsed time (min)	4.0	6.25	9.0	12.25	16	25	36	64	100
Thickness of specimen (mm)	18.40	18.27	18.14	18.05	17.98	17.90	17.85	17.79	17.76

After 24 hr the thickness was 17.58 mm. Using the root-time method, determine (a) the coefficient of consolidation, and (b) the initial and primary compression ratios, for this stress range.

10.13 Using the log-time method and the data given in Exercise 12, determine the coefficient of consolidation and the initial and primary compression ratios.

10.14 A 15 m square raft foundation is founded at a depth of 2 m in a layer of over-consolidated clay, which itself is 8 m thick from the surface. It has been estimated that there will be an average increase in effective stress in the clay of 220 kPa. Laboratory tests corresponding to this stress range have shown:

Coefficient of volume compressibility $(m_v) = 0.20$ m^2/MN
Pore pressure coefficient $(A) = 0.50$

Obtain an estimate of the ultimate consolidation settlement, corrected by the Skempton–Bjerrum method.

Soil Mechanics Spreadsheets and Reference Assignments and Quizzes (available on the accompanying CD)

Assignments A.10   Settlement and consolidation
Quiz Q.10   Settlement and consolidation

Chapter 11

Bearing capacity of foundations

11.1 Design requirements and concepts

A *foundation* is the supporting base of a structure which forms the interface across which the loads are transmitted to the underlying soil or rock. In the majority of cases, foundations for building and civil engineering structures are constructed of plain or reinforced concrete. Some notable exceptions are earth and rock structures, such as roads, embankments and dams; also some piled foundations may be of steel or timber.

Since this book is concerned with basic soil mechanics, rather than foundation engineering in a specialist sense, the topics in this chapter relate principally to the soil mechanics aspects of shallow foundations and simple piled foundations. For more detailed analyses, constructional details and specialist coverage of topics such as deep basements, caissons, etc., readers should consult appropriate texts, e.g. Barker (1981), Tomlinson (1995).

The overall performance and functional viability of a foundation depends largely on the interaction between the structural unit above and the soil/rock unit below. The behaviour of the foundation depends on the nature of the soil and the behaviour of the soil depends on the shape and size of the foundation and also on the distortional behaviour of the superstructure. Thus, soils should not be thought of as having *intrinsic* bearing capacities; rather, the designer should think in terms of the *bearing capacity of specific foundation arrangements*, the parameters of which describe the performance criteria of both the soil and the structure above.

The requirements of the design process must therefore include provision for limiting settlement or other movement, safety against ultimate shear failure, functional serviceability and material durability, as well as economy in both the cost of construction and of future maintenance.

The great majority of foundation design problems are routine and are solved by the combined application of experience and local regulations. For example, in the United Kingdom (excluding Northern Ireland) the design and construction of *low-rise* buildings and their foundations must conform to one of the following:

(a) In England and Wales – The Building Regulations (current)
(b) In Inner London   – The London Building Acts
                             Constructional Bye-laws
(c) In Scotland   – The Building Standards (Scotland) (Consolidation)
                             Regulations

Foundations designed in accordance with BS 8004 and BS 8103, as appropriate, are deemed to satisfy the requirements of the Building Regulations. Under *Eurocode 7: Geotechnical Design*, such problems will fall into *Geotechnical Category 1* (see Appendix 1), for which the advice of a soil mechanics specialist will seldom be required. A significant degree of overdesign may occur, but not usually enough to seriously affect efficiency or performance. Such routine design depends largely on detailed observance of local practice and on experience, including qualitative studies of the performance of earlier similar structures.

11.2   Factors in the design of shallow foundations

Shallow foundations are sometimes called *spread* foundations and include isolated pads, strip footings and rafts. The most usual definition of a *shallow* foundation refers to the founding depth being less than the breadth. For the most part in the case of pad foundations and strip footings this is acceptable, but for wide rafts it is clearly unacceptable. It is sensible, therefore, to limit the term 'shallow' to mean *less than 3 m or less than the breadth of the footing*. From an analytical point of view 'shallow' is taken to mean that the shear strength of the soil above founding level is ignored in the computation of bearing capacity (see Section 11.5).

Leaving aside the structural design of the concrete footing itself, three main design criteria must be considered:

(a) *Adequate depth.* The depth at which a footing is *founded*, i.e. the depth below the surface to its underside, must be sufficient to prevent any adverse effects due to changes in surface conditions. These may include climatic changes in temperature or rainfall, the action of freezing and thawing, temperature changes transmitted from the building and changes in groundwater levels. Adequate depth may also be significant when foundations are subject to horizontal loads or strong overturning moments.

(b) *Limiting settlement.* The amount of total settlement, differential settlement and angular distortion that may be tolerated depends on the required functional performance of the building and the requirements of the user, as well as on economic factors, such as asset values, insurance costs, potential production losses, etc. Immediate (undrained) settlement must be considered separately from time-dependent settlement, most settlement damage may be classified as *architectural* and will be confined to cladding and finishes. Immediate settlement occurs during construction as the dead and structural loading is imposed; subsequent damage will therefore be minimised if the application of finishes, etc. is delayed until the dead loading is complete.

Guidelines to limiting values are suggested by a number of sources, but the following routine limits appear to be conventionally acceptable (Skempton and MacDonald, 1956):

SANDS   Maximum total settlement = 40 mm for isolated footings
= 40–65 mm for rafts
Maximum differential settlement between adjacent columns = 25 mm
CLAYS   Maximum total settlement = 65 mm for isolated footings
= 65–100 mm for rafts
Maximum differential settlement between adjacent columns = 40 mm

See also Section 11.6.

(c)   *Factor of safety against shear failure.* Shear failure occurs when the soil divides into separate blocks or zones which move fully or partially, and tangentially with respect to each other, along *slip surfaces.* Thus, a plastic yielding condition has developed wherein the shear stress along the slip surface has reached a limiting value, i.e. an *ultimate limit state.* Ultimate limit state shear failure is only likely to be a controlling design factor in the case of foundations on hard brittle soils or soil of low or intermediate plasticity; more generally, limiting settlement will be the controlling factor. Foundations most susceptible to shear failure are those having high live-to-dead load ratios, e.g. heavy engineering structures, silos, bridges, masts, water-retaining structures. The principal criterion for design will be that the ratio of the shear strength of the soil to the maximum mobilised shear stress must not be less than an appropriate value, i.e. *factor of safety.* Conventionally, this factor of safety should be at least 2.5 or 3.

Three principal modes of shear failure may be defined:

General shear failure (Fig. 11.1(a)).   This occurs when a clearly defined plastic yield slip surface forms under the footing and develops outward towards one or both sides and eventually to the ground surface. Failure is sudden and will often be accompanied by severe tilting leading to final collapse on one side. This mode of failure is associated with dense or over-consolidated soils of low compressibility.

Local shear failure (Fig. 11.1(b)).   In compressible soils, significant vertical movement may take place before any noticeable development of shear planes occurs. As the soil beneath the footing reaches the yield condition shear planes develop, but fail to extend to the ground surface. Some adjacent bulging may occur, but very little tilting takes place. The settlement which occurs will usually be the principal design criterion.

Punching shear failure (Fig. 11.1(c)).   In weak compressible soils considerable vertical movement may take place with the development of slip surface restricted to vertical planes adjacent to the sides of the footing. Bulging at the surface is usually absent and may even be replaced by *drag down.*

The failure mode depends mainly on the compressibility of the soil. In dense sands and cohesive soils of low compressibility the general shear condition usually occurs. In highly compressible clays and silts a punching shear failure is most likely. Punching failures have also been observed in

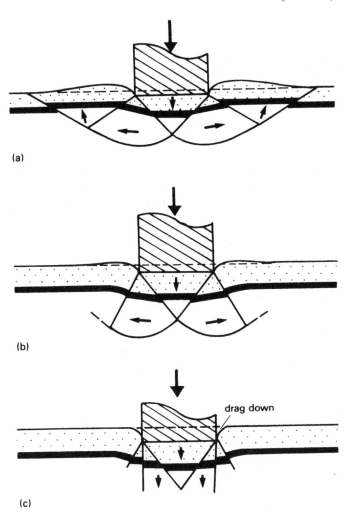

Fig. 11.1   Modes of shear failure
(a) General shear failure (b) Local shear failure (c) Punching shear failure

loose sands, but are more likely to occur at greater depths, or as a result of the compression of an underlying soft layer. In normally consolidated clays drainage conditions may influence the mode of failure: whereas general shear failure might occur under undrained conditions, under longer term drained conditions failure may be due to punching shear.

11.3   Definitions of bearing capacity

The *ultimate bearing capacity* (q_f) is the intensity of bearing pressure at which the supporting ground is expected to fail in shear, i.e. a collapse will take place. In *Eurocode 7* the equivalent value is defined (with considerable more length) as the *ultimate limit state design vertical capacitance* (Q_d) and is quantified as a load (force) and not as a pressure (stress).

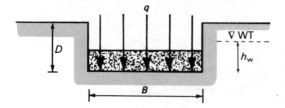

Fig. 11.2

The *net bearing pressure* (q_n) (also known as the *net foundation pressure*) is the net change in total stress experienced by the soil at the base of the foundation; this being the difference between the total applied stress and the stress removed due to excavation.

$$q_n = q - \sigma_o \qquad\qquad [11.1(a)]$$

where q = contact pressure at the base of the foundation
σ_o = overburden pressure adjacent to the foundation
(or overburden relief due to excavation)
= γD
γ = unit weight of soil
D = founding depth (Fig. 11.2)

In terms of effective stress (e.g. when the water table is above the base):

$$q'_n = q - \sigma'_o \qquad\qquad [11.1(b)]$$

where $\sigma'_o = \gamma D - \gamma_w h_w$
h_w = height of water table above foundation base (Fig. 11.2)

The *factor of safety* against shear failure is defined as the ratio between the net ultimate bearing capacity and the net bearing pressure:

$$\text{Factor of safety, } F = \frac{q_{nf}}{q_n} = \frac{q_f - \sigma'_o}{q - \sigma'_o} \qquad\qquad [11.2]$$

The usual value of F is 3.0, with 2.5 being considered a minimum for building foundations, but see also Appendix B.

The *allowable bearing capacity* (q_a) is defined as the bearing pressure that will cause either undrained or drained settlement or creep equal to a specified tolerable design limit (see also Section 11.6). For a given foundation in a given soil, the allowable design value for the applied bearing pressure must satisfy the two criteria given below.

I   An ultimate limit state value (shear strength)

$$\text{Allowable design bearing capacity, } q_a = \frac{q_f - \sigma'_o}{F} + \sigma'_o \qquad\qquad [11.3(a)]$$

II   A serviceability limit state (settlement)

Allowable design bearing capacity, q_a = bearing pressure corresponding to a specified limit value (s_L) of undrained or drained settlement.

II(a)   Immediate or undrained settlement – from eqn [6.74]:

$$q_a = \frac{s_L E_u}{B(1 - v^2)I_\rho} + \sigma_o'$$ [11.3(b)]

II(b)   Consolidation or drained settlement – from eqn [10.2]:

$$q_a = \frac{s_L}{m_v(I + 1)H_o} + \sigma_o'$$ [11.3(c)]

See Section 11.7.

A *presumed bearing pressure* is a conservative value attributed to a rock or soil for preliminary design purposes. Values are obtained using empirical data, and bearing in mind factors such as the width of the foundation, the probable settlement limits and local experience. Foundations on cohesionless soil may be designed using presumed bearing pressure estimated from the results of *in-situ* tests, such as standard penetration tests, cone penetrometer tests, pressuremeter tests, etc. (see Section 11.8). Some values of presumed bearing pressure are given in Table 11.1.

The design procedure usually commences with a notional type, shape and size of foundation being proposed, using the presumed bearing values given in

Table 11.1   Presumed bearing values for rocks and soils

Types of rocks or soils	Presumed bearing value (kPa)	Remarks
Rocks		
Hard igneous or gneissic rocks	10 000	Only sound unweathered
Hard limestones and sandstones	4 000	rocks
Schists and slates	3 000	
Hard shales and mudstones; soft sandstones	2 000	Thin-bedded or shattered
Soft shales and mudstones	600–1000	rocks must be assessed
Hard sound chalk; soft limestone	600	after inspection
Cohesionless soils		
Compact gravel or sand/gravel	>600	Providing:
Medium-dense gravel or sand/gravel	200–600	width $B \nless 1$ m
Loose gravel or sand/gravel	<200	and groundwater level
Compact sand	>300	$\nless B$ below base of
Medium-dense sand	100–300	footing
Loose sand	<100	
Cohesive soils		
Very stiff boulder clays; hard clays	300–600	This group is susceptible
Stiff clays	150–300	to long-term settlement
Firm clays	75–150	
Soft clays and silts	<75	
Very soft clays and silts	Not applicable	

Notes: (a) These values are intended as a guide for *preliminary* design purposes only.
(b) They are *gross* values, with allowance for embedment.
Adapted from Table 1, BS 8004 with permission of the British Standards Institution.

Table 11.1. The next steps involve the evaluation of the loading conditions and the calculation of the expected contact pressure. From laboratory or field tests, values are obtained for both the shear strength and settlement characteristics of the soil. Armed with this information, the designer can now evaluate the bearing capacities and compare these with the expected pressures. The size and depth of the foundation are now adjusted so as to arrive at a design embodying both structural and economic efficiency. In this context, not only must the final performance requirements of the structure be considered, but also safe and efficient construction processes.

11.4 Ultimate bearing capacity of shallow foundations

As with most of the analytical procedures in soil mechanics, the behaviour of an actual foundation is first idealised in the form of a simplified model. The model then provides a basis for a mathematical analysis, which very often requires some empirical modification in order to achieve good correlation between theoretical and observed behaviour.

Initially, in this case, the bearing capacity is considered of an infinitely long strip foundation, placed on the surface of a deep layer of soil, which is assumed to be weightless and in all senses uniform. A general shear failure is proposed and the condition of limiting equilibrium assumed to apply at the point of failure. In Fig. 11.3, such a foundation is shown resting on the surface of a clay soil under undrained loading conditions. The overburden pressure due to an actual founding depth of D is considered as a surcharge on the notional ground surface CABD: $\sigma_o = \gamma D$. The soil is assumed to behave elastically until a state of limiting equilibrium is reached, and then to be in plastic equilibrium along the shear slip surfaces shown.

As the *ultimate bearing capacity* (q_f) is reached, a wedge of soil (ABF) is displaced downward, within which the Rankine active state is developed and where the major principal stress is vertical and the minor principal stress is horizontal. The adjacent sectors (AFE and BGF) are forced sideways to rotate about the edges of the footing. Here the shear planes radiate from the edges A and B and the principal stress directions are rotated through a total angle of 90°.

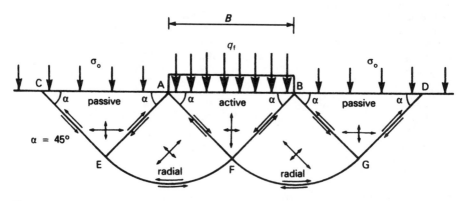

Fig. 11.3 General shear failure in undrained ($\phi_u = 0$) conditions

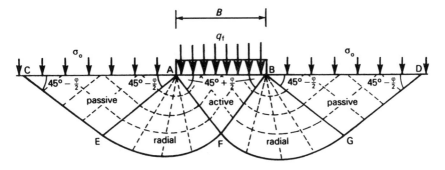

Fig. 11.4 General shear failure in drained conditions

In the outer wedges (CAE and BDG), therefore, the principal stresses are aligned in accordance with the Rankine passive state.

Under *undrained conditions* (i.e. $\tau_f = c_u$), the curves FE and FG are circular arcs, so that a solution may be obtained as follows:

$$q_f = (\pi + 2)c_u + \sigma_o = 5.14c_u + \sigma_o \qquad [11.4]$$

Under *drained conditions* ($\phi' > 0$), the angles α of the shear planes correspond to the active and passive Rankine values of $(45° + \phi'/2)$ and $(45° - \phi'/5)$ respectively (Fig. 11.4). The curves FE and FG now take the form of logarithmic spirals and the solution is expressed in the form:

$$q_f = cN_c + \sigma'_o N_q \qquad [11.5]$$

where N_c and N_q are dimensionless factors dependent on the drained angle of shearing resistance ϕ'.

A number of exact solutions have been obtained for N_c and N_q, but since the numerical disparities are in the main insignificant the original values proposed by Prandtl (1921) and Reissner (1924) are usually used:

$$N_q = \exp^{\pi \tan \phi'} \tan^2(45° + \phi'/2)$$

$$N_c = \cot \phi'(N_q - 1)$$

So far, the soil has been assumed to be weightless, and so a term is required in the q_f expression to take account of soil density. Terzaghi (1943) proposed a three-term expression for an infinitely long *strip footing* of breadth B:

$$q_f = cN_c + \sigma_o N_q + \tfrac{1}{2}B\gamma N_\gamma \qquad [11.6]$$

where N_c and N_q are defined as before and N_γ is a similar coefficient defining the bearing capacity of a soil of density γ, where both c and σ_o are equal to zero. No exact solutions have so far been obtained for N_γ. Terzaghi produced a set of empirical values based on a failure zone angle of $\alpha = \phi'$, but later work has shown $\alpha = 45° + \phi'/2$ to be a more realistic value. No general agreement as yet exists on a set of empirical N_γ values, but several pieces of research have shown that two very similar expressions are as good as any and yield maximum errors of less than 10 per cent:

Table 11.2 Bearing capacity factors ($\phi > 0$)

ϕ	N_c	N_q	N_γ	ϕ	N_c	N_q	N_γ	ϕ	N_c	N_q	N_γ
0	5.14	1.00	0.00	15	11.0	3.94	1.42	30	30.1	18.4	18.1
1	5.38	1.09	0.00	16	11.6	4.34	1.72	31	32.7	20.6	21.2
2	5.63	1.20	0.01	17	12.3	4.77	2.08	32	35.5	23.2	24.9
3	5.90	1.31	0.03	18	13.1	5.26	2.49	33	38.6	26.1	29.3
4	6.19	1.43	0.05	19	13.9	5.80	2.97	34	42.2	29.4	34.5
5	6.49	1.57	0.09	20	14.8	6.40	3.54	35	46.1	33.3	40.7
6	6.81	1.72	0.14	21	15.8	7.07	4.19	36	50.6	37.8	48.1
7	7.16	1.88	0.19	22	16.9	7.82	4.96	37	55.6	42.9	56.9
8	7.53	2.06	0.27	23	18.1	8.66	5.85	38	61.4	48.9	67.4
9	7.92	2.25	0.36	24	19.3	9.60	6.89	39	67.9	56.0	80.1
10	8.34	2.47	0.47	25	20.7	10.7	8.11	40	75.3	64.2	95.5
11	8.80	2.71	0.60	26	22.3	11.9	9.53	41	83.9	73.9	114
12	9.28	2.97	0.76	27	23.9	13.2	11.20	42	93.7	85.4	137
13	9.81	3.26	0.94	28	25.8	14.7	13.10	43	105	99	165
14	10.4	3.59	1.16	29	27.9	16.4	15.40	44	118	115	199
								45	134	135	241
								46	152	159	294
								47	174	187	359
								48	199	222	442
								49	230	266	548
								50	267	319	682

Values of N_c after Prandtl
N_q after Reissner
N_γ after Hansen
For $\tau_f = c_u$ see Fig. 11.5
(values of N_c)

$N_\gamma = 2.0(N_q + 1) \tan \phi'$ (Caquot and Kerisel, 1953)

$N_\gamma = 1.8(N_q - 1) \tan \phi'$ (Hansen, 1961)

Table 11.2 gives values of N_c and N_q, together with Hansen's values of N_γ. Modifications to N_c, N_q and N_γ to take into account inclined and eccentric loading have been considered by de Beer (1965, 1966), Chen (1975), Meyerhof (1953), Hansen (1961) and Sokolovski (1960). For most foundations of adequate depth it is only necessary to consider the shape; exceptions are isolated footings subject to substantial horizontal loading or overturning moments (see Section 11.5).

11.5 Ultimate stability design for shallow foundations

In the analyses summarised in the preceding section, the length of the foundation is assumed to be infinitely long. In order to provide a general method for engineering design purposes three shape categories are usually defined: strip, rectangular and circular. In the light of work by de Beer (1967) and Vesic (1970) the following general expression seems acceptable, with certain limitations, for shallow foundations of adequate depth and loaded vertically.

$q_{nf} = cN_cs_c + \sigma_oN_qs_q + \frac{1}{2}\gamma BN_\gamma s_\gamma - \gamma D$ [11.7]

Table 11.3 Shape factors for shallow foundations ($\phi' > 0$)

Shape of footing	s_c	s_q	s_γ
Strip	1.00	1.00	1.00
Rectangle	$1 + \left(\dfrac{B}{L}\right)\left(\dfrac{N_q}{N_c}\right)$	$1 + \left(\dfrac{B}{L}\right)\tan\phi'$	$1 - 0.4\left(\dfrac{B}{L}\right)$
Circle or square	$1 + \dfrac{N_q}{N_c}$	$1 + \tan\phi'$	0.6

After de Beer (1967) and Vesic (1970)

where N_c, N_q, N_γ are bearing capacity factors (Table 11.2) and s_c, s_q, s_γ are shape factors (Table 11.3).

The main limitations of this general expression are:

(a) No account is taken of soil compressibility and consequent changes in volume.
(b) No account is taken of changes in water table level or pore pressure.
(c) A linear relationship is assumed to exist between ϕ' and normal effective stress, which may not be the case at high values of effective stress.

Foundations on sands and gravels
In soils of high permeability, changes in groundwater level can occur rapidly with consequent effects on soil density and pore pressures. Bearing capacity calculations should be carried out in terms of effective stress and eqn [11.7] written as:

$$q_{nf} = \sigma'_0 N_q s_q + \tfrac{1}{2}\gamma' B N_\gamma s_\gamma - \sigma'_0 \qquad [11.8]$$

Note that the 'cohesion term' is eliminated, since $c' = 0$. The values of N_q and N_γ are usually deduced from *in-situ* tests (see Section 11.8), or they may be taken from Table 11.2 in relation to the drained angle of shearing resistance ϕ'. The value of σ'_0 is the effective overburden pressure at the founding depth.

The bearing capacity of cohesionless soils with moderate to high values of ϕ' is substantially reduced when the water table lies within the zone extending from the surface down to a depth of B below the foundation. When the water level is expected to remain well below a depth of $z_w = D + B$, the unit weight used in each term of eqn [11.8] will be the value γ for the drained soil. If the water table lies at the surface ($z_w = 0$), the buoyant or submerged unit weight must be used: $\gamma' = \gamma_{sat} - \gamma_w$. For water table depths intermediate between $z_w = 0$ and $z_w = D + B$, values of γ' should be used as indicated in Fig. 11.5. When $z_w < D$, it is preferable that the effective overburden stress (σ'_0) be calculated from $\sigma'_0 = \gamma D - \gamma_w h_w$ and $\gamma' = \gamma_{sat} - \gamma_w$ used the $\tfrac{1}{2}B\gamma'N_\gamma$ term. Note that if upward seepage flow takes place, the effective unit weight of the soil is further reduced.

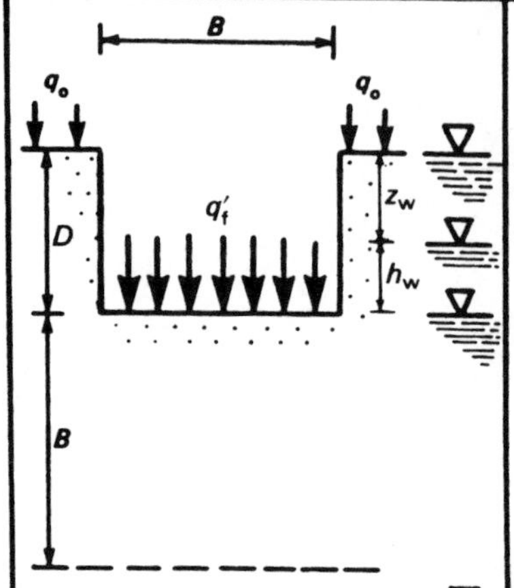

	position of water table	term $\sigma'_o N_q$	term $\frac{1}{2} B \gamma' N_\gamma$
	at ground surface	$\gamma' = \gamma - \gamma_w$	$\gamma' = \gamma - \gamma_w$
	intermediate	Calc. σ'_0	Calc. $\gamma' = \gamma - \gamma_w$
	at base of foundation	$\gamma' = \gamma$	$\gamma' = \gamma - \gamma_w$
	below the passive zone	$\gamma' = \gamma$	$\gamma' = \gamma$

Note: If there is an upward component of seepage flow of hydraulic gradient *i*, the effective unit weight (γ') must be further reduced by an amount equal to $i\gamma_w$
$\gamma_w = 9.81$ kN/m^3

Fig. 11.5 Reduction in γ' due to groundwater

Worked example 11.1   *A shallow strip footing of breadth 2.5 m is to be founded at a depth of 2.0 m in a well-drained compact sand having the following properties:*

$c' = 0$ $\phi' = 34°$ $\gamma = 19.0$ kN/m^3

Determine the safe load per metre length based on the ultimate bearing capacity and applying a factor of safety of 3.0.

First, determine the net ultimate bearing capacity. Assuming there is no surface surcharge adjacent to the foundation:

$$q_{nf} = \gamma' D N_q s_q + \tfrac{1}{2} \gamma' B N_\gamma s_\gamma - \gamma' D$$

From Table 11.2: $N_q = 29.4$, $N_\gamma = 34.5$

From Table 11.3: $s_q = s_\gamma = 1.0$

In well-drained conditions: $\gamma' = \gamma = 19$ kN/m^3

Hence, $q_{nf} = 19 \times 2(29.4 - 1) + \tfrac{1}{2} \times 19 \times 2.5 \times 34.5$

$= 1899$ kPa

Design bearing capacity, $q_d = q_{nf}/F + \gamma' D$

$= 1899/3 + 19 \times 2 = 671$ kPa

Design load per metre run, $Q_d = 671 \times 2.5 = \underline{1678 \text{ kN}}$

Worked example 11.2 *A strip footing at a depth of 0.9 m is required to transmit an inclusive load of 650 kN/m to a compact sand having the following properties:*

$c' = 0$ $\phi' = 38°$ $\gamma_{sat.} = 20.4 \ kN/m^3$

Assume that the water table may rise to the surface and, adopting a factor of safety of 3.0, determine the breadth of footing required.

From Table 11.2: $N_q = 48.9$, $N_\gamma = 67.4$

Then the net ultimate bearing capacity,

$$q_{nf} = \gamma' DN_q + \tfrac{1}{2}\gamma' BN_\gamma - \gamma' D$$
$$= (20.4 - 9.81)0.9 \times 47.9 + \tfrac{1}{2}(20.4 - 9.81)B \times 67.4$$
$$= 456.9 + 356.9B$$

Design loading intensity, $q_d = Q/B = q_{nf}/F + \gamma' D$

Therefore $650/B = 152.3 + 119.0B + 0.9(20.4 - 9.81)$

Giving $0 = 119.0B^2 + 161.8B - 650$

From which $B = \underline{1.75 \ m}$

Worked example 11.3 *A square footing of side dimension 4.2 m is founded at a depth of 2.0 m in a sand having the following properties:*

$c' = 0$ $\phi' = 32°$ $\gamma_{sat.} = 20 \ kN/m^3$

Determine the ultimate bearing capacity of the footing when the water table is: (a) well below the base, (b) at a level equal to that of the base, (c) at the ground surface, and (d) at the ground surface with an upward seepage of hydraulic gradient i = 0.4.

From Table 11.2, for $\phi' = 32°$: $N_c = 35.5$, $N_q = 23.2$, $N_\gamma = 24.9$
From Table 11.3, $s_c = 1 + 23.2/35.5 = 1.653$
                                   $s_q = 1 + \tan 32° = 1.625$
                                   $s_\gamma = 0.6$

Since there is no surface surcharge, $\sigma'_o = \gamma' D$

(a)   When the water table is well below the base:

$q'_f = \gamma DN_q s_q + \tfrac{1}{2}\gamma BN_\gamma s_\gamma$
       $= 20 \times 2.0 \times 23.2 \times 1.625 + \tfrac{1}{2} \times 20 \times 4.2 \times 24.9 \times 0.6$
       $= 20 \times 75.4 + 20 \times 31.4 = \underline{2315 \ kPa}$

(b)   Water table at level of base:

$q'_f = \gamma DN_q s_q + \tfrac{1}{2}(\gamma_{sat.} - \gamma_w)BN_\gamma s_\gamma$
       $= 20 \times 75.4 + (20.0 - 9.81)31.4 = \underline{1828 \ kPa}$

(c)   Water table at ground surface:

$q'_f = (\gamma_{sat.} - \gamma_w)DN_q s_q + \tfrac{1}{2}(\gamma_{sat.} - \gamma_w)BN_\gamma s_\gamma$
       $= (20.0 - 9.81)75.4 + (20.0 - 9.81)31.4 = \underline{1088 \ kPa}$

(d)   With an upwards seepage for which $i = 0.4$:

$q'_f = (20.0 - 9.81 - 0.4 \times 9.81)75.4 + (20.0 - 9.81 - 0.4 \times 9.81)31.4$
       $= \underline{669 \ kPa}$

Worked example 11.4   *Show that the evaluation of net ultimate bearing capacity is very sensitive to variations in the measured or estimated values of ϕ'.*

Consider a shallow strip footing for which:

$B = 1.5$ m $D = 1.0$ m $c' = 0$ $\gamma = 20$ kN/m^3

Then $q_{nf} = \gamma'DN_q + \frac{1}{2}\gamma'BN_\gamma - \gamma'D$

$= 20(N_q - 1) + 15N_\gamma$

The variation of q_{nf} with ϕ' is shown in the table below. The other tabulated values are $R = q_f/q_{f(30)}$, a normalising coefficient R with respect to q_{nf} for $\phi' = 30°$, and the percentage error incurred with an overestimate of $1°$ in the value of ϕ'.

ϕ' *(deg)*	N_q	N_γ	q_{nf} *(kPa)*	$R = \dfrac{q_u}{q_{u(30)}}$	*% error of* $\Delta\phi' = +1°$
20	6.40	3.54	161	0.26	14.2
25	10.7	8.11	316	0.51	14.2
30	18.4	18.1	620	1.0	14.5
35	33.3	40.7	1257	2.03	16.0
40	64.2	95.5	2697	4.35	17.5
45	135	241	6295	10.2	20.2

Medium to dense drained sands and gravels
In these soils the ultimate bearing capacity will be so high, even allowing for movement of the water table, that a *tolerable settlement* criterion will govern the design in most cases (see Sections 11.6 and 11.7).

Saturated medium dense sands and gravels
High water table levels may reduce the ultimate bearing capacities by up to 60 per cent and so predictions must be made of post-construction drainage conditions and 'probable worst cases' evaluated. In some cases of narrow strip footings ultimate bearing capacity may govern the design, but with wide footings and raft foundations settlement will usually be the controlling factor. It will be important also to consider the effects of pumping during construction.

Loose saturated sands and silty sands
In very loose ($N < 5$) saturated fine or medium-fine and silty sands, rapid rises in water table can seriously affect ultimate stability. The possibility must also be considered of liquefaction due to vibrations resulting from traffic, machinery, earthquakes, etc. Either the soil should be compacted to a denser state (e.g. by rolling, ramming, vibrating, dynamic compaction, vibroflotation, etc.) or the foundation should be taken down to a more suitable layer.

Foundations on clays and silts
The ultimate bearing capacity of fine soils with low permeabilities is most critical immediately after construction before the excess pore pressure has had time to

dissipate, i.e. under *undrained* conditions. In the course of time, as consolidation takes place, the stiffness of the soil increases and so does the shear strength. The design of foundations on fine soils should therefore be in terms of undrained or total stress conditions.

Skempton (1951) suggested that for an undrained saturated clay ($\phi_u = 0$), the basic form of Terzaghi's equation should be used, but with values of N_c related to the shape and depth of the foundation.

For $\tau_f = c_u$: $q_f = c_u N_c + \gamma D$ [11.9(a)]

$q_{nf} = c_u N_c$ [11.9(b)]

(since $N_q = 1$ and $N_\gamma = 0$)

Values of N_c may be obtained from the chart shown in Fig. 11.6 or from the following empirical expression:

$$N_c = 5.14\left(1 + 0.2\frac{B}{L}\right)\left[1 + \sqrt{\left(0.053\frac{D}{B}\right)}\right]$$ [11.9(c)]

with maximum values when $D/B \geq 4.0$ as follows:

when $B/L = 0$, $N_c = 7.5$ (i.e. strip footing)
where $B/L = 1.0$, $N_c = 9.0$ (i.e. square or circle)

These values seem to be borne out by some back analyses of known failure except in the case of very sensitive or quick clays.

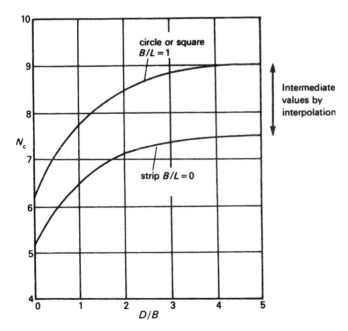

Fig. 11.6 Skempton's values for N_c for undrained conditions

Worked example 11.5 *A rectangular footing of breadth 6 m and length 15 m is founded at a depth of 4.5 m in a fine soil having the following properties:*

$c_u = 40 \ kPa \quad \gamma = 18 \ kN/m^3$

Determine the ultimate bearing capacity for the footing.

$D/B = 4.5/6.0 = 0.75 \quad B/L = 6/15 = 0.40$

Using Skempton's N_c from Fig. 11.6 (interpolating): $\quad N_c = 6.7$.

So $\quad q_f = c_u N_c + \gamma D$

$\qquad = 40 \times 6.7 + 18 \times 4.5 = \underline{349 \ kPa}$

Using equ [11.9(c)]:

$\qquad N_c = 5.14(1 + 0.2 \times 0.40)[1 + (0.053 \times 0.75)^{1/2}] = 6.66$

So $\quad q_f = 40 \times 6.66 + 18 \times 4.5 = \underline{347 \ kPa}$

Using the general eqn [11.7] and factors from Tables 11.2 and 11.3:

$\qquad N_c = 5.14, \quad N_q = 1, \quad N_\gamma = 0$

$\qquad s_c = 1 + 0.40 \dfrac{1}{5.14} = 1.078 \quad s_q = 1.0$

So $\quad q_f = 40 \times 5.14 \times 1.078 + 18 \times 4.5 = 303 \ kPa$

This last result is some 11 per cent lower than the other two and may be considered a conservative value.

Worked example 11.6 *Determine the breadth of a strip footing required to carry an inclusive load of 550 kN/m at a depth of 1.5 m in a fine soil having the following properties:*

$c_u = 90 \ kPa \quad \gamma = 19 \ kN/m^3$

Design bearing capacity, $\quad q_d = \dfrac{Q}{B}$

and $\qquad\qquad\qquad q_{nf} = c_u N_c = 90 N_c$

$\qquad\qquad \therefore \quad q_d = \dfrac{90 N_c}{F} + \gamma D = \dfrac{550}{B}$

Try $B = 2.0 \ m \quad D/B = 0.75 \quad B/L = 0$

$\qquad N_c = 5.14[1 + (0.053 \times 0.75)^{1/2}] = 6.16$

$\qquad q_d = 30 \times 6.16 + 19.0 \times 1.5 = \dfrac{550}{B}$

Giving $\quad B = 2.6 \ m$

Try $B = 2.5 \ m \quad D/B = 0.6$

$\qquad N_c = 5.14[1 + (0.053 \times 0.6)^{1/2}] = 6.06$

$\qquad q_d = 30 \times 6.06 + 19.0 \times 1.5 = \dfrac{550}{B}$

Giving $\quad B = 2.6 \ m$

Hence $\quad B = \underline{2.6 \ m}$

Worked example 11.7 *A strip footing will be required to transmit an inclusive load of 900 kN/m in an overconsolidated soil having the following properties:*

$$c_u = 130 \; kPa \quad \gamma = 19 \; kN/m^3 \quad \phi_c' = 26° \quad c' = 18 \; kPa \quad \phi_p' = 22°$$

It is proposed to found the footing at a depth of 2.0 m and a factor of safety of 3 is required. Calculate the minimum breadth required with respect to the undrained, critical and peak strengths.

Initially assuming a breadth of $B = 3.0$ m:

Undrained

$D/B = 2/3$;   then $N_c = 5.14(1 + \sqrt{(0.053 \times 2/3)}) = 6.106$

Design bearing capacity, $q_d = \dfrac{Q}{B} = \dfrac{c_u N_c}{F} + \gamma D$

Substituting values: $\dfrac{900}{B} = \dfrac{130 \times 6.106}{3} + 19 \times 2.0$

giving $B = \underline{2.97 \text{ m}}$

Drained – critical strength

From Table 11.2: $N_q = 11.9$, $N_\gamma = 9.53$

$q_{nf} = \gamma D (N_q - 1) + \frac{1}{2} \gamma B N_\gamma$

$\quad = 19 \times 2.0(11.9 - 1) + \frac{1}{2} \times 19 \times 3.0 \times 9.53$

$\quad = 414.2 + 271.6 = 685.8 \text{ kPa}$

Design bearing capacity, $q_d = \dfrac{Q}{B} = \dfrac{q_{nf}}{F} + \gamma D$

Substituting values: $\dfrac{900}{B} = \dfrac{685.8}{3} + 19 \times 2.0$

giving $B = \underline{3.38 \text{ m}}$

Drained – peak strength

From Table 11.2: $N_c = 16.9$, $N_q = 7.82$, $N_\gamma = 4.96$

$q_{nf} = c' N_c + \gamma D (N_q - 1) + \frac{1}{2} \gamma B N_\gamma$

$\quad = 18 \times 16.9 + 19 \times 2.0(7.82 - 1) + \frac{1}{2} \times 19 \times 3.0 \times 4.96$

$\quad = 304.2 + 229.2 + 141.4 = 674.8 \text{ kPa}$

Design bearing capacity, $q_d = \dfrac{Q}{B} = \dfrac{q_{nf}}{F} + \gamma D$

Substituting values: $\dfrac{900}{B} = \dfrac{674.8}{3} + 19 \times 2.0$

Giving $B = \underline{3.42 \text{ m}}$

Comments: undrained stability is marginally better than drained here. However, undrained strength is dependent on water content and therefore groundwater conditions. If future groundwater conditions cannot be predicted a drained critical strength solution is preferable: after reiteration a minimum breadth of 3.30 m is found to be required.

Firm to stiff clays

The ultimate bearing capacity criterion will govern design in most cases where the soil has a low compressibility, e.g. firm to stiff normally consolidated clays, non-fissured and unweathered overconsolidated clays (including boulder clay). If a minimum factor of safety of 2.5 is used against ultimate failure, the settlement will usually remain within a tolerable serviceability limit.

Short-term undrained loading conditions during or immediately after construction will be most critical, so use eqn [11.9]. The undrained shear strength may be assumed to be constant for shallow depths and equal to the confidence limit *average* of values obtained in conventional consolidated-undrained triaxial tests. Samples should be taken in the zone between the founding depth and a depth below this equal to the breadth of the foundation.

Soft normally consolidated clays and silts

Consolidation settlement is the most likely design criterion in soils of moderate to high compressibility. It is prudent to evaluate the allowable bearing capacity in terms of both tolerable drained settlement and ultimate shear failure, adopting the lowest as the design value.

Soft to firm overconsolidated clays

The stiffness and strength of overconsolidated clays are seriously affected by weathering, e.g. weathered brown London Clay. Swelling takes place and the undrained shear strength decreases considerably. The presence of fissures will speed up the rate of swelling, especially under certain conditions of exposure, e.g. new sloping cut surfaces, and where fluctuations in water table occur. Both short-term (undrained) and long-term (drained) stability should be checked. For long-term analyses, effective stresses must be evaluated and critical state parameters used, i.e. $c' = 0$ and $\phi' = \phi'_c$. Settlement (or allowable bearing capacity) must be calculated using the properties of the weathered and not fresh clay. The possibility should also be considered of *heave* during construction after deep excavation and after completion of lightly-loaded foundations.

High-rise and heavy buildings

Mostly, these should be considered to be *Geotechnical Category 3* problems (see Appendix A) and necessitate the use of a geotechnical specialist. A full analysis of stresses, pore pressures, vertical and lateral strains, bearing capacities and settlements should be carried out.

Layered deposits

Thin clay layer over stronger deposits (Fig. 11.7(a))

If ultimate bearing capacity would control design in a thicker layer of the same soil, a modified form of eqn [11.9(b)] can be used.

Net ultimate bearing capacity, $q_{nf} = c_u N_{cH}$ [11.10]

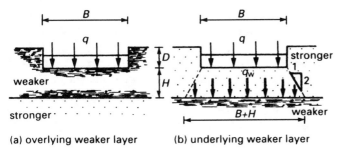

(a) overlying weaker layer (b) underlying weaker layer

Fig. 11.7 Effect of a weaker layer

where N_{cH} = Skempton's bearing capacity factor modified as follows:

B/H	2	2–7	>7
N_{cH}	N_c	$N_c + (B/2H - 1)$	7.6

and H = thickness of the thin layer

Underlying weaker layer (Fig. 11.7(b))
The simplest approach to this problem is to assume that the applied load spreads laterally in a ratio of 2 vertical: 1 horizontal. Therefore, at the top of the weaker layer, the uniform applied load is assumed to be:

Strip footing: $\qquad q_w = q\dfrac{B}{B + H}$ $\qquad\qquad$ [11.11(a)]

Rectangular footing: $\quad q_w = q\dfrac{B^2}{(L + H)(B + H)}$ $\qquad$ [11.11(b)]

Circular footing: $\qquad q_w = q\dfrac{B^2}{(B + H)^2}$ $\qquad\qquad$ [11.11(c)]

These expressions actually underestimate q_w at the centre and overestimate it at the edges, but are considered sufficiently acceptable if the factor of safety is at least 3. For more rigorous and accurate analyses refer to Section 6.10.

11.6 Settlement criteria for foundation design

The majority of foundations are specified for light or medium-sized buildings, including domestic buildings. In the main, these are 'designed' simply on the basis of conventional practice or routine, or they follow guidelines set out in local regulations (see Section 11.1). It is likely that most are over-designed, but the usual extent of this does not impose severe economic penalties. The simplest 'proper' design methods are based only on index properties (see Sections 7.18 and 10.3), but produce results very close to those obtained by more rigorous methods.

The values assumed for design loads will often be much greater than those that will subsequently occur, especially where allowance is being made for live loading. In most buildings, live loading will be about 15–30 per cent of dead loads, and usually all of the dead loads are applied before the final fittings and finishes are complete. Thus, most of the immediate settlement will be due to dead loads and in many cases (e.g. in cohesionless soil and stiff clays) will be virtually complete by the end of construction.

The main settlement-associated problem in the performance of buildings is cracking due to angular distortion. Building components, such as brick walls and concrete floor slabs, are extremely brittle and crack easily when distorted. Although such brittle failure is less likely with ductile components fabricated in steel, timber and reinforced concrete, unacceptable over-stressing may still occur.

Deciding the criteria

The decision as to what constitutes a *tolerable* settlement limit is often taken arbitrarily or by default. The consequences, however, if adverse, will affect many different people: architect, structural designer, geotechnical specialist, contractor, building owner, building user, developer, insurer, etc. Performance limits should be arrived at by those responsible for design, but after due consultation with other interested parties. Three groups of criteria will need to be considered:

(a) *Visual appearance*. Deviations from vertical of greater than 1/250 will be noticeable and may cause subjective feelings of alarm in users and observers. Similarly noticeable are deviations from horizontal of 1/100 and deflection/span ratios of 1/250.

(b) *Visible damage*. Acceptable levels of damage vary subjectively and with the type and part of building affected. Jennings and Kerrich (1962) proposed a classification of damage related to the ease of repair. Table 11.4, which is based largely on this work, was prepared by Burland, Broms and de Mello (1978).

(c) *Damage affecting structural or functional integrity*. Cracking damage may allow penetration of water or other liquids, leading to the subsequent development of other damage, e.g. corrosion of reinforcement, deterioration of jointing materials, promotion of rot or decay, etc. Movement and distortion limits will mostly be dictated by the use and function of a building or a particular component, e.g. lifts, conveyors, overhead cranes, precision machinery, drains, large doors, etc. In modern warehousing the use of automated store-and-retrieve systems requires special consideration. Particular care must be taken to prevent cracking damage in load-bearing brick walls, since loss of structural integrity may alter the wider pattern of loading in the building.

A number of studies have been presented dealing with recommended design limits for settlement, differential settlement and angular distortion: Skempton and MacDonald (1955), Polshin and Tokar (1957), Bjerrum (1963), Burland and Wroth (1975). Guidelines for settlement and differential settlement were given in Section 11.2. The following definitions (after Burland and Wroth) of settlement

Table 11.4   *Classification of visible damage to walls with particular reference to ease of repair*

Degree of damage	Description of typical damage* (ease of repair is underlined)	Approx. crack width (mm)
0. Negligible	Hairline cracks of less than 0.1 mm are classified as negligible	<0.1
1. Very slight	*Fine cracks which can easily be treated during normal decoration.* Perhaps isolated slightly fracturing in building. Cracks in external brickwork visible on close inspection	1–5
2. Slight	*Cracks easily filled. Redecoration probably required.* Several slight fractures showing inside of building. Cracks are visible externally and *some repointing may be required externally* to ensure weathertightness. Doors and windows may stick slightly	>5
3. Moderate	*The cracks require some opening up and can be patched by a mason. Recurrent cracks can be masked by suitable linings. Repointing of external brickwork and possibly a small amount of brickwork to be replaced.* Doors and windows sticking; service pipes may fracture; weathertightness often impaired	5–15 or a number of cracks ≥3
4. Severe	*Extensive repair work involving breaking out and replacing sections of walls, especially over doors and windows.* Windows and door frames distorted, floor sloping noticeably; walls leaning or bulging noticeably, some loss of bearing in beams; service pipes disrupted	15–25 but also depends on no. of cracks
5. Very severe	*Major repair required involving partial or complete rebuilding.* Beams lose bearing; walls lean badly and require shoring; windows broken with distortion; danger of instability	Usually >25 but also depends on no. of cracks

* Crack width is only one aspect of damage and should not be used on its own as a direct measure of it.
Note: In assessing the degree of damage account must be taken of its location in the building or structure.
After Burland, Broms and de Mello (1978)

and distortion quantities are recommended for use in quantitative studies relating to the foundation design (see also Fig. 11.8).

Settlement (s or ρ) is the downward displacement of a given point, e.g. settlement of point $B = s_B$.

Differential settlement (δs or $\delta \rho$) is the displacement of one point with respect to another, e.g. displacement of B with respect to $A = \delta s_{BA}$.

Angular strain (α) at a given point is the change in slope at that point, e.g. $\alpha_A = \delta s_{BA}/L_{BA} + \delta s_{BC}/L_{BC}$.

Tilt angle (ω) is the rigid body rotation angle of a well-defined unit of structure.

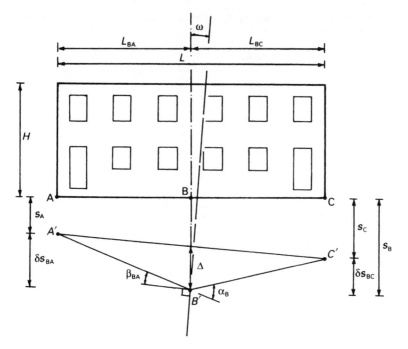

Fig. 11.8 Definitions and symbols for foundation distortion (after Burland and Wroth, 1975)

Relative rotation (β) is the rotation of a straight line between two reference points relative to the tilt.

Relative deflection (Δ) is the maximum displacement occurring between two points relative to the straight line drawn between them.

Angular distortion (Δ/L) is the ratio of the relative deflection between two points and the distance between them. Note that the angular distortion is the same as the relative rotation when applied to simple shear distortion; however, this will not be the case in bending since the rotation occurs at the supports in a simple beam.

Limitation of distortion
Table 11.5 gives guide figures for the angular distortion limits in relation to observed damage. Skempton and MacDonald recommended practical design limits for angular distortion of 1/500 for most conventional structures, but suggested that a figure of 1/1000 would be necessary to avoid visible cracking damage. Structural damage is unlikely to occur with values of Δ/L less than 1/150.

Burland and Wroth examined a number of relationships for use in design, mostly based on the concept of cracking in a brittle uniform beam. Although these were arrived at by a non-rigorous analysis, they do produce values comparable with those relating to observed damage. A classical elasticity analysis was used and, adopting criteria proposed by Skempton and MacDonald and Polshin and Tokar, a set of practical criteria are suggested for cracking and other damage. It is proposed that the onset of cracking is related to the tensile strain induced by the distortion: for brickwork and similarly brittle materials,

Table 11.5 Angular distortion limits

Δ/L	Critical structural situation
1/5000	Hair cracks observed in unreinforced brickwork; load-bearing walls subject to hogging
1/3000	Visible cracks in load-bearing walls
1/1000	Visible cracks in brick in-fill panels
1/750	Practical limit to prevent unbalance of sensitive machinery
1/600	Level of overstress in diagonal members becomes significant
1/500	Practical limit to prevent serious cracking in framed buildings and modern constructions
1/300	Damage to building frames and panel walls; misalignment problems in overhead cranes
1/250	Tilting noticeable in high-rise buildings
1/150	Structural damage to most buildings expected

After Skempton and MacDonald (1956), Bjerrum (1963), Burland and Worth (1975)

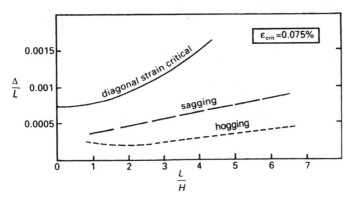

Fig. 11.9 Maximum angular distortion related to component shape
(after Burland and Wroth, 1975)

this appears to lie in the range 0.05 to 0.10 per cent. Thus, a limiting value is proposed of $e_{crit.}$ = 0.075 per cent. Another governing factor is the stiffness modulus ratio (E/G): for isotropic materials, E/G = 2.6, but for structures flexible in shear E/G will be higher, and lower for relatively stiff structures.

The suggested limiting values are given in Fig. 11.9 and are based on $e_{crit.}$ = 0.075 per cent and categorised for different building types as follows:

(a) *Structures of low shear stiffness*, e.g. framed buildings, reinforced walls, etc., where cracking due to diagonal tensile strain will be critical.

(b) *Sagging structures extremely brittle in tension*, e.g. unreinforced brickwork and masonry walls, where tensile bending strains at the lower boundary will be critical.

(c) *Hogging structures extremely brittle in tension*, e.g. unreinforced brickwork and masonry walls, where tensile bending strains at the upper boundary will be critical.

(d) *Structures in combined bending and shear*, e.g. infilled frame panels and structures where cracking is not permissible, where the maximum relative rotation β must not exceed 0.002 (1/500).

11.7   Evaluation of allowable bearing capacity

In the majority of foundation design problems a serviceability limit of tolerable settlement will govern design, but designers should be aware of the overall behavioural probabilities. Short-term and long-term stability may be related to different mechanisms, as may also the behaviour during construction. Generalised arbitrary limits may be inappropriate for complex structures and where variable ground conditions occur. Some parts of structure may be considerably stiffer or less stiff than others, requiring therefore different critical limits. Interaction between the structure and the ground may often give rise to the most critical mechanisms. The essence of good design is that evaluation must be dependent upon and consequent to a full understanding of behaviour in both the structures and the ground.

Allowable bearing capacity of clays
The methods of computing settlement are described in Chapter 10. Once a *tolerable settlement limit* has been established with due regard to structural and functional implications, the intensity of bearing pressure that would produce this may be calculated. The following simplified method can be used for straightforward cases, but where deep or stratified layers of compressible soil are involved, a more rigorous analysis should be used.

For a *tolerable immediate settlement* of s_i (from eqn [6.74]):

$$q_a = \frac{s_i E}{B(1 - v^2)I_\rho} + \gamma D \qquad [11.12(a)]$$

For a *tolerable consolidation settlement* of s_c (from eqn [10.2]):

$$q_a = \frac{s_c}{m_v(I + 1)H} + \gamma D \qquad [11.12(b)]$$

where m_v = coefficient of volume compressibility
$\quad\quad\quad H$ = the thickness of the compressible layer beneath the foundation or $2B$, whichever is least
$\quad\quad\quad I$ = the appropriate stress coefficient obtained from Chapter 6, corresponding to a depth of H below the foundation

Which of these two equations is applicable in drained conditions depends on the level of initial and final effective stresses in the soil below the foundation and their relationship with the preconsolidation stress σ'_p. If the soil is not to be stressed beyond σ'_p, elastic deformation may be assumed and eqn [11.12(a)] used with drained parameters E' and v'. Consolidation settlement takes place once σ'_p has been exceeded; a two-stage calculation will be necessary when σ'_p falls between the initial and final stresses.

Worked example 11.8   *A rectangular raft foundation has a breadth of 9 m and a length of 25 m and is founded at a depth of 3.2 m. The water table is located at a depth of 2.5 m below the ground surface. The soil is a deep layer of clay for which the following properties have been obtained from laboratory tests:*

$c_u = 25\ kPa \quad \phi_u = 0 \quad \gamma = 17.4\ kN/m^3$
$c' = 5\ kPa \quad \phi' = 23° \quad E_o' = 36\ MPa$

Determine the allowable bearing capacity with respect to the following conditions:

(A) Immediately after construction, assuming saturated undrained conditions.
(B) After a longer period of time, assuming the rate of loading to be slow enough so that no excess pore pressure develops.

The allowable bearing capacity will be taken as the least of those obtained respectively from the consideration of a factor of safety against shear failure of 3.0 and an average tolerable settlement of 75 mm.

Firstly, some basic geometric and elastic properties need to be evaluated.

$B/L = 9/25 = 0.360 \quad D/B = 3.2/9.0 = 0.356$

$$v' \approx \frac{K_o}{1 + K_o} = \frac{1 - \sin 23°}{2 - \sin 23°} = 0.38$$

From eqn [6.21]: $E' = E_o' \dfrac{(1 + v')(1 - 2v')}{1 - v'}$

$$= 36.0 \frac{1.38 \times 0.24}{0.62} = 19.2\ MPa$$

From eqn [6.24]: $E_u = E' \times \dfrac{1.5}{1 + v'} = 20.9\ MPa$

Condition A

(i) Shear failure: use the Skempton method.

$$N_c = 5.14(1 + 0.2 \times 0.360)[1 + (0.053 \times 0.356)^{1/2}] = 6.26$$

Then $q_{nf} = c_u N_c = 25 \times 6.26 = 157\ kPa$

Therefore $q_a = \frac{157}{3} + 17.2 \times 3.2 = \underline{108\ kPa}$

(ii) Settlement: from Table 6.4 for $L/B = 2.778$, $I_{p(average)} = 1.477$.
For saturated undrained condition, $v = 0.5$.

Then $q_a = \dfrac{s_i E_u}{B(1 - v^2)I_p} = \dfrac{75 \times 10^{-3} \times 20.9 \times 10^3}{9(1 - 0.5^2)1.477} = \underline{157\ kPa}$

So for condition A the shear strength of the soil is the governing factor and the allowable bearing capacity is $\underline{108\ kPa}$.

Condition B

(i) For long-term or drained conditions effective stress parameters (c', ϕ', v', etc.) are used, together with the Terzaghi equation.

From Table 11.2, for $\phi' = 23°$: $N_c = 18.1$, $N_q = 8.66$, $N_\gamma = 5.85$

From Table 11.3: $s_c = 1 + 0.360 \times 8.66/18.05 = 1.17$

$s_q = 1 + 0.360 \tan 23° = 1.15$

$s_\gamma = 1 - 0.40 \times 0.360 = 0.856$

$\sigma_o' = 17.4 \times 2.5 + (17.4 - 9.81)(3.2 - 2.5) = 48.8\ kPa$

Then
$$q'_{nf} = c'N_c s_c + \sigma'_o N_q s_q + \tfrac{1}{2}\gamma'BN_\gamma s_\gamma - \sigma'_o$$

$$= 5 \times 18.1 \times 1.17 + 48.8 \times 8.66 \times 1.15$$
$$+ \tfrac{1}{2}(17.4 - 9.81)9.0 \times 5.85 \times 0.856 - 48.8$$

$$= 714 \text{ kPa}$$

Therefore
$$q_a = \tfrac{714}{3} + 48.8 = 287 \text{ kPa}$$

(ii) Settlement: since the preconsolidation stress is not known the elastic deformation under drained conditions will be used to provide an estimated value of q_a:

$$q_a = \frac{s_t E'}{B(1 - v^2)I_\rho} = \frac{75 \times 10^{-3} \times 19.2 \times 10^3}{9(1 - 0.38^2)1.447} = \underline{129 \text{ kPa}}$$

In the longer term, then, the tolerable settlement limit will be reached before the safe shear strength condition. The allowable bearing capacity is therefore 129 kPa. It is possible, however, that the tolerable settlement limit of 75 mm may be tempered by a consideration of the rate at which settlement proceeds. For example, if the 50 per cent value were to be taken as a more realistic long-term criterion (assuming a reasonably slow rate) the allowable bearing capacity would be nearer the value obtained in (i).

Allowable bearing capacity of sands

In the case of sands, the settlement is almost entirely immediate and should be calculated for the maximum functional (dead + live) loading intensity. It has been found that where footings are of equal size, differential settlement is unlikely to exceed 40 per cent of the total amount, and that even for wide foundations it rarely exceeds 75 per cent.

The settlement characteristics, and therefore the allowable bearing capacity, of a sand depend mainly on its relative density and the position of the water table; stress history and the degree of cementation can also be significant. The majority of sand deposits, including silty and gravelly mixtures, tend to be non-homogeneous in terms of size, grading and relative density. It is both difficult and expensive attempting to obtain good representative undisturbed samples. These factors have consequently led to the development of several *in situ* testing procedures from which allowable bearing capacity calculations and settlements predictions can be made (see Section 11.8).

11.8 Interpretation of in situ *bearing capacity tests*

Descriptions of the apparatus and methods used in a variety of *in situ* tests are given in Chapter 12. In this section, the intention is to explain the interpretation of the results of tests and the associated methods of estimating settlement and allowable bearing capacity. It is important to note that, as in the case of all empirical tests, the reliability of the information gathered is not intrinsic, but requires confirmation from other sources, e.g. laboratory tests, alternative *in situ* tests, past experience. A most useful review of *in situ* test methods has been published by Schmertmann (1975).

The plate bearing test

The procedure for the plate bearing test is described in Section 12.5. In essence, the soil is loaded at the bottom of a trial pit or borehole through a steel plate at least 300 mm square. In a series of papers, Marsland (e.g. 1974) has reported comparisons of penetration and pressuremeter tests with tests using a 865 mm plate down to depths of 25 mm in London Clay. The reliability of the results depends largely on the homogeneity of the soil in the area of stress influence under the plate compared with that under the actual footing. The biggest danger in interpretation results from over-extrapolation, the information then becoming erratic and misleading. Ideally, a number of tests need to be carried out using plates of different diameters and located at different depths, both above and below the water table. This, of course, can be expensive and time consuming.

In order to estimate the immediate settlement of a foundation using data from this test, Terzaghi and Peck (1967) have suggested the following relationship:

$$s_B = s_b \left(\frac{2B}{B + b} \right)^2 \qquad\qquad [11.13]$$

where s_b = settlement of a test plate of side dimension b
s_B = settlement of a foundation of side dimension B at the same intensity of loading

Considerable evidence has been found that the values given by this test are widely variable; underestimates of 20 per cent may be common and in some cases they have been found as high as 100 per cent. Reliability, especially in fissured clays, is improved by using larger diameter plates in carefully prepared holes. A recent development is the *screw plate*, which eliminates surface irregularity errors and can be screwed into the ground continuously.

The standard penetration test (SPT)

This is a dynamic test carried out in boreholes during site investigations. The procedure is specified in BS 1377 and described briefly in Section 12.5 of this book. A split-barrel sampler is driven into the soil at the bottom of a hole using a standard hammer. After an initial drive of 150 mm, the number of blows required to drive the sampler a further 300 mm is recorded. This number of blows is referred to as the *standard penetration resistance* or *N-value*.

When the test is carried out below the water table in fine sands or silty sands, the pore pressure tends to be reduced in the vicinity of the sampler, resulting in a transient increase in effective stress. A corrected *N*-value may be obtained from the following expression suggested by Terzaghi and Peck (1948):

$$N_{corr.} = 15 + \tfrac{1}{2}(N - 15) \qquad\qquad [11.14]$$

The measured *N*-value is also influenced by the confining pressure at the depth of measurement, and so a correction should be made with respect to the effective overburden stress:

$$N = C_N N \qquad\qquad [11.15]$$

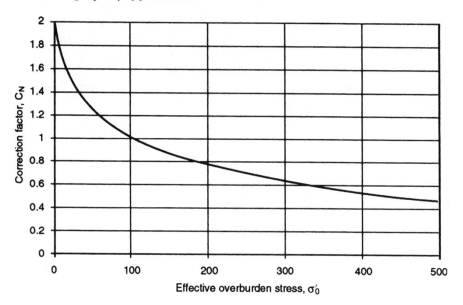

Fig. 11.10 Correction of SPT *N*-values for overburden stress

where C_N = a correction factor. A number of correction proposals have been published (Gibbs and Holtz, 1957; Tomlinson, 1994; Peck and Bazaraa, 1969; Peck *et al.*, 1974), some of which suggest correction factors as high as four, which is considered by many to be too high. The chart given in Fig. 11.10 follows the Peck, Hanson and Thornburn proposal in which $C_N = 0.77 \log(2000/\sigma_0')$ with a maximum value of two.

Skempton (1986), in considering the influence of test procedure on measured *N*-values, suggested that the type of anvil, the method of releasing the hammer and the length of boring rods, variously affected the actual amount of energy available for producing penetration. The rod energy ratio (delivered energy/free-fall energy of the hammer) was found to vary between about 45 per cent and 80 per cent in different countries using different methods. In Great Britain, using the British Standard method and equipment, the rod energy ratio is estimated at 60 per cent for rods over 10 m long. It is therefore recommended that for comparative purposes test values obtained by other than the BS method should be normalised to a rod energy ratio of 60 per cent and described as N_{60} values. Corrections with respect to drill rod length and borehole diameter may be made as follows:

Drill rod correction:

for $1 < L_R < 10$, $C_R = 1 - 0.04(10 - L_R)$ [11.16]
for $L_R \geq 10$ $C_R = 1$

where L_R = total length of drill rods

Borehole diameter (mm)	115	150	200
Correction factor (C_B)	1.0	1.05	1.15

Table 11.6 Classification using SPT N-values
(a) Cohesionless soils

SPT N-value	Relative density	
	Description	$I_d(\%)$
<4	Very loose	0–15
4–10	Loose	15–35
10–30	Medium dense	35–65
30–50	Dense	65–85
>50	Very dense	85–100

(b) Cohesive soils

SPT N-value	Consistency	Approximate undrained shear strength (kPa)
<2	Very soft	<20
2–4	Soft	20–40
4–8	Firm	40–75
8–15	Stiff	75–150
>15	Very stiff or hard	>150

Burland and Burbidge (1985) have suggested that in the case of gravels and gravel-sands a further correction is necessary, especially when using the N-value to assess compressibility:

Correction for gravel: $C_G = 1.25$ [11.17]

Thus, if all of these corrections are applied a representative N-value (N') is derived from the measured data. Corrected N-values may be used to provide estimates of relative density, peak angle of shearing resistance and allowable bearing capacity. A correlation between relative density (density index) and the N-value was proposed by Terzaghi and Peck (1948) and provides the current standard for classification purposes (Table 11.6(a)). The use of N-values to estimate the undrained shear strength of cohesive soils is far from reliable, but for classification purposes the figures given in Table 11.6(b) may be used; the values given for undrained shear strength should be regarded as being very approximate.

Peck *et al.* (1974) proposed approximate relationships between the N-value and the peak angle of shearing resistance and also the bearing capacity factors N_q and N_γ (Fig. 11.11).

Allowable bearing capacity using N-values

Terzaghi and Peck (1948) proposed a correlation between allowable bearing capacity and the corrected N-value in the form of a chart (Fig. 11.12). The breadth of the footing and the N-value are used as entry data and the allowable

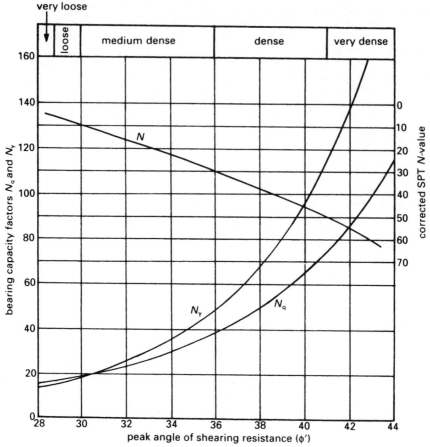

Fig. 11.11   Relationship between *N*-value and ϕ', N_q and N_γ

Notes: 1. ϕ'/*N*-value relationship after Peck, *et al.* (1974)
2. N_q/ϕ' and N_γ/ϕ' relationship from Table 11.2

bearing capacity (q_{TP}) is read off the left vertical axis. The effect of the water table may be taken into account by applying the following correction:

Water table correction, $C_w = \dfrac{1}{2}\left(1 + \dfrac{D_w}{D + B}\right)$ [11.18]

where D_w = depth of water table below surface
D = founding depth below surface
B = footing breadth

Thus $q_a = C_w q_{TP}$ [11.19]

The Terzaghi and Peck method yields quite conservative values of q_a, since it attempts to ensure that the settlement is nowhere greater than 25 mm. For wide footings and rafts the limiting values may be raised to 50 mm. Meyerhof (1965) suggested that the q_{TP} value could be increased by 50 per cent and that no correction should be made for the water table since the effect would be incorporated

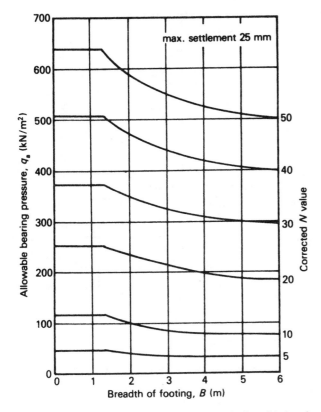

Fig. 11.12 Relationship between *N*-value and allowable bearing pressure (after Terzaghi and Peck, 1967)

in the measured *N*-values. Meyerhof also proposed a set of simple design relationships as follows:

For $B < 1.25$ m: $\quad q_a = \dfrac{s_L N}{1.9}$ [11.20(a)]

For $B > 1.25$ m: $\quad q_a = \dfrac{s_L N}{2.84}\left(\dfrac{B + 0.33}{B}\right)^2$ [11.20(b)]

For rafts: $\quad q_a = \dfrac{s_L N}{2.84}$ [11.20(c)]

where s_L = permitted settlement limit (mm)
  N = average *N*-value between $z = D$ and $z = D + B^*$
  B = breadth of footing

Burland and Burbidge (1985), using a large number of settlement observations, plotted the ratio of settlement per unit bearing pressure (s/q) against breadth (B) (Fig. 11.13). In each case, the soil (sand) is broadly classified as loose, medium dense or dense according to the measured *N*-value. No account is taken of the

---

* Some opinions suggest that this range should be between $z = D$ and $z = D + 2B$.

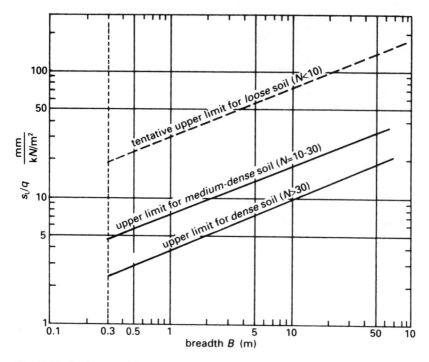

Fig. 11.13 Settlement of footings on sand (after Burland and Burbidge, 1985)

water table, founding depth or shape of the footing. The strong linear trend of the results supports the view that settlement is proportional to bearing pressure, e.g. Meyerhof's equations ([11.20]). Upper limit lines have been drawn for each relative density class which can be used as design guidelines as follows:

(a) Let the *probable* settlement = half the appropriate upper-limit value.
(b) Assume that the *maximum* value will not exceed *1.5 times the probable value*.
(c) Only use the *loose* upper-limit line as a preliminary assessment in problems such as large rafts or storage tanks on loose sands.

In the same work, detailed analyses were carried out on over 200 records of settlements of foundations, tanks and embankments on sands and gravels. Factors such as shape, founding depth, grain size and time were included and a relatively simple relationship established between the N-value, bearing pressure, breadth of foundation and settlement. The compressibility of the soil is stated in terms of the average N-value as a grade of compressibility and a *compressibility index* (I_c). Figure 11.14 illustrates these relationships and the following empirical expression is derived:

Compressibility index, $I_c = \dfrac{1.71}{N^{1.4}}$ [11.21]

where N = average N-value measured between a depth of $z = D$ and z
          $= D + B$, corrected for fine-sand/silt or gravel (eqns [11.14] and [11.17]), or the equivalent value related to grain size derived from the cone penetration test

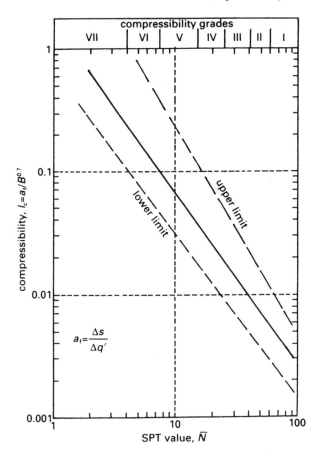

Fig. 11.14 Relationship between compressibility (I_c) and N-value (after Burland and Burbidge, 1985)

For normally consolidated sands, the immediate settlement at the end of construction, corresponding to an average effective foundation pressure of q', is given by:

Immediate average settlement, $s_i = q'B^{0.7}I_c$ [11.22]

or where s_L is the limiting settlement,

Allowable bearing capacity, $q_a = \dfrac{s_L}{B^{0.7}I_c}$ [11.23]

where I_c may be obtained from eqn [11.21] or from Fig. 11.14.

   For overconsolidated sands, or loading at the base of an excavation for which the maximum previous overburden pressure (or overconsolidation stress) is σ'_o, the average immediate settlement is given by:

$s_i = \frac{1}{3}\sigma'_oB^{0.7}I_c + (q' - \sigma'_o)B^{0.7}I_c$

in which the second term cannot be negative, so that:

When $q' > \sigma'_o$: $s_i = (q' - \frac{2}{3}\sigma'_o)B^{0.7}I_c$ [11.24(a)]

When $q' \leq \sigma'_o$: $s_i = \frac{1}{3}q'B^{0.7}I_c$ [11.24(b)]

or where s_L is the limiting settlement,

Allowable bearing capacity, $q_a = \dfrac{s_L}{B^{0.7}I_c} + \dfrac{2}{3}\sigma'_o$ [11.25]

It was concluded that the water table level did not have a statistically significant effect on these values, but that a significant correlation did exist with respect to the shape of the foundation. Equations [11.24(a) and (b)] should therefore be multiplied by the following *shape correction factors*:

$$f_S = \left[\frac{1.25L/B}{(L/B) + 0.25} \right]^2$$ [11.26(a)]

$$f_L = \frac{s_L}{s_i} = \frac{H_s}{z_i}\left(2 - \frac{H_s}{z_i} \right)$$ [11.26(b)]

where z_i = depth of the zone of influence in which 75 per cent of the settlement will occur and which may be taken approximately as:

$$z_i = e^{0.77 \ln B}$$ [11.27]

   H_s = thickness of layer beneath foundation
   L = length of foundation
   B = breadth of foundation

In the case of foundations subject to fluctuating loads, e.g. bridges, silos, tall chimneys, machinery bases, etc., immediate settlements will be higher than for static loading, and time-dependent settlement may also be significant. The following conservative correction is suggested when $t > 3$ years:

$$f_t = \frac{s_t}{s_i} = [1 + R_3 + R_t \log(t/3)]$$ [11.28]

where R_3 = proportional increase in settlement taking place during the first 3 years
= $0.3s_i$ (static loading) or $0.7s_i$ (fluctuating loading)
   R_t = proportional increase in settlement taking place in each log-cycle of time after 3 years
= $0.2s_i$ (static loading) or $0.8s_i$ (fluctuating loading)

Equation [11.28] gives the following values:

No. of years after completion (t):	3	5	10	20	30
f_t for static loading:	1.32	1.36	1.42	1.47	1.50
f_t for fluctuating loading:	1.80	1.94	2.18	2.39	2.50

The cone penetration test (CPT)

The equipment and test method for the cone penetration test (CPT) are described in Section 12.5. The cone is pushed into the soil at a rate of 20 mm/s and the *cone resistance* (q_c) measured as the maximum force recorded during penetration divided by the end area. Standard cones have an end (or tip) with an apex angle of 60° and a diameter of 35.7 mm, giving an end area of 1000 mm²; when the tip incorporates a friction sleeve, this has a (cylindrical) area of 1500 mm². The *local side friction* (f_s) is then measured as the frictional resistance per unit area on the friction sleeve.

A compressibility coefficient (similar to the compression index of fine soil described in Section 6.2) was suggested by de Beer and Martens (1951):

$$C = \frac{1.5q_c}{\sigma_o'} \qquad [11.29]$$

where
q_c = cone resistance (MPa)
σ_o' = effective overburden pressure (MPa)

The settlement s_i at the centre of a layer of thickness H is then given by:

$$s_i = \frac{H}{C} \log\left(\frac{\sigma_o' + \Delta q}{\sigma_o'}\right) \qquad [11.30]$$

where
Δq = increase in stress at the centre of the layer due to a foundation pressure of q

This method is considered to overestimate the value of s_i. A rapid conservative method was suggested by Meyerhof (1974):

$$s_i = \frac{q_n B}{2\bar{q}_c} \qquad [11.31]$$

where
q_n = net applied loading = $q - \sigma_o'$
$\bar{q}_c$ = average cone resistance over a depth below the footing equal to the breadth B

Probably the most thorough and reliable method for computing settlement from CPT results was proposed by Schmertmann (1970) and modified in the following form by Schmertmann *et al.* (1978). The sand mass below the foundation is divided into a number of layers down to a depth of $2B$ below the base for a square footing and down to $4B$ for a long footing ($L \geq 10B$). The strain within each layer is then calculated as:

$$e_L = \frac{q_n I_z}{E}$$

where
q_n = net applied loading = $q - \sigma_o'$
E = deformation modulus
I_z = a strain influence obtained from Fig. 11.15

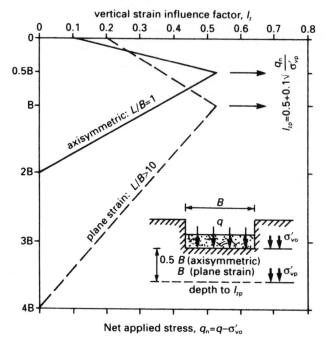

Fig. 11.15   Vertical strain influence factor (after Schertmann *et al.*, 1978)

The immediate settlement is then given by:

$$s_i = C_1 C_2 q_n \sum \frac{I_z}{E} \Delta z \qquad [11.32]$$

where C_1 and C_2 are correction factors as follows:

Embedment correction factor, $C_1 = 1 - 0.5(\sigma'_0/q_n)$ for $q_n \leq \sigma'_0$, $C_1 = 0.5$ [11.33(a)]

Creep correction factor, $C_2 = 1 + 0.2 \log(10t)$ [11.33(b)]

where t = time in years after loading

The stiffness modulus E is derived from the cone resistance as indicated in Table 11.7.

Estimates of undrained shear strength may be obtained from the following relationship:

$$c_u = \frac{q_c - \sigma_0}{N_k} \qquad [11.34]$$

where N_k = the *cone factor*, which is analogous to the bearing capacity factor N_c. The value of N_k varies with the shape and type of penetrometer tip, the rate of penetration and also the stress history and micro-fabric of the soil: recommended values are:

NC clays: $N_k = 15-21$

Stiff and OC clays: $N_k = 24-30$

UK glacial clays: $N_k = 14-22$

Table 11.7 Relationship between stiffness modulus and cone resistance

	Normally consolidated sands	Overconsolidated sands
Square foundation ($L/B = 1.0$)	$E = 2.5q_c$	$E = 5q_c$
Long foundation ($L/B > 1.0$)	$E = 3.5q_c$	$E = 7q_c$

Table 11.8 Confined modulus indicated by cone resistance

Soil type	Classification	$\alpha_M = E'_o/q_c$	
NC clays and silts[1]			
Highly plastic clays and silts	CH. MH	2–7.5	
Clays of intermediate or low plasticity	CI. CL		
$q_c < 0.7$ MPa		3–10	
$q_c > 0.7$ MPa		2–6	
Silts	MI. ML	3–7.5	
Organic silts	OL	2–10	
Peat	Pt. OH		
50% < w < 100%		1.5–5.0	
100% < w < 200%		1.0–1.9	
w > 200%		0.4–1.25	
		$q_c < 2.0$ (MPa)	$q_c > 2.0$ (MPa)
OC clays and silts[2]			
Highly plastic clays and silts	CH. MH	2–6	
Clays of intermediate or low plasticity	CI. CL	2–5	1–2.5
Silts	MI. ML	3–6	1–3

Notes: (1) α_M *range includes mantle cone and reference tip data; reference tip values are (on average) 25% higher.*
(2) Mantle cone data only.
After Sanglerat (1979)

The confined modulus E'_o for clay can be expressed as:

$$E'_o = \frac{1}{m_v} = \alpha_M q_c \qquad [11.35]$$

where α_M = a coefficient for which values were given by Sanglerat (1979) (Table 11.8). For better estimates of E'_o it is preferable to use oedometer test results and index properties, but local correlations between q_c and E'_o are useful when assessing variations in compressibility.

CPT/SPT relationship
It is possible to utilise the various methods of calculating allowable bearing capacity based on *N*-values by converting cone resistances. For example Meyerhof (1956) proposed that

$$q_c(\text{MPa}) = 0.4N \qquad [11.36]$$

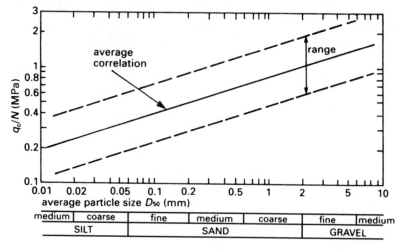

Fig. 11.16 Relationship between CPT and SPT (after Burland and Burbidge, 1985)

Subsequent observations by Meigh and Nixon (1961) have shown a variable relationship between silty fine sand ($q_c = 0.25N$) and coarse gravels ($q_v = 1.2N$). More recently, Burland and Burbidge (1985) produced correlations based on a large number of observations between q_c/N and the average grain size D_{50}. Figure 11.16 shows the average correlation line, together with the zone into which the results fall.

Worked example 11.9   *A square foundation of 3.5 m side is to be founded at a depth of 1.5 m in a medium sand ($\gamma = 19.4 \ kN/m^3$). The water table is located at a depth of 3.6 m. During site investigations a standard penetration test produced the following values.*

Depth BS (m):	1.5	2.2	3.0	3.8	4.6	5.4
N-value:	7	9	12	12	17	20

Determine an estimate for the allowable bearing capacity based on an immediate settlement limit of 25 mm.

Terzaghi and Peck method
The measured N-values are first corrected with respect to overburden pressure and then the average calculated between the founding depth and a depth equal to the breadth below this.

Depth z (m)	$\sigma_o = \gamma z$ (kPa)	u (kPa)	σ_o' (kPa)	N	C_N (Fig. 11.10)	N'	Average N'
1.5	29.1	0	29.1	9	1.5	13.5	
2.2	42.7	0	42.7	11	1.35	14.9	
3.0	58.2	0	58.2	14	1.24	17.4	16.6
3.8	73.7	2.0	71.7	14	1.15	16.1	
4.6	89.2	9.8	79.4	19	1.10	20.9	
5.4	104.8	17.7	87.1	23			

Thus, from Fig. 11.12, for $B = 3.5$ m: $q_{TP} = 164$ kPa.

Correction for water table,

$$C_w = \frac{1}{2}\left(1 + \frac{3.6}{1.5 + 3.5}\right) = 0.86$$

Hence, allowable bearing capacity, $q_a = 164 \times 0.86 = \underline{141 \text{ kPa}}$.

Meyerhof method (a)
The average measured N-value between $z = D$ and $z = D + B$ is 13.4.

From Fig. 11.12: $q_{TP} = 125$ kPa.

This is increased by 50 per cent, hence, $q_a = 125 \times 1.5 = \underline{188 \text{ kPa}}$.

Meyerhof method (b)
Using Meyerhof's expression (eqn [11.20(b)]) with a limiting settlement of $s_L = 25$ mm:

$$q_a = \frac{25 \times 16.6}{2.84}\left(\frac{3.5 + 0.33}{3.5}\right)^2 = \underline{175 \text{ kP}}$$

Burland and Burbidge method
The zone of influence depth corresponding to $B = 3.5$ is obtained from:

$$z_i = e^{0.77 \ln B} = \exp(0.77 \times \ln 3.5) = 2.6$$

The average measured N-value is therefore taken between $z = 1.5$ and $z = 4.1$: average $N = 12$.

Then compressibility index, $I_c = 1.71/12^{1.4} = 0.0527$.

If the limiting immediate settlement is 25 mm, then from eqn [11.22]:

$$q_a = \frac{s_L}{B^{0.7}I_c} = \frac{25}{35^{0.7} \times 0.527} = \underline{197 \text{ kPa}}$$

This is the *net* value, therefore the original overburden pressure is added, i.e. eqn [11.24(a)]:

$$q_a = 197 + \tfrac{2}{3} \times 29.1 = \underline{216 \text{ kPa}}$$

Worked example 11.10 *A square foundation of 3.0 m side is to be founded at a depth of 1.5 m in a layer of medium-fine sand ($\gamma = 18.4$ kN/m³ and $\gamma_{sat.} = 20.2$ kN/m³). The water table is located at a depth of 3.5 m. The net loading intensity is to be 160 kPa. Using the cone penetration resistances given in Fig. 11.17, obtain an estimate of the probable settlement.*

The Schmertmann, Hartman and Brown method is used.

At $z = D$, $\sigma_o' = 18.4 \times 1.5 = 27.6$ kPa

Then $q_n = 160 - 18.4 \times 1.5 = 132.4$ kPa

At $B/2$ below the footing,

$z = 3.0$ m and $\sigma_o' = \Sigma \gamma z = 18.4 \times 3 = 55.2$ kPa

Refer to Fig. 11.15: $I_p = 0.5 + 0.1\sqrt{(132.4/55.2)} = 0.655$

For a square footing on NC soil (Table 11.7): $E = 2.5q_c$

Note that the top layer is thinner because of the founding depth.

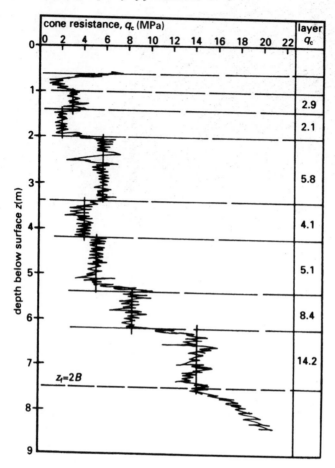

Fig. 11.17

Depths

z (m)	z_f (m)	Δz (m)	q_c (MPa)	$\frac{z_f}{B}$	I_z (Fig. 11.15)	$\frac{I_z \Delta z}{E}$ (m^3/MN)
1.2	2.9					
1.7	0.25	0.5	2.1	0.083	0.192	0.0182
2.7	1.2	1.4	5.8	0.400	0.544	0.0525
3.8	2.3	0.8	4.1	0.767	0.566	0.0442
4.8	3.3	1.2	5.1	1.100	0.455	0.0428
5.8	4.3	0.8	8.4	1.433	0.344	0.0131
6.85	5.35	1.3	14.2	1.783	0.277	0.0083

$\Sigma = 0.1792$

Embedment correction, $C_1 = 1 - 0.5(27.6/132.4) = 0.896$
Creep correction, $C_2 = 1.0$ {immediately after construction}

Hence, immediate settlement,

$s_i = C_1 C_2 q_n \Sigma(I_z \Delta z/E) = 0.896 \times 1.0 \times 132.4 \times 0.1792 = 21$ mm

Applying the creep correction, the future probable settlement will be:

After 2 years: $C_2 = 1.26$ $\therefore$ $s_t = $ 26 mm

After 5 years: $C_2 = 1.34$ $\therefore$ $s_t = $ 28 mm

After 10 years: $C_2 = 1.40$ $\therefore$ $s_t = $ 29 mm

11.9 Pile foundations

The use of pile foundations in poor or waterlogged ground goes back to prehistoric times. Evidence of the use of timber piles in the form of straight tree trunks and later parallel- and taper-sawn baulk piles may be found among archaeological evidence from the earliest civilisations up to comparatively recent times. Modern piles and piling techniques make use of modern science and technology, and some of the processes, as well as some design techniques, are impressively sophisticated. Nevertheless, the basic reasons for using piles, as opposed to other foundation types, remain much the same. Here are some typical foundation problems for which a pile system might provide a solution:

(a) Where a soil layer of reliable adequate bearing capacity lies too deep for the economic use of conventional footings.
(b) Where the soil layer(s) immediately underlying a structure are soft or poorly compacted.
(c) Where the soil layer(s) immediately underlying a structure are moderately to highly variable in nature.
(d) On sites where the soil strata, and in some cases the ground surface, are steeply inclined.
(e) On river and shore-zone sites where tidal or wave action, or scouring, may vary the amount of material near the surface.
(f) For structures transmitting very high concentrated loads.
(g) For structures transmitting significant horizontal or inclined loading.
(h) For structures which structurally or functionally may be sensitive to differential settlement.

For a detailed treatment of pile types and piling methods, the reader should refer to books relating to foundation engineering practice, such as Tomlinson (1995) or Whittaker (1976).

From the point of view of both design and construction, piles are classified into two types:

(a) *Driven or displacement* piles, which are usually pre-formed before being driven, jacked, screwed or hammered into ground.
(b) *Bored or replacement* piles, which require a hole to be first bored into which the pile is then formed, usually of reinforced concrete.

A useful tabular guide to the selection of pile types was published by Armstrong (1973).

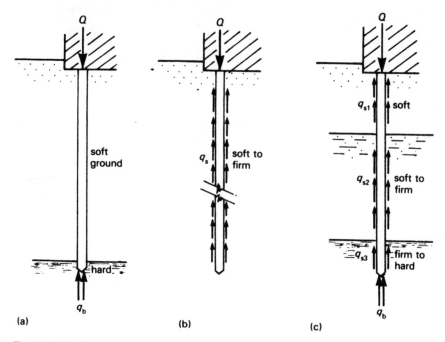

Fig. 11.18 Load-carrying capacity of piles
(a) End-bearing (b) Friction (c) Combined end-bearing and friction

Another method of classifying piles in terms of the way in which they derive their load-carrying capacity is:

End-bearing piles (Fig. 11.18(a)) Here the pile acts as a laterally restrained column, the load being transmitted to the toe and resisted by the bearing capacity of the soil.

Friction piles (Fig.11.18(b)) Here the load is transmitted to the soil through the adhesion or frictional resistance along the shaft of the piles. Pure friction piles tend to be quite long, since the load-carrying capacity is a function of the shaft area in contact with the soil. In cohesionless soils, such as sands of medium to low relative density, friction piles are often used to increase the density and thus the shear strength.

In the majority of cases, however, the load-carrying capacity is dependent on both end-bearing and shaft friction (Fig. 11.18(b)). Other types of piles for specialist problems are *tension piles* (Fig. 11.19(a)), e.g. to resist overturning in tall buildings; *raking piles* (Fig. 11.19(b)), e.g. in harbour and river installations; and *shear piles* (Fig. 11.19(c)), e.g. to resist horizontal forces or movement.

It is most convenient, from a design point of view, to divide piles into the first two classes, *driven* and *bored*, since the approach to obtaining a value for load-carrying capacity is significantly different for each. For driven piles *dynamic* formulae are used and for bored piles *static* formulae; static formulae may also be used for driven piles, especially in cohesionless soils.

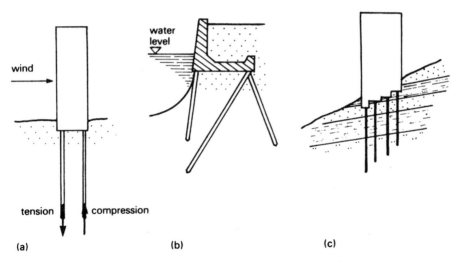

Fig. 11.19 Special types of pile
(a) Tension piles (b) Raking piles (c) Shear piles

11.10 Dynamic pile formulae

The ultimate static resistance of the piles is predicted from the dynamics of the pile-driving operation itself. The kinetic energy imparted by the hammer is equated to the work done by the pile in penetrating into the soil.

Net kinetic energy = work done during penetration

Basically for a hammer of weight W, falling from a height h and causing a penetration or *set* of s, the resistance load could be obtained from:

$$Wh - (\text{energy losses}) = R\,s$$

The energy losses may be due to hammer rebound, pile and pile cap compression, friction, heat and to the strain caused in the soil. A number of formulae have been proposed to take account of all of these effects. Varying degrees of reliability are claimed for a particular equation in particular circumstances; but in all cases, particularly those involving cohesionless soils, the results should be correlated with suitable static load tests. Ramey and Johnson (1979) have published the results of a survey in which the results of load tests are compared statistically with load-carrying capacities predicted by five of the most commonly used formulae.

11.11 Static pile capacity equations

The static equilibrium approach to pile design is much the same as that for shallow foundations. Conventional elasticity/plasticity theory is the basis, with either undrained or undrained shear strength parameters. The main differences arise from the fact that the soil at the base and along the shaft of a pile is usually well below the surface.

The ultimate axial load of a single pile is considered to be the sum of its end-bearing and its shaft friction resistances that are mobilised by the applied load.

$$Q_u = Q_b + Q_s - W$$

$$= A_b q_b' + \Sigma A_s q_s' - W \qquad [11.37]$$

where A_b and A_s = area of base and shaft respectively

$\quad\quad q_b'$ = ultimate net bearing capacity of the soil at the end of the pile

$\quad\quad q_s'$ = mobilised adhesion or frictional resistance along the shaft of the pile

$\quad\quad W$ = weight of the pile – weight of soil replaced

$\quad\quad\quad = 0.25 \pi d^2 L(\gamma_p - \gamma)$

$\quad\quad d$ = pile diameter L = pile length

$\quad\quad \gamma_p$ = average density of pile

Note that W will be negligible in most cases, especially with timber piles, but may be significant in the case of steel piles.

Cohesionless soils

The ultimate bearing capacity of sands depends mainly on their density index, but if a pile is driven-in this will have the effect of increasing I_d. Since this may have a marked effect on the subsequent shear strength properties, the prediction of load-carrying capacity using parameters measured prior to driving is often difficult. In such cases, correlation with load test results is essential.

The net ultimate bearing capacity at the base of the pile is:

$$q_b' = \sigma_o' N_q \qquad [11.38]$$

where σ_o' = effective overburden pressure

$\quad\quad N_q$ = a bearing capacity factor

Values of N_q have been proposed by several, including Meyerhof (1976) and Berezantsev *et al.* (1961), these later values correlating well with acceptable settlement limitations (Fig. 11.20).

The frictional resistance developed along the shaft comprises the shaft area multiplied by the average skin friction, summated for a series of layers. For a given layer the skin friction resistance is given by:

$$q_s = K_s \sigma_o' \tan \delta = \beta_s \sigma_o' \qquad [11.39]$$

where K_s = a coefficient of earth pressure dependent largely on the relative density of the soil

$\quad\quad \sigma_o'$ = average effective overburden pressure in the layer

$\quad\quad \delta$ = angle of friction between the pile and the soil

The base and shaft friction resistances do not develop linearly with depth below certain depths. This is probably mainly due to arching effects in the soil related to its relative density and compressibility. It is therefore recommended that the effective overburden pressure (in eqns [11.38] and [11.39]) should be calculated linearly with depth only down to a *limiting depth* (z_L) and then assumed to

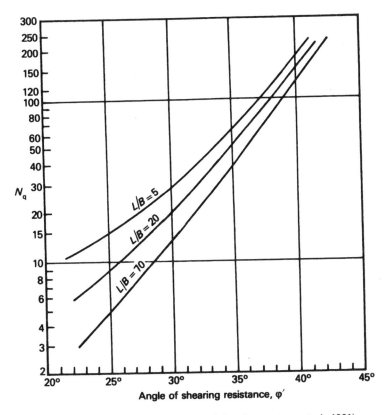

Fig. 11.20 Values of N_q for pile formulae (after Berezantsev *et al.*, 1961)

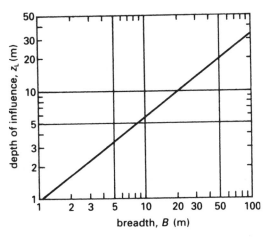

Fig. 11.21 Depth of influence (after Burland and Burbidge, 1985)

remain constant below this. Values of z_L are given in Fig. 11.21 correlating approximately with density index (I_d per cent) and ϕ'.

When $z < z_L$: $\quad \sigma'_o = \gamma z - u_z$

Where $z \geq z_L$: $\quad \sigma'_o = \gamma z_L - u_{zL}$

Table 11.9   Skin friction coefficients

	K_s	δ	$\beta_s = K_s \tan \delta$	
			$\phi' = 25°$	$\phi' = 40°$
Concrete piles				
Loose sand	1.0	$0.75\phi'$	0.34	0.58
Dense sand	2.0		0.68	1.15
Steel piles				
Loose sand	0.5	$20°$		0.18
Dense sand	1.0			0.36

Table 11.10   Ultimate pile bearing capacities and SPT N-values

Pile type*	Soil type	Ultimate base resistance q_b (kPa)	Ultimate shaft resistance q_s (kPa)
Driven	Gravelly sand and sand	$40 (L/d) N \leq 400N$	
	Sandy silt and silt (ML)	$30 (L/d)N \leq 300N$	$2\bar{N}$
Bored	Gravels and sands	$13 (L/d)N \leq 130N$	
	Sandy silt and silt (ML)	$10 (L/d)N \leq 100N$	$\bar{N}$

where L = embedded length
       d = shaft diameter
       N = N-value near pile toe
       $\bar{N}$ = average N-value along embedded length
* For piles with a taper exceeding 1%, add 50%.
After Meyerhof (1976)

Table 11.9 gives suggested empirical values for K_s, δ and the combined factor β_s for concrete and steel piles in sands. Since it is difficult in practice to obtain undisturbed samples for laboratory testing, empirical values of ϕ' are usually obtained from loading tests or *in situ* tests.

Pile bearing capacity using SPT N-values
Meyerhof (1976) produced correlations between base and friction resistances and N-values (Table 11.10); it is recommended that the N-values are first normalised with respect to an overburden pressure (at measurement level) of 100 kPa, using the following correction factor:

$$C_{NP} = 0.77 \log(1920/\sigma_o') \tag{11.40}$$

where σ_o' = effective overburden pressure at the depth of measurement

Pile bearing capacity using cone resistance values
Because of the inherent variability of sand deposits, it is necessary to determine an average of the measured cone resistances above and below the toe of the pile. A method for doing this was suggested by Heijnen (1974):

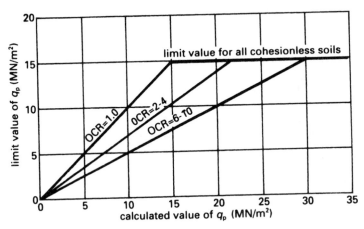

Fig. 11.22 Limit values of pile end-bearing capacity in sands and gravels
(after te Kamp, 1977)

Ultimate end-bearing capacity, $Q_b = q_p A_b$ [11.41(a)]

where $q_p = \frac{1}{2}(\bar{q}_{c1} + \bar{q}_{c2})$ [11.41(b)]

and $\bar{q}_{c1}$ = least average value of q_c below the pile toe down to a depth below
the toe between $0.7d$ and $4.0d$ (calculated in a series of trials)

$\bar{q}_{c2}$ = average value of q_c between the depth of the pile toe and $8d$ above it.

Typical Dutch practice is to set limit values for q_p related to the value calculated
using eqns [11.41(a) and (b)] and the stress history of the soil; recommended
limit values (te Kamp, 1977) are given in Fig. 11.22.

Although shaft resistance can be determined from values of local side friction,
more accurate estimates can be obtained from cone resistances.

Ultimate shaft resistance, $Q_s = S_p \sum_o^L q_c \pi d \Delta L$ [11.42]

where L = embedded length
and S_p = a coefficient dependent on the type of pile: recommended values are
0.005–0.012 for solid timber, pre-cast concrete and steel displacement piles and 0.003–0.008 for open-ended steel piles

Single piles in fine soils
A number of factors conspire to reduce the shear strength and consequent
adhesion around a pile, such as remoulding, softening due to water in the hole
or from fresh concrete, or due to porewater seepage. It is common to assume
undrained conditions exist at the time of loading, but recent studies have shown
that dissipation of excess pore pressure takes place fairly quickly. In this case,
by the time the pile is fully loaded, conditions may be nearer to being drained.
A method of applying an effective stress analysis using empirical coefficients is
suggested by Burland (1973).

In terms of undrained strength c_u:

End-bearing capacity, $Q_b = c_u N_c A_b$ [11.43]

Shaft resistance capacity, $Q_s = \Sigma(\alpha c_u \Delta A_s)$ [11.44]

where $N_c = 9.0$ for intact clays or 6.75 for fissured clays
  α = adhesion factor, usually taken as 0.45, but may vary from 1.0 for
    soft clays to 0.3 for overconsolidated clays
  ΔA_s = shaft surface area within a given layer of UD strength c_u

The shaft resistance capacity in terms of effective stresses is:

$$Q_s = K_o \sigma'_o \tan \delta = \beta_s \sigma'_o$$ [11.45]

The value of K_o depends on the soil type, the method of installing the pile and the stress history of soil.

Correlations with loading tests have shown that for soft clays β_s falls within a narrow range of values (0.25–0.4), irrespective of the clay type. For stiff and overconsolidated clay, a more complex situation exists due mainly to the difficulty of estimating K_o which varies with depth. Skempton and others have shown that in London Clay, for example, K_o may be as high as 3 near the surface, decreasing to less than 1 at depths over 30 m.

A lower bound value of β_s is:

$$\beta_s = (1 - \sin \phi'_d) \tan \phi'_d$$ [11.46]

where ϕ'_d = remoulded drained angle of shearing resistance

Negative skin friction
When piles are driven through a layer of fill material which slowly compacts or consolidates due to its own weight, or if the layers underlying the fill consolidate under the weight of the fill, a downward drag is imposed on the pile shaft (Fig. 11.23). The skin friction between the pile and soil therefore acts in a downward direction, and so *increases* the load on the pile. As consolidation proceeds, the magnitude of negative skin friction increases, since the effective overburden pressure σ'_o increases as the excess pore pressure dissipates. This is counterbalanced to some extent by the reduction in effective overburden pressure due to the transfer of load to the pile. Burland (1973) has proposed that a value of β of 0.25 is appropriate, at least for preliminary design purposes, in normally consolidated clays.

The general expression for the load-carrying capacity of a pile subject to negative skin friction is:

$$Q + \Sigma q_N A_{SN} = q_b A_b + \Sigma q_s A_{SR}$$ [11.47]

11.12   *Bearing capacity of pile groups*

Piles are usually placed in groups with centre-to-centre spacings typically between 2 and 5 shaft diameters. The pile heads are incorporated into a rigid

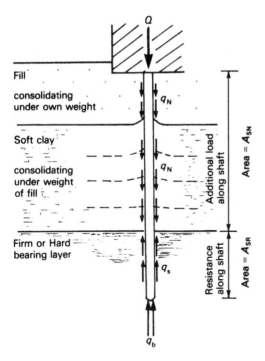

Fig. 11.23 Negative skin friction

slab or *pile cap*, which has the effect of constraining the group to act as a single monolithic unit, rather than as individual piles. This tendency is increased with closer spacing.

The overall load-bearing capacity of a pile group is not generally equal to the sum of the individual pile capacities. The ratio of the average load in a pile group to the ultimate single pile load is referred to as the efficiency (ζ) of the group:

$$\zeta = \frac{\text{ultimate group load}}{N \times \text{ultimate individual load}} \qquad [11.48]$$

Although it is generally assumed that the loading distribution is uniform for an axially loaded situation, it is clear from experimental work that the piles on the perimeter of a group are more heavily loaded than those in the centre.

Cohesionless soils
The action of driving piles into sands increases the relative density and thus the group bearing capacity may be greater than the sum of the individual capacities (i.e. $\zeta > 1.0$). It is recommended that in this case the design value of $\zeta = 1$ be taken. In the case of bored piles in sands, the efficiency is lower because the shear zones of adjacent piles overlap, giving values of ζ as low as 0.6.

Cohesive soils

If the spacing in a group is $2B$ to $3B$ the efficiency will be about 0.7. The failure mode is likely to be a *block failure* in which the piles and the soil between them may be considered as a large pier. The basic ultimate load equation [11.37] is modified as follows:

$$Q_{u(group)} = N_c c_u B_g L_g + \bar{c}_u 2D(B_g + L_g) \tag{11.49}$$

where N_c = Skempton's values taken from Fig. 11.6
c_u = undrained shear strength at depth D
B_g = breadth of group 'block'
L_g = length of group 'block'
D = depth to base of group
$\bar{c}_u$ = average undrained shear strength between 0 and D m below the surface

Since the dissipation of excess pore pressure will take longer in the case of a pile group, the undrained condition may be assumed for design. If the pile cap rests on the ground the design load should be taken as the least of the ultimate group load ($Q_{u(group)}$) and the sum of the individual pile capacities ($Q_u \times N$). When the piles are free-standing the design load should not exceed $\frac{2}{3}Q_u N$.

Worked example 11.11   *A 400 mm square-section concrete pile is driven to an embedded depth of 12 m in a cohesive soil, which has the following properties:*

$c_u = 35 + 4.2 \ kPa \ (where \ z = depth \ below \ surface)$

$\gamma = 20 \ kN/m^3$

The groundwater level is at a depth of 3 m. Calculate: (a) the design load capacity for the pile, adopting the following factors of safety: 3.0 against end-bearing, 1.5 against shaft resistance failure and 2.5 against overall shear failure; and (b) the safe load capacity for a square group of nine of these piles in this soil, when spaced at 2.25 m centres and adopting a factor of safety of 2.5 against group failure. What is the efficiency of the group?

At $z = 12$ m: $c_u = 35 + 4.2 \times 12 = 85.4$ kPa

(a)   End bearing (eqns [11.37] and [11.43]):

$$Q_b = A_b q_b = A_b c_u N_c$$

$$= 0.400^2 \times 9.0 \times 85.4 = 123 \ kN$$

Shaft resistance (eqns [11.37] and [11.45]):

$$Q_s = A_s \bar{q}_s = A_s \beta \bar{\sigma}_{vo}$$

Average overburden stress, $\bar{\sigma}_o' = 20 \times 6 - 9.81 \times 3 = 90.6$ kPa

Assume $\beta = 0.4$

$\therefore \quad Q_s = 12 \times 0.40 \times 4 \times 0.4 \times 90.6 = 696$ kN

Applying the factors of safety:

Safe load $Q \not> \dfrac{123}{3} + \dfrac{696}{2.5} = 319$ kN

Hence design $Q = \underline{319 \text{ kN}}$

or $Q \not> \dfrac{123 + 696}{2.5} = 328$ kN

(b) Group failure: $\dfrac{z}{B_g} = \dfrac{12}{2.25} = 5.33$

$N_c = 5.14 \times 1.2[1 + (0.053 \times 5.33)^{1/2}] = 9.45$ $\therefore$ $N_c = 9.0$ (max. value)

Average shear strength, $\bar{c}_u = \frac{1}{2}(35 + 35 + 4.2 \times 12) = 60.2$ kPa

Equation [11.49]:

$Q_{u(group)} = 9.0 \times 85.4 \times 2.25^2 + 60.2 \times 2 \times 12(2.25 + 2.25)$

$\qquad = 10\,393$ kN

Design load: $Q_s = \dfrac{10\,393}{2.5} = \underline{4157 \text{ kN}}$

Equation [11.48]:

Efficiency $\zeta = \dfrac{10\,393}{9(123 + 696)} = \underline{1.41}$

Settlement of pile groups

The settlement of a pile group in both cohesionless and cohesive soils will be greater than that due to an individual pile. In order to estimate the settlement it is usual to consider an *equivalent raft* located at a depth of $\frac{2}{3}D$, where $D =$ the embedded depth of the piles (Fig. 11.24). The area of the equivalent raft is determined by assuming that the load spreads from the underside of the pile cap in the ratio of 1 horizontal to 2 vertical. The settlement due to this uniformly-loaded equivalent raft is then calculated by conventional means, with parameters obtained either from laboratory tests or from *in situ* penetration tests.

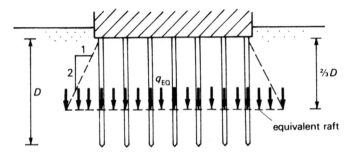

Fig. 11.24 Settlement of pile groups

Exercises

11.1   Determine the net ultimate bearing capacity of a strip footing of breadth 2.5 m which is to be founded 1.25 m below the surface of a thick stratum of clay, having the following properties:

$c_u = 88$ kPa

11.2   A strip footing is 2.4 m wide and is founded at a depth of 2.8 m in a soil having the following properties:

$c' = 12$ kPa $\phi' = 20°$ $\gamma = 19$ kN/m³

Determine: (a) the net ultimate bearing capacity, and (b) the design bearing capacity, adopting a factor of safety of 3.0.

11.3   The foundation for a circular tank is to be of diameter 18 m and founded at a depth of 2.5 m. The soil properties are:

$c' = 0$, $\phi' = 34°$, saturated $\gamma = 21$ kN/m³, drained $\gamma = 19$ kN/m³

Determine the net ultimate bearing capacity when the groundwater level is: (a) well below the foundation, (b) at the base of the foundation, and (c) at the ground surface.

11.4   A foundation 12 m long × 3 m wide is to be constructed in a coffer dam at a depth of 1.8 m below the dredge line. The soil has the following properties:

$c' = 0$ $\phi' = 32°$ $\gamma_{sat.} = 20$ kN/m³

Calculate the gross ultimate bearing capacity for a condition where the water level is 1.5 m above the dredge line and there is a vertical upwards flow of water with a hydraulic gradient of 0.25.

11.5   Determine the breadth of strip footing required to carry a uniform load 600 kN/m with a factor of safety of 3.0 against shear failure in the following situations. (The groundwater level is well below the base.)

(a) Founding depth = 0.5 m:	$c_u = 80$ kPa		$\gamma = 18$ kN/m³
(b) Founding depth = 4.5 m:	$c_u = 80$ kPa		$\gamma = 18$ kN/m³
(c) Founding depth = 0.5 m:	$c' = 0$	$\phi' = 30°$	$\gamma = 20$ kN/m³
(d) Founding depth = 4.5 m:	$c' = 0$	$\phi' = 30°$	$\gamma = 20$ kN/m³

11.6   A square footing is to be founded at a depth of 2.6 m and will transmit a uniform load of 2.6 MN. Adopting a factor of safety of 3.0, determine the size of footing required.

$c_u = 64$ kPa $\gamma = 19$ kN/m³

11.7   A rectangular foundation 10 × 5 m is to be designed, using a factor of safety of 3.0 to carry a uniform load of 86.6 MN, in fully-drained conditions in the following soil:

$c' = 0$ $\phi' = 35°$ $\gamma = 20$ kN/m³ $\gamma_{sat.} = 22$ kN/m³

(a) Determine a suitable founding depth.
(b) Determine the percentage reduction in load-carrying capacity that results from the groundwater level rising:
   (i) To the underside of the footing.
   (ii) To the ground surface.
(c) Calculate the factor of safety if, when the load is 86.6 MN, the two conditions in (b) occur.

11.8 A spoil heap 120 m long and 60 m wide is to be constructed on a thick layer of alluvial silt, the properties of which are:

$$c_u = 24 \text{ kPa} \quad \gamma = 18 \text{ kN/m}^3$$

Assuming that the maximum slope angle for the side batters is 45°, a unit weight for the spoil of 20 kN/m³ and adopting a factor of safety of 2.0, determine the maximum safe height to which the tip may be built.

11.9 A strip footing of breadth 2.5 m is to be founded at a depth of 2.0 m. The factor of safety against shear failure is 3.0 and the settlement must not exceed 50 mm. Determine the allowable bearing capacity when the footing is founded in the following soils. (Assume a thickness below the footing of 2B.)

(a) A soft alluvial clay:
$$c_u = 35 \text{ kPa} \quad \gamma = 18 \text{ kN/m}^3 \quad m_v = 0.92 \text{ m}^2/\text{MN} \quad E_u = 5 \text{ MPa}$$
(b) A stiff boulder clay:
$$c' = 25 \text{ kPa} \quad \phi' = 22° \quad \gamma = 20 \text{ kN/m}^3 \quad m_v = 0.10 \text{ m}^2/\text{MN} \quad E'_o = 22 \text{ MPa}$$
(c) A dense gravelly sand:
$$c' = 0 \quad \phi' = 32° \quad \gamma = 20 \text{ kN/m}^3 \quad E'_o = 145 \text{ kPa}$$

11.10 A circular raft foundation of diameter 15 m is founded at a depth of 4.5 m. The groundwater level is located at a depth of 2.5 m below ground level the soil is a deep layer of a lightly overconsolidated clay having the following properties:

$$c_u = 32 \text{ kPa} \quad \gamma = 18 \text{ kN/m}^3$$
$$c' = 10 \text{ kPa} \quad \phi' = 24° \quad E_o = 55 \text{ kPa}$$

Determine the allowable bearing capacity with respect to both immediate undrained and long-term drained conditions, if a factor of safety of 3.0 against shear failure is required and the average settlement must not exceed 60 mm.

11.11 In Fig. 11.25 the cross-section is shown of a strip footing. The relevant soil properties are as follows:

sand: $c' = 0 \quad \phi' = 32° \quad \gamma = 19 \text{ kN/m}^3 \quad \gamma_{sat.} = 21 \text{ kN/m}^3$
clay: $c_u = 60 \text{ kPa} \quad \gamma = 19 \text{ kN/m}^3$

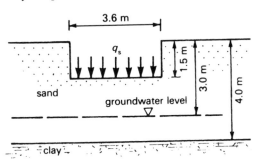

Fig. 11.25

If the factor of safety against shear failure in either the sand or the clay is to be 3.0, calculate the design bearing capacity (q_d).

11.12 A strip footing of breadth 4 m is to be founded at a depth of 4 m. A standard penetration test carried out at this depth yielded an uncorrected N value of 16. Obtain a corrected value for N' and hence estimate the allowable bearing capacity for the footing when the footing is: (a) well below the base, and (b) 1.6 m below the ground surface.

(Unit weight of soil = 19 kN/m³)

11.13 A raft foundation 5 m square is to be founded at a depth of 3.5 m in a deep layer of sand; the groundwater level is known to remain at approximately 1.8 m below the surface. During site exploration the following uncorrected N values were recorded in a standard penetration test:

Depth below surface (m)	1.5	2.5	3.5	4.5	5.5	6.5	7.5	8.5	9.5	10.5
N value		10	11	15	18	24	26	29	31	34	36

drained $\gamma = 19$ kN/m³   saturate $\gamma = 21$ kN/m³

Determine the allowable bearing capacity.

11.14 A circular raft foundation of diameter 8 m is founder at a depth of 2.4 m and transmits a contact pressure of 150 kPa to a deep layer of silty sand. The ground-water level is located at a depth of 3.5 m. The unit weights of the silty sand are: drained $\gamma = 17.5$ kN/m³; saturated $\gamma = 19.4$ kN/m³. Using the cone resistances given below, obtain an estimate of settlement of the foundation using the Schmertmann, Hartman and Brown method.

Depth (m)	2.5	3.2	4.0	4.8	5.6	6.6	7.6	8.6
q_c (MPa)	2.1	2.8	2.6	2.9	3.5	4.2	4.1	6.2

Depth (m)	9.6	10.6	11.6	12.6	13.6	14.6	16.6	17.6
Q_C (MPa)	6.8	7.5	8.6	8.1	10.2	10.9	11.4	13.2

11.15 Bored piles of 600 mm diameter are to be installed at a depth of 10 m in a firm clay for which the following properties may be used:

Depth (m)	2.0	4.0	5.5	7.0	9.0	10.5	12.5
Undrained shear strength (kPa)	34	36	42	49	61	17	85

$\beta = 0.6$ $\gamma = 19.6$ kN/m³

The groundwater level is at a depth of 2.5 m.
The factors of safety required are as follows:

Overall = 2.5, end bearing = 3.0, shaft resistance = 1.5, group failure = 2.5

(a) Calculate the estimated safe load capacity of a single pile.
(b) Calculate the estimated safe load capacity of a 4 × 3 group of piles spaced at 1.2 m centres.
(c) Calculate the efficiency of the group.

Soil Mechanics Spreadsheets and Reference Assignments and Quizzes (available on the accompanying compact disc):

Assignments A.11   Bearing capacity of foundations
Quiz Q.11   Bearing capacity of foundations

Chapter 12

Site investigations and *in situ* testing

12.1 Scope and purpose of site investigation

A thorough and comprehensive site investigation is an essential preliminary to the design and construction of a civil engineering project. The size and type of project will influence the scope of an investigation, but not its necessity: even the smallest job warrants some form of site investigation. The British Standard Code of Practice BS 5930:1991 gives a list of the primary objectives of site investigation as follows:

(a) To assess the general *suitability* of the site and environs for the proposed works including implications of previous use or contamination.

(b) To enable an adequate and economic *design* to be prepared, including the design of temporary works.

(c) To plan the best method of *construction*; to foresee and provide against difficulties and delays that may arise during construction due to ground and other local conditions.

(d) To determine the *changes* that may arise in the ground and environmental conditions, either naturally or as a result of the works, and the effect of such changes on the works, on adjacent works, and on the environment in general.

(e) Where alternatives exist, to advise on the relative suitability of *different sites*, or different parts of the same site.

In addition, site investigations are necessary when reporting on the safety of existing works, where alteration to existing works is being planned, when investigating cases where failure has occurred, and when looking for sources of constructional materials.

Much of the investigation needs to be completed prior to the design stage of a project, although some overlap is likely, especially in the case of major works and where *in situ* testing is required.

A site investigation in connection with a moderate to large project will consist of several stages. The stages will all be interrelated and may overlap or occur in different sequences for different types of site and project. The overall process may be recursive, i.e. data from one stage may need further work in other stages, and this may continue throughout and even beyond the completion of the contract.

(a)   *Desk study*: this is essentially the collection of a wide variety of information relating to the site, e.g. maps, drawings; details of existing or historic development, local authority information; geological maps, memoirs, records; details of utilities, services, restrictions, rights-of-way, ownership of adjacent property; aerial photographs. (A comprehensive list of information and sources is given in BS 5930 and Dumbleton and West (1976).)

(b)   *Site reconnaissance*: an early examination of the site by appropriate experts is most desirable, e.g. geologist, land surveyor, soils engineer, hydrologist, etc. Information should be collected on the overall site layout, topography, basic geology; details of access, entry and height restrictions. Local conditions should be examined, such as climate, stream flows, groundwater conditions, site utilisation related to weather and time of year. Where possible photographic records should be kept.

(c)   *Detailed site exploration and sampling*: investigation of detailed geology and sub-surface soil conditions using surface surveys, trial pits, headings, boreholes, soundings, geophysical methods, as appropriate; survey of groundwater conditions over a significant period of time (maybe even after completion of works); examination of existing and adjacent structures for signs of cracking or settlement; location of underground structures or cavities, buried pipes, services, etc.; provision of samples for further examination and laboratory testing.

(d)   *Laboratory testing of samples*: tests on disturbed and undisturbed samples submitted from the site team; tests on soils (as specified) for classification, quality, permeability, shear strength, compressibility, etc.; tests on rock cores and samples for strength and durability; tests on constructional materials, such as compaction tests; tests on groundwater; chemical and petrographic analyses.

(e)   *In situ testing*: texts carried out on the site either prior to or during the construction process; ground tests such as shear-vane, standard penetration, cone penetration, plate bearing, pressuremeter; structure loading tests, such as tests on piles, proof loading; displacement observations.

(f)   *Reporting results*: details of geological study, including structures, stratigraphy and mapping; results of borings, etc., including log, references for samples and stratigraphy interpretations as requested; comments and recommendations relating to the design and construction of the proposed works; recommendations relating to further investigating or testing, and to ongoing or post-completion monitoring.

The extent and scope of a site investigation will depend partly on the nature of the site itself and partly on the type of structure. The number of variations and permutations is, of course, endless, but the following examples will serve to show how wide the range of variables is:

Types of site

(a)   *Compact*: enclosed urban, open country, mountain region, alluvial flood plain, river valley, coastal, harbour, off-shore.

(b)   *Extended*: road, railway routes; tunnels, sewers, pipelines, canals, transmission lines, coastal defences.

Types of structure New or existing; high-rise or low-rise; buildings – framed construction, unit construction, lightweight, etc. – flats, offices, factories, warehouses, power stations, etc.; materials – steel, reinforced concrete, pre-stressed concrete, brickwork, timber, etc.

The cost of an investigation is obviously a function of the extent and scope of the work to be done, and will be strongly influenced by the type of site. Taken as a percentage of the total contract price, the cost will generally be somewhat less for larger projects and where good ground conditions exist: as a general rule the figure will lie between 0.2 and 2 per cent of the total expenditure. Guidelines for specification and procurement of site investigation have been published by the Institution of Civil Engineers (Site Investigation Steering Group, 1993).

12.2 Extent and depth of site exploration

Most of the information collected during site exploration will relate to sub-surface deposits of rocks and soils. The overall objective is to build up a three-dimensional picture of the site, which extends laterally and vertically to include all of the strata likely to be affected by the changes in loading, etc. brought about by the proposed construction.

In order to achieve adequate coverage, the spacing of trial pits or boreholes needs to be based on considerations of both the type of structure (e.g. narrow or wide; low or tall; heavy or light, etc.) and the nature of the ground conditions (e.g. rock or soil; firm or soft; homogeneous or stratified, etc.). Borings should be sunk at strategic points related to the layout of buildings, such as near corners, near to points of heavy loading, near to locations where settlement is to be limited. In uniform and homogeneous conditions, boreholes may be located up to 100 m apart; in conditions of lateral or vertical variability the spacing is accordingly reduced, to as close as 5 m in severe cases.

The depth of exploration is related mainly to the types of material present and their susceptibility to compression under load. A useful guide (strictly a *guide* and not a rule) may be obtained by computing the predicted increases in vertical stress. Strata likely to be subject to an increase in vertical effective stress of 10 per cent or more should be included in the investigation. Another simple, but perhaps not so reliable, rule of thumb is to carry borings down to at least 1.5 times the breadth of the foundation; care is required in interpreting this guideline where multiple foundations are to be installed (see Section 6.12).

Where rock strata are encountered, it is good practice to continue all borings for at least 1.5 m into sound unweathered rock. In all cases, a small number of borings should be taken down significantly deeper, so that no unexpectedly weak layers are missed.

12.3 Methods of site exploration

In choosing methods of site exploration there are four main sets of factors to consider.

(a) *Geological nature of the site*: in clay soils the most practicable form of exploration is the sinking of boreholes; in sandy soils, lined boreholes may be required, together with special sampling equipment; trial pits are only practicable in firm or compact soils, or in soft rocks, above the water table; in hard rocks core borings can be taken down to sufficient depth, but tend to give little information on structural features, such as joint geometry.

(b) *Topographical nature of the site*:   both the type of terrain and access to the site are important from the point of view of moving exploration equipment about; in hilly or steeply-inclined sites headings driven more or less horizontally may be more convenient than vertical borings; in waterlogged or swampy sites some preliminary work may be necessary to form roads, hardstandings, etc.

(c) *Type of information required*:   in the main, both designer and constructor will be interested in the nature and sequence of the sub-surface rocks and soils, but specialist information is sometimes required: details of joint geometry, groundwater flows, presence of buried features (either natural or man-made), location of previous failure surfaces, etc.

(d) *Cost and time*:   usually, the deeper the exploration the more costly and time consuming it will be; most costs are reduced as the number of holes (or pits, etc.) increases; time factors should be considered in the project planning stage.

Trial pits

In cohesive soils and soft rocks, above the water table, trial pits are often preferable to borings; they are easily dug with a mechanical excavator, or even by hand, and they have the advantage of exposing the succession of strata for easy visual examination. The main disadvantage is that they are limited to depths of 2–3 m; perhaps a little deeper by additional hand-digging. The sides of trial pits must be adequately supported, even when they are to be left open only for short periods; it may also be necessary to install some form of water pumping, especially in permeable rocks and soils. Samples may be taken by hand from the bottom and sides of the pit. Trial pits are often particularly useful in soils containing boulders or cobbles, for groundwater observations and for locating buried pipes and services. Long trench pits can be useful on filled sites and where pre-existing works and services are expected to exist.

Headings

Headings (also called *adits*) are excavated almost horizontally, either from the surface of steeply-inclined ground, from cliff or quarry faces, or from the bottom of shafts. The cost of driving headings (and sinking deep shafts) is very high, and in addition the operation is difficult in loose ground or below the water table. Headings are therefore seldom used, except for specialist investigations, such as pilot tunnels, mineral exploitation surveys, or where other methods are unsuitable.

Hand auger

The hand auger (also called a *post-hole* or *Iwan auger*) is a very simple hand-tool used for drilling into soft soils down to a maximum of 5–6 m. The usual form consists of a 100 mm diameter half-cylinder clay auger, which is attached, through

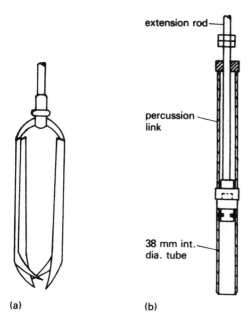

extension rod

percussion
link

38 mm int.
dia. tube

(a)   (b)

Fig. 12.1 Hand auger and sampler
(a) Iwan auger (b) 38 mm sample tube

a series of 1 m extension rods, to a cross-piece, that may be turned manually at the surface (Fig. 12.1(a)). To obtain samples, the clay auger is replaced with a 38 mm sample tube attached to a sliding percussion link (Fig. 12.1(b)). By raising the extension rod/percussion link assembly and forcing it downwards, the sample tube is driven into the ground at the bottom of the hole. The cross-piece is then rotated to shear off the bottom end of the sample and the sample tube driven upwards using the percussion link.

Percussion rig boring

The light cable percussion or shell auger method is the most widely used method in the United Kingdom; the equipment is relatively simple and easily portable, and in soft to very firm soils free from boulders or cobbles a good rate of progress can be maintained. The most usual borehole diameter is 150 mm, but others up to 300 mm can be drilled; the maximum depth of exploration, although dependent on soil type to some extent, is around 50–60 m.

The equipment consists of a derrick, a power-winch (Fig. 12.2(a)) and a set of drilling tools. A percussion method is used, whereby the tool assembly is raised by the winch to about 1 m above the bottom of the hole and then allowed to fall under its own weight, thus driving the cutting tool into the soil. When the tool becomes full of soil, it is raised to the surface, where disturbed samples may be taken from its contents.

The *clay cutter* (Fig. 12.2(c)) is an open-ended cylinder with slots in the sides and a detachable cutting shoe at the lower end; this tool is particularly used for clay soils in dry boreholes. A *clay auger* is occasionally used; this is rotated by

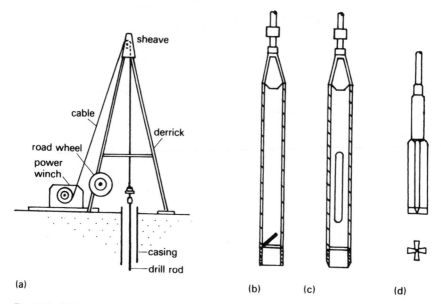

Fig. 12.2 Light percussion drilling equipment
(a) Derrick and winch (b) Bailer (c) Clay cutter (d) Chisel

hand to advance the drilling; it is also used to clean the bottom of the hole before sampling.

For boring in sands and gravels, the *sand shell* or *bailer* is used (Fig. 12.2(b)). This is also an open-ended (non-slotted) cylinder, with a detachable cutting-shoe which incorporates a flap valve. The flap valve ensures that loose material or slurry is retained in the shell as it is raised to the surface. In compact cohesionless soils, or where boulders or cobbles are encountered, the *chisel* (Fig. 12.2(d)) is used to break up hard materials; fragments and slurry are then removed using the bailer.

In wet conditions and in loose soils, and for very deep holes, a *casing* must be installed near the surface. This usually consists of steel tubes, screwed together in as many lengths as appropriate, and jacked or knocked into the drilled hole as drilling proceeds. They can be hauled out after completion of drilling or left in place if further observations are required.

Rotary augers
Power-operated rotary augers mounted on vehicles provide an extremely portable and versatile method of drilling. As well as drilling for exploration purposes, they are also used in piling operations; a variety of types and sizes are available, from small units mounted on ordinary tractors, to larger machines capable of drilling holes to depths of 50 m and, with certain types of auger, up to 2 m in diameter.

Augers are usually referred to as *flight augers* or *bucket augers*. Short-flight augers (Fig. 12.3(a)) consist of a short-length helical blade attached to the shaft, with cutter teeth and a small diameter pilot bit at the lower end; these are

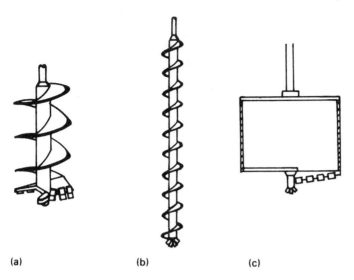

Fig. 12.3 Rotary augers
(a) Short-flight auger (b) Continuous-flight auger (c) Bucket auger

available in diameters between 75 and 500 mm, and with some special devices as large as 1 m. For some larger augers the helix comprises only one revolution: these are called *single-flight* augers. *Continuous-flight* augers (Fig. 12.3(b)) are used to enable the cut soil to rise, up the spiral, to the top of the hole; thus removing the necessity to lift and clear the tool a number of times; as the drilling advances, additional spiral sections are added.

A more recent development of the continuous-flight auger is the *hollow-stem* auger, which consists of a long flight of spiral attached to a hollow tube and an adaptor cap at the top. A central drill rod passes through the hollow stem and attaches to the cutting head at the lower end and adaptor at the top. As drilling proceeds, additional lengths of hollow-stem and drilling rods are added as required; the adaptor and drilling rod may be removed at any time to facilitate core sampling, the hollow-stem serving as a casing. The use of the hollow-stem auger is particularly recommended in loose or soft deposits, especially below the water table.

Bucket augers (Fig. 12.3(c)) consist of an open-top metal cylinder; with cutters mounted on a baseplate; as the soil is cut it passes into the bucket, which is then raised and emptied at intervals. Bucket augers can be used to drill holes rapidly in firm soil (they are not suitable for drilling in cohesionless soils below the water table) with diameters ranging from 300 mm to 2 m. Inspection shafts of about 1 m diameter can be conveniently drilled using a bucket auger; using a special cage and with suitable ventilation a person can be lowered into the hole to carry out a visual examination.

Core-drilling

In hard soils and rocks power-operated *core-drills* are used, consisting of small-diameter hollow tube, fitted at the lower end with a *coring bit* (Fig. 12.4). The core barrel is rotated at speeds ranging between 600 and 1200 rpm, a controlled

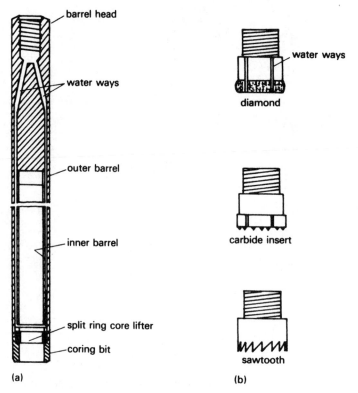

Fig. 12.4 Rotary coring equipment
            (a) Double-tube core barrel (b) Coring bits

pressure applied and water circulated through the bit. The fragments removed in the annular cut are brought to the surface with the circulating water as the core fills the barrel. A drilling run of 1–3 m is usually made before raising the barrel and removing the core. The more usual standard sizes of core barrel used in site investigation range between 30 and 100 mm (hole diameter), although larger-diameter equipment is available for special uses.

12.4   Methods of sampling

There are two main categories of soil samples:

Undisturbed samples: in which the structure and water content is preserved, as far as possible, to truly represent site conditions; undisturbed samples are required for tests of shear strength, consolidation and permeability; these are usually obtained by a suitable coring method.

Disturbed samples: these should be collected as drilling or digs proceed, where possible attempting to preserve the *in situ* water content; disturbed samples are mainly required for soil identification and for classification and quality tests; as samples are collected they are placed and sealed into glass or plastic containers, or tins, or plastic bags.

Sample disturbance

It is virtually impossible to obtain totally *undisturbed* samples, especially from moderate to deep holes. The process of boring, driving the coring tool, raising and withdrawing the coring tool and extruding the sample from the coring tool, all conspire to cause some disturbance. In addition, samples taken from holes may tend to swell as a result of stress relief. Samples should be taken only from a newly drilled or newly extended hole, with care being taken to avoid contact with water. As soon as they are brought to the surface, core tubes should be labelled inside and outside, the ends sealed with wax and capped, and then stored away from extremes of heat or cold and vibration.

Soft silts and clays of low plasticity are particularly sensitive to sampling disturbance, especially on the outside where in contact with the sample tube. In samples raised from deep holes, the reduction in total stress produces a negative pore pressure in the centre; and as swelling takes place water will be drawn from the outer zone to dissipate the excess negative pore pressure. The outer zone is therefore consolidated, but as the reduction in negative pore pressure also results in a reduction of effective stress, the shear strength of the soil tends to be reduced.

Sample disturbance may be reduced by using an appropriate type of sample tube. The types of sampler in common use are: open-drive, thin-walled, split-barrel and piston sampler.

Open-drive sampler

The simplest and most common form of sample tube is called an *open-drive* sampler (Fig. 12.5(a)), consisting of a steel tube with a screw thread at each end. A cutting shoe, and sometimes an extension piece, is screwed on to the lower end; a sampler head, which incorporates a non-return valve, is screwed-on to the upper end. As the sample is driven into the tube the valve allows air and water to escape but it remains closed as the sampler is raised to the surface, thus holding the sample within the tube.

It is recommended (British Standard BS 5930) that the diameter of the cutting (D_C) edge be typically 1 per cent less than that of the inside of the tube (D_S), so as to reduce the frictional resistance between the sample and the tube. Also the external diameter of the cutting shoe (D_W) should be slightly greater than that of the tube, so as to reduce the force required to withdraw the tube from the hole. The volume of soil displaced by the sampler in proportion to the volume of the sample is given by the *area ratio* (A_r) as follows:

$$A_r = \frac{D_W^2 - D_C^2}{D_C^2} \times 100 \qquad [12.1]$$

To reduce sample disturbance, the area ratio should be kept as low as possible consistent with maintaining sufficient strength in the wall of the tube.

The most common type of open-drive sampler has an internal diameter of 100 mm and a length ranging between 380 and 450 mm, giving an area ratio of about 30 per cent. This is suitable for all types of cohesive soils, but a flap-valved extension piece may be needed for loose or wet sands.

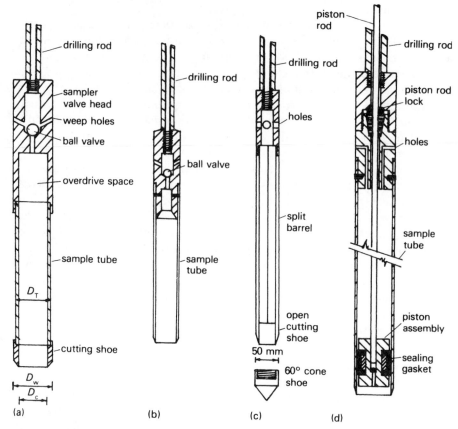

Fig. 12.5 Soil samplers
(a) Open-drive sampler (b) Thin-walled sampler (c) Split-barrel sampler
(d) Piston sampler

Thin-walled sampler

In soft silts and clays, which are sensitive to sample disturbance, a *thin-walled* (or *Shelby tube*) sample is preferred to an open-drive type (Fig. 12.5(b)). The end of the tube is machined into a cutting edge, with an inward reduction in diameter to give an area ratio of 10 per cent. A valved sampler head is used, but with a screw-out stud attachment method, rather than a screw thread. Thin-walled samplers are available in diameters ranging from 35 to 100 mm and in a variety of lengths.

Split-barrel sampler

The *split-barrel* sampler is a dual-purpose tool (Fig. 12.5(c)), which may be fitted with a cylindrical cutting shoe for taking samples, or a solid conical shoe may be fitted, and the tool may also be used to perform the standard penetration test (see Section 12.5). As the name suggests, the barrel is split longitudinally into two hemicylindrical halves. The sampler head and shoe are attached by means of screw threads at the upper and lower ends; when these have been removed, the two halves of the barrel can be separated to allow removal of the sample. The main disadvantage of the split-barrel sampler is that, since the internal diameter is 35 mm and the outside diameter is 50 mm, the area ratio is about 100 per cent.

Piston sampler

For very soft alluvial silts and clays it will be necessary to use a *piston sampler* (Fig.12.5(d)), consisting of a thin-walled tube fitted with a piston. The sampler is attached to the lower end of a hollow boring rod, through which passes an inner rod which operates the piston. To begin with, the sampler is lowered to the bottom of the borehole with the piston locked in the lower position.

The piston incorporates a seal which prevents water and debris from entering the tube. As the piston is held against the soil at the bottom of the hole, it is unlocked and the tube driven down into the soil for the full length of travel of the piston. The piston is now locked at the top of the tube and the whole assembly withdrawn to the surface, where the sampler head and the piston are removed before waxing and sealing the tube.

The piston sampler is generally available in sizes ranging from 35 to 100 mm (internal diameter), producing sample lengths of up to 600 mm.

Swedish foil sampler

This is a special variation of the piston sampler designed to secure long lengths (up to 25 m) of continuous undisturbed samples. The basic principle is the complete elimination of frictional resistance between the sample and the sides of the tube. This is achieved by housing rolls of metal foil in recesses built into the sampler head; these unroll and form a lining between the soil and the tube as the core is driven up the sampler. The piston is drawn upwards on a cable and lengths of sample tube (68 mm internal diameter) are attached as the hole is driven deeper.

As the tubes are raised, they are uncoupled and a cut made through foil and soil together.

This type of sampler is primarily used for sampling in soft silts and clays, in thinly stratified soil, and in very soft or liquid soils, such as lake bed and harbour deposits.

12.5 In situ *testing*

In certain soils, such as soft sensitive silts and clays and some coarse non-cohesive soils, it is difficult (almost impossible occasionally) to obtain good undisturbed samples. It is difficult also to accurately model in the laboratory truly representative conditions of structure and/or pore pressure under certain site conditions; e.g. in very soft alluvium. A number of relatively simple *in situ* testing procedures have therefore been devised which will enable good estimates of soil properties to be made under actual site conditions. Although in *in situ* testing the degree of accuracy and control possible is lower than would be expected in the laboratory, this is often compensated for by a large number of tests being carried out.

Table 12.1 In situ *testing*

Required to measure	Test	Required to measure
	Core cutter ⎫	Soil density
	Sand replacement ⎬	
	⎧ Standard penetration ⎫	Relative
Shear strength	⎨ Cone penetration ⎬	density
	⎪ Shear vane	
	⎩ Pressuremeter ⎫	
	⎧ California bearing ⎪	
Allowable	⎨ ratio ⎬	Compressibility
bearing capacity	⎪ Plate bearing ⎪	
	⎩ Pile loading ⎭	
Drainage pattern ———————	Groundwater level ⎫	
In situ stress ———————	Piezometer ⎬	Pore pressure
	Pumping test ———————	Permeability

In Table 12.1 a list is given of the most important *in situ* tests, together with an indication of what each is primarily intended to measure. The methods of interpretation and application of test results are given (in most cases) elsewhere.

Core cutter test
The core cutter is also referred to in Section 3.6 and is fully detailed in BS 1377. The purpose of the test is to determine the bulk density of soil that has been placed by compaction, or of natural soil; after measuring the water content the dry density can then be calculated. The apparatus (Fig. 3.8) consists of steel cylinder internal diameter 100 mm and length 130 m, machined to a cutting edge at one end. With the upper end protected by the dolly, the cutter is knocked into the ground using a specially designed rammer; afterwards it is dug out, and with the ends of the soil struck off level, it is weighed.

Sand replacement test
This is another method for determining soil density; it also is referred to in Section 3.6 and is detailed in BS 1377. A cylindrical hole approximately of diameter 100 mm and depth 150 mm is dug out through a hole in a special tray; excavated material is then carefully weighed. The sand-pouring cylinder (Fig. 3.9) is placed over the hole and sand run out to fill it, and so determine its volume. Two sizes of sand-pouring cylinder are available for soils of fine-medium and coarse grain size.

Standard penetration test
This test is widely used during the course of sinking test boreholes as a means of estimating the relative density and shear strength characteristics. A standard 50 mm diameter split-barrel sampler (Fig. 12.5(c)) is driven into the ground at the bottom of the hole by repeated blows from a drop-hammer of mass 65 kg, falling a distance of 0.76 m. The sample is driven a total of 450 mm into the soil and the number of blows recorded for the last 300 mm of penetration. The interpretation of the test results is dealt with in Section 11.8. For sands and

cohesive soils, the cylindrical cutting shoe can be used and samples taken at the same time; for coarser-grained soils a solid conical shoe is preferred, giving identical results. The test is fully detailed in BS 1377.

Cone penetration test
The cone penetration test is a *static penetration* test in which the device (Fig. 12.6) is pushed, rather than being driven by blows, into the soil. The cone,

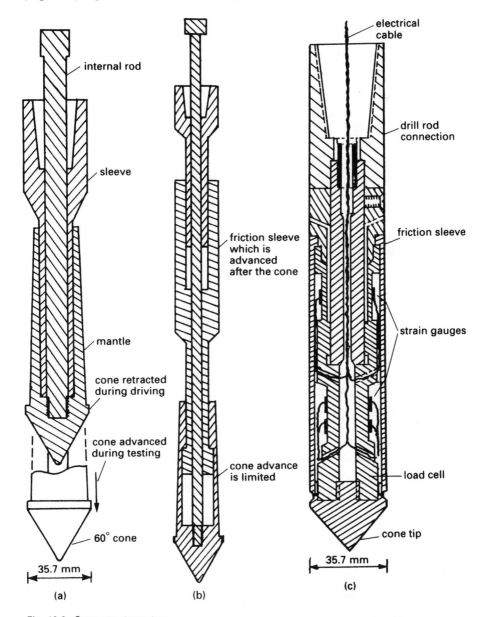

Fig. 12.6 Cone penetrometers
(a) Dutch mantle cone (b) Friction sleeve cone (c) Electric penetrometer

which has an apex angle of 60° and an end diameter of 35.7 mm (giving an end area of 1000 mm^2), is attached to a rod; this latter being protected by an outer sleeve. By applying a measured force to the rod, the cone is pushed for about 80 mm into the soil at a uniform rate of penetration of 20 mm/s. The ratio of the force required to the end area is called the cone penetration resistance (q_c). The interpretation of the test results is dealt with in Section 11.8.

The results appear to be most reliable for sands and silts, that are less than 85 per cent saturated. A recent development is the 'electric' cone penetrometer in which a load cell is incorporated, thus enabling a continuous charted readout of penetration resistance against depth. A detailed description of the test and its interpretation is given in Meigh (1987).

Shear vane test

Very often silts and clays, particularly those of alluvial or shallow-water origin, are notoriously difficult to sample. The shear vane is used to measure the *in situ* undrained shear strength of these soils. Interpretation of the test results is dealt with in Section 7.17, and the test procedure is fully detailed in BS 1377.

A four-bladed vane (Fig. 12.7) is driven into the soil at the end of a rod and the vane then rotated at a constant rate between 6 and 12 deg./min. until the cylinder of soil contained by the blades shears. The maximum torque required for this is recorded. For weak soils ($c_u < 50$ kPa) the blade size should be 75 mm wide by 150 mm long, and for slightly stronger soils ($50 < c_u < 100$ kPa) 50×100 mm. Remoulded strength can be measured by rotating the vane at a faster rate

The vane rod and extension pieces are enclosed in a sleeve to prevent soil adhesion during the application of the torque. Depending on the nature of the soil, vane tests may be carried out down to depths of 60–70 m.

Plate bearing test

The object of the plate bearing test is to obtain a load/settlement curve. A series of such tests is often time consuming and expensive, and quite often will give unreliable results. They can be useful under certain circumstances, where other test procedures are difficult or impossible to apply: in testing, for example, weathered rocks, chalk or hardcore fill.

A trial pit is first excavated to the required depth, the bottom levelled and a steel plate set firmly on the soil. A static load is then applied to the plate in a series of increments, and the amount and rate of settlement (vertical displacement of the plate) measured. Loading is continued until the soil under the plate yields. A number of tests will be required using different plate diameters and at different depths; the diameters range between 300 mm and 1 m. Errors in the resulting settlement estimates are mainly due to the difference between the plate area and that of the actual foundation and the fact that the much larger structure will affect a larger volume of soil: deep layers may be stressed by the actual structure, but not by the plate bearing test. Another source of error often arises from carelessness in bedding-in the plate; ideally this should be done by setting the plate in plaster of Paris.

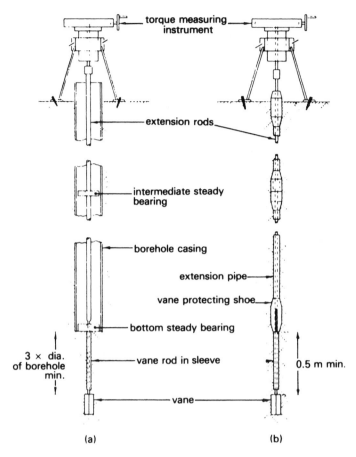

torque measuring instrument

extension rods

intermediate steady bearing

borehole casing

extension pipe

vane protecting shoe

bottom steady bearing

3 × dia. of borehole min.

vane rod in sleeve

0.5 m min.

vane

(a)   (b)

Fig. 12.7 Shear vane apparatus
(a) Borehole vane test (b) Penetration vane test
(Reproduced from BS 1377 by permission of the British Standards Institution)

California bearing ratio test

The California bearing ratio test is basically a laboratory penetration test, but it can also be carried out *in situ*. It is not easy to obtain reliable reproducible results with the wet cohesive soils commonly found in the United Kingdom; the test is much more applicable in tropical and sub-tropical regions where drier soils occur. The test procedure is fully detailed in BS 1377, and is required for some road and airfield design procedures.

A circular area of ground is first trimmed flat and the test carried out by forcing a standard circular plunger of 1935 mm² end area to penetrate the soil at a constant rate of 1 mm/min., while readings of the load are taken at every 0.25 mm of penetration. A plot is drawn of force/penetration, and the forces corresponding to a penetration of 2.5 and 5.0 mm read off. The California bearing ratio (CBR) is given by the larger ratio:

$$CBR = \frac{\text{force required on test soil}}{\text{force required for same penetration on standard soil}}$$

Pile loading test

Loading tests are usually carried out on one of the piles to be used in the structure installed at an appropriate depth. The object may be either to deter- mine the ultimate load or to load the pile to a 'proof load', this latter exceeding the proposed working load by a specified percentage. The pile may be loaded by means of a hydraulic jack or by adding kentledge to a platform built on top of the pile (Fig. 12.8).

Two test procedures are recommended and detailed in British Standard BS 8004: the *maintained load method* and the *constant rate of penetration method*.

(a)   *Maintained load method*: the test load is applied in equal increments, each load being maintained until all observable settlement has ceased. Curves are then plotted of settlement against load, and settlement against time. This procedure is recommended when estimates of settlement are required for working loads, the results are less reliable when ultimate loads are being determined.

(b)   *Constant rate of penetration method*: in this method, the pile is forced into the ground using the hydraulic jack at a constant rate: 0.75 mm/min. in clay soils and 1.5 mm/min. in sands and gravels. The load is measured either by using a calibrated jack, or a load cell or proving ring. This method is quite rapid and usually requires several observers, briefed to take read- ings simultaneously, or an electronic logging system.

Interpretation of load/settlement curves

In firm to stiff soils the curve often becomes parallel to the penetration axis (Fig. 12.9) or even falls back; the ultimate load may then be taken as the max- imum load value. In looser or softer soils, the curve does not flatten out; in these cases it is recommended that the ultimate load is taken as the value corresponding to a settlement equal to 10 per cent of the pile diameter. An improved method of interpretation has been suggested by Chin (1978): when the settlement/load ratio is plotted against settlement, a straight-line curve results for unbroken piles. The ultimate load can then be evaluated from the inverse slope of the curve (Fig. 12.10(a)).

$$\text{Ultimate base capacity} = \frac{\mathrm{d}s}{\mathrm{d}(s/P)} = \frac{15.4 - 0}{20.0 - 3.2} \times 10^3 = 917 \text{ kN}$$

Broken piles can easily be detected when the plot is curved or has a distinct change in slope (Fig. 12.10(b)).

Pressuremeter test

The pressuremeter is a device developed by Menard in the 1950s (Menard, 1956, 1969; Baguelin *et al.*, 1978) for the purpose of measuring *in situ* strength and compressibility. The apparatus consists of two basic components: a probe which is inserted in an unlined borehole, and a pressure-transducer volumeter.

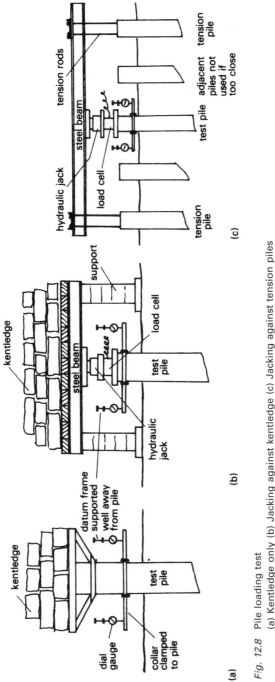

Fig. 12.8 Pile loading test
(a) Kentledge only (b) Jacking against kentledge (c) Jacking against tension piles

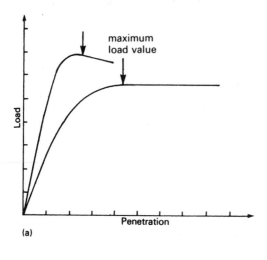

(a)

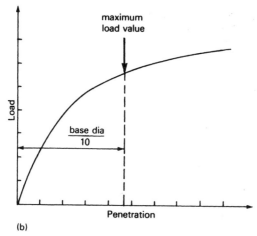

(b)

Fig. 12.9 Interpretation of pile loading tests

The probe is made up of a measuring cell, with a guard cell above and below, enclosed in a rubber membrane (Fig. 12.11(a)). The membrane is inflated using water under an applied gas pressure (CO_2) and the pressure and volume readings taken continuously; the two guard cells ensure that a purely radial pressure is set up on the sides of the borehole. A pressure/volume-change curve is then plotted, from which shear strength and strain characteristics may be evaluated.

Modern instruments, such as the Cambridge pressuremeter (Fig. 12.10(b)), are not only self-boring but will also provide measurements of a horizontal strain and pore pressure, thus allowing a complete effective-stress/strain curve to be drawn. From this curve may be determined the undrained or drained shear strength parameters as well as the shear modulus and Poisson's ratio. It appears (Windle and Wroth, 1977; Hughes *et al.*, 1977) that, although the strength values measured by the pressuremeter tend to be on the high (unsafe) side, the measurements of horizontal effective stress and elastic constants agree well with the

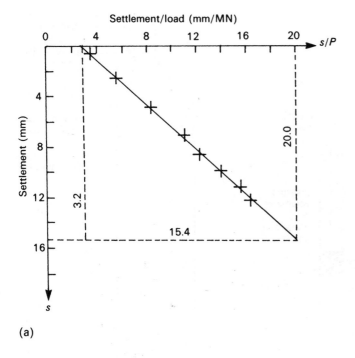

(a)

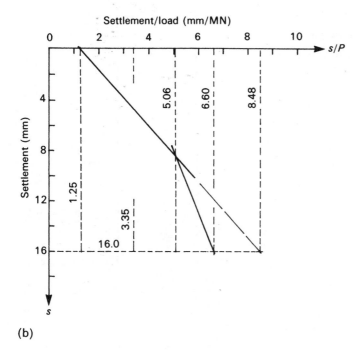

(b)

Fig. 12.10 Pile test interpretation using Chin's method
(a) Ultimate load from inverse slope (b) Curved plot showing broken pile

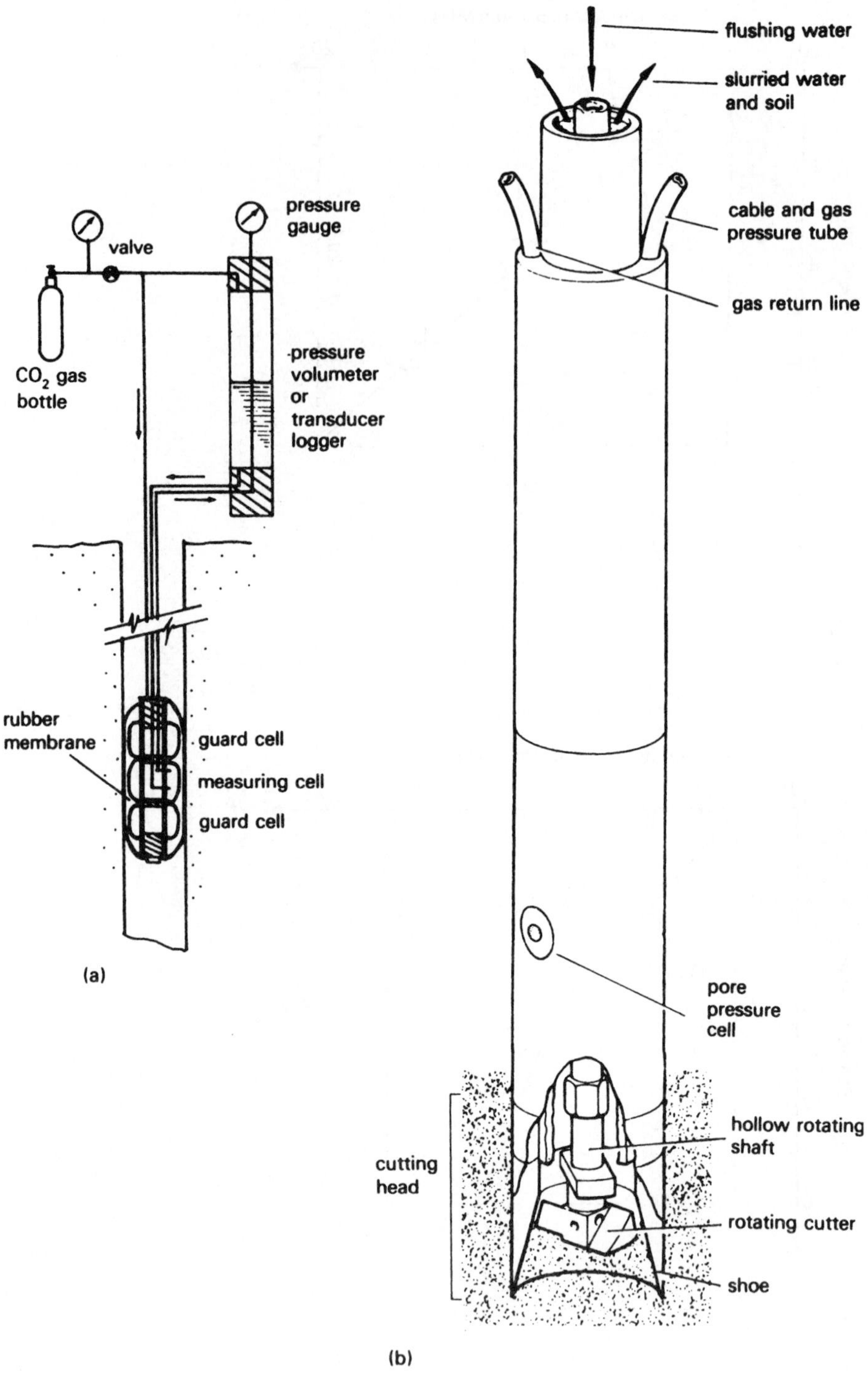

Fig. 12.11 Pressuremeters
(a) Menard type (b) Cambridge type

results of other tests. A detailed account of the test and its interpretation is given in Mair and Wood (1987).

12.6   Groundwater observations

There are two main groups of groundwater observations: determinations of levels and pore pressures, and permeability tests. Observations of groundwater level made during borehole drilling are often found to be unreliable and should be taken as a guide to conditions, rather than as accurate assessments. Variations in climate, stream flow, vegetation growth, etc. may lead to quite different groundwater regimes at different times of the year. It is essential, therefore, that a series of readings is taken over an appropriate period of time.

Permeability pumping tests were dealt with in Section 5.9 and will not be discussed further here. Observations of water table and piezometric surface level are of considerable value, especially when deep excavations are planned.

Borehole observations
Boreholes drilled during the course of site investigations can be left open and cased if necessary to provide an observation point. A minimum period of 24 hr after drilling has ceased should be allowed to pass before observations are commenced. The level of water in a hole is determined usually by means of a chalked tape, or a tape with a float, or by lowering an electrical switching device which is actuated on contact with the water.

It may be of advantage for a long-term series of observations to insert into the borehole a smaller diameter observation tube. The observation tube terminates at a convenient height above the ground surface and is fitted with a cap; the top of the boreholes can then be sealed to prevent blockage by debris and the ingress of rainwater or surface run-off water.

In fine soils the stabilisation of the water table may take some time; in this event, its location may be found using the *fill and bail* method. The hole is filled with water and a quantity bailed out; if the water level is then observed to be falling more is bailed out; if it is rising a small quantity is poured back. This process is continued until a stationary level is found – this being the true groundwater level.

An alternative extrapolation method may be used in which observation of water level is plotted against time. The true groundwater level is found by extrapolating the curve until it becomes parallel to the time axis (Fig. 12.12(a)). If several levels are noted at equal time intervals, the following computational method may be used:

Let: h_1 = rise in water level from time t_0 to t_1
       h_2 = rise in water level from time t_1 to t_2
       h_3 = rise in water level from time t_2 to t_3

Also let: $t_1 - t_0 = t_2 - t_1 = t_3 - t_2$ (Fig. 12.12(b))

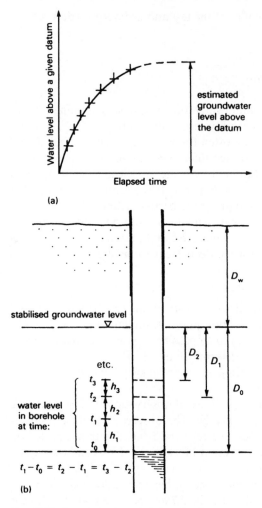

Fig. 12.12   Location of groundwater level

Then the depths of the observed water level below the stabilised groundwater level are:

$$D_0 = \frac{h_1^2}{h_1 - h_2}; \quad D_1 = \frac{h_2^2}{h_1 - h_2}; \quad D_2 = \frac{h_3^2}{h_2 - h_3}; \quad \text{etc.}$$

Observations are continued until sufficient estimates of the location of the stabilised groundwater level have been made.

Worked example 12.1   *The following water levels were recorded at intervals of 24 hr in a borehole in a cohesive soil:*

Depth of water level from surface (m)	8.62	7.74	7.07	6.57	6.18	5.89
Day	0	1	2	3	4	5

Calculate the estimated stabilised groundwater level.

Referring to Fig. 12.12(b): $h_1 = 8.62 - 7.74 = 0.88$ m $h_4 = 6.5 - 6.18 = 0.39$ m
$\qquad\qquad\qquad\qquad\quad h_2 = 7.74 - 7.07 = 0.67$ m $h_5 = 6.18 - 5.89 = 0.29$ m
$\qquad\qquad\qquad\qquad\quad h_3 = 7.07 - 6.57 = 0.50$ m

Then $D_0 = \dfrac{0.88^2}{0.88 - 0.67} = 3.69$ m $\quad D_3 = \dfrac{0.39^2}{0.50 - 0.39} = 1.38$ m

$\qquad D_1 = \dfrac{0.67^2}{0.88 - 0.67} = 2.14$ m $\quad D_4 = \dfrac{0.29^2}{0.39 - 0.29} = 0.84$ m

$\qquad D_2 = \dfrac{0.50}{0.67 - 0.50} = 1.47$ m

Estimate 1:	$D_w = 8.62 - 3.69$	$= 4.93$ m
Estimate 2:	$D_w = 8.62 - 2.14 - 0.88 - 0.67$	$= 4.93$ m
Estimate 3:	$D_w = 8.62 - 1.47 - 0.88 - 0.67 - 0.50$	$= 5.10$ m
Estimate 4:	$D_w = 8.62 - 1.38 - 0.88 - 0.67 - 0.50 - 0.39$	$= 4.80$ m
Estimate 5:	$D_w = 8.62 - 0.84 - 0.88 - 0.67 - 0.50 - 0.39 - 0.29$	$= \underline{5.05}$ m
		Average $= 4.96$ m

Stabilised groundwater level = *5.0 m below surface*

Pore pressure measurements

Pore pressures may be measured directly by sinking *standpipes* or *open piezometers* to the required level in the required stratum. In an unconfined aquifer the water level in the standpipe should correspond to the free water table, so that the pore pressure at the tip of the standpipe is given by the height to the standing water level (Fig. 12.13(a)). When a standpipe penetrates a confined aquifer, the artesian pressure forces the water level up to a level corresponding to the piezometric surface representing the pore pressure in the confined aquifer.

The artesian piezometric surface level may occur below (Fig. 12.13(b)) or above (Fig. 12.13(c)) the water table level in an unconfined surface layer.

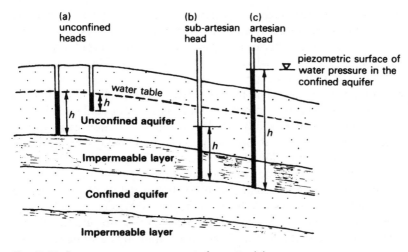

Fig. 12.13 Pore pressure measurements from standpipes

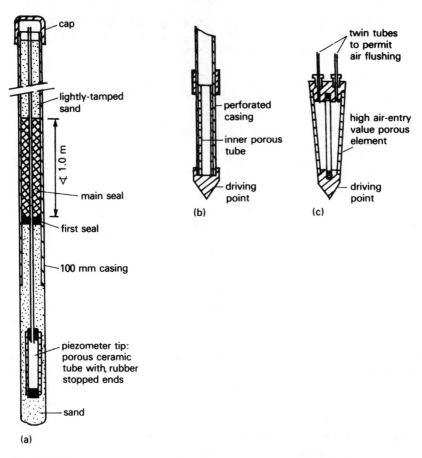

Fig. 12.14   Piezometers
(a) Casagrande type. Driven types: (b) Geonor (c) Bishop

Open-ended standpipe tubes have several disadvantages: they tend to become clogged with silt and sometimes clay and their response to pore pressure changes is rather slow. For soils of low permeability and where a more rapid response is required a *closed piezometer* is used, of which there are two types: the *hydraulic* type and the *pressure transducer* type.

In the hydraulic-type arrangement, the piezometer tip is installed at the bottom of a borehole or (in sands) driven into the ground. These are usually hollow and consist mainly of a porous body and two hydraulic tube connections (Fig. 12.14). The hydraulic tubes lead back to a Bourdon-type gauge or a manometer panel, and are usually made of stiff nylon tubing. It is essential to keep the system air-free. The water inside the piezometer tip responds to changes in pore pressure in the surrounding soil and this is reflected at the observation panel; a correction has to be made for the difference in position between the observation panel and the tip.

Among recent developments are piezometer tips incorporating electrical load-cells and vibrating-wire gauges, offering even more rapid and accurate response, and quite free of the problems of air contamination.

12.7 Geophysical methods

Certain geophysical properties, such as electrical resistance, elasticity, magnetic susceptibility, etc., vary from stratum to stratum; a stratum boundary is therefore indicated by an anomaly in the measurements of the particular property.

Geophysical methods do not actually measure engineering properties, hence they provide indirect methods of soil exploration; they can be used economically to determine soil stratum boundaries, to locate bedrock and water table levels, to detect organic soil areas and the presence of sub-surface cavities. In all cases, geophysical data needs to be correlated with information gathered from borings or trial pits.

There are two types of geophysical method in common use, so far as civil engineering is concerned: *electrical resistivity surveys* and *seismic methods*.

Electrical resistivity methods

Although the mineral constituents of soils are in the main poor conductors of electricity, the presence of groundwater containing dissolved salts does allow the passage of measurable currents over short to moderate distances. Significant variations in resistivity can be detected between different strata: above and below the water table; between unfissured rocks and soils; between voids and soil/rock.

The usual method of carrying out a resistivity survey consists of driving four equally-spaced electrodes into the ground in a straight line (Fig. 12.15(a)). As a current (I amp) is passed through the outer pair of electrodes, the potential difference (E volts) between the inner pair is measured. This arrangement is known as the *Wenner configuration*. For a soil of uniform apparent resistivity (ρ_α), it can be shown that:

$$\rho_\alpha = \frac{2\pi dE}{I} \qquad [12.2]$$

For the location of vertical boundaries, the spacing is kept constant and the configuration traversed across the site; a significant change in resistivity is recorded as each electrode passes across the boundary. To determine the depth of a horizontal boundary, a number of readings are taken with different electrode spacings and a field curve plotted of ρ/d. The field curve is then compared with a set of theoretical resistivity-ratio curves representing a 2-, 3- or 4-layer system of strata as appropriate. Figure 12.15(b) shows a set of two-layer curves; from which the depth of layer A may be found from the distance factor X.

Seismic refraction method

Seismic waves travel at different velocities through different types of material; several factors affect the velocity of shock-wave propagation, such as density, moisture content, texture, presence of voids or discontinuities and elasticity. The method of seismic refraction involves generating a sound wave in the rock or soil, using a sledgehammer, a falling weight or a small explosive charge, and then recording its reception at a series of *geophones* located at various distances

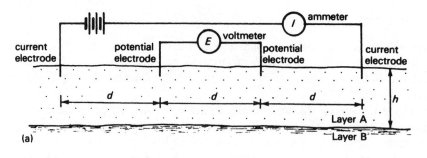

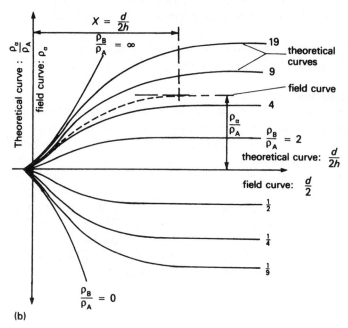

(b)

Fig. 12.15 Resistivity method
(a) Wenner configuration (b) Two-layer resistivity curves

from the shotpoint (Fig. 12.16(a)). The time of the first sound arrival at each geophone is noted from the pen-trace of a continuous recorder.

For geophones close to the shotpoint, the direct wave, travelling with velocity V_A in the upper layer A, arrives first. If the velocity in the lower layer is greater, then at some distance away from the shotpoint, the refractive wave, travelling with velocity V_B in the lower layer B and with velocity V_A in the upper layer, will arrive first. The first arrival times are plotted against distance from the shotpoint (Fig. 12.16(b)), to give two time/distance straight-line curves of slopes $1/V_A$ and $1/V_B$ respectively. If the intersection of the two slopes occurs at 'distance' X, it can be shown that the depth of layer A is given by:

$$h_A = \tfrac{1}{2}X\left(\frac{V_B - V_A}{V_B + V_A}\right)^{1/2}$$

[12.3]

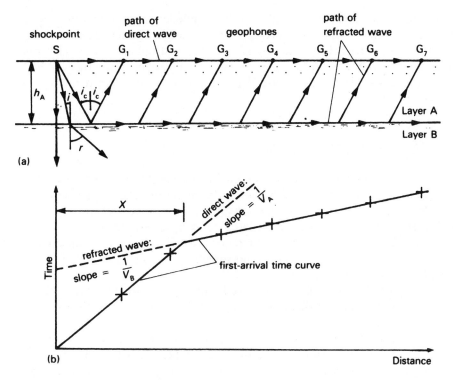

Fig. 12.16 Seismic refraction method

The seismic refraction method can be used for explorations down to about 300 m, but it is a prerequisite that the wave velocity in the upper layer must be less than that in the lower layer(s), i.e. $V_B > V_A$. This condition arises from the fact that for the waves to be refracted along the boundary they have to arrive along a path which lies at the critical angle i_c to the boundary normal, such that $\sin i_c = V_A/V_B$. In a multilayered series of stratum a 'blind' layer will occur when its wave velocity is less than that in the overlying layer: the direct wave will pass through a blind layer perpendicular to the boundary and is then not refracted. Other errors may be caused due to the fact that lithological boundaries do not always correspond with boundaries between strata of different wave velocity; also in anisotropic strata there may be a difference between the vertical and horizontal velocities.

Seismic reflection method

Reflection methods are not suitable for shallow-depth exploration in rocks and soils, since the arrival of reflected waves is obscured by much higher amplitude surface and shear waves. Seismic reflection profiling is, however, widely used for deep exploration (>300 m) and for sub-water exploration in rivers, lakes and shallow seas. The method is particularly applicable in marine surveys, because water cannot transmit shear waves or high-velocity waves.

The technique involved is similar to that employed in echosounding: a pulse of sound is emitted which is reflected from the sea bed or sub-bed boundaries and the sound arrivals recorded by an array of hydrophones. The wave velocities of the water and sub-bed layer need to be determined by other methods. From the results, lateral and vertical profiles may be constructed rapidly and economically.

12.8   Site investigation reports

A site investigation report is usually the culmination of the investigation, exploration and testing programme, although intermediate reports may sometimes be required where long-term ongoing observations are involved. The report will be addressed to the client or whomever has commissioned the investigation: it may be purely factual or may contain (if requested) advice and recommendations relating to design and construction, and sometimes suggestions relating to post-construction monitoring.

Although individual reports vary according to the particular brief received and conditions encountered, a typical report will normally include the following:

(a) *Introduction*: a brief summary of the proposed works, the investigations carried out, the location of the site and significant names and dates.

(b) *Description of site*: a general description of the site: its topography and main surface features; details of access; details of previous development or relevant history; details of existing works, underground openings, drainage, etc.; a map showing site location, adjoining land and borehole locations.

(c) *Geology of the site*: commencing with a description of overall geology, related to the regional geology of the area; description of main soil and rock formations and structures; comments on the influences of geology on design and construction.

(d) *Soil conditions*: a detailed account of the soil conditions encountered, related to the design and construction of the proposed works; description of all relevant layers, together with results of laboratory and *in situ* tests; details of groundwater and drainage conditions.

(e) *Construction materials*: a detailed account of the nature, quantity, availability and significant properties of materials considered for construction purposes.

(f) *Comments and construction review*: comments are necessary on the validity and reliability of the information being presented; where further work is required this should be mentioned; if the brief is also to make recommendations, these should include consideration of alternative methods of both design and construction.

(g) *Appendices*: it is convenient to assemble most of the collected data into a series of appendices: borehole logs; laboratory test details and results; results of *in situ* tests; geophysical survey records; references; relevant literature extracts.

Copies of the factual report, together with any associated factual records, should be made available with tender documents and subsequently to all interested parties throughout the contract.

Soil Mechanics Spreadsheets and Reference Assignments and Quizzes (available on the accompanying CD)

Assignments A.12   Site investigations and *in situ* testing

Quiz Q.12   Site investigations and *in situ* testing

Answers to exercises

Chapter 2

2.1 0.042 mm, 2.45, 1.3; SM
2.2 0.3, 20, 1.25; GW
2.3 0.005, 280, 0.128; GM/SM
2.4 A: $D_{60} = 0.42$ mm, $D_{30} = 0.32$ mm
    B: $D_{60} = 1.75$ mm, $D_{30} = 0.35$ mm
    C: $D_{60} = 1.50$ mm, $D_{30} = 0.40$ mm
2.5 (a) 54 per cent
    (b) 21, MH
2.6 (a) 46 per cent
    (b) 24, CL
2.7 E: CL, 0.80, 31 per cent
    F: MH, 4.8, 36 per cent
    G: CL–ML, 0.64, –0.33 per cent
    H: CH, 1.56, 78 per cent

Chapter 3

3.1 (a) 2.029 Mg/m³, 1.796 Mg/m³
    (b) 0.504, 0.335
    (c) 70 per cent
    (d) 10.2 per cent
3.3 (a) 0.724
    (b) 1.913 Mg/m³, 1.560 Mg/m³
    (c) 22.6 per cent
    (d) 1.980 Mg/m³
3.4 2.020 Mg/m³, 1.663 Mg/m³,
    21.4 per cent, 0.630, 3.1 per cent
3.5 2.009 Mg/m³, 24.8 per cent
3.6 1.872 Mg/m³, 0.920, 1.032
3.7 528 litres/m³
3.8 (a) 1.818 Mg/m³, 73.2 per cent
    (b) 2.066 Mg/m³, 0.594
3.9 (a) 0.613
    (b) 16.36 kN/m³
    (c) 20.09 kN/m³
    (d) 19.14 kN/m³

3.10 (a) 0.504, 17.48 kN/m³
    (b) 1892 g, 303 ml
3.11 19.2 kN/m³, 16.7 kN/m³
3.12 (a) 1.911 Mg/m³, 13.3 per cent
    (b) 3.8 per cent
3.13 (a) 1.867 Mg/m³, 14.1 per cent
    (b) 4.5 per cent, 0.446,
        85.4 per cent
    (c) 1.76 Mg/m³

Chapter 4

4.2 0.6 m
4.3 17.7 kPa
4.4 $\sigma = \sigma' + u$
4.5 0.9983
4.6 26.4 per cent, swelling will
    occur
4.8 (a) at $z = 6$ m: $\sigma'_z = 67.1$ kPa
                $\sigma_z = 126.0$ kPa
        at $z = 11$ m: $\sigma'_z = 113.1$ kPa
                $\sigma_z = 221.0$ kPa
    (b) at $z = 6$ m: $\sigma'_z = 108.0$ kPa
                $\sigma_z = 108.0$ kPa
        at $z = 11$ m: $\sigma'_z = 153.9$ kPa
                $\sigma_z = 203.0$ kPa
4.9 (a) at $z = 3$ m: $\sigma'_z = 24.6$ kPa
                $\sigma_z = 54.0$ kPa
        at $z = 5$ m: $\sigma'_z = 45.0$ kPa
                $\sigma_z = 94.0$ kPa
    (b) at $z = 0$ m: $\sigma'_z = 55.0$ kPa
                $\sigma_z = 55.0$ kPa
        at $z = 3$ m: $\sigma'_z = 79.6$ kPa
                $\sigma_z = 109.0$ kPa
        at $z = 5$ m: $\sigma'_z = 100.0$ kPa
                $\sigma_z = 149.0$ kPa
4.10 $\sigma'_{AA} = 6.87H - 9.75$; 1.42 m

4.11 (a) 73.0 kPa, 162.0 kPa
(b) 8.3 m
(c) 11.0 m

Chapter 5
5.1 1.06
5.2 1.72×10^{-3} m/s
5.3 0.633×10^{-3} m/s
5.4 1.04×10^{-6} m/s
5.5 3.7×10^{-5} m/s
5.6 0.49×10^{-3} m/s
5.7 1.36×10^{-4} m/s
5.8 0.25×10^{-3} m/s
5.9 2.49×10^{-3} m/s, 7.31×10^{-7} m/s
5.10 7.36 kPa, 5.7 kPa;
−5.89 kPa, 32.16 kPa
5.11 0.23 m³/hr (per m)
5.12 0.89 m³/hr (per m)
5.13 (a) 12.0 m³/hr (per m)
(b) 1.11
5.14 (a) 0.044 m³/hr (per m)
(b) at 4 m intervals, from piling:
24, 23, 18, 12 and 0 kPa
5.15 (a) 0.10 m³/hr (per m)
(b) 0.077 m³/hr (per m)
0.077 m³/hr (per m)
(c) piling failure would be likely
5.16 0.26 m³/hr (per m)
5.17 (a) 2.1 m³/day (per m)
(b) 1.8 m³/day (per m)

Chapter 6
6.1 $21\sigma'$, $24\sigma'$
6.2 0.5, ∞
6.3 34.5 kPa, 14.5 kPa, 26.0 kPa
6.4 (a) 124 kPa
(b) 123 kPa
(c) 124 kPa
6.5 (a) 76.4, 77.3, 73.3, 71.2 and 66.9 kPa
(b) 78.1, 13.8 and 23.3 kPa; at a
point 0.3 m from 100 kN/m load
6.6 2.65 m
6.7 120 kPa, 101 kPa, 66 kPa
6.8 (a) 6.7 m
(b) 102.8 kPa, 18.5 kPa
6.9 107 kPa, 124 kPa, 74 kPa
6.10 158, 142, 77, 14 and 6 kPa

Chapter 7
7.2 (a) 29°
(b) 41°

7.3 (a) 30°
(b) 37°
7.4 (a) 492 kPa, 575 kPa, 0.614,
0.226
(b) −182 kPa, 213 kPa, 189 kPa
(c) 29°
7.6 124 kPa
7.7 (a) 0, 23°; 0, 33°
(b) normally consolidated
7.8 40 kPa, 19°; 0, 28°
7.9 (a) 103 kPa
(b) 83 kPa
(c) 164 kPa
7.10 (a) 65 kPa, 79 kPa, 101 kPa
(b) 827 kPa
7.11 (a) A: 116 kPa, 16°; B: 52 kPa, 16°
(b) 0.18, 16°
7.12 (b) −204 kPa
(c) 67 kPa; 83 kPa; 303 kPa
7.13 (a) 0, 0
(b) 160 kPa, 230.5 kPa
7.14 (a) 730 kPa, 518 kPa
(b) 0.92
(c) 0.21
Below: strains elastic
7.15 (b) -1.22×10^{-3}
(c) 8.33 kPa

Chapter 8
8.1 432 kN/m, 4.0 m
8.2 519 kN/m, 4.34 m
8.3 366 kN/m, 4.15 m
8.4 484 kN/m, 3.90 m
8.5 510 kN/m, 4.0 m
8.6 441 kN/m, 2.33 m
8.7 408 kN/m, 3.21 m
8.8 529 kN/m, 3.65 m
8.9 736 kN/m, 4.27 m
8.10 655 kN/m, 3.34 m
8.11 458 kN/m, 3.61 m
8.12 (a) 187 kN/m
(b) 286 kN/m
(c) 404 kN/m
8.13 (a) 632 kN/m
(b) 662 kN/m
(c) 899 kN/m
8.14 (a) 108 kN/m, 98 kN/m
(horizontal comp.)
(b) 1.28 (horizontal comp.)
8.15 308, 415 and 450 kN/m (at 24° to
wall surface normal)

8.16 (a) 290 kN/m, 330 kN/m
     (b) 11.3 m
8.17 (a) 293 kN/m
     (b) 683 kN/m
8.18 (a) 2.94, 1.53, 82 kPa
     (b) 2.28, 0.61, 104 kPa
8.19 (a) 6.8 m
     (b) 6.9 m
     (c) 6.8 m
8.20 (a) 1.41
     (b) 2.5
8.21 11.7 m
8.22 (a) 2.38 m
     (b) 213 kN
8.23 (a) 0.35
     (b) 654 kN

Chapter 9
9.1 2.27, 1.98, 1.81
9.2 (a) 2.04
     (b) 1.04
     (c) 1.51
9.3 22.6°
9.4 (a) 1.30
     (b) 1.03
     (c) 0.98
9.5 23 kPa
9.6 1.29 (ignoring tension crack), 1.18
     (with tension crack); $X = 8$ m,
     $Y = 14.5$ m
9.7 (a) 1.43
     (b) 1.39
     (c) 1.36
9.8 (a) 1.45
     (b) 1.79
9.9 (a) 1.20
     (b) 1.78
     (c) 2.67
9.10 29°
9.11 3.9 m
9.12 1.74
9.13 1.20
9.14 (a) 1.66
     (b) 1.1
9.15 (a) 1.62
     (b) 1 vertical: 3.5 horizontal

Chapter 10
10.1 13 mm
10.2 (a) 5.2 MN/m^2
     (b) 48 mm
10.3 17 mm

10.4 (a) 0.798, 0.772
     (b) 7.23×10^{-2} m^2/MN
10.5 (a) 0.844, 0.806, 0.792, 0.767,
         0.740, 0.719, 0.772
     (b) (i) 67 mm (ii) 82 mm
10.6 (a) 61 kPa, 124 kPa
     (b) 68 mm
10.7 103 mm
10.8 213 mm
10.9 (a) 198 mm
     (b) 2.14 yr
10.10 (a) 67 mm
      (b) 11.0 yr, 27.1 yr
10.11 (a) 9.7 yr
      (b) 175 mm
10.12 (a) 3.82 mm^2/min.
      (b) 0.029, 0.779
10.13 3.82 mm^2/min.; 0.029, 0.779
10.14 203 mm

Chapter 11
11.1 528 kPa
11.2 (a) 546 kPa
     (b) 235 kPa
11.3 (a) 5830 kPa
     (b) 4376 kPa
     (c) 3415 kPa
11.4 292 kPa
11.5 (a) 3.8 m
     (b) 2.3 m
     (c) 2.65 m
     (d) 0.9 m
11.6 3.5 m
11.7 $z = 3.8$ m
     (a) 13.7 per cent, 37.9 per cent
     (b) 2.57, 1.81
11.8 3.7 m
11.9 (a) 50 kPa (consolidation)
     (b) 113 kPa (elastic compression)
     (c) 544 kPa (shear failure)
11.10 155 kPa, 168 kPa
11.11 162 kPa
11.12 (a) 320 kPa
      (b) 365 kPa
11.13 445 kPa
11.14 (a) 77 mm
      (b) 46 mm
11.15 (a) 515 kN
      (b) 4540 kN
      (c) 0.73
11.16 1.24, 1.67

References and bibliography

AASHTO (1986) *Standard Specifications for Transportation Materials and Methods of Sampling and Testing: Part II. Methods of Sampling and Testing* (14th edn). American Association of State Highway and Transportation Officials, Washington.

Ahlvin, R. G. and Ulery, H. H. (1962) Tabulated values for determining the complete pattern of stresses, strains and deflections beneath a uniform load on a homogeneous half-space. *Highway Research Board, Bulletin 342*.

Anon. (1991) *Specification for Highway Works*. Department of Transport, HMSO, London.

Armstrong, J. H. (1973) 'Pile selection', *Ground Engineering*, **6**(6).

ASTM (1986) *Annual Book of ASTM Standards: Part 19*. American Society for Testing Materials, Philadelphia.

Atkinson, J. H. (1981) *Foundations and Slopes*. McGraw-Hill, London.

Atkinson, J. H. and Bransby, P. L. (1978) *The Mechanics of Soils – An Introduction to Critical State Soil Mechanics*. McGraw-Hill, London.

Atkinson, J. H. and Clinton, D. B. (1986) 'Stress paths on 100 mm diameter samples', *Eng. Geology Special Publication No. 2*. Geological Society, London.

Atkinson, J. H., Evans, J. S. and Richardson, D. (1986) 'Effects of stress path and stress history on the stiffness of reconstituted London Clay', *Eng. Geology Special Publication No. 2*. Geological Society, London.

Baguelin, F., Jezequel, J.-F. and Shields, D. H. (1978) *The Pressuremeter and Foundation Engineering*. Transtech Publications, Basle.

Baird, H. G. (1988) 'Earthworks control – "assessment of suitability"'. *Ground Engineering*, **22**(4, May), pp. 22–8.

Berezantsev, V. G., Kristoforov, V. S. and Golubkov, V. N. (1961) 'Load-bearing capacity and deformation of piled foundations', *Proceedings*, International Conference on Soil Mechanics and Foundation Engineering, Paris 2.

Bishop, A. W. (1955) 'The use of the slip circle in the stability analysis of slopes', *Geotechnique*, **4**(2).

Bishop, A. W. (1958) 'Test requirements for measuring the coefficient of earth pressure at rest', *Proceedings*, Conference on Earth Pressure Problems, Brussels, **1**, 2–14.

Bishop, A. W. (1966) 'Strength of soils as engineering materials, 6th Rankine Lecture', *Geotechnique*, **16**, 89–130.

Bishop, A. W. (1967) 'Progressive failure with special reference to the mechanism causing it', *Proceedings*, European Conference on Soil Mechanics and Foundation Engineering, Oslo.

Bishop, A. W., Green, G. W., Garga, V. K., Andresen, A. and Brown, J. D. (1971) 'A new ring shear apparatus and its application to the measurement of residual strength', *Geotechnique*, **21**.

Bishop, A. W. and Henkel, D. J. (1962) *The Measurement of Soil Properties in The Triaxial Test* (2nd edn). Edward Arnold, London.

Bishop, A. W. and Morgenstern, N. R. (1960) 'Stability coefficients for earth slopes', *Geotechnique*, **10**.

Bjerrum, L. (1954) 'Geotechnical properties of Norwegian clays', *Geotechnique*, **4**(2), pp. 46–69.

Bjerrum, L. and Simons, N. E. (1960) 'Comparison of Shear Strength Characteristics of Normally Consolidated Clays', *Proc. ASCE Research Conf. on the Shear Strength of Cohesive Soils*, Boulder, pp. 711–26.

Bjerrum, L. (1963) 'Discussion', *Proceedings*, European Conference on Soil Mechanics and Foundation Engineering, Weisbaden **3**.

Bowden, F. P. and Tabor, D. (1950, 1964) *The Friction and Lubrication of Solids, Parts I and II.* Oxford University Press, London.

Brackley, I. J. A. (1980) 'Prediction of soil heave from suction', *Proc. 7th Regional African Conf. on Soil Mechanics*, Accra, **1**.

British Standards Institution (1973) *CP 102 Protection of Buildings Against Water from the Ground.* BSI, London.

British Standards Institution (1981) *BS 5930: Code of Practice for Site Investigations.* BSI, London.

British Standards Institution (1986) *BS 8004: Code of Practice for Foundations.* BSI, London.

British Standards Institution (1986) *BS 8103: Part 1: Structural Design of Low-rise Buildings.* BSI, London.

British Standards Institution (1989) *BS 8081: Code of Practice for Ground Anchorages.* BSI, London.

British Standards Institution (1990) *BS 1377: Methods of Test for Soils for Civil Engineering Purposes.* BSI, London.

British Standards Institution (1994) *BS 8002: Code of Practice for Earth Retaining Structures.* BSI, London.

British Standards Institution (1995) *BS 8006: Code of Practice for Strengthened/Reinforced Soils and Other Fills.* BSI, London.

British Standards Institution (1995) *DDENV 1997–1:1995: Geotechnical Design.* BSI, London.

British Standards Institution (1999) *BS 6031: Code of Practice for Earthworks.* BSI, London.

Bromhead, E. N. (1979) 'A simple ring-shear apparatus', *Ground Engineering*, **12**(5), pp. 40–4.

Bromhead, E. N. (1986) *The Stability of Slopes.* Surrey University Press, London.

Bromhead, E. N. and Curtis, R. D. (1983) 'A comparison of alternative methods of measuring the residual strength of London clay', *Ground Engineering*, **16**, pp. 39–41.

Bromhead, E. N. and Dixon, N. (1986) 'The field residual strength of London clay: its correlation with laboratory measurements, especially ring-shear tests', *Geotechnique*, **36**.

Broms, B. (1971) 'Lateral earth pressures due to compaction of cohesionless soils', *Proc. 4th Int. Conf. Soil Mechanics*, Budapest, pp. 373–84.

Brooker, E. W. and Ireland, H. O. (1965) 'Earth pressures at rest related to stress history', *Canadian Geotechnical Journal*, **2**, pp. 1–15.

Building Research Establishment (1980) *Low-Rise Buildings on Shrinkable Clays: Part 1: BRE Digest No. 240*. HMSO, London.

Building Research Establishment (1981) *Concrete in Sulphate-Bearing Soils: BRE Digest No. 250*. HMSO, London.

Building Research Establishment (1981) *Low-Rise Buildings on Shrinkable Clays: Part 2: BRE Digest No. 241*. HMSO, London.

Burland, J. B. (1973) 'Shaft friction of piles in clay – a simple fundamental approach', *Ground Engineering* **6**(3).

Burland, J. B., Broms, B. B. and de Mello, V. F. B. (1978) Behaviour of foundations and structures (*BRE Current Paper CP51/78*) HMSO, London.

Burland, J. B. and Burbidge, M. C. (1985) Settlement of foundations on sand and gravel. *Proceedings*, Institution Civil Engineers, **78**(Dec), pp. 1325–81.

Burland, J. B., Butler, F. G. and Duncan, P. (1966) The behaviour and design of large-diameter bored piles in stiff clay. *Proceedings*, Symposium on Large Bored Piles, Institution Civil Engineers, London.

Burland, J. B. and Fourie, A. B. (1985) 'The testing of soils under conditions of passive stress relief', *Geotechnique*, **35**(2), pp. 193–8.

Burland, J. B. and Wroth, C. P. (1975) 'Settlement of buildings and associated damage', *Proceedings*, British Geotechnical Society Conference on Settlement of Structures, Cambridge, pp. 611–54,

Calhoon, M. L. (1972) 'Discussion on D'Appolonia, Poulos and Ladd (1971)', *Proceedings*, American Society of Chartered Engineers, **98**(SM3), pp. 306–8.

Calladine, C. R. (1971) 'A micro-structural view of the mechanical properties of saturated clay', *Geotechnique*, **21**(4).

Caquot, A. and Kerisel, J. (1948) *Tables for the Calculation of Passive Pressure, Active Pressure and Bearing Capacity of Foundations*, Gauthier-Villars, Paris.

Caquot, A. and Kerisel, J. (1953) 'Sur le terme de surface dans le calcul des fondations en milieu pulvérulent', *Proceedings*, 3rd International Conference on Soil Mechanics and Foundation Engineering, Zurich **1**.

Casagrande, A. (1936) 'The determination of pre-consolidation load and its practical significance', *Proceedings*, 1st International Conference on Soil Mechanics, Harvard, **3**.

Casagrande, A. (1937, 1940) 'Seepage through dams'. Reprinted in *Contributions to Soil Mechanics*, Boston Society of Civil Engineers, Boston.

Cedergren, H. R. (1987) 'Drainage and dewatering techniques', in Bell, G. F. (ed.) *Ground Engineer's Reference Book*. Butterworths, London.

Chandler, R. J. (1969) 'The effect of weathering on the shear strength properties of Keuper Marl', *Geotechnique*, **19**, pp. 321–4.

Chandler, R. J. and Peiris, T. A. (1989) 'Further extensions to the Bishop and Morgenstern slope stability charts', *Ground Engineering*, May.

Chandler, R. J. and Skempton, A. W. (1974) 'The design of permanent cutting slopes in stiff fissured clays', *Geotechnique*, **24**, pp. 438–43.

Chen, W-F. (1975) *Limit Analysis and Soil Plasticity*. Elsevier, Amsterdam.

Chen, R. H. and Chameau, J.-L. (1983) 'Three-dimensional limit equilibrium of slopes', *Geotechnique*, **33**(1), pp. 31–40.

Chin, Fung Kee (1978) 'Diagnosis of Pile Condition', *Geotechnical Engineering*, **9**, pp. 85–104.

Civil Engineering Code of Practice CP 2 (1951) *Earth Retaining Structures*. Institute of Structural Engineers, London (now superseded by BS 8002:1994).

Clarke, K. E. (1957) 'Tropical black clays', *Proceedings*, Institution of Civil Engineers, London.

Clayton, C. R. I. and Symons, I. F. (1992) 'The pressure of compacted fill on retaining walls' (Technical Note), *Geotechnique*, **42**(1), pp. 127–30.

Clayton, C. R. I., Symons, I. F. and Heidra Cobo, J. C. (1991) 'The pressure of clay backfill against retaining structures', *Canadian Geotechnical Journal*, **28**(April), pp. 282–97.

Clayton, C. R. I., Simons, N. E. and Matthews, M. C. (1995) *Site Investigation – A Handbook for Engineers*. 2nd edn, Granada Publishing, London.

Coatsworth, A. M. (1986) 'A rational approach to consolidated undrained testing', *Engineering Geology Special Publication No. 2*. Geological Society, London.

Corbett, B. O. (1987) 'Exclusion techniques', in Bell, F. G. (ed.) *Ground Engineer's Reference Book*. Butterworths, London.

Corbett, B. O. and Stroud, M. A. (1975) 'Temporary retaining wall constructed by Berlinoise system at Centre Beaubourg, Paris', *Proceedings*, Conference on Diaphragm Walls and Anchorages. London, pp. 95–101.

Coulomb, C. A. (1776) 'Essai sur une application des regeles des maximus et minimus a quelque problemes de statique relatif à l'architecture', *Memoirs Divers Savants*, Academie Science, Paris **7**.

Crawford, C. B. (1986) State of the art: evaluation and interpretation of soil consolidation tests', in Young, R. N. and Townsend, F. C. (eds) *Consolidation of Soils: Testing and Evaluation*. ASTM STP, Philadelphia.

Crawford, C. B. and Burn, K. N. (1962) 'Settlement studies at Mount Sinion Hospital, Toronto', *Engineering Journal*, **45**(17), Ottawa.

Croney, D. and Jacobs, J. C. (1967) *The Frost Susceptibility of Soils and Road Materials*. Road Research Laboratory. HMSO, London.

D'Appolonia, D. J., Poulos, H. G. and Ladd, C. C. (1971) 'Initial settlement of structures on clay', *Proceedings*, American Society of Civil Engineers, **97**(SM10).

Davis, E. H. and Raymond, G. P. (1965) 'A non-linear theory on consolidation', *Geotechnique*, **15**(2).

de Beer, E. E. (1965) 'Bearing capacity and settlement of shallow foundations on sand', *Proceedings*, Symposium on Bearing Capacity and Settlement of Foundations. Duke University, N. Carolina.

de Beer, E. E. (1966) 'Computation of the failure in the bearing capacity of shallow foundations with inclined and eccentric loads', *Trans. No. 66.13*. US Army Engineers Waterways Experimental Station, Vicksburg.

de Beer, E. E. and Martens, A. (1951) 'Method of computation of an upper limit for the influence of heterogeneity of sand layers on the settlement of bridges', *Proceedings*, 4th International Conference on Soil Mechanics and Foundation Engineering. London **1**, pp. 275–8.

Driscoll, R. (1983) 'The influence of vegetation on the swelling and shrinkage of clay soils in Britain', *Geotechnique*, **33**, pp. 141–50.

Drucker, D. C. and Prager, W. (1952) 'Soil mechanics and plastic analysis or limit design', *Quarterly Applied Mathematics*, **10**, pp. 157–65.

Dumbleton, M. J. and West, G. (1976) *Preliminary Sources of Information for Site Investigations in Britain*. Transport and Road Research Laboratory Report. DOE, London.

Fadum, R. F. (1948) 'Influence values for estimating stresses in elastic foundations', *Proceedings*, 2nd International Conference on Soil Mechanics and Foundation Engineering. Rotterdam, **3**.

Feld, J. (1967) *Lessons from Failures of Concrete Structures*. American Concrete Institute, Detroit.

Feld, J. (1968) *Construction Failure*. John Wiley and Sons, New York.

Gibbs, H. J. and Holtz, W. G. (1957) 'Research on determining the density of sand by spoon penetration testing', *Proceedings*, 4th International Conference on Soil Mechanics and Foundation Engineering. London, **1**.

Giroud, J. P. (1968) 'Settlement of a linearly-loaded rectangular area', *Proceedings*, American Society of Civil Engineers, **94**(SM4).

Giroud, J. P. (1970) 'Stresses under a linearly-loaded rectangular area', *Journal of the Soil Mechanics and Foundations Div.*, ASCE, **98**, No. SM1, pp. 263–8.

Hamilton, J. J. and Crawford, C. B. (1959) *Papers in Soils – 1959 Meetings*, ASTM STP254, ASTM Philadelphia, pp. 254–71.

Hammond, R. (1956) *Engineering Structural Failures*. Odhams Press, London.

Hansen, J. B. (1961) 'A general formula for bearing capacity', *Bulletin* Danish Geotechnical Institute, **II**.

Hawkins, A. B. and Privett, K. D. (1985) 'Measurement and use of residual strength of cohesive soils', *Ground Engineering*, **18**, p. 229.

Hawkins, A. B. and Privett, K. D. (1986) 'Residual strength: does BS 5930 help or hinder?' *Engineering Geology Special Publication No. 2*. Geological Society, London.

Head, K. H. (1992) *Manual of Soil Laboratory Testing: Vols 1, 2 and 3*. Pentech Press, London.

Heijnen, W. J. (1973) 'Penetration testing in the Netherlands: a state-of-the-art report', *Proceedings*, 1st European Symposium on Penetration Testing, Stockholm 1974, **1**, pp. 79–83.

Henkel, D. J. (1982) 'The design of sheet pile walls', *Arup Journal*, **17**(1), pp. 22–3.

Holtz, W. G. and Gibbs, J. H. (1956) 'Engineering properties of expansive clays', *Transactions*, American Society of Civil Engineers, **121**, 641–6.

Holtz, W. G. (1959) 'Expansive clay properties and problems', *Col. School of Mines Quarterly* **54**(4), pp. 89–125.

Hooper, J. A. (1973) 'Observations on the behaviour of a piled foundation on London Clay', *Proceedings*, ICE. London.

Hovlund, H. J. (1977) 'Three-dimensional slope stability analysis method', *Proceedings*, American Society of Civil Engineers, **103**(GT9), pp. 971–86.

Hughes, J. M. O., Wroth, C. P. and Windle, D. (1977) 'Pressuremeter tests on sands', *Geotechnique*, **27**(4).

Humphrey, N. F. and Dunne, T. (1982) 'Pore pressures in debris failure initiation', *Report 45*. University of Washington, Seattle.

Hungr, O. (1987) 'An extension of Bishop's simplified method of slope stability analysis to three dimensions', *Geotechnique*, **37**, pp. 113–17.

Hutchinson, J. N. and Sarma, S. K. (1985) 'Discussion on three-dimensional limit equilibrium analysis of slopes', *Geotechnique*, **35**(2), p. 215.

Hvorslev, M. J. (1973) 'Uber die fesigkeitseigenshaften gestorter bindiger boden', *Ingvidensk. Skr. A*, No. 45.

Ingold, T. S. (1979) 'The effects of compaction on retaining walls', *Geotechnique*, **29**(3), pp. 265–83.

Institution of Civil Engineers (1976) *Manual of Applied Geology*. Institution of Civil Engineers, London.

International Society for Soil Mechanics and Foundation Engineering (1985), 'Report by the Technical Committee on Symbols, Units, Definitions and Correlations', *ISSMFE Executive Meeting*, San Francisco, 9/10 Aug.

Jaky, J. (1944) 'The coefficient of earth pressure at rest', *Journal of Society of Hungarian Architects and Engineers*, pp. 355–8.

Janbu, N. (1954) *Stability Analysis of Slopes With Dimensionless Parameters*. Harvard Soil Mechanics Series, No. 46.

Janbu, N. (1973) 'Slope stability computations', in Hirschfield and Poulos (eds) *Embankment Dam Engineering, Casagrande Memorial Volume*. John Wiley, New York.

Janbu, N., Bjerrum, L. and Kjaernsli, B. (1956) 'Veiledring ved losning av fundermentering-soppgaver', *Norwegian Geotechnical Institute Publication No. 16*, Oslo.

Janbu, N., Tokheim, O. and Senneset, K. (1981) 'Consolidation tests with continuous loading', *Proceedings*, 10th International Conference on Soil Mechanics and Foundation Engineering, Stockholm 1, pp. 645–54.

Jennings, J. E. and Kerrich, J. E. (1962) 'The heaving of buildings and associated economic consequences with particular reference to the Orange Free State Goldfields', *The Civil Engineer in S. Africa*, 5(5), p. 122.

Jones, C. J. F. P. (1996) *Earth reinforcement and soil structures* Thomas Telford, London.

Kerisel, J. and Absi, E. (1990) *Active and Passive Earth Pressure Tables* (3rd edn). A. A. Balkema, Rotterdam.

King, C. J. W. (1989) 'Revision of effective-stress method of slices', *Geotechnique*, 39(3), 497–502.

Ladd, C. C. and Foott, R. (1974) 'New design procedures for stability of soft clays', *Proceedings*, American Society of Civil Engineers, 100(G17), pp. 776–86.

Ladd, C. C. *et al.* (1977) 'Stress-deformation and strength characteristics', *Proceedings*, 9th International Conf. on Soil Mechanics and Foundation Engineering, Tokyo, 1977, 2, pp. 421–4.

Lambe, T. W. (1962) 'Pore pressures in a foundation clay', *Proceedings*, American Society of Civil Engineers, 88(SM2).

Leonards, G. A. and Ramiah, B. K. (1960) *Papers in Soils – 1959 Meetings* ASTM STP 254. ASTM, Philadelphia, pp. 116–30.

Lerouriel, S., Kabbaj, M., Tavenas, F. and Bouchard, R. (1985) 'Stress–strain rate relation for the compressibility of sensitive clays', *Geotechnique*, 35(2), pp. 159–80.

Littlejohn, G. S. and Bruce, D. A. (1977) *Rock Anchors – State of the Art*. Foundation Publications, Brentford, Essex.

Littlejohn, G. S. (1970) 'Soil anchors', *Proceedings*, Conference on Ground Engineering. Institution of Civil Engineers, London, pp. 33–44.

Lupini, J. F., Skinner, A. E. and Vaughan, P. R. (1981) 'The drained residual strength of soils', *Geotechnique*, 31, pp. 181–213.

McKaig, T. H. (1962) *Building Failures: Case Studies in Construction and Design*. McGraw-Hill, New York.

Mair, R. J. and Wood, D. M. (1987) *Pressuremeter Testing: Methods and Interpretation*. CIRIA, Butterworths, London.

Marsland, A. (1971) 'The shear strength of stiff fissured clay', *Proceedings*, Roscoe Memorial Symposium, Cambridge.

Marsland, A. (1974) 'Comparisons of the results from static penetration tests and large *in-situ* plate tests in London Clay', *BRE Current Paper CP87/74*. BRE, Watford.

Masch, F. D. and Denny, K. J. (1966) 'A statistical approach to the characterising of the permeability of a mass', *Proceedings of the Fourth International Conference on Application of Statistics and Probability in Soil and Structural Engineering*, University of Florence.

Massarch, K. R. (1979) 'Lateral Earth Pressure in Normally Consolidated Clay', *Proc. 17th Conf. on Soil Mechs. and Foundation Eng.*, Brighton, England, 2, pp. 245–50.

Matheson, G. D. (1983, 1988) 'The use and application of the moisture condition apparatus in testing soil for suitability for earth working', *SDD Guide No. 1*, TRRL, Scottish Branch, Livingston.

Medhat, F. and Whyte, I. L. (1986) 'An appraisal of soil index tests', *Engineering Geology Special Publication No. 2*. Geological Society, London.

Meigh, A. C. (1987) *Cone Penetrations Methods and Interpretation*. CIRIA/Butterworths, London.

Meigh, A. C. and Nixon, I. K. (1961) 'Comparison of in-situ tests on granular soils', *Proceedings*, 5th International Conf. on Soil Mechanics and Foundation Engineering, Paris, Vol. 1, pp. 499–507.

Menard, L. F. (1956, 1959) An apparatus for measuring the strength of soil in place. MSCE thesis, University of Illinois.

Menard, L. F. (1969) *The Geocel Pressuremeter*. Geocel Inc., Colorado.

Mesri, G. and Godlewski, P. M. (1977) Time- and stress-compressibility interrelationship. *Proceedings*, American Society of Civil Engineers, 103(GT5), pp. 417–30.

Mesri, G., Rokshar, A. and Bohor, B. F. (1975) 'Composition and compressibility of typical samples of Mexico City clay', *Geotechnique*, 25(3), pp. 527–54.

Meyerhof, G. G. (1953) 'The bearing capacity of foundations under eccentric and inclined loads', *Proceedings*, 3rd International Conference on Soil Mechanics and Foundation Engineering, Zurich 1, pp. 440–45.

Meyerhof, G. G. (1956) 'Penetration tests and bearing capacity of cohesionless soils', *Proceedings*, ASCE–Journal of Soil Mech and FE Div., 82(SM1), Jan., pp. 1–19.

Meyerhof, G. G. (1965) 'Shallow foundations', *Proceedings*, American Society of Civil Engineers, 91(SM2).

Meyerhof, G. G. (1976) 'Bearing capacity and settlement of pile foundations', *Proceedings*, American Society of Civil Engineers, 102(GT3), pp. 195–228.

Morgenstern, N. R. and Price, V. E. (1965) The analysis of the stability of general slip surfaces, *Geotechnique*, 15, 79–93.

Morgenstern, N. R. and Price, V. E. (1967) A numerical method for solving the equations of stability general slip surfaces. *Computer Journal*, 9 pp. 388–92.

Morrison, I. M. and Greenwood, J. R. (1989) 'Assumptions in simplified slope stability analysis by the method of slices', *Geotechnique*, 39(3), pp. 503–9.

Muir Wood, D. (1990) *Soil Behaviour and Critical State Soil Mechanics*. Cambridge University Press, Cambridge.

Murray, R. T. (1980) 'Fabric reinforced earth walls: development of design equations', *Ground Engineering*, 13(7), 29–36.

O'Connor, M. J. and Mitchell, R. J. (1977) An extension to the Bishop and Morgenstern slope stability charts. *Canadian Geotechnical Journal*, 14, pp. 144–51.

Packshaw, S. (1946) 'Earth pressure and earth resistance', *Journal of the Institution of Civil Engineers*, 25, p. 233.

Padfield, C. J. and Mair, R. J. (1984) 'Design of retaining walls embedded in stiff clay', *CIRIA Report 104*. CIRIA, London.

Parry, R. H. G. (1960) Triaxial compression and extension tests on remoulded saturated clay. *Geotechnique*, 10, 166–80.

Parry, R. H. G. (ed.) (1972) 'Stress–strain behaviour of soils', *Proceedings*, Roscoe Memorial Symposium. Cambridge University.

Parsons, A. W. (1987) 'Shallow compaction', in Bell, F. G. (ed.) *Ground Engineer's Reference Book*. Butterworths, London.

Peck, R. W. and Bazaraa, A. S. (1969) 'Discussion on settlement of spread footings on sand', *Proceedings*, American Society of Civil Engineers, 95(SM3), pp. 905–9.

Peck, R. B., Hanson, W. E. and Thornburn, T. H. (1974) *Foundation Engineering*. John Wiley, New York.

Polshin, D. E. and Tokar, R. A. (1957) 'Maximum allowable non-uniform settlement of structures', *Proceedings*, 4th International Conference on Soil Mechanics and Foundation Engineering, London, 1, p. 402.

Poulos, G. H. and Davis, E. H. (1974) *Elastic Solutions for Soil and Rock Mechanics.* John Wiley, New York.

Powers, J. P. (1976) *Construction Dewatering: A Guide to Theory and Practice.* John Wiley, New York.

Prandtl, L. (1921) 'Uber die endringungsfestigkeit plastischer baustoffe und die festigkeit von Schnieden', *Zeitschrift fur Angewandte Mathematik und Mechanik*, **1**(1).

Ramey, G. E. and Johnson, R. C. (1979) 'Relative accuracy and modification of some dynamic pile capacity prediction equations', *Ground Engineering*, Sept.

Rankine, W. J. M. (1857) 'On the stability of loose earth', *Phil. Trans. Royal Society*, **147**.

Reissner, H. (1924) 'Zum erddruckproblem', *Proceedings*, 1st International Conference on Applied Mechanics, Delft.

Reynolds, O. (1866) 'Experiments showing dilatancy, a property of granular materials', *Proceedings*, Royal Institute, **2**, 354–63.

Road Research Laboratory (1970) *Road Note 29: A Guide to the Structural Design of Pavements.* HMSO, London.

Roscoe, K. H. and Burland, J. B. (1968) 'On the generalised stress–strain behaviour of "wet" clay', in Heyman, J. and Leckie, F. A. (eds) *Engineering Plasticity.* Cambridge University Press, 535–609.

Rowe, P. W. (1952) 'Anchored sheet pile walls', *Proc. Part 1*, Inst. of Civil Engineers, London.

Rowe, P. W. and Peaker, K. (1965) 'Passive earth pressure measurements', *Geotechnique*, **15**(1).

Rowe, P. W. and Barden, L. (1966) 'A new consolidation cell', *Geotechnique*, **16**(2).

Samuels, S. G. (1975) 'Some properties of the Gault clay from the Ely–Ouse Essex water tunnel', *Geotechnique*, **25**, pp. 239–64.

Sanglerat, G. (1979) *The Penetrometer and Soil Exploration.* Elsevier, Amsterdam.

Sangrey, D. A. (1972) 'Naturally sensitive soils', *Geotechnique*, **22**, pp. 139–52.

Schlosser, F. *et al.* (1972) 'Etude de la terre armée a l'appareil triaxial', *Rapport de Recherché* No. 17. LCPC, Paris.

Schmertmann, J. H. (1970) 'Static cone to compute settlement over sand', *Proceedings*, American Society of Civil Engineers, **96**(SM3).

Schmertmann, J. H., Hartman, J. P. and Brown, P. R. (1978) 'Improved strain influence factor diagrams', *Proceedings*, American Society of Civil Engineers, **104**(GT8), pp. 1131–5.

Schofield, A. N. and Wroth, C. P. (1968) *Critical State Soil Mechanics.* McGraw-Hill, London.

Schofield, C. P. and Wroth, D. M. (1978) 'The correlation of index properties with some basic properties of soils', *Canadian Geotechnical Journal*, **15**(2), May, pp. 137–45.

Shields, D. H. and Tolunary, A. Z. (1973) Passive pressure by method of slices. *Proceedings*, American Society of Civil Engineers, **99**(SM12), 1043–53.

Site Investigation Steering Group (1993) *Site Investigation in Construction: Parts 1, 2, 3 and 4.* Thomas Telford, London.

Skempton, A. W. (1951) 'The bearing capacity of clays', *Proceedings*, Building Research Congress, London.

Skempton, A. W. (1954) 'The pore pressure coefficients A and B', *Geotechnique*, **4**(4).

Skempton, A. W. (1960) 'Pore pressure and suction in soils', in *Selected Papers.* Butterworths, London.

Skempton, A. W. (1964) 'The long-term stability of clay slopes', *Geotechnique*, **14**(2).

Skempton, A. W. (1970) 'The consolidation of clays by gravitational compaction', *QJGS* London, **125**, pp. 373–412.

Skempton, A. W. (1977) 'Slope stability of cuttings in Brown London Clay', *Proceedings*, 9th International Conference on Soil Mechanics and Foundation Engineering, Tokyo, **3**: pp. 261–70.

Skempton, A. W. (1986) 'Standard penetration test procedures and the effects in sands of overburden pressure, relative density, particle size, ageing and overconsolidation', *Geotechnique*, **36**(3).

Skempton, A. W. and Bjerrum, L. (1957) 'A contribution to the settlement analysis of foundations on clay', *Geotechnique*, **7**(4).

Skempton, A. W. and MacDonald, D. H. (1956) 'Allowable settlement of buildings', *Proceedings*, Institution Civil Engineers, **5**(3).

Skempton, A. W. and Northey, R. D. (1953) 'The sensitivity of clays', *Geotechnique*, **3**(1), pp. 30–53.

Sokolovski, V. V. (1960) *The Statics of Granular Media*. Pergamon Press, Oxford.

Steinbrenner, W. (1934) 'Tafeln zur Setzungsberechnung', *Die Strasse*, **1**, pp. 121–4.

Taylor, D. W. (1948) *Fundamentals of Soil Mechanics*. John Wiley, New York.

te Kamp, W. L. (1977) 'Static cone penetration and testing on piles in sand', *Fugro Sounding Symposium*, Utrecht.

Terzaghi, K. (1936) 'The shearing resistance of saturated soil and the angle between planes of shear', *Proceedings*, 1st International Conference on Soil Mechanics, Harvard.

Terzaghi, K. (1943) *Theoretical Soil Mechanics*. John Wiley, New York.

Terzaghi, K. and Peck, R. B. (1948, 1967, 1969) *Soil Mechanics in Engineering Practice*. John Wiley, New York.

Terzaghi, K., Peck, R. B. and Mesri, G. (1996) *Soil Mechanics in Engineering Practice*. 3rd edn, John Wiley, New York.

Tomlinson, M. J. (1995) *Foundation Design and Construction* (6th edn). Longman, London.

Tomlinson, M. J., Driscoll, R. and Burland, J. B. (1978) 'Foundations for low-rise buildings', *The Structural Engineer*, **56A**(6), pp. 161–73.

Trenter, N. A. (1999) 'A note on the estimation of permeability of granular soils', *Quarterly Journal of Engineering Geology*, **32**, 383–388.

US Army Corps of Engineers (1965) *Soils and Geology – Pavement Design for Frost Conditions*. Technical Manual TM 5-852-6.

Van de Merwe, D. H. (1964) 'Determination of potential expansiveness of soils', *The Civil Engineer in South Africa*, **6**, pp. 103–16.

Vesic, A. S. (1970) *Research on bearing capacity of soils*. Research thesis (unpublished).

Vickers, B. (1983) *Laboratory Work in Civil Engineering: Soil Mechanics*. Granada Publishing, London.

Vidal, H. (1969) 'The principal of reinforced earth', *Highway Research Record*, Washington, DC (282), pp. 1–16.

Vidal, H. (1966) 'Le terre armée', *Annales de l'Institute Technique du Batiment et de Travaut Publiques*, France, July–Aug, pp. 888–938.

Ward, W. H. (1955) 'Experiences with some sheet-pile coffer dams at Tilbury', *Geotechnique*, **5**, p. 327.

Warren, S. J. (1982) 'The pressuremeter test in a homogeneous linearly elastic cross-anisotropic soil', *Geotechnique*, **32**(2), pp. 157–9.

Weltman, A. J. and Head, J. M. (1983) *Site Investigation Manual*. CIRIA Special Publication, PSA. DOE, London.

Whitlow, R. (2000) *Soil Mechanics Spreadsheets and Reference* (on a CD-ROM accompanying *Basic Soil Mechanics*, 4th edn.). Pearson Education, Harlow, Essex.

Whitman, R. V. and Bailey, W. A. (1967) 'Use of computers for slope stability analysis', *Proceedings*, American Society of Civil Engineers, **93**(SM12), pp. 475–98.

Whittaker, T. (1976) *The Design of Piled Foundations*. Pergamon Press, Oxford.

Whyte, I. L. (1982) 'Soil plasticity and strength – a new approach using extrusion', *Ground Engineering*, **15**(1), pp. 16–24.

Williams, A. A. B. and Pidgeon, J. T. (1983) 'Evapotranspiration and heaving clays in South Africa', *Geotechnique*, **33**, pp. 141–50.

Windle, D. and Wroth, C. P. (1977) 'The use of a self-boring pressuremeter to determine the undrained properties of clays', *Ground Engineering*, Sept.

Wissa, A. E., Christian, J. T., Davis, E. H. and Heiberg, S. (1971) 'Consolidation at constant rate of strain', *Proceedings*, American Society of Civil Engineers, **97**(SM10), 1393–1414.

Wroth, C. P. and Wood, D. M. (1978) 'The correlation of index properties with some basic engineering properties of soils', *Canadian Geotechnical Journal*, **15**(2), pp. 137–45.

Youssef, M. S., El Ramli, A. H. and El Demery, M. (1965) 'Relationships between shear strength, consolidation, liquid limit and plastic limit for remoulded clays', *Proceedings*, 6th International Conference on Soil Mechanics and Foundation Engineering, Montreal, **1**, pp. 126–9.

Appendix A:
Eurocode 7: Geotechnical Design

The first draft model of *Eurocode 7: Foundations* was prepared in 1981 by an *ad-hoc* committee comprising representatives of the Geotechnical Societies of the nine European Community members. A second version was published in 1986 when the committee also included members from Portugal and Spain. A 'Draft for public comment' was published in 1995 by the British Standards Institution, (DDENV 1997–1:1995), the title now changed to Eurocode 7: Geotechnical Design.

Purpose and scope
The purpose of the code is to provide a set of principles and procedures intended to ensure an adequate technical quality for foundations, retaining structures and earthworks. Examples of geotechnical engineering work included in the scope of the code are:

Buildings, bridges, retaining walls, excavations, embankments, coffer dams, dykes, small dams.

Site evaluation, field and laboratory investigations, design and evaluation procedures, observations during and after construction, evaluation of material resources.

Geotechnical categories
It is a requirement of the code that projects must be supervised at all stages by personnel with *geotechnical knowledge*. In order to establish minimum requirements for the extent and quality of geotechnical investigation, design and construction three *Geotechnical Categories* are defined (see Table A.1).

The following factors must be considered in arriving at a classification of a structure or part of a structure:

Nature and size of the structure
Local conditions, e.g. traffic, utilities, hydrology, subsidence, etc.
Ground and groundwater conditions
Regional seismicity

Table A.1 Eurocode 7: Foundations – Geotechnical Categories

Geotechnical category	GC.1	GC.2	GC.3
Geotechnical risk	Low	Moderate	High
Expertise required	Experience and QUALITATIVE geotechnical investigations	Qualified ENGINEERS with relevant experience	Experienced GEOTECHNICAL ENGINEERS
Nature and size	Small and relatively simple structures or construction	Conventional structures and foundations with no abnormal risks	Large or unusual structures and abnormal risks
Ground conditions	NOT: abnormal, sloping, loose fill, swelling clay, soft/loose/highly compressible soil	CAN be determined (e.g. properties for design) using ROUTINE field and laboratory procedures	Unusual or exceptionally difficult; requiring NON-ROUTINE tests, calculations, interpretations, etc.
Surroundings	NO risk of damage to or from neighbouring buildings, utilities, public areas, etc.	POSSIBLE risk to neighbouring structures from excavation, piling, groundwater lowering or drainage	Possible SEVERE risk to neighbouring structures
Groundwater problems	NO excavation below water table, except where local experience indicates low risk	NO RISK of damage without prior warning to structures or load-bearing strata due to groundwater lowering or drainage	Possible HIGH RISK of groundwater problems; e.g. multi-layered strata of variable permeability, variable water table levels

	Seismically INACTIVE areas or INSENSITIVE structures	Where NATIONAL SEISMIC CODES *only* are adequate for design; NOT areas of high risk	Areas of HIGH SEISMIC risk and VERY SENSITIVE structures
Regional seismicity			
Influence of environment	If NO significant problems of hydrology, vegetation, surface water, subsidence, landslip, etc. exist	Where ROUTINE procedures are adequate to deal with the problem	DIFFICULT or COMPLEX environmental problems
Examples of structures	Light buildings: max column loads of 250 kN; wall load of 100 kN; no settlement problems; conventional foundations. Retaining walls and excavations not exceeding 2 m deep; no surcharge. Earthworks not exceeding 3 m of fill below trafficked area or 1 m below ground slabs. Ground slabs empirically designed, requiring no detailed analysis. One- and two-storey houses and agricultural buildings on conventional piled foundations. Small excavations for drainage works, pipe laying, etc.	Spread footings. Raft foundations. Piled foundations. Earth and water retaining walls. Moderate excavations. Bridge piers and abutments. Embankments and earthworks. Ground anchors and tie-backs	Buildings with exceptional loads. Multistorey basements. Retaining dams and other structures acted upon by great differential water pressures. Where temporary or permanent water table lowering occurs. Earthworks and pavement subject to abnormal traffic loads. Large buildings and tunnels. Dynamically loaded foundations. Offshore structures. Power stations. Chemical plants, especially if hazardous. Structures sensitive to seismic activity or where seismicity is high. Coastal and other major landslip problems.

Classification must take place *prior to* the geotechnical investigations. The category may be changed at a later stage and different parts of a project may be placed into different categories. The procedures for higher categories may be used to justify more economic design solutions, where suitably qualified and experienced engineers consider them appropriate.

Performance criteria and limit states

The performance criteria which must be considered are given for each type of structure, together with suggested calculation models and some prescriptive measures. Alternative approaches are also allowed provided that they can be justified.

The code is based on the *limit state method*, in which each possible performance criterion is considered as a *limit state*. The design process must demonstrate that the occurrence of each possible limit state, considered separately, is either eliminated or is sufficiently improbable.

The two main classes of limit state considered are:

Type 1   an *ultimate limit state*
   1(A)   a *mechanism* is formed in the *ground*
   1(B)   a *mechanism* is formed in the *structure or severe structural damage occurs*
Type 2   a *serviceability limit state*
            at which *deformation in the ground* will cause *loss of serviceability in the structure*

Although *durability* must be considered when selecting design parameters, it should not be defined as a limit state as such. Durability should be secured by paying attention to the detailed aspects of design, with adequate provision being made for protection, maintenance, etc.

Prescriptive measures

For certain limit states, calculation models are either not available or unnecessary. Instead, the limit state can be avoided by adopting conventional and generally conservative details in design, and by attention to specification and control of materials, workmanship, protection and maintenance procedures. For example, prescriptive measures can be used to ensure protection from frost action or chemical attack, or to avoid unnecessary calculations in very familiar design situations.

Appendix B:
Geotechnical design: factors of safety and partial factors

Factors and design values

A conventional *factory of safety* is defined as the ratio of ultimate and design (working) actions, e.g.:

$$\text{Factor of safety, } F_s = \frac{\text{Ultimate load}}{\text{Design load}} \quad \text{or} \quad \frac{\text{Shear strength}}{\text{Mobilised shear stress}}$$

When a performance criterion is expressed in terms of more than one parameter, *partial factors* are used, e.g.:

$$F_s = \frac{\text{shear strength}}{\text{mobilised shear stress}} = \frac{c' + \sigma' \tan \phi'}{c'_{mob} + \sigma' \tan \phi'_{mob}}$$

Therefore, the design values are: $\quad c'_{mob} = \dfrac{c'}{F_c} \quad$ and $\quad \tan \phi'_{mob} = \dfrac{\tan \phi'}{F_\phi}$

where F_c and F_ϕ = *partial factors*

Similarly, Design load = $G_D + Q_D = \gamma_G G_R + \gamma_Q Q_R$

where γ_G and γ_Q = partial factors, G_R and Q_R = representative values of permanent and variable actions respectively (e.g. dead load and live load).

When applied to loads, partial factors are usually called *load factors*.

Deciding values for factors of safety

The value chosen for a factor of safety or partial factor is normally arrived at in the light of experience; however, the following considerations are important:

(1) The governing limit state: *ultimate limit state* – shear slip in slopes, overturning or sliding in retaining walls; *serviceability limit state* – settlement in foundations.
(2) The relationship between strength and stiffness for soil, which is not constant but varies with such things as depth, level of stress, stress history and drainage condition.

(3) The variation of soil, both laterally and vertically, and the reliability of measured soil parameters resulting from tests on samples or from field tests.
(4) Whether temporary or permanent safety conditions are being considered.
(5) The effect of time and changes in conditions on measured values of parameters.
(6) The consequences of a limit state occurring, and the degree of risk therefore incurred.
(7) The mathematical form of the calculation entailing a factor of safety, and the relative importance of key parameters within such a calculation.

Some typical factors of safety in geotechnical design

Foundation design

Mechanism and failure state		Typical F_s range
Bearing capacity failure (ultimate LS)	Shear slip at critical strength along a shear-slip surface	
(i) Shallow foundation		Permanent works 2.3–3.0
(ii) Piled foundation		Temporary works 1.5–2.0
		End-bearing 2.5–3.0
		Shaft friction 1.5–2.0
		Overall 2.0–2.5
Settlement (serviceability LS)	Vertical displacement due to soil compression	Value of F_s above or a modified value is used, or computation of allowable bearing capacity at design
(i) Immediate: at the time of loading		
(ii) Consolidation: at some time following completion of loading		
Durability: LS is defined in term of adequate depth required to prevent damage due to frost, scouring, shrinking/swelling soil, effect of tree roots, etc.		

Slope design

Mechanism and failure state	Typical F_s range
Shear slip Discontinuous movement along a shear-slip surface which may be plane, cylindrical or log-spiral in form.	
(i) Temporary cuttings and embankments: using undrained strength (c_u) and TS	1.1–1.3
(ii) Permanent cuttings: using critical strength (ϕ'_c) and ES	1.2–1.4
(iii) Embankment – foundation: undrained (c_u) or drained (ϕ')	1.2–1.5
(iv) Embankment – fill: drained (ϕ') for compacted soil and ES	1.2–1.4
(v) Reactivated landslip: residual strength (ϕ'_r)	(natural value)

TS = total stresses, ES = effective stresses.

Seepage design

Mechanism and failure state	Typical F_s range
Uplift and heave; due to vertical seepage pressure	1.5–2.0
Exit velocity gradient, piping	2.0–3.0

Retaining wall design
Choice of strength parameters must be related to the amount of wall *yield* expected:

(i) No yield (i.e. rigid wall): at-rest pressure (K_o) and peak strength (c_u or ϕ'_f)

(ii) Small yield: active/passive pressures (K_a, K_p, K_{ac}, K_{pc}) and peak strength (c', ϕ'_f), but higher F_s against passive pressure.

(iii) Large yield (i.e. flexible wall): active/passive pressures (K_a, K_p, K_{ac}, K_{pc}) and critical strength (c', ϕ'_f), but lower F_s against passive pressure.

Mechanism and failure state	Typical F_s range	
	Permanent	*Temporary*
(a) Strength factor: drained analysis – stiff clays	1.1–1.5	1.0–1.2
– other soils	1.5–2.0	1.0–1.3
undrained analysis	2.0	1.5
(b) Net passive/active pressure moment factor:		
drained analysis	1.5–2.0	1.1–1.5
undrained analysis	2.0	1.5
(c) Overturning stability of gravity walls	2.0–3.0	1.5
(d) Sliding along base	1.5–2.5	1.5
(e) Overall slip stability	1.2–1.5	1.0–1.2
(f) Anchor rod failure	1.5–2.0	1.2–1.5
(g) Internal stability (e.g. bending of sheet piles)	1.5 (e.g. $f_b = 0.65 f_y$)	

Range values: use HIGHER values when $\phi' > 30°$ or for *conservative* design; use LOWER values when $\phi' < 20°$ or for *worst credible* design.

Limit state design
BS 8002 (1994) adopts the philosophy of limit state design and makes the following recommendations with respect to soil parameters to be used in design. *Parameters for effective stress design* – either for peak state or critical state:

$$\text{Design } \tan \phi' = \frac{\text{representative } \tan \phi'}{M} \qquad \text{Design } c' = \frac{\text{representative } c'}{M}$$

where M = mobilisation factor = 1.2 (minimum value)

Design $\tan \delta = 0.75 \times$ design $\tan \phi'$ (incorporating $M = 1.2$)

or

design $\tan \delta = 0.667 \times$ representative $\tan \phi'$

If the wall is prevented from sinking on the passive side, or may move upward, wall friction must be excluded from passive quantities.
Parameters for total stress design:

$$\text{Design undrained strength (design } c_u) = \frac{\text{Representative } c_u}{M}$$

If wall displacements are required to be <0.5 per cent of wall height, $M = 1.5$; M should be >1.5 when using peak strengths mobilised at high strains; M should be 2.5–3.0 for bearing capacity calculations.

Design $c_w = 0.75 \times$ design c_u (incorporating $M = 1.5$)

or

design $c_w = 0.50 \times$ representative c_u

Ground anchors

BS 8081 (1989) Temporary works < 6 months: $F \geq 2.0$
Temporary works ≥ 6 months: $F \geq 2.5$
Permanent works: $F \geq 3.0$

Other minimum partial factors

Mechanism and failure state	Load factor (non-beneficial)	Resistance factor (beneficial)
Loads: dead loads	1.2	0.85
live loads, wind or earthquake	1.5	
water pressures	1.25	0.85
Shear strength: undrained (c_u) – stability, foundations		0.5–0.75
drained (c') – foundations		0.5
wall adhesion		0.5
wall friction		0.5–0.8

Appendix C:
Guide to *Soil Mechanics Spreadsheets and Reference* (on the CD-ROM)

About Soil Mechanics Spreadsheets and Reference

Soil Mechanics Spreadsheets and Reference is a package of teaching/learning materials for use by students of civil, structural and geotechnical engineering, engineering geology, building technology and engineering, architecture and related disciplines. The package contains:

* Sets of interactive spreadsheets (Microsoft Excel) in soil mechanics and related topics
* An on-screen *Soil Mechanics Reference Manual* (Windows Help files)
* An on-screen *Soil Mechanics Glossary of Terms and Symbols* (Windows Help file)
* On-screen *Soil Mechanics Quizzes* for student self-assessment (Windows Help files)
* A *Tutorial Manual* (Word file) of study assignments incorporating the use of the spreadsheets
* A *Soil Mechanics Spreadsheets and Reference User Guide* (Word file) which includes guides to using the spreadsheets, assignments and quizzes and suggestions for editing and customising by the user.

Soil Mechanics Spreadsheets and Reference has been written and developed by Roy Whitlow and is published on a CD-ROM by Pearson Education as part of, and in conjunction with, *Basic Soil Mechanics* (4th edn) by Roy Whitlow.

Soil Mechanics Spreadsheets and Reference is designed to be an interactive, editable teaching/learning resource. It is not meant to replace conventional methods, but to supplement them; it is not intended as a substitute for textbooks or other reference material. The components of the package are simple and easy to use either on single PCs or networks. Students will be able to work at their own pace and according to their own style of learning; they will be able to study material in depth or simply browse. Numerous calculations can be completed at very short study sessions; repetition and experimentation can assist in the exploration of ideas and concepts; a number of laboratory routines and material tests can be simulated; some spreadsheets are designed to receive student's own data.

The spreadsheets are written in standard *Microsoft Excel* and as such will run via Windows 95/98 without conversion. All of the spreadsheets are initially saved as read-only files or are otherwise protected from accidental corruption. However, all are editable using the tools available in *Excel*. Teachers are invited to customise some or all of the spreadsheets to suit their own purposes: modified versions can be saved easily under different filenames. Similarly, students and teachers can rename and save spreadsheets containing their own data.

Hardware and software required

All of the *Soil Mechanics Spreadsheets and Reference* software runs on conventional desktop, laptop or notebook computers via *Microsoft Windows*. A screen resolution of 800×600 pixels or 640×480 pixels is recommended and there should be at least 4 Mbyte of RAM and 5 Mbyte of hard disk space available. All of the programs and files can be accessed and run directly from the CD-ROM, or they can be downloaded to a hard disk. All downloaded files should be held in a folder named C:\SMSS.

Word document files

Some of the material on the CD-ROM is in the form of Word document files (i.e. those appended **.doc**). These can be viewed in Word or a number of other file-viewing utilities (e.g. WordPad), or they can be printed for future use. The individual licence restriction applies also to these printed files: they must be for the personal use of the licence holder. The following is a summary.

Soil Mechanics Spreadsheets and Reference User Guide (SMSRuserGuide.doc)

The *User Guide* provides information and guidelines to assist users in navigating and using the *Soil Mechanics Spreadsheets and Reference* materials. It is assumed that users are familiar with the Microsoft software of Windows 95/98, Word and Excel. In particular the basic conventions and protocols in Excel need to be known. For those with more advanced knowledge it will be possible to rewrite or tailor some spreadsheets to suit specific purposes and needs. The following sections are included:

1. About Soil Mechanics Spreadsheets and Reference
2. Getting started – installing the software
3. Using Soil Mechanics Spreadsheets and Reference in teaching
4. Using the spreadsheets
5. Soil Mechanics Reference
6. Soil Mechanics Glossary
7. Study assignments and learning outcomes
8. Self-assessment quizzes and monitoring

Study assignments in soil mechanics

The main basis of this component is the provision of study exercises. The range of subjects and topics is given below. ***Study Assignment Sheets*** are sets of instructions, tasks and hints to student users; some pages will be in pro-forma style with spaces for entry of values, recovered information, etc. It is intended that students users will study using a combination of the *Study Assignment Sheets*, the *Soil Mechanics Spreadsheets*, the *Soil Mechanics Reference Manual*, the *Soil Mechanics Glossary* and the *Soil Mechanics Quizzes.* Students will be advised to keep additional notes, to record important key values and to work in methodical manner. Tutors will be advised that they can easily devise programmes of work using this package. The *Study Assignment Sheets* and the *Soil Mechanics Spreadsheets* are editable. The *Soil Mechanics Reference Manual, Glossary* and *Quizzes* are 'standalone' components and therefore can be used within other study arrangements, e.g. laboratory and other practical work.

The *Study Assignment Sheets* do not constitute a textbook, nor are they intended to be a fully comprehensive work scheme for soil mechanics. Tutors must provide a curriculum and programmes of study according to course and individual needs, and also engage students in other aspects of tuition, e.g. lectures, tutorials, laboratory programmes and field exercises. The following sections are included; theses reflect the chapter headings in *Basic Soil Mechanics*:

A.1   Origins and composition

A.2   Soil classification and description

A.3a   Basic physical properties of soils

A.3b   Compaction of soil

A.4   Groundwater, pore pressure and effective stress

A.5   Soil permeability and seepage

A.6a   Stresses and strains in soils

A.6b   Compression and volume change

A.6c   Stresses due to applied loading

A.7   Measurement of shear strength

A.8   Earth pressure and retaining walls

A.9   Stability of slopes

A.10   Settlement and consolidation

A.11   Bearing capacity of foundations

A.12   Site investigations and *in situ* testing

Guide to self-assessment quizzes

In trials of the software many users expressed a desire to see numerous self-testing routines included. However, there are limitations associated with computer-based testing. For example, should it be totally user-orientated or should there be built-in monitoring. To monitor progress and to seek out problem areas for treatment are reasonable objectives in operational teaching/learning programmes, and so too may be attainment grading. However, the author feels that mechanisms for such purposes are best devised by course tutors and examiners, and not all will be best placed in computer routines. The *Soil Mechanics Quizzes* are therefore designed principally for students to use as a means of self-testing their own current understanding and progress; some quizzes or parts may be adaptable for other purposes if the need is perceived.

Index